William Benjamin Carpenter

The Microscope and it's Revelations

Vol. II

William Benjamin Carpenter

The Microscope and it's Revelations
Vol. II

ISBN/EAN: 9783744690249

Printed in Europe, USA, Canada, Australia, Japan

Cover: Foto ©berggeist007 / pixelio.de

More available books at **www.hansebooks.com**

THE MICROSCOPE

AND ITS

REVELATIONS

BY

WILLIAM B. CARPENTER, C.B. M.D. LL.D.

F.R.S. F.G.S. F.L.S.

CORRESPONDING MEMBER OF THE INSTITUTE OF FRANCE,
AND OF THE AMERICAN PHILOSOPHICAL SOCIETY,
ETC., ETC.

SIXTH EDITION

ILLUSTRATED BY TWENTY-SIX PLATES
AND FIVE HUNDRED WOOD ENGRAVINGS

VOLUME II.

NEW YORK
WILLIAM WOOD & COMPANY
56 & 58 LAFAYETTE PLACE
1883

TABLE OF CONTENTS.

CHAPTER X.

MICROSCOPIC FORMS OF ANIMAL LIFE:—PROTOZOA.

	PAGE		PAGE
Protozoa,	1	Heliozoa,	11
Monerozoa,	2	Lobosa,	14
Rhizopoda,	7	Coccoliths and Coccospheres	19
Reticularia,	7	Gregarinida,	21

CHAPTER XI.

ANIMALCULES:—INFUSORIA AND ROTIFERA.

INFUSORIA,	24	Infusoria continued:—	
Flagellata,	26	Ciliata,	41
Cilio-flagellata,	37	ROTIFERA,	53
Suctoria,	39	Tardigrada,	62

CHAPTER XII.

FORAMINIFERA AND POLYCYSTINA.

FORAMINIFERA,	64	RADIOLARIA,	109
Porcellanea,	70	Discida,	112
Arenacea,	77	Polycystina,	113
Vitrea,	85	Acanthometrina,	113
Eozoön Canadense,	101	Colloza,	115

CHAPTER XIII.

SPONGES AND ZOOPHYTES.

SPONGES,	117	Zoophytes continued:—	
ZOOPHYTES,	122	Acalephæ,	132
Hydrozoa,	123	Actinozoa,	134
Production of Medusoids,	126	Ctenophora,	137

CHAPTER XIV.

ECHINODERMATA.

Structure of Skeleton,	140	Echinoderm-Larvæ,	150

CHAPTER XV.

POLYZOA AND TUNICATA.

POLYZOA,	157	TUNICATA,	163

CHAPTER XVI.

MOLLUSCOUS ANIMALS GENERALLY.

Structure of Shells,	171	Ciliary motion on Gills,	189
Palate of Gasteropods,	180	Organs of Sense of Mullusks,	190
Development of Mollusks,	183	Chromatophores of Cephalopods,	191

CHAPTER XVII.

ANNULOSA OR WORMS.

	PAGE		PAGE
Entozoa,	192	Annelida,	195
Turbellaria,	194	Development of Annelids,	197

CHAPTER XVIII.

CRUSTACEA.

	PAGE		PAGE
Pycnogonida,	205	Cirrhipeda,	213
Entomostraca,	207	Malacostraca,	214
Suctoria,	212	Metamorphosis of Decapods,	215

CHAPTER XIX.

INSECTS AND ARACHINDA.

	PAGE		PAGE
Number and variety of Objects afforded by Insects,	218	Wings,	241
		Feet,	243
Structure of Integument,	219	Stings and Ovipositors,	245
Scales and Hairs,	220	Eggs,	246
Eyes,	229	Agamic Reproduction,	246
Antennæ,	232	Embryonic Development,	248
Mouth,	233		
Circulation of the Blood,	237	Acarida,	248
Respiratory Apparatus,	238	Parts of Spiders,	250

CHAPTER XX.

VERTEBRATED ANIMALS.

	PAGE		PAGE
Elementary Tissues,	253	Epidermis,	275
Cells and Fibres,	253	Pigment-Cells,	275
Bone,	255	Epithelium,	276
Teeth,	258	Fat,	277
Scales of Fish,	261	Cartilage,	278
Hairs,	263	Glands,	279
Feathers,	266	Muscle,	281
Hoofs. Horns, etc.,	267	Nerve,	284
Blood,	267	Circulation of the Blood,	286
White and Yellow Fibres,	271	Injected Preparations,	292
Skin, Mucous and Serous Membranes,	274	Vessels of Respiratory Organs,	299

CHAPTER XXI.

APPLICATION OF THE MICROSCOPE TO GEOLOGY.

	PAGE		PAGE
Fossilized Wood, Coal,	302	Fossil Bones, Teeth, etc.,	310
Fossil Foraminifera; Chalk,	304	Inorganic materials of Rocks,	312
Organic Materials of Lime-stones,	308	Nachet's Mineralogical Microscope,	315

CHAPTER XXII.

INORGANIC OR MINERAL KINGDOM.—POLARIZATION.

	PAGE		PAGE
Mineral Objects,	318	Organic Structures suitable for Polariscope,	323
Crystallization of Salts,	319		
Molecular Coalescence,	323	Micro-Chemistry,	326

APPENDIX.

	PAGE		PAGE
"Numerical Aperture" and "Angular Aperture,"	327	Swift's "Wale" Model Students' Microscope,	332
Watson's New Model Microscopes,	331	Nachet's Objective-carrier,	333

THE MICROSCOPE.

CHAPTER X.

MICROSCOPIC FORMS OF ANIMAL LIFE:—PROTOZOA.

391. PASSING-ON, now, to the Animal Kingdom, we begin by directing our attention of those minute and simple forms, which correspond in the Animal series with the *Protophyta* in the Vegetable (Chap. VI.); and this is the more desirable, since the formation of a distinct group to which the name of PROTOZOA (first proposed by Siebold) may be appropriately given, is one of the most interesting results of Microscopic inquiry. This group, which must be placed at the very base of the Animal scale, beneath the great Sub-Kingdoms marked-out by Cuvier, is characterized by the extreme simplicity that prevails in the structure of the beings composing it; the lowest of them being single protoplasmic particles or 'jelly specks;' whilst even among the highest, however numerous their units may be, these are (as among *protophytes*, § 227) mere repetitions of one another, each capable of maintaining an independent existence. In this there is a very curious and significant parallelism to the earliest embryonic stage of higher Animals. For the fertilized germ of any one of these first shapes itself as a single cell; and then, by repeated binary subdivisions, develops itself into a *morula* or 'mulberry-mass' of cells (Fig. 403), corresponding to the 'multicellular' organisms met with among the higher Protozoa (Fig. 350). There is, so far, in neither case, any sign of that 'differentiation' of organs which is characteristic of the higher Animals; but whilst, in the Protozoon, each cell is not merely similar to its fellows, but is independent of them, the *morula*, in such as go on to a higher stage, becomes the subject of a series of developmental changes, tending to the production of a single whole, whose parts are mutually-dependent. The first of these changes is its conversion into a *gastrula* or primitive stomach, whose wall is formed of a double membrane,—the outer lamella, or *ectoderm*, being derived directly from the external cell-layer of the morula, whilst the inner, or *endoderm*, is formed by the 'invagination' of that layer into the space left void by the dissolution of the central cells of the 'morula.'¹ This *gastrula-stage*, as we

---

[1] It has not yet been certainly ascertained that the endoderm is formed by invagination in all cases; but as several of the supposed exceptions have disappeared under the light of fuller investigation, it seems probable that the remainder will be found conformable to the general rule.

shall see hereafter (§ 513), remains permanent in the great group of *Cœlenterata;* though the endoderm and ectoderm are separated from each other in its higher forms by the development of generative and other organs between them. But in all Classes above the Cœlenterates, the primitive stomach has only a transitory existence, being superseded by the permanent structures that have their origin in its walls.—Thus the whole Animal Kingdom may be divided, in the first place, into the PROTOZOA, which are either single cells, or aggregates of similar cells corresponding to the *morula*-stage of higher types; and the METAZOA, in which the morula takes-on the condition of an individualized organism, the life of every part of which contributes to the general life of the whole.

392. The lowest of the *Protozoa*, however, like the simplest Protophytes, do not even attain the rank of a true *cell*,—understanding by that designation a definite protoplasmic unit, limited by a cell-wall, and containing a 'nucleus.' For they consist of particles of protoplasm, termed ('cytodes' or 'plastids') of indefinite extent, which have neither cell-wall nor nucleus, but which yet take-in and digest food, convert it into the material of their own bodies, cast out the indigestible portions, and reproduce their kind, with the regularity and completeness that we have been accustomed to regard as characteristic of higher Animals. Between some of these *Monerozoa* (as they have been designated by Prof. Haeckel, who first drew attention to them and the *Myxomycetes* (§ 222) the *Chlamidomyxis* (§ 324) already described, no definite line of division can be drawn; the only justification for the separation here adopted being that the affinities of the former seem to be rather with the lowest forms of Vegetation, whilst the whole life-history of the types now to be described and the connected gradation by which they pass into undoubted Rhizopods, leave no doubt of *their* claim to a place in the Animal Kingdom.

MONEROZOA.

393. A characteristic example of this lowest Protozoic type is presented by the *Protomyxa aurantiaca* (Fig. 279), a marine 'Moner' of an orange-red color, found by Professor Haeckel upon dead shells of *Spirula* near the Canary Islands. In its active state is has the stellar form shown at F; its arborescent extensions dividing and inosculating so as to form a constantly changing network of protoplasmic threads, along which stream in all directions orange-red granules obviously belonging to the body itself, together with foreign organisms (*b c*)—such as marine Diatoms, Radiolarians, and Infusoria,—which, having been entrapped in the pseudopodial network, are carried by the protoplasmic stream into the central mass, where the nutrient matter of their bodies is extracted, the hard skeletons being cast out. Neither nucleus nor contractile vesicle is to be discerned; but numerous floating and inconstant vacuoles (*a*) are dispersed through the substance of the body.— After a time, the currents become slower; the ramified extensions are gradually drawn inwards; and, after ejecting any indigestible particles it may still include, the body takes the form of an orange-red sphere, round which a cyst soon forms itself, as shown at A. After a period of quiescence, the protoplasmic substance retreats from the interior of the cyst, and breaks up into a number of small spheres (B), which, at first inactive, soon begin to move within the cyst, and change their shape to that of a pear with the small end drawn out to a point. The cyst then

bursts, and the red pear-shaped bodies issue forth into the water (c), moving freely about by the vibrations of *flagella* formed by the drawing-out of their small ends,—just as do the flagellated zoöspores of protophytes (§ 231). These bodies, being without trace of either nucleus, contractile vesicle, or cell-wall, are to be accounted **as** particles of simple homogeneous protoplasm, to which the designation *plastidules* has been appropriately given. After about a day the **motions** cease; the flagella are drawn in, and the plastidules take the **form** and lead the life of *Amœbæ* (§ 403), putting forth inconstant pseudopodial processes, and engulfing nutrient particles in their substance (D). Two or more of these amœbiform bodies unite to form a 'plasmodium' (as in the *Myxo-*

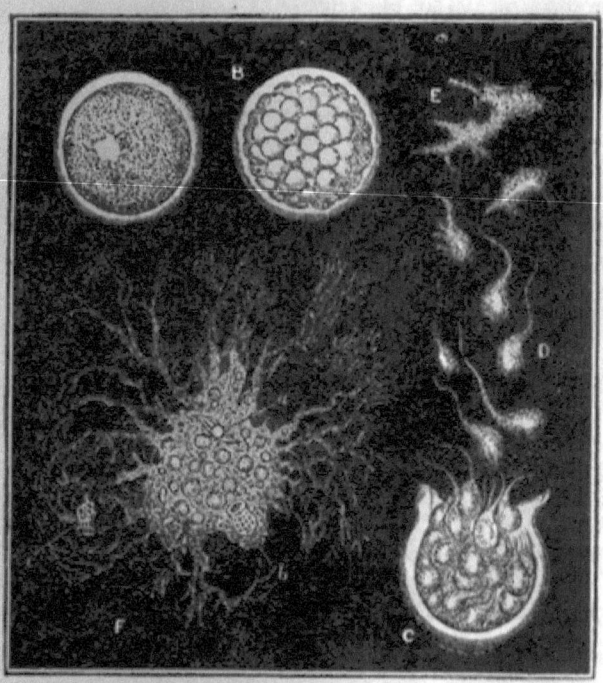

Protomyxa aurantiaca:—A, encysted statospore; B, incipient formation of swarm-spores, shown at C escaping from the cyst, at D swimming freely by their flagellate appendages, and at E creeping in the amœboid condition; F, fully-developed reticulate organism, showing numerous vacuoles, *a*, and captured prey, *b*, *c*.

mycetes, § 222); its pseudopodial extensions send out branches which inosculate to form a network; and the body grows, by the ingestion of nutriment, to the size of the original.—In this cycle of change there seems no intervention of a generative act, the coalescence of the amœbiform plastidules having none of the characters of a true 'conjugation.' But it is by no means improbable that after a long course of multiplication by successive subdivisions, a sexual act of some kind may intervene.

394. Another very interesting 'moneric' type is, the *Vampyrella;* of which one form (Fig. 280, B) has long been known in its encysted con-

dition as a minute brick-red sphere attached to the filaments of the Conjugate *Spirogyra;* whilst another (Fig. 281, *a, a*) similarly attaches itself to the branches of *Gomphonema* (§ 294). The walls of the cysts are composed of two membranes; of which the interior gives the characteristic reaction of cellulose, whilst the softer external layer is nitrogenous. After remaining some time in the quiescent condition, the encysted protoplasm breaks up into two or four 'tetraspores' (Fig. 281, *b, d*); these escape by openings in the cyst (Fig. 280, c); and soon take the spherical form, emitting very slender pseudopodial filaments (Figs. 280, D, 281, B) like those of an *Actinophrys,* but possessing neither nucleus nor contractile vesicle. In this condition they show great activity; moving about in search of the special nutriment they require, drawing themselves out in strings and fine filaments which tear asunder and again unite to send off branches and form fine fan-like expansions, and these occasionally contract-

Fig. 280.

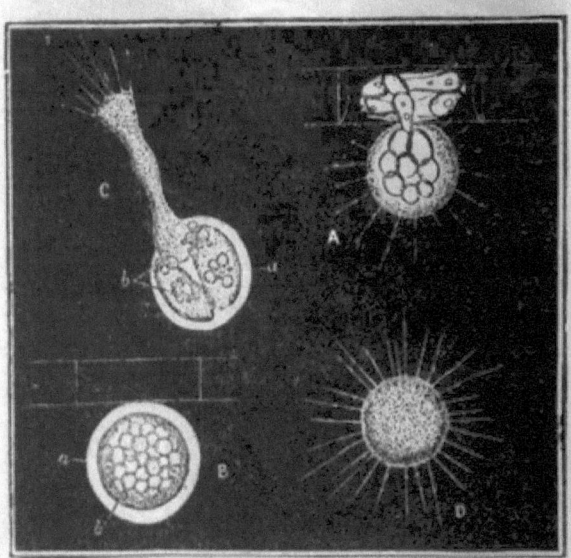

Vampyrella spirogyræ, as seen at A sucking out contents of *Spirogyra*-cell: at B in encysted condition, the cyst a inclosing granular protoplasm *b;* at C, division of contents of cyst into tetraspores, of which one is escaping in the amœboid condition, to develop itself into the adult form shown at D.

ing again into minute spheres. When the *V. spirogyræ* is watched in water containing some filaments of *Spirogyra,* it may be seen to wander until it meets one of these filaments, to which, if it be healthy and loaded with chlorophyll, it attaches itself. It soon begins to perforate the wall of the filament; and when the interior of this has been reached, its endoplasm, carrying with it the chlorophyll-granules it includes, passes slowly into the body of the *Vampyrella.* In this manner, cell after cell is emptied of its contents; and the plunderer, satiated with food, resumes its quiescent spherical form to digest it. The chlorophyll granules which it has ingested become diffused through the body, but gradually cease to be distinguishable, the protoplasmic mass assuming a brick-red color.

The first layer it exudes to form its cyst is the outer or nitrogenous investment, within which the cellulose layer is afterwards formed.—The *V. gomphonematis* in like manner creeps over the stems and branches of the *Gomphonema* (Fig. 281, *e*), adapting itself to the form of its support; and as soon as it has reached one of the terminal siliceous cells of the Diatom, it extends itself over it so as completely to envelop the cell in a thin layer of protoplasm. From the surface of this, a number of fine pseudopodia radiate into the surrounding water (*f*); whilst another portion of the protoplasm finds its way between the two siliceous valves into the interior,

Fig. 281.

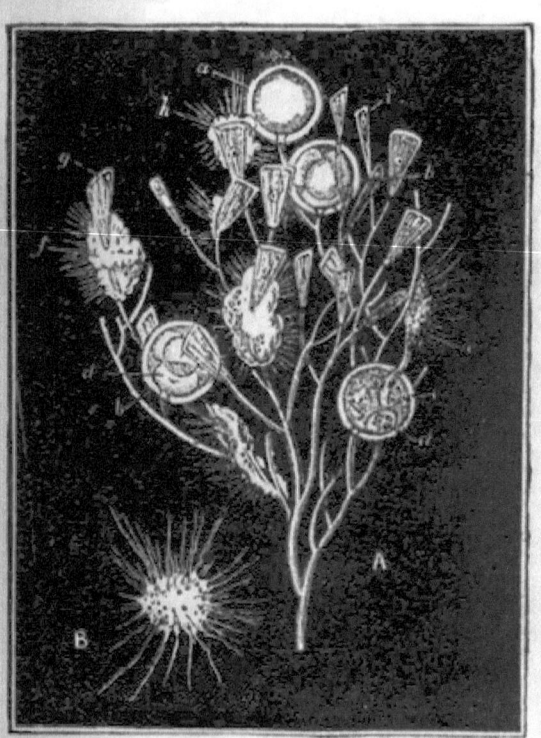

ampyrella gomphonematis:—A, colony of *Gomphonema* attacked by *Vampyrellæ*; *a*, encysted state; *b, b*, cysts with contents breaking-up into tetraspores, *d, d*, seen escaping at *e*; at *f* is shown a *Vampyrella* sucking-out contents of *Gomphonema*-cells, the emptied frustules of which, *g, h*, are cast forth.—B, isolated *Vampyrella*, creeping about by its extended pseudopodia.

and appropriates its contents. The valves, when emptied, break off from their support, and are cast out of the body of the *Vampyrella*, which soon proceeds to another *Gomphonema*-cell and plunders it in the same manner. After thus ingesting the nutriment furnished by several cells, and acquiring its full size, it passes, like *V. spirogyræ*, into the encysted condition, to recommence—after a period of quiescence—the same cycle of change.

395. Intermediate between the foregoing and the 'reticularian'

Rhizopods to be presently described, is another simple Protozoön discovered in ponds in Germany by M.M. Claparède and Lachmann, and named by them *Lieberkühnia Wageneri*.[1] The whole substance of the body of this animal and its pseudopodial extensions (Fig. 282) is composed of a homogeneous, semifluid, granular protoplasm; the particles of which, when the animal is in a state of activity, are continually performing a circulatory movement, which may be likened to the rotation of the particles in the protoplasmic network within the cell of a *Tradescantia* (§ 355). It is a marked peculiarity of the pseudopodial extension of this type, that it does not take place by radiation from all parts of the body indifferently; but that it proceeds entirely from a sort of trunk that soon divides into branches, which, again, speedily multiply by further subdivision, until at last a multitude of finer and yet finer threads are spun-out, by whose continual inosculations a complicated network is produced, which may be likened to an animated Spider's web. The entire absence of anything like a membranous envelope is clearly evidenced by the readiness with which the subdivision and the coalescence of the pseudopodia alike take place. Any small alimentary particles that may come into contact with the glutinous surface of the pseudopodia, are retained in adhesion by it, and speedily partake of the general movement going-on in their substance. This movement takes place in two principal directions; from the body towards the extremities of the pseudopodia, and from these extremities back to to the body again. In the larger branches a double current may be seen, two streams passing at the same time in opposite directions; but in the finest filaments the current is single, and a granule may be seen to move in one of them to its very extremity, and then to return, perhaps meeting and carrying back with it a granule that was seen advancing in the opposite direction. Even in the broader processes, granules are sometimes observed to come to a stand, to oscillate for a time, and then to take a retrograde course, as if they had been entangled in the opposing current,—just as often is to be seen in *Chara*. When a granule arrives at a point where a filament bifurcates, it is often arrested for a time, until drawn into one or the other current; and when carried across one of

Lieberkühnia Wageneri.

---

[1] "Etudes sur les Infusoires et les Rhizopods;" Geneva, 1850–1861. The beautiful figure of *Lieberkühnia*, given by M. Claparède, has been reproduced by the Author in Plate 1 of his 'Introduction to the Study of the Foraminifera.'—A Rhizopod of the same type has been discovered by Mr. Siddall (of Chester) in *Sea*-water from the North and South Coasts of Wales, which he regards as especially identical with *L. Wageneri* ("Quart. Microsc. Journ.," N. S., Vol. xx., p. 144), but which the Author (who has great confidence in the accuracy of the excellent observers by whom the latter was described) must regard as differentiated from it (1) by the existence of a pellucid flexible investment (foreshadowing the 'test' of Gromia), having a definite orifice bordered by four infolded lips, through which the sarcodic trunk issues forth; and (2) by the presence of a number of highly refractive, short, rod-like spicules set at various angles on the external surface.

the bridge-like connections into a different band, it not unfrequently meets a current proceeding in the opposite direction, and is thus carried back to the body without having proceeded very far from it. The pseudopodial network along which this 'cyclosis' takes place, is continually undergoing changes in its own arrangement; new filaments being put forth in different directions, sometimes from its margin, sometimes from the midst of its ramifications, whilst others are retracted. Not unfrequently it happens that to a spot where two or more filaments have met, there is an afflux of the protoplasmic substance that causes it to accumulate there as a sort of secondary centre, from which a new radiation of filamentous processes takes place. Occasionally the pseudopodia are entirely retracted, and all activity ceases; so that the body presents the appearance of an inert lump. But if watched sufficiently long, its activity is resumed; so that it may be presumed to have been previously satiated with food, which is undergoing digestion during its stationary period. No encysting process has been noticed in *Lieberkühnia;* and the manner in which this type reproduces itself is at present entirely unknown. As the marine type of it occurs on our own coasts, the fresh-water type may very likely be found in our ponds; and either may be recommended as a most worthy object of careful study.

RHIZOPODA.

396. We now arrive at the group of *Rhizopods,* or 'root-footed' animals, first established by Dujardin for the reception of the *Amœba* (§ 403) and its allies, which had been included by Prof. Ehrenberg among his Infusory Animalcules, but which Dujardin separated from them as being mere particles of *sarcode* (protoplasm), having neither the definite body-wall nor the special mouth of the true *Infusoria,* but putting forth extensions of their sarcodic substance, which he termed *pseudopodia* (or false feet), serving alike as instruments of locomotion, and as prehensile organs for obtaining food. According to Dujardin's definition of this group, the *Monerozoa* already described would be included in it; but it seems on various grounds desirable to limit the term *Rhizopoda* to those Protozoa in which the presence of a *nucleus,* the differentiation of an *ectosarc* (or firmer superficial layer of protoplasm) from the semi-fluid *endosarc,* together with the more definite form and restricted size, indicate a distinct approach to the condition of true cells.—Many different schemes for the classification of the Rhizopods have been proposed; but none of them can be regarded as entirely satisfactory, our knowledge of the Reproductive processes, and of other important parts of the life-history of these creatures, being still extremely imperfect. And as some parts of the scheme proposed by the Author twenty years ago,[1] based on the characters of the pseudopodial extensions, have been accepted by more recent systematists, he thinks it best still to adhere to it, as seeming to him to be on the whole most natural.

I. In the First division, *Reticularia,* the pseudopodia freely ramify and inosculate, so as to form a network, exactly as in *Lieberkühnia;* from which they are distinguished by the possession of a nucleus, and by the investment of their sarcodic bodies in a firm envelope. This is most commonly either a *calcareous* shell of very definite shape, or a *test* built up of sand-grains or other minute particles more or less firmly united by

---

[1] "Natural History Review," 1861, p. 456; and "Introduction to the Study of the Foraminifera" (1862), Chap. II.

a calcareous cement exuded from the sarcodic body. These testaceous forms, which are exclusively marine, constitute the group of *Foraminifera;* whose special interest to the microscopist entitles it to separate consideration (Chap. XII.). And it is only for convenience, that two *Reticularia* which inhabit fresh water also, and the envelopes of whose bodies are usually membranous, are here separated from the Foraminifera (to which they properly belong) for description as types of the group. The *Reticularia* have little locomotive power, and only seem to exercise it to find a suitable situation for their attachment; the capture of their food being effected by their pseudopodial network.

II. The Second division, *Heliozoa*,[1] consists of the Rhizopods whose pseudopodia extend themselves as straight radiating rods, having little or no tendency to subdivide or ramify, though they are still sufficiently soft and homogeneous (at least in the lower types, § 399), to coalesce when they come into contact with each other. These have usually (probably always) a contractile vesicle as well as a nucleus; and the higher forms of them are characterized by the inclosure of peculiar yellow corpuscles (whose import is unknown) in the substance of their endosarc. By far the larger number of this group also have skeletons of Mineral matter, which are always *siliceous;* and these are sometimes perforated casings of great regularity of form, as in the marine *Polycystina;* sometimes internal frameworks of marvellous symmetry, as in the marine *Radiolaria*. These two groups, also, will be reserved for special notice (Chap. XII.); the simple *Heliozoa* which are among the commonest inhabitants of fresh water, furnishing the best illustrations of the essential characters of the type. They seem for the most part to have but little locomotive power, capturing their prey by their extended pseudopodia.

III. The Third group, *Lobosa*, contains the Rhizopods which most nearly approach the condition of true Cells, in the differentiation of their almost membranous ectosarc and their almost liquid endosarc, and in the non-coalescence of their pseudopodial extensions, which, instead of being either thread-like or rod-like, are *lobate*, that is, irregular projections of the body, including both ectosarc and endosarc, which are continually undergoing change both in form and number. The *Lobosa* are comparatively active in their habits, moving freely about in search of food, which is still received into the substance of their bodies through any part of their surface,—unless this is inclosed in envelopes, such as are formed by many of them, either by exudation from the surface of their bodies of some material (probably chitinous) which hardens into a membrane, or by aggregating and uniting grains of sand or other small solid particles, which they build up into 'tests.' A large proportion of them are inhabitants of fresh water, and some are even found in damp earth.

397. *Reticularia*.—This type is very characteristically represented by the genus *Gromia* (Fig. 283); some of whose species are marine, and are found, like ordinary *Foraminifera*, among tufts of Corallines, Algæ, etc.; whilst others inhabit fresh water, adhering to Confervæ and other Plants of running streams. It was in this type that the presence of a nucleus formerly supposed to be wanting in Reticularia generally, was first estab-

---

[1] To this group the Author formerly extended the name *Radiolaria* given by Müller to one section of it; but he now thinks it preferable to employ the general term *Heliozoa* given to it by Hertwig and Lesser, restricting the term *Radiolaria* to the group to which it was originally applied.

lished by Dr. Wallich. The sarcode-body of this animal is incased in an egg-shaped, brownish-yellow, chitinous envelope, which may attain a diameter of from 1-12th to 1-10th of an inch, looking to the naked eye so like the egg of a Zoophyte or the seed of an aquatic Plant, that its real nature would not be suspected as long as it remains quiescent. The 'test' has a single round orifice, from which, when the Animal is in a state of activity, the sarcodic substance streams forth, speedily giving off ramifying extensions, which, by further ramification and inosculation, form a network like that of Lieberkühnia. But the sarcode also extends itself so as to form a continuous layer over the whole exterior of the 'test;' and from any part of this layer fresh pseudopodia may be given off. By the alternate extension and contraction of these, minute Protophytes and Protozoa are entrapped and drawn into the interior of the test, where their nutritive material is extracted and assimilated; and if the 'test' (as happens in some species) be sufficiently transparent, the indigestible hard parts (such as the siliceous valves of Diatoms, shown in Fig. 283) may be distinguished in the midst of the sarcodic substance. By the same agency, the *Gromia* sometimes creeps up the sides of a glass vessel. In the intervals of quiescence, on the other hand, the whole sarcodic body, except a film that serves for the attachment of the test, is withdrawn into its interior.

398. Another example of the Reticularian group is afforded by the curious little *Microgromia socialis* (Fig. 284), first discovered by Mr. Archer, and further investigated with great care by Hertwig;[1] which has the curious habit of uniting with neighboring individuals, by the fusion of the pseudopodia, into a common 'colony;' the individuals sometimes remaining at a distance from one another as at A, but sometimes aggregating themselves into compact masses as at B. The nearly globular thin calcareous shell is prolonged into a short neck having a circular orifice,

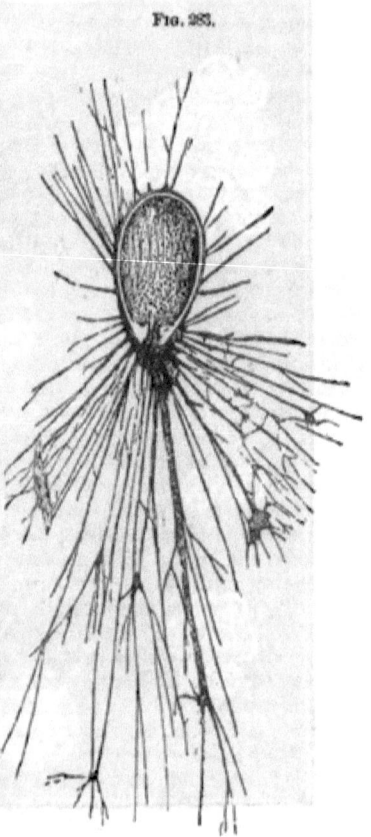

Fig. 283.

Gromia oviformis, with its pseudopodia extended.

from which the sarcode-body extends itself, giving off very slender pseudopodia which radiate in all directions. A distinct nucleus can be seen in the deepest part of the cavity; while a contractile vesicle lies imbedded in the sarcodic substance nearer the mouth. Multiplication by duplicative subdivision has been distinctly observed in this type; but with a peculiar

---

[1] 'Ueber *Microgromia*;' in "Archiv für Mikr. Anat.," Bd. x., Supplement.

departure from the usual method. A transverse constriction divides the body into two halves—as shown in two individuals of colony A,—each half possessing its own nucleus and contractile vesicle; the posterior segment, which at first lies free at the bottom of the cell, then presses forwards towards its orifice, as shown at C, and finally, by amœboid movements, escapes from it, sometimes stretching itself out like a worm (as seen at D), sometimes contracting itself into a globe, and sometimes

FIG. 284.

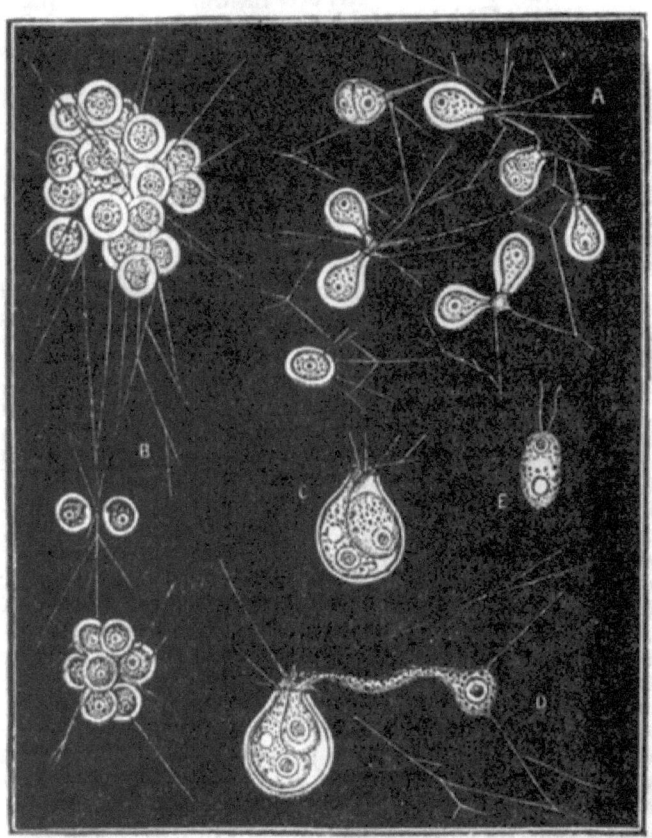

Microgromia socialis:—A, colony of individuals in extended state, some of them undergoing transverse fission; B, colony of individuals (some of them separated from the principal mass) in compact state; C, D, formation and escape of swarm-spore, seen free at E.

spreading itself out irregularly over the pseudopodia of the colony. But it finally gathers itself together and takes an oval form; and either develops a pair of flagella, and forsakes the colony as a free swimming *Monad* (§ 416), or assumes the form of an *Actinophrys*, moving about by three or four pointed pseudopodia,—probably in each case coming after a time to rest, excreting a shell, and laying the foundation of a new colony. There is reason to think that a multiplication by longitudinal fission also takes place, in which the escaping segment and the one left behind in the old shell remain attached by their pseudopodia, and the former develops a new shell without undergoing any change of condition.

399. *Heliozoa.*—The *Actinophrys sol*, sometimes termed the 'sun-animalcule' (Fig. 285), is one of the commonest examples of this group; being often met-with in lakes, ponds, and streams, amongst Confervæ and other aquatic plants, as a whitish-gray spherical particle distinguishable by the naked eye, from which (when it is brought under a sufficient magnifying power) a number of very pellucid, slender, pointed rods are seen to radiate. The central portion of the body is composed of homogeneous sarcode, inclosing a distinct nucleus with a large nucleolus (as in Fig. 287, N); but the peripheral part has a 'vesicular' aspect, as in the type next to be described (Fig. 286). This appearance is due to the number of 'vacuoles' filled with a watery fluid, which are included in the sarcodic substance, and which may be artificially made either to coalesce into larger ones, or to subdivide into smaller. A 'contractile vesicle,' pulsating rhythmically with considerable regularity, is always to be distinguished, either in the midst of the sarcode-body, or (more commonly) at or near its surface; and it sometimes projects considerably from this, in the form of a flattened sacculus with a delicate membranous wall, as shown at *o*. The cavity of this sacculus is not closed externally, but communicates with the surrounding medium; not, however, by any distinct and permanent orifice, the membraniform wall giving way when the vesicle contracts, and then closing-over again. This alternating action seems to serve a respiratory purpose, the water thus taken-in and expelled being distributed through a system of channels and vacuoles excavated in the substance of the body; some of the vacuoles which are nearest the surface being observed to undergo distention when the vesicle contracts, and to empty themselves gradually as it re-fills. The body of this animal is nearly motionless, but it is supplied with nourishment by the instrumentality of its pseudopodia; its food being derived not merely from Vegetable particles, but from various small Animals, some of them (as the young of Entomostraca) possessing great activity as well as a comparatively high organization. When one of these happens to come into contact with one of the pseudopodia (which have firm axis-filaments clothed with a granular sarcode), this usually retains it by adhesion, but the mode in which the particle thus taken captive is introduced into the body, differs according to circumstances. If the prey is large and vigorous enough to struggle to escape from its entanglement, it may usually be observed that the neighboring pseudopodia bend over and apply themselves to it, so as to assist in holding it captive, and that it is slowly drawn by their joint retraction toward the body of its captor. Any small particle not capable of offering active resistance, on the other hand, may be seen after a little time to glide towards the central body along the edge of the pseudopodium, without any visible movement of the latter, much in the same manner as in Gromia. When in either of these modes the food has been brought to the surface of the body, this sends over it on either side a prolongation of its own sarcode-substance; and thus a marked prominence is formed (Fig. 285 c), which gradually subsides as the food is drawn more completely into the interior. The struggles of the larger Animals, and the ciliary action of *Infusoria* and *Rotifera*, may sometimes be observed to continue even after they have been thus received into the body; but these movements at last cease, and the process of digestion begins. The alimentary sub-

stance is received into one of the vacuoles of the endosarc (Fig. 287, F), where it lies in the first instance surrounded by liquid; and its nutritive portion is gradually converted into an undistinguishable gelatinous mass, which becomes incorporated with the material of the sarcode-body, as may be seen by the general diffusion of any coloring particles it may contain. Several vacuoles may be thus occupied at one time by alimentary particles; frequently four to eight are thus distinguishable, and occasionally ten or twelve; Ehrenberg, in one instance, counted as many as sixteen, which he described as multiple stomachs. Whilst the digestive process, which usually occupies some hours, is going on, a kind of slow circulation takes place in the entire mass of the endosarc with its included vacuoles. If, as often happens, the body taken-in as food possesses some hard indigestible portion (as the shell of an Entomostracan or Rotifer), this, after the digestion of the soft parts, is gradually pushed towards the surface, and is thence extruded by a process exactly the converse of that by which it was drawn in. If the particle be large, it usually escapes at once by an opening which (like the mouth) extemporizes itself for the occasion (D); but, if small, it sometimes glides along a pseudopodium from its base to its point, and escapes from its extremity.

Fig. 288.

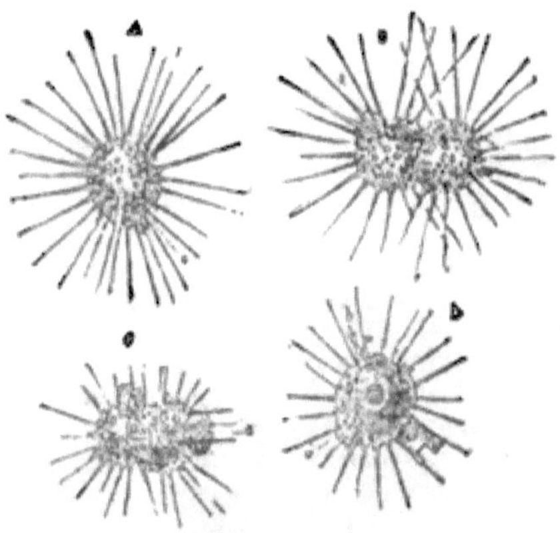

Actinophrys sol, in different states:—A, in its ordinary sun-like form, with a prominent contractile vesicle *o*; B, in the act of division or of conjugation, with two contractile vesicles *o, o*; C, in the act of feeding; D, in the act of discharging fæcal (?) matters, *a* and *b*.

400. The ordinary mode of Reproduction in *Actinophrys* seems to be by binary subdivision: its spherical body showing an annular constriction, which gradually deepens so as to separate its two halves by a sort of hour-glass contraction; and the connecting band becoming more and more slender, until the two halves are completely separated. This process of fission, which may be completed within half an hour from its commencement, seems to take place first in the contractile vesicle; for each segment very early shows itself to be provided with its own (B, *o*, *o*), and the two vesicles are commonly removed to a considerable distance from one another. The segments thus divided are not always equal, and sometimes their difference in size is very considerable. A junction of two individuals, on the other hand, has been seen to take place in *Actinophrys*, and has been supposed to correspond to the 'conjugation' of Protophytes; it is very doubtful, however, whether this junction really involves a complete fusion of the substance of the bodies which take part in it; and there is not sufficient evidence that it has any true generative character. Certain it is that such a junction or 'zygosis' may take place, not between two only, but between several individuals at once, their number being recognized by that of their contractile vesicles; and that, after remaining thus united for several hours, they may separate again without having undergone any discoverable change.

401. Under the generic name *Actinophrys* was formerly ranked the larger but less common Heliozoon now distinguished as *Actinosphærium Eichornii* (Fig. 286); one important difference consisting in the structure of the radiating pseudopodia, each of which has here a firm axis-filament or 'spine,' which passing through the superficial zone, rests on the surface of the central sphere, as shown at *a a*, Fig. 287. This axis is clothed with a layer of soft sarcode derived from the superficial or cortical zone of the body. Several nuclei (*n, n*) are usually to be seen embedded in the protoplasmic mass.—The general life-history of this type corresponds with that of the preceding; but its mode of reproduction presents some marked peculiarities. The binary segmentation is preceded by a withdrawal of the pseudopodia, even their clearly-defined axis becoming indistinct and finally disappearing; the body becomes enveloped,

FIG. 286.

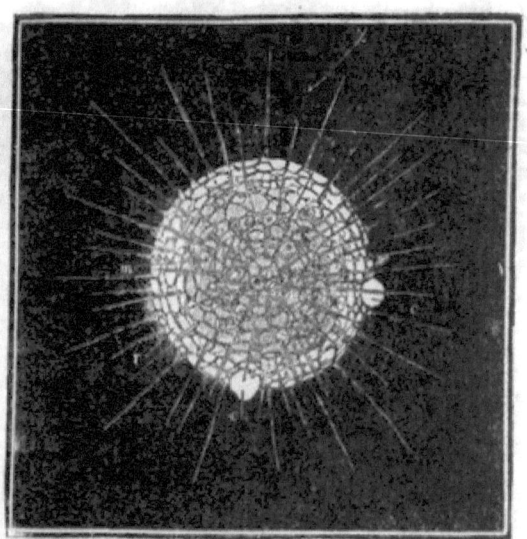

Actinosphærium Eichornii:—*m*, endosarc; *r*, ectosarc; *c, c,* contractile vacuoles.

by a clear gelatinous exudation, which forms a kind of cyst; and within this the process of binary subdvision is repeatedly performed, until the original single mass is replaced by a sort of *morula* (§ 391), each spherule of which shows the distinction between the central and cortical regions, the former including a single nucleus, whilst the latter is strengthened by siliceous deposit into a firm investment. After remaining in this state during the winter, the young *Actinosphæriæ* come forth in the spring without this siliceous investment; and gradually grow into the likenesses of their parent.

402. A large number of new and curious fresh-water forms of this type have been recently brought under notice; of which the *Clathrulina elegans* (Fig. 288) may be specially mentioned as presenting an obvious transition to the *Polycystine* type (§ 504). This has been found in various

parts of the Continent, and also (by Mr. Archer[1]) in Wales and Ireland; occurring chiefly in dark ponds shaded by trees and containing decaying leaves. Its soft sarcode body is incased by a siliceous capsule of spherical form, regularly perforated with oval apertures, and supported on a long silicified peduncle. The body itself, and the pseudopodia which it puts forth through the apertures of the capsule, seem closely to correspond with those of *Actinophrys*.—Reproduction here takes place not only by binary fission, but by the formation of 'swarm spores.' In the first mode, one of the two segments remains in possession of the siliceous capsule, whilst the other finds its way out through one of the apertures, lives for some hours in a free condition as an Actinophrys, and ultimately produces the capsule and stem characteristic of its type. In the second mode, numerous small rounded sarcode-masses, each possessing a nucleus, are produced within the capsule, in what manner cannot

FIG. 287.

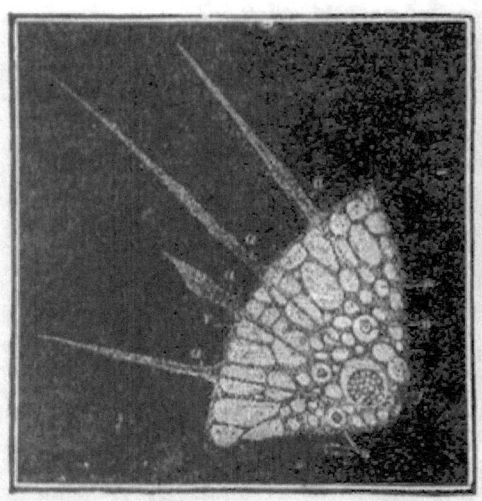

Marginal portion of *Actinosphærium Eichornii*, as seen in optical section under a higher magnifying powder:—*m*, endosarc; *r*, ectosarc; *a, a, a*, pseudopodia; *n, n*, nuclei with nucleoli; *f*, ingested food-mass.

be clearly made-out; and every one of these is enveloped in a firm envelope, set round with short spines, probably siliceous. These cysts remain for months within the common capsule; and when the time arrives for their further development, the sarcode-corpuscles slip out of their cysts, and escape through the orifices of the capsule as flagellated Monads of oval form (Fig. 288, B,) each having a nucleus, *n*, near the base of the flagella, and two contractile vesicles near its opposite end. After swarming for some hours in this condition, they change to the free *Actinophrys* form, and finally acquire the siliceous capsule and stem of the Clathrulina.

403. *Lobosa*.—No example of the Rhizopod type is more common in

---

[1] See his Memoir on Fresh-water Radiolaria in "Quart. Journ. of Micros. Sci.," N.S., Vol. ix. (1869), p. 250.

streams and ponds, vegetable infusions, etc., than the *Amœba* (Fig. 289); a creature which cannot be described by its form, for this is as changeable as that of the fabled Proteus, but may yet be definitely characterized by peculiarities that separate it from the two groups already described.

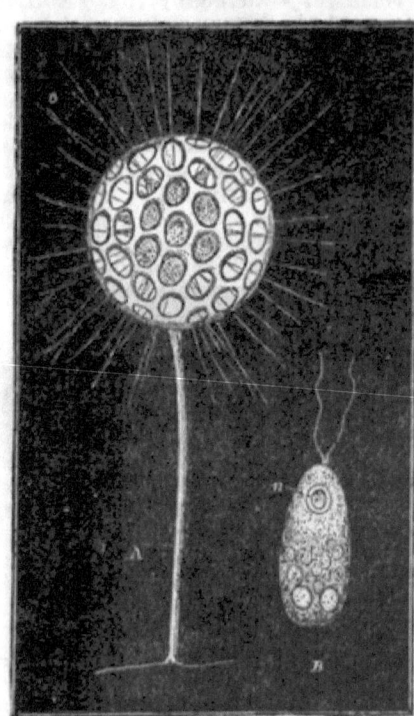

Clathrulina elegans:—A, complete organism; B, swarm-spore, showing nucleus, N, and two contractile vesicles near its opposite end.

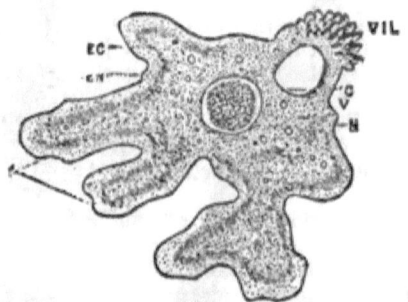

Diagrammatic representation of *Amœba proteus;*—E c, ectosarc; EN, endosarc; CV, contractile vesicle; N, nucleus; P, pseudopodia; VIL, villous tuft.

The distinction between 'ectosarc' and 'endosarc' is here clearly marked, so that the body approaches much more closely in its characters to an ordinary 'cell' composed of cell-wall and cell-contents. It is through the 'endosarc' alone, EN, that those colored and granular particles are diffused, on which the hue and opacity of the body depend; its central portion seems to have an almost watery consistence, the granular particles being seen to move quite freely upon one another with every change in the shape of the body; but its superficial portion is more viscid, and graduates insensibly into the firmer substance of the 'ectosarc.' The ectosarc, EC, which is perfectly pellucid, forms an almost membranous investment to the endosarc; still it is not possessed of such tenacity as to oppose a solution of its continuity at any point, for the introduction of alimentary particles, or for the extrusion of effete matters; and thus there is no evidence, in *Amœba* and its immediate allies, of the existence of any more definite orifice, either oral or anal, than exists in other Rhizopods. The more advanced differentiation of the ectosarc from the endosarc of *Amœba*, is made evident by the effects of re-agents. If an *Amœba radiosa* be treated with a dilute alkaline solution, the granular and molecular endosarc shrinks together and retreats towards the centre, leaving the radiating extensions of the ectosarc in the condition of cæcal tubes, of which the walls are not soluble at the ordinary temperature, either in acetic or mineral acids, or in dilute alkaline solutions; thus agreeing with the envelope noticed by Cohn as possessed by *Paramecium* and other ciliated *Infusoria*, and with the containing membrane of ordinary animal cells. A 'nucleus,' N, is always distinctly visible in *Amœba*, adherent to the inner portion of the ectosarc, and projecting from this into the cavity occupied by the endosarc; when most perfectly seen, it presents the aspect of a clear flattened vesicle surrounding a solid and usually spherical nucleolus; it is readily soluble in alkalies, and first expands and then dissolves when treated with acetic or sulphuric acid of moderate strength; but when treated with dilute acid it is rendered darker and more distinct, in consequence of the precipitation of a finely granular substance in the clear vesicular space that surrounds the nucleolus. A 'contractile vesicle,' CV, seems also to be uniformly present, though it does not usually make itself so conspicuous by its external prominence as it does in *Actinophrys*; and the neighboring part of the body is often prolonged into a set of villous processes VIL, the presence of which has been thought by some to mark a specific distinction, but which seems too variable and transitory to be so regarded.

404. The pseudopodia, which are not so much appendages, as lobate extensions of the body itself, are few in number, short, broad, and rounded; and their outlines present a sharpness which indicates that the substance of which their exterior is composed possesses considerable tenacity. No movement of granules can be seen to take place along the surface of the pseudopodia; and when two of these organs come into contact, they scarcely show any disposition even to mutual cohesion, still less to fusion of their substance. Sometimes the protrusion seems to be formed by the ectosarc alone, but more commonly the endosarc also extends into it, and an active current of granules may be seen to pass from what was previously the centre of the body into the protruded portion, when the latter is undergoing rapid elongation; whilst a light current may set towards the centre of the body from some other protrusion which is being withdrawn into it. It is in this manner that an *Amœba* moves from place to place; a protrusion like the finger of a glove being first formed, into which the substance of the body itself is gradually transferred; and another protrusion being put forth, either in the same or in some different direction, so soon as this transference has been accomplished, or even before it is complete. The kind of progression thus executed by an *Amœba* is described by most observers as a 'rolling' movement, this being certainly the aspect which it commonly seems to present; but it is maintained by M.M. Claparède and Lachmann that the appearance of rolling is an optical illusion, for that the nucleus and contractile vesicle always maintain the same position relatively to the rest of the body, and that 'creeping' would be a truer description of their mode of progression. It is in the course of this movement from place to place, that the *Amœba* encounters particles which are fitted to afford it nourishment: and it appears to receive such particles into its interior through any part of the ectosarc, whether of the body itself or of any of its lobose expansions; insoluble particles which resist the digestive process being got rid of in the like primitive fashion.

405. It may often be seen that portions of the sarcode-body of an *Amœba*, detached from the rest, can maintain an independent existence; and it is probable that such separation of fragments is an ordinary mode of increase in this group. When a pseudopodial lobe has been put-forth to a considerable length, and has become enlarged and fixed at its extrem-

ity, the subsequent contraction of the connecting portion, instead of either drawing the body towards the fixed point, or retracting the lobe into the body, causes the connecting band to thin-away until it separates; and the detached portion speedily shoots out pseudopodial processes of its own, and comports itself in all respects as an independent Amœba. Multiplication also takes place by regular binary subdivision. And an issue of 'swarm-spores,' which swim about for a time like Infusoria, has been witnessed by a competent observer.[1] In the *A. terricola* discovered by Greef in earth and dry sand, this process is seen to commence in the nucleus, which breaks-up into rounded corpuscles that diffuse themselves through the substance of the endosarc. The creature then ceases to take-

Fig. 290.

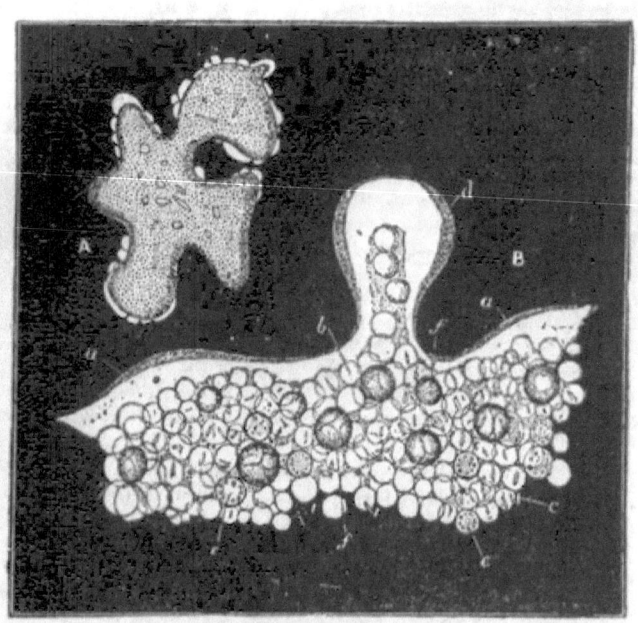

Pelomyxa palustris:—A, as it appears when in amœboid motion:—B, portion more highly magnified; showing *a, a*, the hyaline ectosarc; *b*, one of the vacuoles of the endosarc; *c*, rod-like bodies scattered through the endosarc; *d*, protruded extension of ectosarc, with endosarc passing into it; *e, e*, nuclei; *f, f*, globular hyaline bodies.

in food; its motions become less active, and its functions seem to be entirely confined to the nurture of the germs, which finally make their way out, and soon attain the size and aspect of their parent.—No sexual act has been certainly recognized as part of the life-history of *Amœba*, the union of two or more individuals, which may be occasionally witnessed, having more the character of the 'zygosis' of Actinophrys (§ 400).

406. A sarcodic organism discovered by Greef, and named by him *Pelomyxa palustris* (Fig. 290), which spreads over the bottom of stagnant ponds in the condition of slimy masses of indefinite form, exhibits a

---

[1] Prof. A. M. Edwards (U. S.) in "Monthly Microsc. Journ.," Vol. viii. (1872), p. 29

further advance upon the Amœban type. The substance of its body exhibits a very clear differentiation between the homogeneous hyaline ectosarc (B, *a*, *d*), and the contained endosarc, which contains such a multitude of spherical vacuoles, *b*, as to have a 'vesicular' or frothy aspect. When it feeds upon the decomposing vegetable matter at the bottom of the pools it inhabits, its body acquires a blackish hue; but in other situations it may be colorless. Besides the vacuoles, there are seen in the endosarc a great number of nucleus-like bodies, *e*, *e*, and also many hyaline globular brilliant bodies, *f*, *f*, which are regarded by Greef as germs or swarm-spores, developed from nucleoli set free within the general cavity of the body by the bursting of the nuclei. This creature, during the active period of its life, moves like an Amœba, either by general undulations of its surface, or by special pseudopodial extensions *d*. After a time, however, its movements cease, and it looks as if dead; but by the giving-way of its ectosarc, a multitude of minute amœbiform bodies break forth, each having its nucleus and contractile vesicle. These at first live as *Amœbæ*, but afterwards pass into a resting state, assuming a spherical or oval shape, and then put forth flagella, by which they swim actively for a time,—probably then settling-down to develop themselves into the parental form.

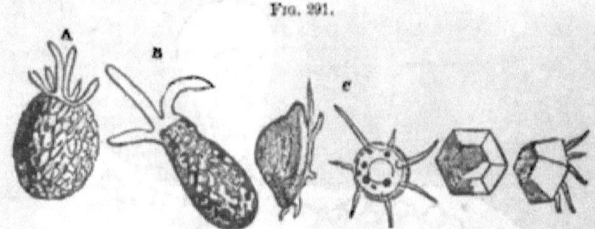

Fig. 291.

Testaceous forms of *Amœban* Rhizopods:—A, *Difflugia proteiformis*; B, *Difflugia oblonga*; C, *Arcella acuminata*; D, *Arcella dentata*.

407. The Amœban like the Actinophryan type shows itself in the testaceous as well as in the naked form; the commonest examples of this being known under the names *Arcella* and *Difflugia*. The body of the former is inclosed in a 'test' composed of a horny membrane, apparently resembling in constitution the *chitine* which gives solidity to the integuments of Insects: it is usually discoidal (Fig. 291, C, D) with one face flat and the other arched, the aperture being in the centre of the flat side; and its surface is often marked with a minute and regular pattern. The test of *Difflugia*, on the other hand, is more or less pitcher-shaped (A, B), and is chiefly made up of minute particles of gravel, shell, etc., cemented together. In each of these genera, these sarcode-body resembles that of *Amœba* in every essential particular; the contrast being very marked between its large, distinct lobose extensions, and the ramifying and inosculating pseudopodia of *Gromia* (Fig. 283). In each case a detached portion of the sarcodic body will put forth pseudopodia of its own type; and the separation of a bud or gemmule put forth from the mouth of the test seems to be an ordinary mode of propagation among the Amœbans thus inclosed. In *Arcella* it has been observed that the pseudopodia of two or more individuals unite by bridges of protoplasm, and afterwards separate; but it seems doubtful whether this is a true generative 'conjugation,' or a mere 'zygosis.' It has been observed by Bütschli, however, that after the separation of three individuals which had been thus united, the sar-

codic body of one of them had withdrawn itself for a considerable space from the wall of the test, and that in the liquid which filled the interval a number of Vibrio-like bodies (spermatozoids?) swarmed; while numerous disk-shaped masses of protoplasm lay on the surface of the body. After some time these showed lively amœboid movements, creeping about between the body of the parent and the wall of the test, and ultimately escaping through its orifice. Each of them contained a nucleus and contractile vesicle, and moved by means of blunt pseudopodia; and it seems probable that they were embryoes which would in time form the characteristic *Arcella*-test.

408. Many testaceous *Amœbans* have been recently discovered, which form tests of remarkable regularity and sometimes of singular beauty; and it is difficult to determine, in many cases, whether the minute plates of which they are composed have been formed by exudation from their own bodies, or have been picked up from the surface over which the animals crawl.[1] There can be no doubt of this kind, however, in regard to the *Quadrula symmetrica* represented in Fig. 292, whose sarcode-body is encased in a pear-shaped test of glassy transparence, made up of a great number of square plates which touch each other by their edges. The sarcode body does not usually fill the test; the intervening space being occupied by a clear liquid, and traversed by bands of protoplasm. In the posterior part of the body is seen a large clear spherical nucleus, with a distinct dark nucleolus; and in front of this are contractile vesicles, usually two in number.

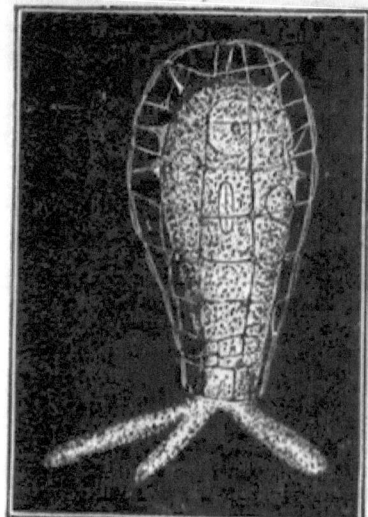

Fig. 292.

Quadrula symmetrica, with extended pseudopodia.

409. *Coccoliths and Coccospheres.*—This would seem the most appropriate place for the description of certain peculiar little bodies found very extensively diffused over the deep-sea bottom, especially abounding in the Globigerina-mud (§ 480), which may be considered as Chalk in process of formation. It was in the specimens of this mud brought up by the 'Cyclops' soundings in 1857, that Prof. Huxley first found the *Coccoliths* (Fig. 293, 1, 2) which Dr. Wallich in 1860 found aggregated in the spherical masses which he designated as 'coccospheres' (3). Regarding the gelatinous matrix in which they were imbedded as a new type of the *Monerozoa* described by Haeckel, having the condition of an indefinitely extended *plasmodium*, Prof. Huxley proposed to designate it by the name *Bathybius*, indicative of its habitat in the depths of the sea; and this idea was

[1] See especially the recent admirable work of Prof. Leidy on the Freshwater Rhizopods of the United States (1880).—It is to be regretted that its able Author's time and opportunities did not permit him to follow-out the life-histories of the many interesting forms which he has described and figured.

accepted by Haeckel, whose representation of a living specimen of *Bathybius*, with imbedded coccoliths, is given in Fig. 293, 4. The observations made in the 'Challenger' Expedition, however, have not confirmed this view; the supposed *Bathybius* being a gelatinous precipitate, consisting of sulphate of lime, slowly deposited in water to which strong spirit has been added. Whatever be their nature, Coccoliths and Coccospheres are bodies of great interest; since their **occurrence in** Chalk and in very early Limestones (§ 699) is an additional **link in the** evidence of the similarity of the conditions under which they were **formed, to those at** present prevailing on the sea-bed of the Atlantic and others oceans.—Two distinct types are recognizable among the Coccoliths, which Prof. Huxley has designated respectively *discoliths* and *cyatholiths*. The former are round or oval disks, having a thick strongly-refracting rim and a thinner internal portion, the greater part of which is occupied by a slightly-opaque, cloud-like patch lying round a central corpuscle (Fig. 293, 5). In general, the 'discoliths' are slightly convex on one side, slightly concave on the other, and the rim is raised into a prominent ridge on the more convex side;

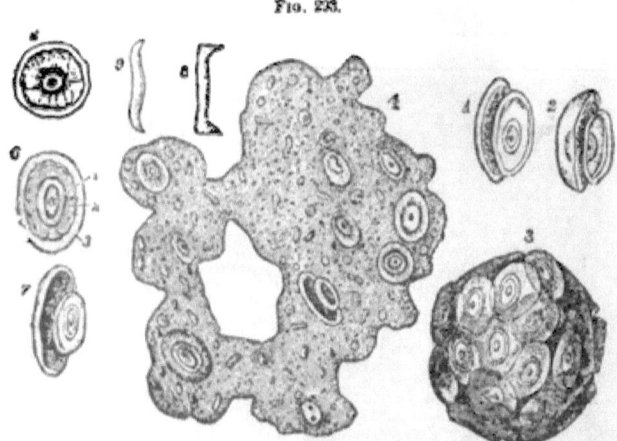

Fig. 293.

Coccoliths and Coccospheres:—1, 2, 7, Cyatholiths seen obliquely;—3, Coccosphere, with imbedded cyatholiths;—4, Coccoliths imbedded in supposed protoplasmic expansion;—5, Discolith seen in front view;—6, Cyatholith seen in front view, showing (1) central corpuscle, (2) granular zone, (3) transparent outer zone;—8, 9, Discoliths seen edgeways.

so that when **viewed** edgewise, they present the appearances shown in figs. 8, 9. Their length is ordinarily between 1-4000th and 1-5000th of an inch; but it ranges from 1-2700th to 1-11,000th. The largest are commonly free; but the smallest are generally found imbedded among heaps of granular particles, of which some are probably discoliths in an early stage of development.—The 'cyatholiths,' also, when full grown, have an oval contour; though they are often circular when immature. They are convex on one face and flat or concave **on the other; and** when left to themselves, they lie on one or other of **these two faces.** In either of these aspects, they seem to be composed of two **concentric zones** (fig. 6, 2, 3) surrounding an oval thick-wall central corpuscle **(1)**, in the centre of which is a clear space sometimes divided into two. The zone (2) immediately surrounding the central corpuscle is usually more or less distinctly granular, and sometimes has an almost bead-like margin. The narrower outer zone (3) is generally clear, transparent, and structureless; but sometimes shows radiating striæ. When viewed sideways or obliquely, however, the 'cyatholiths' are found to have a form somewhat resembling that of a shirt-stud (figs. 1, 2, 7). Each consists of a lower plate, shaped like a deep saucer or watch-glass; of a smaller upper plate, which is sometimes flat, sometimes more or less concavo-convex; of the oval, thick-wall, flattened corpuscle, which connects these two plates together at their centres; and of an intermediate granular substance, which more or less completely fills up the interval between the two plates. The length of these cyatholiths ranges from about 1-1600th to 1-8000th of an inch, those of 1-3000th of an inch and under being always circular.—It appears from the action of dilute acids upon the Coccoliths, that they must mainly consist of calcareous matter, as they readily dissolve, leaving scarcely a trace behind. When the cyatholiths are treated with very weak acetic acid, the central corpuscle rapidly loses its strongly refracting character; and there remains an extremely delicate, finely-granular membranous framework. When treated with iodine, they are stained, but not very strongly; the intermediate substance being the most affected. Both discoliths and cyatholiths are completely destroyed by strong hot solutions of caustic potass or soda.—The Coccospheres (Fig. 3) are made up by the aggregation of bodies resembling 'cyatholiths' of the largest size in all but the absence of the granular zone; they sometimes attain a diameter of 1-760th of an inch.—What is their relation to the Coccoliths, and under what conditions these bodies are formed, are questions whereon no positive judgment can be at present given. (See § 710.)

GREGARINIDA.

410. A very curious animal parasite is often to be met with in the intestinal canal of Earthworms, Insects, etc., and sometimes in that of higher animals, the simplicity of whose structure requires that it should be ranked among the Protozoa. Each individual *Gregarina* (Fig. 294, A) essentially consists of a large single cell, usually more or less ovate in form, and sometimes attaining the extraordinary length of *two-thirds of an inch*.[1] A sort of beak or proboscis frequently projects from one extremity; and in some instances this is furnished with a circular row of hooklets, closely resembling that which is seen on the head of Tænia. There is here a much more complete differentiation between the cell-membrane and its contents, than exists either in *Actinophrys* or in *Amœba;* and in this respect we must look upon *Gregarina* as representing a decided advance in organization. Being nourished upon the juices already prepared for it by the digestive operations of the animal which it infests, it has no need of any such apparatus for the introduction of solid particles into the interior of its body, as is provided in the 'pseudopodia' of the Rhizopods and in the oral cilia of the Infusoria. Within the cavity of the cell, whose contents are usually milk-white and minutely granular, there is generally seen a pellucid nucleus; and when, as often happens, the cell undergoes duplicative subdivision, the process commences in a constriction and cleavage of this nucleus. The membrane and its contents, except the nucleus, are soluble in acetic acid. Cilia have been detected both upon the outer and the inner surface; but these would seem destined, not so much to give motion to the body, as to renew the stratum of fluid

---

[1] See Prof. Ed. Van Beneden on *Gregarina gigantia*, in "Quart. Journ. Microsc. Sci.," N. S., Vol. x. (1870), p. 51, and Vol. xi., p. 242.

in contact with it; for such change of place as the animal does exhibit, is effected by the contractions and extensions of the body generally, as in Amœba (§ 403). An 'encysting process,' very much resembling that of the lower Protophytes, is occasionally observed to take place in *Gregarinæ*, and seems to be preparatory to their multiplication. Whatever the original form of the body may be, it becomes globular, ceases to move, and becomes invested by a structureless 'cyst,' within which the substance of the body undergoes a singular change. The nucleus disappears; and the sarcodic mass breaks up into a series of globular particles, which gradually resolve themselves (as shown at B, C) into forms very like those of *Naviculæ*. These 'pseudo-navicellæ' are set-free, in time, by the bursting of the capsule that incloses them; and they develop themselves into a new generation of Gregarinæ, first passing through an Amœba-like stage.—A sort of 'conjugation' has been seen to take place between two individuals, whose bodies, coming into contact with each

FIG. 294.

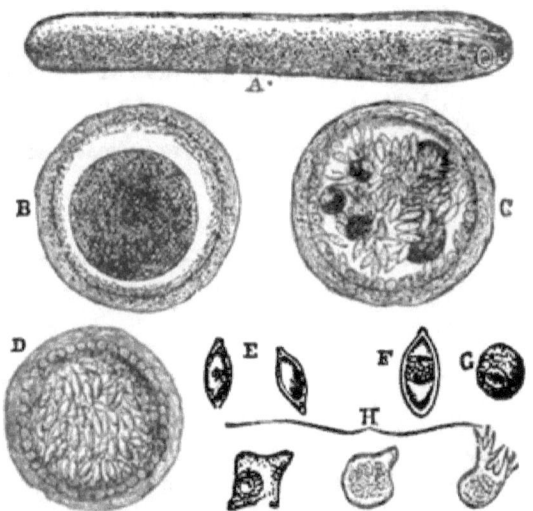

Gregarina of the Earthworm:—A, in its ordinary aspect; B, in its encysted condition; C, D, showing division of its contents into pseudo-navicellæ; E, F, free pseudo-navicellæ; G, H, free amœboids produced from them.

other by corresponding points, first become more globular in shape, and are then encysted by the formation of a capsule around them both; the partition-walls between their cavities disappear; and the substance of the two bodies becomes completely fused together. But as the product of this 'zygosis' is the same as that of the ordinary encysting process, there seems no sufficient reason for regarding it, like the 'conjugation' of Protophytes, as a true Generative act.

Prof. Haeckel's Memoirs on *Monera* and the *Gastræa Theory* will be found in the successive Nos. of the "Jenaische Zeitschrift" beginning with 1868; and in a collected form, in the two parts of his "Biologische Studien." The first of his Memoirs on *Monera* is translated in "Quart. Journ. Microsc. Sci.," N.S., Vol. ix. (1869); and the first of his Papers on the *Gastræa Theory* in Vol. xiv. (1874)

of the same Journal. See also the valuable series of papers on the *Freshwater Rhizopods* by Mr. Wm. Archer, in the current series of the "Quart. Journ. Microc. Sci.;" the important Memoirs of Hertwig and Lesser in the "Archiv für Mikr. Anat." (especially the Suppl. Heft to Bd. x., 1874), and the Presidential Addresses of Prof. Allman to the Linnæan Society for 1876 and 1877 (in Nos. 69 and 71 of its Journal) on "Recent Researches on some of the more simple Sarcode-Organisms," of which the Author has freely availed himself.

CHAPTER XI.

ANIMALCULES.—INFUSORIA AND ROTIFERA.

411. Nothing can be more vague or scientifically inappropriate than the title Animalcules; since it only expresses the small dimensions of the beings to which it is applied, and does not indicate any of their characteristic peculiarities. In the infancy of Microscopic knowledge, it was natural to associate together all those creatures which could only be discerned at all under a high magnifying power, and whose internal structure could not be clearly made out with the instruments then in use; and thus the most heterogeneous assemblage of Plants, Zoophytes, minute Crustaceans, larvæ of Worms, Mollusks, etc., came to be aggregated with the true Animalcules under this head. The Class was being gradually limited by the removal of all such forms as could be referred to others; but still very little was known of the real nature of those that remained in it, until the study was taken up by Prof. Ehrenberg, with the advantage of instruments which had derived new and vastly improved capabilities from the application of the principle of Achromatism. One of the first and most important results of his study, and that which has most firmly maintained its ground, notwithstanding the overthrow of Prof. Ehrenberg's doctrines on other points, was the separation of the entire assemblage into two distinct groups, having scarcely any feature in common except their minute size; one being of very *low*, and the other of comparatively *high* organization. On the lower group he conferred the designation of *Polygastrica* (many-stomached), in consequence of having been led to form an idea of their organization which the united voices of the most trustworthy observers now pronounces to be erroneous; and as the retention of this term must tend to perpetuate the error, it is well to fall back on the name *Infusoria*, or Infusory Animalcules, which simply expresses their almost universal prevalence in infusions of organic matter. To the higher group, Prof. Ehrenberg's name *Rotifera* or *Rotatoria* is on the whole very appropriate, as significant of that peculiar arrangement of their cilia upon the anterior parts of their bodies, which, in some of their most common forms, gives the appearance (when the cilia are in action) of wheels in revolution; the group, however, includes many members in which the ciliated lobes are so formed as not to bear the least resemblance to wheels. In their general organization, these 'Wheel-animalcules' must certainly be considered as members of the *Articulated* division of the Animal Kingdom; and they seem to constitute a Class in that lower portion of it, to which the designation *Worms* is now commonly given.—Notwithstanding the wide zoological separation between these two kinds of Animalcules, it seems most suitable to the plan of the present work to treat of them in connection with one another; since the Microscopist continually finds them associated together, and studies them under similar conditions.

Section I.—Infusoria.

412. This term, as now limited by the separation of the *Rhizopoda* on the one hand, and of the *Rotifera* on the other, is applied to a far smaller range of forms than was included by Prof. Ehrenberg under the name of 'polygastric' animalcules. For a large section of these, including the *Desmidiaceæ, Diatomaceæ, Volvocineæ,* and many other Protophytes, have been transferred, by general (though not universal) consent, to the Vegetable kingdom. And it is not impossible that many of the reputed *Infusoria* may be but larval forms of higher organisms, instead of being themselves complete animals. Still an extensive group remains, of which no other account can at present be given, than that the beings of which it is composed go through the whole of their lives, so far as we are acquainted with them, in a grade of existence which is essentially Protozoic (§ 391); each individual apparently consisting of but *a single cell*, though its parts are often so highly differentiated, as to represent (only, however, by way of *analogy*) the 'organs' of the higher animals after which they are usually named.

413. Among the *ciliate* Infusoria, which form not only by far the largest, but also the most characteristic division of the group, there is probably none which has not a *mouth*, or permanent orifice for the introduction of food, which is driven towards it by ciliary currents; while a distinct *anal* orifice, for the ejection of the indigestible residue, is also generally present. The mouth is often furnished with a *dental* armature; and leads to an *œsophageal canal*, down which the food passes into the digestive cavity. This cavity is still occupied, however, as in Rhizopods (403), by the *endosarc* of the cell; but instead of lying in mere vacuoles formed in the midst of this, the food-particles are usually aggregated, during their passage down the œsophagus, into minute pellets, each of which receives a special investment of firm protoplasm, constituting it a digestive vesicle (Fig. 299); and these go through a sort of circulation within the cell-cavity.

414. The 'contractile vesicles' again, attain a much higher development in this group, and are sometimes in a connection with a network of canals channelled-out in the 'ectosarc' while their rhythmical action resembles that of the *circulatory* and *respiratory* apparatuses of higher animals. There is ample evidence, also, of the presence of a specially contractile modification of the protoplasmic substance, having the action (though not the structure) of *muscular* fibre; and the manner in which the movements of the active free-swimming Infusoria are directed, so as to avoid obstacles and find-out passages, seems to indicate that another portion of their protoplasmic substance must have to a certain degreee the special endowments which characterize the *nervous* systems of higher animals. Altogether, it may be said that in the Ciliate Infusoria *the Life of the Single Cell finds its highest expression*.[1]

---

[1] The doctrine of the unicellular nature of the *Infusoria* has been a subject of keen controversy among Zoologists, from the time when it was first definitely put forward by Von Siebold ("Lehrbuch der vergleich. Anat.," Berlin, 1845) in opposition to the then paramount doctrine of Ehrenberg as to the complexity of their organization, which had as yet been called in question only by Dujardin "Hist. Nat. des Infusoires," Paris, 1841). Of late, however, there has been a decided convergence of opinion in the direction above indicated; which has been brought about in great degree by the contrast between the *Protozoic* simplicity of the reproductive and developmental processes in Infusoria, and the com-

415. Before proceeding to the description of the *ciliate* Infusoria, however, it will be of advantage to notice two smaller groups—the *flagellate* and the *suctorial*—which, on account of the peculiarities of their structure and actions, are now ranked as distinct, and of whose 'unicellular' character there can be no reasonable doubt, since they are for the most part 'closed' cells, scarcely distinguishable morphologically from those of Protophytes.

416. FLAGELLATA.—Our knowledge of this tribe has been greatly augmented in recent years, not only by the discovery of a great variety of new forms, but still more by the careful study of the life history of several among them. The *Monads*, properly so called,[1] which are the smallest animals at present known, are its simplest representatives; but it also includes organisms of much greater complexity; and some of its composite forms have a very remarkable relation to Sponges (§ 508). The *monas lens*, long familiar to Microscopists as occurring in stagnant waters and infusions of decomposing organic matter, is a spheroidal particle of protoplasm, from 1-2000th to 1-5200th of an inch in diameter, inclosed in a delicate hyaline investment or 'ectosarc' and moving freely through the water by the lashing action of its slender *flagellum*, whose length is from three to five times the diameter of the body. Within the body may be seen a variable number of vacuoles; and these are occasionally occupied by particles distinguishable by their color, which have been introduced as food. These seem to enter the body, not by any definite mouth (or permanent opening in the ectosarc), but through an aperture that forms itself in some part of the oral region near the base of the flagellum. In the smallest *Monadinæ*, neither nucleus nor contractile vesicle is distinguishable; but in larger forms a nucleus can be clearly seen. The life-history of several simple *Monadinæ*, presenting themselves in infusions of decaying animal matter (a cod's head being found the most productive material), has been studied with admirable perseverance and thoroughness by Messrs. Dallinger and Drysdale, of whose important observations a general summary will now be given.[2]

417. The Monad-form most recently and completely studied by Mr. Dallinger—with all the advantages derived from trained experience, and under objectives of the highest quality and greatest magnifying power—is the *Dallingeria Drysdali* (Kent) represented in Plate xiii. Its normal shape, as seen in fig. 1, is a long oval, slightly constricted in the middle, and having a kind of pointed neck (*a*), from which proceds a flagellum about half as long again as the body. From the shoulder-like projections behind this (*b, c*) arise two other long and fine flagella, which are directed backwards. The sarcode body is clear, and apparently structureless, with minute vacuoles distributed through it; and in its hinder part a nucleus (*d*) is distinguishable. The extreme length of the body is seldom more than the 1-4,000th of an inch, and is often less. This Monad swims with

---

plexity of the like processes as seen even in the lowest of the *Metazoa* (§ 391) which has been specially and forcibly insisted on by Haeckel ("Zur Morphologie der Infusorien," Jenaische Zeitschr., Bd. vii., 1873).—An excellent summary of the whole discussion was given by Prof. Allman, in his Presidential Address to the Linnæan Society in 1875.

[1] The Family *monadina* of Ehrenberg and Dujardin consists of an aggregate of forms now known to be of very dissimilar nature, many of them belonging to the Vegetable Kingdom.

[2] See their successive Papers in the "Monthly Microsc. Journ.," Vol. x. (1873), pp. 53, 245; Vol. xi. (1874), pp. 7, 69, 97; Vol. xii. (1874), p. 261; and Vol. xiii. (1875), p. 185;—and "Proceed. Roy. Soc.," Vol. xxvii. (1878), p. 332.

PLATE XIII.

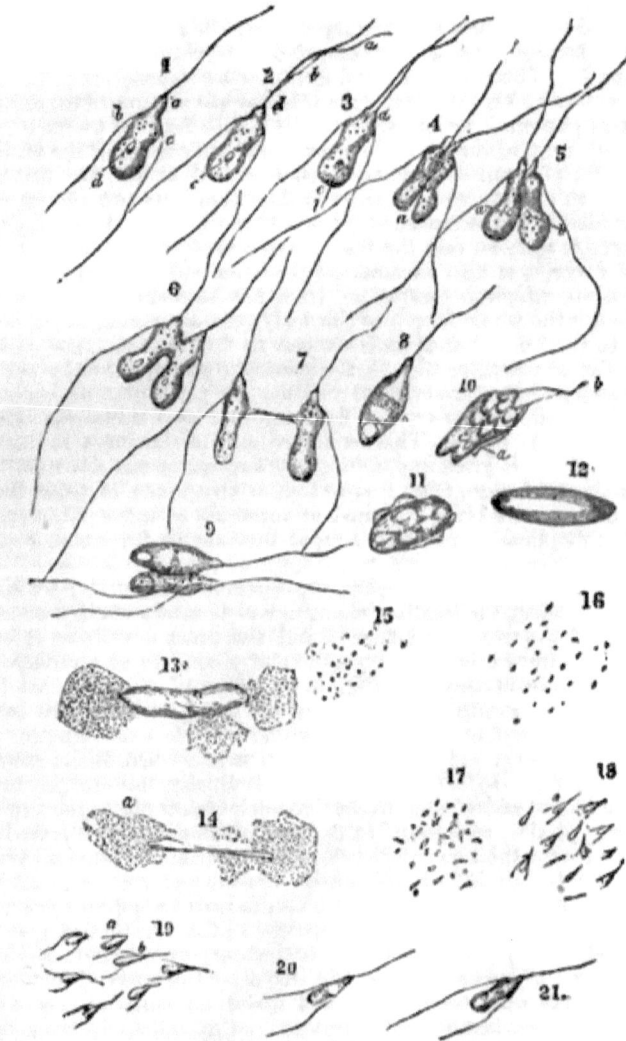

LIFE-HISTORY OF FLAGELLATE INFUSORIUM (after Dallinger).

Fig. 1. Normal form, showing three flagella, a, b, c, and nucleus d.
2. Anterior flagellum, a, b, double ; c, nucleus.
3. Fission commencing in nucleus c, and in anterior portion of body, a.
4. Fission more advanced, and showing itself also in posterior portion of body, a.
5. Fission still more advanced, both in nucleus, a, b, and in body.
6, 7. Fission proceeding to completion.
8. Change to amœboid condition, with single flagellum and granular band a.
9. Conjugation of this with free-swimming form.
10, 11. Stages of progressive fusion, terminating in production of still sac, 12 which afterwards opens and pours out spores, as at 13, 14, the progressive growth of which is shown in figs. 15-21.

great rapidity; its movements, which are graceful and varied, being produced by the action of the flagella, which can not only impel it in any direction, but can suddenly reverse its course or check it altogether. But besides this free-swimming movement, a very curious 'springing' action is performed by this Monad when the decomposing organic matter of the infusion is breaking up, the process of disintegration being apparently assisted by it. The two posterior flagella anchor themselves and coil into a spiral, and the body then darts forwards and upwards, until the anchored flagella straighten out again, when the body falls forward to its horizontal position, to be again drawn back by the spiral coiling of the anchored flagella. This Monad multiplies by longitudinal fission; the first stage of which is the splitting of the anterior flagellum into two (fig. 2, a, b); and a movement of the nucleus (c) towards the centre. In the course of *from thirty to sixty seconds* the fission extends down the neck fig. 3, a; a line of division is also seen at the posterior end (c), and the nucleus (b) shows an incipient cleavage. In a few seconds the cleavage-line runs through the whole length of the body, the separation being widest posteriorly (fig. 4, a); and in from one to four minutes the cleavage becomes almost complete (fig. 5), the posterior part of the body, with the two halves (a and b) of the original nucleus, being now quite disconnected, though the anterior parts are still held together by a transverse band of sarcode, as seen in fig. 6. This soon narrows and elongates, as shown in fig. 7; and at last it gives way, setting the two bodies entirely free. The whole process of fission, from first to last, is completed in from four to seven minutes; and being repeated at intervals of a few minutes, this mode of multiplication produces a rapid increase in the number of the Monads.

418. Such fission does not, however, continue indefinitely; for certain individuals undergo a peculiar change, which shows itself first in the absorption of the two lateral flagella and the great development of the nucleus, and afterwards in the formation of a transverse granular band across the middle of the body (fig. 8, a). One of these altered forms swimming into a group in the 'springing' state, within a few seconds firmly attaches itself to one of them, which at once unanchors itself, and the two swim freely and vigorously about, as shown in fig. 9, generally for from thirty-five to forty-five minutes. Gradually, however, a 'fusion' of the two bodies and of their respective nuclei takes place, the two trailing flagella of the 'springing' form being drawn-in (fig. 10); and in a short time longer the two anterior flagella also disappear, and all trace of the separate bodies is lost, the nuclei vanish, and the resultant is an irregular amœboid mass (fig. 11), which gradually acquires the smooth, distended, and 'still' condition represented in fig. 12. This a cyst filled with reproductive particles of such extraordinary minuteness, that, when emitted from the ends of the cyst (fig. 13) after the lapse of four or five hours, they can only be distinguished under an amplification of 5,000 diameters, with perfect central illumination through an aperture in the diaphragm of from 1-80th to the 1-100th of an inch in diameter. Yet these particles, when continuously watched, are soon observed to enlarge and to undergo elongation (figs. 15–17); and within two hours after their emission from the sac, the anterior flagellum, and afterwards the two lateral flagella (fig. 18) can be distinguished. Slight movements then commence; the neck-like protrusion shows itself (fig. 19, a, b), and in about half an hour more the regular swimming action begins. About four hours after the escape of its germ from the sac, the Monad acquires

its characteristic form (fig. 20), though still only one-half the length of its parent; but this it attains (passing through the stage shown in fig. 21) in another hour, and the process of multiplication by fission, as already d scribed, commences very soon afterwards.—There can be no reasonable doubt that the 'conjugation' of two individuals, followed by the transformation of their fused bodies into a sac filled with reproductive germs, is to be regarded (as in protophytes) in the light of a true *generative* process; and it is interesting to observe the indication of sexual distinction here marked by the different states of the two conjugating individuals. —There is every reason to believe that *the entire life-cycle* of this Monad has thus been elucidated; and it will now be sufficient to notice the principal diversities observed by Messrs. Dallinger and Drysdale in the life-cycles of the other Monadine forms which they have studied.

419. Their simple *uniflagellate* Monad (*Monas Dallingeri*, Kent). having an ovate form with a long diameter never exceeding 1-4000th of an inch, and advancing slowly with a straight, uniform motion like that of *Monas termo*, differs from the preceding in its mode of multiplication; for this takes place, not by duplicative fission, but by the breaking-up of the sarcodic substance (as in the production of 'swarm spores' by Protophytes) into from thirty to sixty segments, which, at first lying closely packed together, make their escape as free-swimming Monads, each provided with its *flagellum*. Conjugation, in this type, occurs between the ordinary forms and certain individuals distinguished between their somewhat larger size, and by the granular aspect of their sarcode towards the flagellate end; and there is reason to think that the latter have never undergone the segmentation by which the former have been multiplied. The smaller are absorbed, as it were, into the larger; and the latter passes after a time into the encysted state, corresponding in its subsequent history with the preceding type.—The *bi-flagellate* or 'acorn' Monad of the same observers (identified by Kent with the *Polytoma uvella* of Ehrenberg) presents some remarkable peculiarities in its mode of reproduction. Its binary fission extends only to the protoplasmic substance of its body, leaving its envelope entire; and by a repetition of the process, as many as 16 segments, each attaining the likeness of the parent, are seen thus inclosed, their flagella protruding through the general investment. This compound state being supposed by Ehrenberg to be the normal one, he named it accordingly. But the parent-cyst soon bursts, and sets free the contained 'macro-spores,' which swim about freely, and soon attain the size of the parent. Again, the posterior part of the body of certain individuals shows an accumulation of granular protoplasm, giving to that region a roughened acorn-cup-like aspect; the bursting of the projection, while the creature is actively swimming through the water, sets free a multitude of shapeless granular fragments, within each of which a minute bacterium-like corpuscle is developed; and this, on its release, acquires in a few hours the size and form of the original monad. This process seems analogous to the development of 'micro-spores' among Protophytes, by the direct breaking-up of the protoplasm. It is, like the previous process, non-sexual or *gonidial;* the true generative process consisting here, as in the preceding cases, in the 'conjugation' of two individuals, with the usual results.

420. A *Cercomonas* (*C. typicus*, Kent), characterized by the possession of a flagellum at each end, was found to multiply, during eight days (and nights) of continuous observation, by *transverse* duplicative subdivision alone. But certain individuals then exhibited a remarkable change,

becoming amœboid and less active; and when two of these came into contact, they underwent a complete fusion, the product of which was a globular cyst, with a very definite investment, filled with reproductive germs.—The 'springing Monad' of the same observers (*Heteromita rostrata*, Kent) is of a long ovate form, with an average length of about 1-3000th of an inch. From its narrower extremity a sort of beak arises, from which proceeds a fine flagellum about half as long again as the body; and at a little distance behind this, another and longer flagellum arises, with which the Monad anchors itself to the covering-glass, constantly springing backwards and forwards by its recurrent coil and uncoil. A nucleus shows itself near the rounded posterior end of the body. This Monad multiplies by *longitudinal* fission, commencing at the beaked end, and completed in six or seven minutes; and the process may be repeated continuously for many days. Among enormous numbers, there are a few distinguishable from the others by a slight excess of size, and by the power to swim freely; these become 'still'—for a time amœboid—then round; a small cone of sarcode pushes out, dividing and increasing into another pair of flagella; the disk splits, each part becomes possessed of a nuclear body, and two well-formed free-swimming Monads are set free. These conjugate with individuals of the ordinary form which have just undergone fission, the nuclei of the two approximating to each other; a complete fusion of sarcode and nuclei takes place; the body, at first motile, comes to rest, assumes a triangular form, and loses its flagella; it then becomes clear and distended, and emits its contained reproductive granules at the angles.—The 'hooked Monad' (*Heteromita uncinata*, Kent) is another bi-flagellate form, usually ovate with one end pointed, and from 1-3000th to 1-4000th of an inch in length; being distinguished from the preceding by the peculiar character of its flagella, of which the one that projects forward is not more than half the length of the body, and is permanently hooked, while the other, whose length is about twice that of the body, is directed backwards, flowing in graceful curves. Its motion consists of a succession of springs or jerks rapidly following each other, which seems produced by the action of the hooked flagellum. Multiplication takes place by *transverse* fission, and continues uninterruptedly for several days. A difference then becomes perceptible between larger and smaller individuals; the former being further distinguished by the presence of what seems to be a contractile vesicle in the anterior part of the body. Conjugation occurs between one of the larger and one of the smaller forms, the latter being, as it were, absorbed into the body of the larger; and the resulting product is a spherical cyst, which soon begins to exhibit a cleavage-process in its interior. This continues until the whole of its sarcodic substance is subdivided into minute oval particles, which are set free by the rupture of the cyst, and of which each is usually furnished with a single flagellum, by whose lashing movement it swims freely. These germs speedily attain the size and form of the parent, and then begin to multiply by transverse fission—thus completing the 'genetic' cycle.

421. The 'calycine Monad' of the same observers (*Tetramitus rostratus*, Perty), has a length of from 1-900th to 1-1000th of an inch, and a compressed body tapering backwards to a point. Its four flagella (which constitute its generic distinction) arise nearly together from the flattened front of the body; and its swimming movement is a graceful gliding. Near the base of the flagella is a pair of contractile vesicles; and further behind is a large nucleus. Multiplication takes place by longitudinal

fission, which is preceded by a change to a semi-amœboid state. This gives place to a more regular pear-like form, the four flagella issuing from the large end; and the fission commences at their base, two pairs being separated by the cleavage-plane. The nucleus also undergoes cleavage, and its two halves are carried apart by the backward extension of the cleavage. The two half-bodies at last remain connected only by their hinder prolongations, which speedily give way, and set them free. Each, however, has, as yet, only two flagella; but these speedily fix themselves by their free extremities, undergo a rapid vibratory movement, and in the course of about two minutes split themselves from end to end. A still more complete change into the amœboid condition, in which the creature not only moves, but also feeds, like an *Amœba* (devouring all the living and dead Bacteria in its neighborhood), occurs previously to 'conjugation;' and this takes place between two of the amœboid forms, which begin to blend into one another almost immediately upon coming into contact. The conjugated bodies, however, swim freely about for a time, the two sets of flagella apparently acting in concert. But by the end of about eighteen hours, the fusion of the bodies and nuclei is complete, the flagella are retracted, and a spherical distended sac is then formed, which, in a few hours more, without any violent splitting or breaking up, sets free innumerable masses of reproductive particles. These, under a magnifying power of 2,500 diameters, can be just recognized as oval granules, which rapidly develop themselves into the likeness of their parents, and in their return multiply by duplicative fission,—thus completing the 'genetic' cycle.

422. One of the most important researches thus ably prosecuted by Messrs. Dallinger and Drysdale, has reference to the Temperatures respectively endurable by the adult or developed forms of these Monads, and by their reproductive germs. A large number of experiments upon the several forms now described, indubitably led to the conclusion that all the *adult* forms, as well as all those which had reached a stage of development in which they can be distinguished from the reproductive granules, are utterly destroyed by a temperature of 150° Fahr. But, on the other hand, the reproductive granules emitted from the cysts that originate in 'conjugation' were found capable of sustaining a *fluid* heat of 220°, and a dry heat of about 30° more,—those of the Cercomonad surviving exposure to a *dry* heat of 300° Fahr. This is a fact of the highest interest in its bearing on the question of 'spontaneous generation' or Abiogenesis; since it shows (1) that germs capable of surviving desiccation may be everywhere diffused through the air, and may, on account of their extreme minuteness (as they certainly do not exceed 1-200,000th of an inch in diameter), altogether escape the most careful scrutiny and the most thorough cleansing processes; while (2) their extraordinary power of resisting heat will prevent these germs from being killed either by boiling, or by dry-heating up to even 300° Fahr.[1]

423. The structural resemblance of these simple Flagellate Infusoria to the 'Monads' of *Volvox* and its allies (§ 237), is so close that no other than physiological reasons can be assigned for separating them. Whilst the *Volvocineæ* grow and multiply under conditions which seem to justify our regarding them as members of the Vegetable Kingdom (§ 220),

---

[1] Descriptions of the special apparatus used by Messrs. Dallinger and Drysdale in their researches will be found in "Monthly Micr. Journ.," Vol. xi. (1874), p. 97; ibid., Vol. xv. (1876), p. 165; and "Proceed. Roy. Soc.," Vol. xxvii. (1878), p. 343.

the 'flagellated' agree with the 'ciliated' *Infusoria* in ordinarily drawing their nutriment from organic compounds; and it seems clear that, although unpossessed of a mouth, they can introduce solid food-particles into the interior of their bodies. It is, however, not a little remarkable that (according to the statement of Messrs. Dallinger and Drysdale)[1] these Flagellata—like *Bacteria* and other forms referred to the group of Fungi—can be cultivated in Cohn's 'nutritive fluid' (§ 303, *note*), which consists **only** of tartrate of ammonia and mineral salts, without any albuminous matter.

424. A large series of more complex forms of Flagellate Infusoria has been recently brought to our knowledge by the researches of the late Prof. James-Clark (U. S.),[2] followed by those of Stein and Saville Kent. In some of these, a sort of collar-like extension of what appears to be the sarcodic ectosarc, proceeds from the anterior extremity of the body (Fig. 295, *cl*), forming a kind of funnel, from the bottom of which the flagellum arises; and by its vibrations a current is produced within the funnel, which brings down food-particles to the 'oral disk' that surrounds its origin, where the ectosarc seems softer than that which envelops the rest of the body. Towards the base of the collar, a nucleus (*n*) is seen; while, near the posterior termination of the body, is a single or double contractile vesicle *cv*. The body is attached by a pedicle proceeding from its posterior extremity, which also seems to be a prolongation of the ectosarc.—These Animalcules multiply by longitudinal fission; and this, in some cases (as in the genus *Monosiga*), proceeds to the extent of a complete separation of the two bodies, which henceforth, as in the ordinary *Monadina*, live quite independently of each other. But in other forms, as *Codosiga*, the fission does not extend through the pedicel; and the twin bodies being thus held together at their bases, and themselves undergoing duplicative fission, clusters are produced which spring from common pedicels (Fig. 296). And by the extension of the division down the pedicels, themselves, composite arborescent fabrics, like those of Zoophytes, are produced.

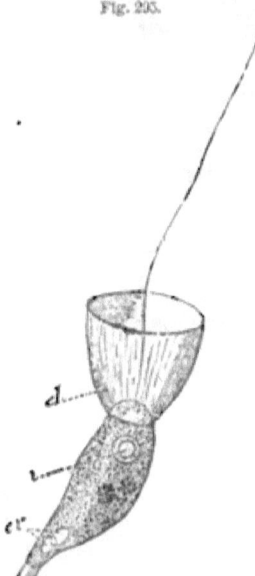

Fig. 295.

Single zooid of *Codosiga umbellata*:—*cl*, collar; *n*, nucleus; *cv*, double contractile vesicle.

425. In another group, a structureless and very transparent horny calyx, closely resembling in miniature the polype-cell of a *Campanularia* (Plate xx.), forms itself around the body of the Monad, which can retract itself into the bottom of it. And in the genus *Salpingœca* both calyx and collar are present. In some forms of this group, multiplication seems to take place, not by fission, but by gemmation; and, as among Hydroid Polypes, the *gemmæ* may either detach themselves and live independently, or may remain in connection with their parent-stocks, form-

---

[1] "Monthly Microscopical Journal," Vol. xiii. (1875), p. 190.
[2] See his Memoirs in "Ann. Nat. Hist.," Ser. 3, Vol. xviii. (1866); ibid., Ser. 4, Vol. i. (1868); Vol. vii. (1871); and Vol. ix. (1872).

ing composite fabrics, in some of which the calyces follow one another in linear series, whilst in others they take on a ramifying arrangement. While some of these composite organisms are sedentary, others, as *Dinobryon*, are free-swimming.

426. Two solitary Flagellate forms, *Anthophysa* and *Anisonema*, may be specially noticed as presenting several interesting points of resemblance to the peculiar type next to be described; the most noticeable being the presence of a distinct mouth, and the possession of two different motor organs—one a comparatively stout and stiff bristle of uniform diameter throughout, which moves by occasional jerks; and the other a very delicate tapering flagellum, which is in constant vibratory motion. If, as appears from the recent observations of Bütschli, the well-known *Astasia* —of which one species has a blood-red color, and sometimes multiplies to such an extent as to tinge with it the water of the ponds it inhabits—has

Fig. 226.

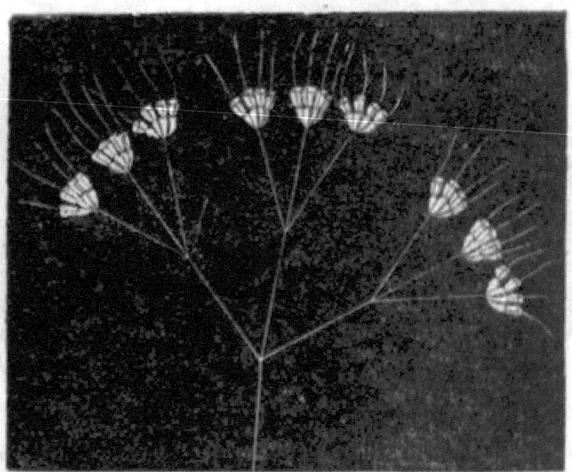

osiga umbellata:—colony-stock, springing from single pedicel tripartitely branched.

a true mouth for the reception of its food, it must be regarded as an Animal, and separated from the *Euglena* (with which it has been generally associated), the latter being pretty certainly a Plant belonging to the same group as *Volvox*.[1]

427. There can be no longer any doubt that the well-known *Noctiluca miliaris*—to which is attributable the *diffused* luminosity that frequently presents itself in British seas—is to be regarded as a gigantic type of the 'unicellular' *Flagellata*. This animal, which is of spheroidal form, and has an average diameter of about 160th of an inch, is just large enough to be discerned by the naked eye when the water in which it may be swimming is contained in a glass jar held up to the light; and its tail-like appendage, whose length about equals its own diameter, and which serves as an instrument of locomotion, may be discerned with a hand-

---

[1] See the Memoir by Prof. Bütschli, in "Zeitschrift f. Wissensch. Zool.," Bd. xxx.; of which an abridgment (with Plate) is given in "Quart. Journ. Microsc. Sci.," Vol. xix. (1879), p. 63.

magnifier. The form of *Noctiluca* is nearly that of a sphere, so compressed that while on one aspect (Fig. 297, A) its outline, when projected on a plane, is nearly circular, it is irregularly oval in the aspect (B) at right angles to this. Along one side of this body is a meridional groove, resembling that of a peach; and this leads at one end into a deep depression of the surface, *a*, termed the *atrium*, from the shallower commencement of which the *tentacle*, *d*,[1] originates, whilst it deepens down at the base of the tentacle to the mouth, *e*. Along the opposite meridian there extends a slightly elevated ridge, *c*, which commences with the appearance of a bifurcation at the end of the atrium farthest from the tentacle; this is of a firmer consistence than the rest of the body, and has somewhat the appearance of a rod imbedded in its walls. The mouth opens into a short œsophagus, which leads directly down to the great central protoplasmic mass; on the side of this canal farthest from the

Fig. 297.

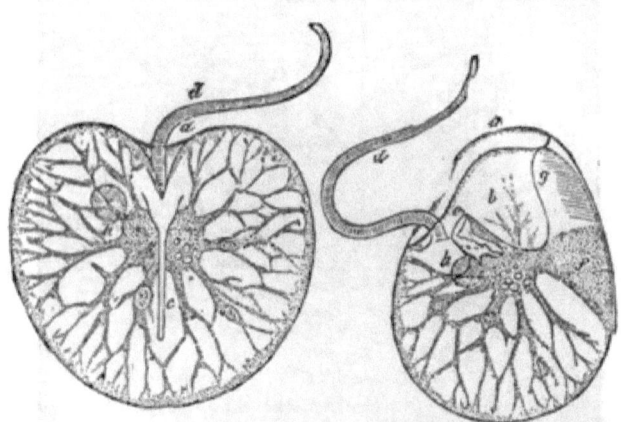

Noctiluca miliaris, as seen at A on the aboral side, and at B on a plane at right angles to it:—*a*, entrance to atrium; *b*, atrium; *c*, superficial ridge; *d*, tentacle; *e*, mouth leading to œsophagus within which are seen the flagellum springing from its base, and the tooth-like process projecting into it from above; *f*, broad process from the central protoplasmic mass, proceeding to superficial ridge; *g*, duplicature of wall; *h*, nucleus.—Magnified about 90 diameters.

tentacle, is a firm ridge that forms a tooth-like projection into its cavity; whilst from its floor there arises a long *flagellum*, which vibrates freely in its interior. The central protoplasmic mass sends off in all directions branching prolongations of its substance, whose ramifications inosculate; these become thinner and thinner as they approach the periphery; and their ultimate filaments, coming into contact with the delicate membranous body-wall, extend themselves over its interior, forming a protoplasmic

---

[1] The organ here termed 'tentacle' is commonly designated *Flagellum;* while what is here termed the *flagellum* is spoken of by most of those who have recognized it, as a *cilium*. The Author agrees with M. Robin in considering the former organ, which has a remarkable resemblance to a single fibrilla of striated muscle (§ 678), as one peculiar to *Noctiluca;* and the latter as the true homologue of the flagellum of the ordinary Flagellata.—It is curious that several observers have been unable to discover the so-called cilium, which was first noticed by Krohn. Prof. Huxley sought for it in at least fifty individuals without success; and out of the great number which he afterwards examined, did not get a clear view of it in more than half-a-dozen.

network of extreme tenuity (Fig. 298). Besides these branching prolongations there is sent off from the central protoplasmic mass a broad, thin, irregularly quadrangular extension (Fig. 297 B, f), which extends to the superficial rod-like ridge, and seems to coalesce with it; its lower free edge has a thickened border; whilst its upper edge becomes continuous with a plate-like striated structure, g, which seems to be formed by a peculiar duplicature of the body-wall. At one side of the protoplasmic mass is seen a spherical vesicle, h, of about 3-2000ths of an inch in diameter, having clear colorless contents, among which transparent oval corpuscles may usually be detected. This, from the changes it undergoes in connection with the reproductive process, must be regarded as a *nucleus*.

428. The particles of food drawn into the mouth (probably by the vi-

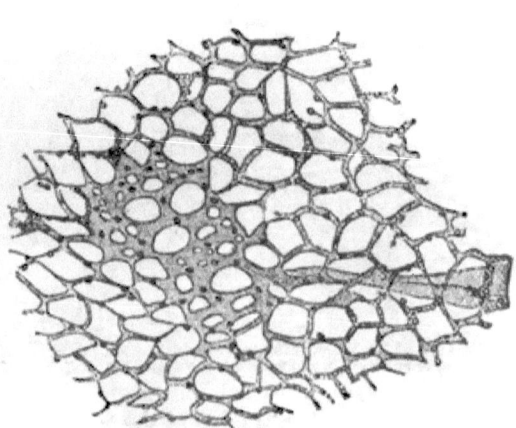

Fig. 298.

Portion of superficial protoplasmic reticulation, formed by ramification of an extension a, of central mass.—Magnified 1000 diameters.

brations of the flagellum) seem to be received into the protoplasmic mass at the bottom of the œsophagus by the extensions of its substance, which envelop them in filmy envelopes that maintain themselves as distinct from the surrounding protoplasm, and thus constitute extemporized digestive vesicles. These vesicles soon find their way into the radiating extensions of the central mass (as shown in Fig. 297, A, B), and are ensheathed by the protoplasmic substance which goes-on to form the peripheral network (Fig. 299). Their number and position are alike variable; sometimes only one or two are to be distinguished; more commonly from four to eight can be seen; and even twelve or more are occasionally discernible. The place of each in the body is constantly being changed by the contractions of the protoplasmic substance; these in the first place carrying it from the centre towards the periphery of the body, and then carrying it back to the central mass, into whose substance it seems to be fused as soon as it has discharged any indigestible material it may have contained, which is got rid of through the mouth. Every part of the protoplasmic reticulation is in a state of incessant change, which serves to distribute the nutrient material that finds its way into it through the walls of the digestive vesicles; but no regular *cyclosis* (like that of plants) can be observed in it. Besides the 'digestive vesicles,' vacuoles filled with clear fluid may be distinguished, alike in the central protoplasmic mass, and in its extensions, as is shown in the centre of Fig. 297. There is no contractile vesicle.

429. The peculiar 'tentacle' of *Noctiluca* is a flattened whip-like filament, gradually tapering from its base to its extremity; the two flattened faces being directed respectively towards and away from the oral aperture. When either of its flattened faces is examined, it shows an alternation of light and dark spaces, in every respect resembling those of striated muscular fibre, except that the clear spaces are not subdivided. But when looked-at in profile, it is seen that between the striated band and the aboral surface is a layer of granular protoplasm. The tentacle slowly bends over towards the mouth about five times in a minute, and straightens itself still more slowly; the middle portion rising first, while the point approaches the base, so as to form a sort of loop, which presently straightens. It seems probable that the contraction of the substance form-

FIG. 299.

Pair of Digestive Vesicles of *Noctiluca*, lying in a course of extension of central protoplasmic mass *a*, to form peripheral reticulation *b*, and containing remains of Algæ.—Magnified 480 diameters.

ing the dark bands, produces the bending of the filament; whilst, when this relaxes, the filament is straightened again by the elasticity of the granular layer.[1]

430. The extreme transparence of *Noctiluca* renders it a particularly favorable subject for the study of the phenomena of phosphorescence. When the surface of the sea is rendered luminous by the general diffusion of *Noctilucæ*, they may be obtained by the tow-net in unlimited quantities; and when transferred into a jar of sea-water, they soon rise to the surface, where they form a thick stratum. The slightest agitation of the jar in the dark causes an instant emission of their light, which is of a beautiful greenish tint, and is vivid enough to be perceptible by ordinary lamp-light. This luminosity is but of an instant's duration, and a short rest is required for its renewal. A brilliant, but short-lived display of luminosity, to be followed by its total cessation, may be produced by electric or chemical stimulation. Professor Allman found the addition of a drop of alcohol to the water containing specimens of *Noctiluca*, on the stage of the microscope, produce a luminosity strong enough to be visible under a half-inch objective, lasting with full intens-

---

[1] According to Robin, the 'tentacle' of *Noctiluca* is derived conjointly from the cell-wall and from its contained protoplasm; being thus differentiated alike from the 'flagellum,' which he regards as an extension of the *latter* alone, and from a 'cilium,' which is an extension of the *former*.

ity for several seconds, and then gradually disappearing. He was thus able to satisfy himself that the special seat of the phosphorescence is the peripheral protoplasmic reticulation which lines the external structureless membrane.

431. The reproduction in this interesting type is effected in various ways. According to Cienkowsky, even a small portion of the protoplasm of a mutilated *Noctiluca* will (as among Rhizopods) reproduce the entire animal. Multiplication by fission or binary sub-division, beginning in the enlargement, constriction, and separation of the two halves of the nucleus, has been frequently observed. Another form of non-sexual reproduction, which seems parallel to the 'swarming' of many Protophytes, commences by a kind of encysting process. The tentacle and flagellum disappear, and the mouth gradually narrows, and at last closes up; the meridional groove also disappears, so that the animal becomes a closed hollow sphere. The nucleus elongates, and becomes transversely constricted, and its two halves separate, each remaining connected with a portion of the protoplasmic network. This duplicative subdivision is repeated over and over again, until as many as 512 'gemmules' are formed, each consisting of a nuclear particle enveloped by a protoplasmic layer, and each having its flagellum. The entire aggregate forms a disk-like mass projecting from the surface of the sphere; and this mass sometimes detaches itself as a whole, subsequently breaking up into individuals; whilst, more commonly, the gemmules detach themselves one by one, the separation beginning at the margin of the disk, and proceeding towards its centre.—The gemmules are at first closed monadiform spheres, each having a nucleus, contractile vesicle, and flagellum; the mouth is subsequently formed, and the tentacle and permanent flagellum afterwards make their appearance.—A process of 'conjugation' has also been observed alike in ordinary *Noctilucæ* and in their closed or encysted forms, which seems to be sexual in its nature. Two individuals, applying their oral surfaces to each other, adhere closely together, and their nuclei become connected by a bridge of protoplasmic substance. The tentacles are thrown off, the two bodies gradually coalesce, and the two nuclei fuse into one. The whole process occupies about five or six hours, but its results have not been followed out.[1]

432. Intermediate between the proper *flagellate*, and the true *ciliate* Infusoria, is the small group of *Cilio-flagellata*, in which, while the body is furnished with rows of cilia, a flagellum is also present. Although this group does not contain any great diversity of forms, yet it is specially worthy of notice, on account of the occasional appearance of some of them in extraordinary multitudes. This is the case, for example, with the *Peridinium* observed by Prof. Allman, in 1854, to be imparting a brown color to the water of some of the large ponds in Phœnix Park, Dublin; this color being sometimes uniformly diffused, and sometimes showing itself more deeply in dense clouds, varying in extent from a few square yards to upwards of a hundred. The animal (Fig. 300, A, B) has a form approaching the spherical, with a diameter of from 1-1000th to 1-5000th of an inch; and is partially divided into two hemispheres, by

---

[1] *Noctiluca* has been the subject of numerous Memoirs, of which the following are the most recent: Cienkowski, "Arch. f. Micr. Anat.," Bd. vii. (1871), p. 131, and Bd. ix. (1873), p. 47; Allman, "Quart. Journ. Micr. Sci.," N.S., Vol. xii. (1872), p. 327; Robin, "Journ. de l'Anat. et de Physiol.," Tom. xiv. (1878), p. 586; and Vignal, "Arch. de Physiol.," Ser. 2, Tom. v. (1878), p. 415.

a deep equatorial furrow, *a*, whilst the flagellum-bearing hemisphere, A, has a deep meridional groove on one side, *b*, extending from the equatorial groove to the pole; the flagellum taking its origin from the bottom of this vertical groove, near its junction with the equatorial. The cilia, in this form, do not seem to be disposed in special bands, but are distributed

FIG. 300.

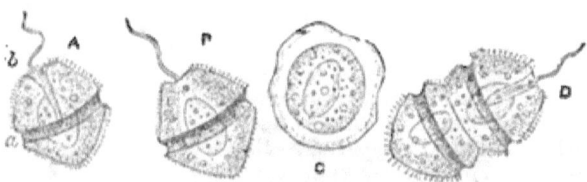

Peridinium uberrimum;—A, B, Front and back views; C, Encysted stage; D, Duplicative subdivision.

over the general surface of the body; but in several other *Peridinians* (Fig. 301), whose bodies are partially invested by a firm *lorica*, the cilia are arranged in special zones. It is questionable whether any definite mouth exists in this type; but it seems certain that alimentary particles

FIG. 301.

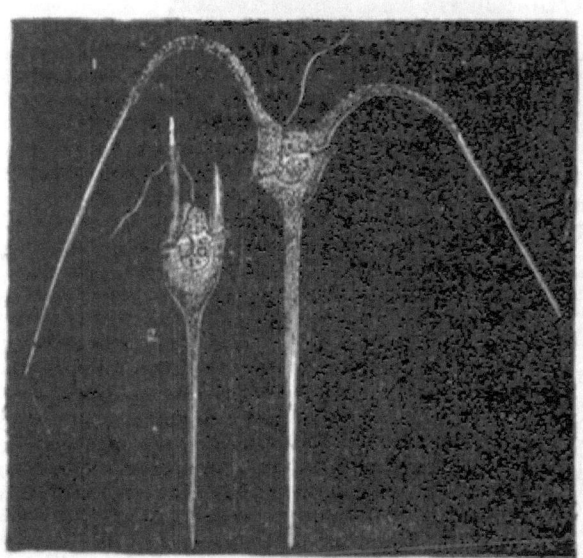

1, *Ceratium tripos;* 2, *Ceratium furca.*

are received into the interior of the body, becoming inclosed in 'digestive vesicles.' A 'contractile vesicle' has been rarely observed; but a large nucleus, sometimes oval, and sometimes horseshoe-shaped, seems always present.—The *Peridinia* multiply by transverse fission (Fig. 300, D), which commences in the subdivision of the nucleus, and then shows itself externally in a constriction of the ungrooved hemisphere, par

to the equatorial furrow. They pass into a quiescent condition, subsiding towards the bottom of the water; and the loricated forms appear to throw off their envelopes. But whether these changes are preparatory to any process of conjugation, is not known.—Some of the *Peridinia* are found in sea-water; but the most remarkable marine forms of the cilioflagellate group belong to the genus *Ceratium* (Fig. 301), in which the cuirass extends itself into long horny appendages. In the *Ceratium tripos* (1), there are three of its appendages; two of them curved, proceeding from the anterior portion of the cuirass, and the third, which is straight or nearly so, from its posterior portion. They are all more or less jagged or spinous. In *Ceratium furca* (2), the two anterior horns are prolonged straight forwards, one of them being always longer than the other; whilst the posterior is prolonged straight backwards. The anterior and posterior halves of the cuirass are separated by a ciliated furrow, from one point of which the flagellum arises; and at the origin of this is a deep depression, into which the flagellum may be completely and suddenly withdrawn. Whether this is, or is not, a true mouth leading into the cell-cavity, has not yet been ascertained.—The Author has found the *Ceratium tripos* extremely abundant in Lamlash Bay, Arran; where it constitutes a principal article of the food of the *Comatulæ* that inhabit its bottom.[1]

433. *Suctoria*.—The *suctorial* Infusoria constitute a well-marked group,—all belonging to one family, *Acinetina*,—the nature of which has been until recently much misunderstood, chiefly on account of the parasitism of their habit. Like the typical *Monadina*, they are closed cells, each having its nucleus and contractile vesicle; but instead of freely swimming through the water, they attach themselves by flexible peduncles, sometimes to the stems of *Vorticellinæ*, but also to filamentous Algæ, stems of Zoophytes, or the bodies of larger animals. Their nutriment is obtained through delicate tubular extensions of the ectosarc, which act as suctorial tentacles (Fig. 302); the free extremity of each being dilated into a little knob, which flattens out into a button-like disk when it is applied to a food-particle. Free-swimming Infusoria are captured by these organs, of which several quickly bend over towards the one which was at first touched, so as firmly to secure the prey; and when several have thus attached themselves, the movements of the imprisoned animal become feebler, and at last cease altogether, its body being drawn nearer to that of its captor. Instead, however, of being received into its interior like the prey of *Actinophrys* (§ 399), the captured Animalcule remains on the outside; but yields up its soft substance to the suctorial power of its victor. As soon as the sucking disk has worked its way through the envelope of the body to which it has attached itself, a very rapid stream, indicated by the granules it carries, sets along the tube, and pours itself into the interior of the Acineta-body. Solid particles are not received through these suctorial tentacles, so that the *Acinetina* cannot be fed with indigo or carmine; but, so far as can be ascertained by observation of what goes on within their bodies, there is a general protoplasmic *cyclosis*, without the formation of any special 'digestive vesicles.'—The ordinary forms of this group are ranked under the two genera *Acineta* and *Podophrya;* which are chiefly distinguished by the presence of a firm envelope or *lorica* in the former, while the body of the latter is

---

[1] See **Allman** in "Quart. Micr. Journ." Vol. iii. (1855), p. 24; and H. James-Clark in "Ann. Nat. Hist.," Ser. 3, Vol. xviii. (1866), p. 429.

naked. In one curious form, the *Ophiodendron*, the suckers are borne in a brush-like expansion on a long retractile proboscis like organ. And the rare *Dendrosoma*, whose size is comparatively gigantic, forms by continuous gemmation an arborescent 'colony,' of which the individual members remain in intimate connection with one another.

434. Multiplication in this group seems occasionally to take place by longitudinal fission; but this is rare in the adult state. Sometimes external *gemmæ* are developed by a sort of pinching-off of a part of the free end of the body, which includes a portion of the nucleus; the tentacula of this bud disappear, but its surface becomes clothed with cilia; and, after a short time, it detaches itself and swims away—comporting itself subsequently like the internal embryos, whose production seems the more ordinary method of propagation in this type. These originate in the breaking-up of the nucleus into several segments, each of which incloses itself in a protoplasmic envelope; and this becomes clothed with cilia, by

Fig. 302.

Suctorial Infusoria:—1. Conjugation of *Podophrya quadripartita*; 2. Formation of embryos by enlargement and subdivision of the nucleus; 3, Ordinary form of the same; 4, *Podophrya elongata*.

the vibrations of which the embryos are put in motion within the body of the parent (Fig. 302, 2), from which they afterwards escape by its rupture. In this condition (*a*) they swim about freely, and seem identical with what has been described by Ehrenberg as a distinct generic form, *Megatricha*. And according to the recent observations of Mr. Badcock,[1] these *Megatricha*-forms multiply freely by self-division. After a short time, however, they settle down upon filamentous Algæ or other supports, lose their cilia, put forth suctorial tentacles (which seem to shoot out suddenly in the first instance, but are afterwards slowly retracted and protruded with a kind of spiral movement), and assume a variety of amœbiform shapes (Fig. 303, 1, 2, 3), some of them corresponding to that of the genus *Trichophrya*. In this stage they become quiescent at

[1] "Journ. of Roy. Microsc. Soc.," Vol. iii. (1880), p. 563.

the approach of winter, the suctorial tentacles and the contractile vesicles disappearing; they do not, however, seem to acquire any special envelope, remaining as clear, motionless, protoplasmic particles. But with the return of warmth their development recommences, a footstalk is formed, and they gradually assume the characteristic form of *Podophrya quadripartita*.—A regular 'conjugation' has been observed in this type, the body of one individual bending down so as to apply its free surface to the corresponding part of another, with which it becomes fused (Fig. 302, 1); but whether this always precedes the production of internal embryos, or is any way preparatory to propagation, has not yet been ascertained.[1]

435. *Ciliata*.—As it is in this tribe of Animalcules that the action of the organs termed *Cilia* has the most important connection with the vital functions, it seems desirable here to introduce a more particular notice of them. They are always found in connection with *cells*, of whose protoplasmic substance they may be considered as extensions, endowed in

Fig. 303.

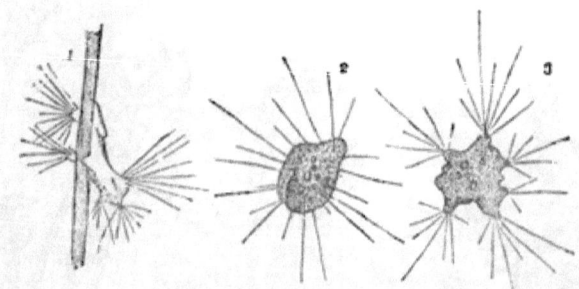

Immature forms of *Podophrya quadripartita*:—1, Amœboid state (*Trichophrya* of Claparède and Lachmann); 2, The same more advanced; 3, Incipient division into lobes.

a special degree with its characteristic contractility. The form of the filaments is usually a little flattened, tapering gradually from the base to the point. Their size is extremely variable; the largest that have been observed being about 1-500th of an inch in length, and the smallest about 1-13,000th. When in motion, each filament appears to bend from its root to its point, returning again to its original state, like the stalks of corn when depressed by the wind; and when a number are affected in succession with this motion, the appearance of progressive waves following one another is produced, as when a corn-field is agitated by successive gusts. When the ciliary action is in full activity, however, little can be distinguished save the whirl of particles in the surrounding fluid; but the *back*-stroke may often be perceived, when the *forward*-

---

[1] The *Acinetina* were described both by Ehrenberg and Dujardin; but the first full account of their peculiar organization was given by Stein in his "Organismus der Infusionsthierchen." Misled, however, by their parasitic habits, Stein originally supposed them not to be independent types, but to be merely transitional stages in the development of *Vorticellinæ* and other Ciliate Infusoria. This doctrine he has long since abandoned; but it is not a little singular that the young of several true *Ciliata* come forth provided with suctorial tentacles as well as with cilia, losing the former as they approximate with advancing growth towards the parental type. Much information as to this group will also be found in the beautiful "Etudes sur les Infusoires et les Rhizopodes" of MM. Claparède and Lachmann, Geneva, 1858-61.

stroke is made too quickly to be seen; and the real direction of the movement is then opposite to the apparent. In this back-stroke, when made slowly enough, a sort of 'feathering' action may be observed; the thin edge being made to cleave the liquid, which has been struck by the broad surface in the opposite direction. It is only when the rate of movement has considerably slackened, that the shape and size of the cilia, and the manner in which their stroke is made, can be clearly seen. Their action has been observed to continue for many hours, or even days, after the death of the body at large.—As *cilia* are not confined to Animalcules and Zoophytes, but give motion to the zoospores of many Protophytes (§ 248), and also clothe the free internal surfaces of the respiratory and other passages in all the higher Animals, including Man (our own experience thus assuring us that their action takes place, not only without any exercise of *will*, but even without *consciousness*), it is clear that

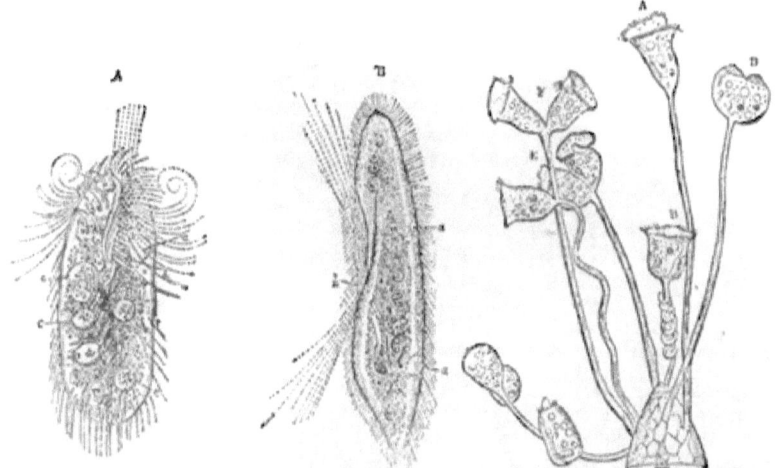

FIG. 304.   FIG. 305.

A, *Kerona silurus*:—a, contractile vesicle; b, mouth; c, c, Animalcules swallowed by the Kerona, after having themselves ingested particles of indigo. B, *Paramecium caudatum*:—a, a, contractile vesicles; b, mouth.

Group of *Vorticella nebulifera* showing A, the ordinary form; B, the same with the stalk contracted; C, the same with the bell closed; D, E, F, successive stages of fissiparous multiplication.

to regard Animalcules as possessing a 'voluntary' control over the action of their Cilia, is altogether unscientific.

436. In the Ciliated Infusoria, the differentiation of the sarcodic substance into 'ectosarc' or cell-wall, and 'endosarc' or cell-contents, becomes very complete; the ectosarc possessing a membranous firmness which prevents it from readily yielding to pressure, and having a definite internal limit, instead of graduating insensibly (as in Rhizopods) into the protoplasmic layer which lines it. A 'nucleus' seems always present; being sometimes 'parietal' (or adherent to the interior of the ectosarc), in other cases lying in the midst of the endosarc. In many *Ciliata* a distinct 'cuticle' or exudation-layer may be recognized on the surface of the ectosarc; and this cuticle, which is studded with regularly arranged markings like those of Diatomaceæ, seems to be the representative of the carapace of *Arcella*, etc. (Fig. 291), as of the cellulose coat of

Protophytes. It is sometimes hardened, so as to form a 'shield' that protects the body on one side only, or a 'lorica' that completely invests it; and there are other cases in which it is so prolonged and doubled upon itself, as to form a sheath resembling the 'cell' of a Zoophyte, within which the body of the Animalcule lies loosely, being attached only by a stalk at the bottom of the case, and being able either to project itself from the outlet or to retract itself into the interior. In a curious group lately described by Haeckel, consisting of Infusoria that spend their lives in the open sea, the body is inclosed in a siliceous lattice-work shell, usually bell-shaped or helmet-shaped, which bears so strong a resemblance to the shells of many Radiolaria as to be easily mistaken for them. The form of the body is usually much more definite than that of the naked Rhizopods; each species having its characteristic shape, which is only departed from, for the most part, when the Animalcule is subjected to pressure from without, or when its cavity has been distended by the ingestion of any substance above the ordinary size. The *cilia* and other mobile appendages of the body are extensions of the outer layer of the 'ectosarc' proper; and this layer, which retains a high degree of vital activity, is sometimes designated the 'cilia-layer.' Beneath this is a layer in which (or in certain bands of which) regular, parallel, fine striæ may be distinguished; and as this striation is also distinguishable in the eminently contractile footstalk of *Vorticella* (Fig. 305, B), there seems good reason to regard it as indicating a special modification of protoplasmic substance, which resembles muscle in its endowments. Hence this is termed the 'myophan-layer.' Beneath this, in certain species of Infusoria, there is found a thin stratum of condensed protoplasm, including minute 'trichocysts,' which resemble in miniature the 'thread-cells' of Zoophytes (§ 528); and this, where it exists, is known as the 'trichocyst-layer.'

437. The vibration of ciliary filaments,—which are either disposed along the entire margin of the body, as well as around the oral aperture, (Fig. 305, A, B), or are limited to some one part of it, which is always in the immediate vicinity of the mouth (Fig. 304),—supplies the means in this group of *Infusoria*, both for progression through the water, and for drawing alimentary particles into the interior of their bodies. In some, their vibration is constant, whilst in others it is only occasional. The modes of movement which Infusory Animalcules execute by means of these instruments, are extremely varied and remarkable. Some propel themselves directly forwards, with a velocity which appears, when thus highly magnified, like that of an arrow, so that the eye can scarcely follow them; whilst others drag their bodies slowly along like a leech. Some attach themselves by one of their long filaments to a fixed point, and revolve around it with great rapidity; whilst others move by undulations, leaps, or successive gyrations: in short, there is scarcely any kind of animal movement which they do not exhibit. But there are cases in which the locomotive filaments have a bristle-like firmness, and, instead of keeping themselves in rapid vibration, are moved (like the spines of Echini) by the contraction of the integument from which they arise, in such a manner that the Animalcule crawls by their means over a solid surface, as we see especially in *Trichoda lynceus* (Fig. 308, P, Q). —In *Chilodon* and *Nassula*, again, the mouth is provided with a circlet of plications or folds, looking like bristles, which, when imperfectly seen, received the designation of 'teeth;' their function, however, is rather that of laying hold of alimentary particles by their expansion

and subsequent drawing-together (somewhat after the fashion of the tentacula of Zoophytes), than of reducing them by any kind of masticatory process.—The curious contraction of the foot-stalk of the *Vorticella* (Fig. 305), again, is a movement of a different nature, being due to the contractility of the tissue that occupies the interior of the tubular pedicle. This stalk serves to attach the bell-shaped body of the Animalcule to some fixed object, such as a leaf or stem of duck-weed; and when the animal is in search of food, with its cilia in active vibration, the stalk is fully extended. If, however, the Animalcule should have drawn to its mouth any particles too large to be received within it, or should be touched by any other that happens to be swimming near it, or should be 'jarred' by a smart tap on the stage of the Microscope, the stalk suddenly contracts into a spiral, from which it shortly afterwards extends itself again into its previous condition. The central cord, to whose contractility this action is due, has been described as muscular, though not possessing the characteristic structure of either kind of muscular fibre; it possesses, however, the special irritability of muscle; being instantly called into contraction (according to the observations of Kühne) by electrical excitation. The only special 'impressionable' organs[1] for the direction of their actions, with the possession of which Infusoria can be credited, are the delicate bristle-like bodies which project in some of them from the neighborhood of the mouth, and in *Stentor* from various parts of the surface. The red spots seen in many *Infusoria*, which have been designated as eyes by Prof. Ehrenberg from their supposed correspondence with the eye-spots of *Rotifera* (§ 447), really bear a much greater resemblance to the red spots which are so frequently seen among Protophytes (§ 230).

438. The interior of the body does not always seem to consist of a simple undivided cavity occupied by soft sarcode; for the tegumentary layer appears in many instances to send prolongations across it in different directions, so as to divide it into chambers of irregular shape, freely communicating with each other, which may be occupied either by sarcode, or by particles introduced from without. The alimentary particles which can be distinguished in the interior of the transparent bodies of Infusoria, are usually protophytes of various kinds, either entire or in a fragmentary state. The Diatomaceæ seem to be the ordinary food of many; and the insolubility of their *loricæ* enables the observer to recognize them unmistakably. Sometimes entire Infusoria are observed within the bodies of others not much exceeding them in size (Fig. 308, B); but this is only when they have been recently swallowed, since the prey speedily undergoes digestion. It would seem as if these creatures do not feed by any means indiscriminately, since particular kinds of them are attracted by particular kinds of aliment; the crushed bodies and eggs of Entomostraca, for example, are so voraciously consumed by the *Coleps*, that its body is sometimes quite altered in shape by the distention. This circumstance, however, by no means proves that such creatures possess a sense of taste and a power of determinate selection; for many instances might be cited, in which actions of the like apparently-conscious nature are performed without any such guidance.—The ordinary process of feeding, as well as the nature and direction of the ciliary currents, may

---

[1] The term 'organs of sense' implies a *consciousness* of impressions, with which it is difficult to conceive that unicellular Infusoria can be endowed. The component cells of the Human body do their work without themselves knowing it.

be best studied by diffusing through the water containing the Animalcules a few particles of indigo or carmine. These may be seen to be carried by the ciliary vortex into the mouth, and their passage may be traced for a little distance down a short (usually ciliated) œsophagus. There they commonly become aggregated together, so as to form a little pellet of nearly globular form; and this, when it has attained the size of the hollow within which it is moulded, seems to receive an investment of firm sarcodic substance, resembling the 'digestive vesicles' of *Noctiluca* (§ 428), and to be then projected into the softer endosarc of the interior of the cell, its place in the œsophagus being occupied by other particles subsequently ingested. (This 'moulding,' however, is by no means universal; the aggregations of colored particles in the bodies of Infusoria being often destitute of any regularity of form.) A succession of such pellets being thus introduced into the cell-cavity, a kind of circulation is seen to take place in its interior; those that first entered making their way out after a time (first yielding up their nutritive materials), generally by a distinct anal orifice, but sometimes by the mouth. When the pellets are thus moving round the body of the Animalcule, two of them sometimes appear to become fused together, so that they obviously cannot have been separated by any firm membranous investment. When the animalcule has not taken food for some time, 'vacuoles,' or clear spaces, extremely variable both in size and number, filled only with a very transparent fluid, are often seen in its sarcode; and their fluid sometimes shows a tinge of color, which seems to be due to the solution of some of the vegetable chlorophyll upon which the Animalcule may have fed last.

439. Contractile Vesicles (Fig. 304, *a, a*), usually about the size of the 'vacuoles,' are found, either singly or to the number of from two to sixteen, in the bodies of most ciliated Animalcules; and may be seen to execute rhythmical movements of contraction and dilatation at tolerably regular intervals; being so completely obliterated, when emptied of their contents, as to be quite undistinguishable, and coming into view again as they are refilled. These vesicles do not change their position in the individual, and they are pretty constant, both as to size and place, in different individuals of the same species; hence they are obviously quite different in character from the 'vacuoles.' In *Paramecium* there are always to be observed two globular vesicles (Fig. 304, B, *a, a*), each of them surrounded by several elongated cavities, arranged in a radiating manner, so as to give to the whole somewhat of a star-like aspect (Plate XIV., fig. 1, *v, v*); and the liquid contents are seen to be propelled from the former into the latter, and *vice versa*. Further, in *Stentor*, a complicated network of canals, apparently in connection with the contractile vesicles, has been detected in the substance of the 'ectosarc;' and traces of this may be observed in other Infusoria. In some of the larger Animalcules, it may be distinctly seen that the contractile vesicles have *permanent* valvular orifices opening outwards, and that an expulsion of fluid from the body into the water around it is effected by their contraction. Hence it appears likely that their function is of a respiratory nature; and that they serve, like the gill-openings of Fishes, for the expulsion of water which has been taken in by the mouth, and which has traversed the interior of the body. (See § 399.)

440. Of the Reproduction of the Ciliated Infusoria, our knowledge is still very imperfect; for although various modes of multiplication have been observed among them, it still remains doubtful whether any process takes place, that can be regarded—like the conjugation of the *Monadina*

(§ 418)—as analogous to the sexual Generation of higher organisms. Binary subdivision would seem to be universal among them; and has in many instances been observed (as elsewhere) to commence in the nucleus. The division takes place in some species longitudinally, that is in the direction of the greatest length of the body (Fig. 305, D, E, F), in other species transversely (Fig. 308, C, D), whilst in some, as in *Chilodon cucullulus* (Fig. 306), it has been supposed to occur in either direction indifferently. But it may be questioned whether, in this latter case, one set of the apparent 'fissions' is not really 'conjugation' of two individuals.—This duplication is performed with such rapidity, under favorable circumstances, that, according to the calculation of Prof. Ehrenberg, no fewer than 268 *millions* might be produced in a month by the repeated subdivisions of a single *Paramecium*. When this fission occurs in *Vorticella* (Fig. 305), it extends down the stalk, which thus becomes double for a greater or less part of its length; and thus a whole bunch of these Animalcules may spring (by a repetition of the same process) from one base. In some members of the same family, arborescent structures are produced resembling that of *Codosiga* (Fig. 296), by the like process of

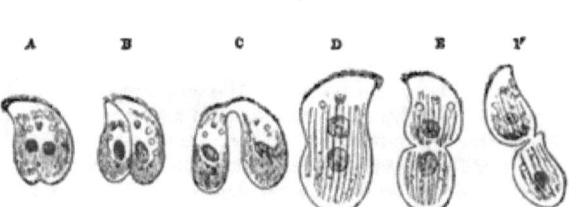

Fig. 306.

A B C D E F

Fissiparous multiplication of *Chilodon cucullulus*:—A, B, C, successive stages of longitudinal fission (?); D, E, F, successive stages of transverse fission.

continuous subdivision.—Another curious result of this mode of multiplication presents itself in the family *Ophrydina*; masses of individuals, which separately resemble certain *Vorticellina*, being found imbedded in a gelatinous substance of a greenish color, sometimes adherent, and sometimes free. These masses, which may attain the diameter of four or five inches, present such a strong general resemblance to a mass of *Nostoc* (§ 247), or even of Frogs' spawn, as to have been mistaken for such; but they simply result from the fact, that the multitude of individuals produced by a repetition of the process of self-division, remain connected with each other for a time by a gelatinous exudation from the surface of their bodies, instead of at once becoming completely isolated. From a comparison of the dimensions of the individual *Ophrydia*, each of which is about 1-120th of an inch in length, with those of the composite masses, some estimate may be formed of the number included in the latter; for a cubic inch would contain nearly *eight millions* of them, if closely packed; and many times that number must exist in the larger masses, even making allowance for the fact that the bodies of the Animalcules are separated from each other by their gelatinous cushion, and that the masses have their central portions occupied by water only. Hence we have, in such clusters, a distinct proof of the extraordinary extent to which multiplication by duplicative subdivision may proceed, without the interposition of any other operation. These Animalcules, however, free themselves at times from their gelatinous bed, and have been observed to undergo an 'encysting process' corresponding with that of the *Vorticellina*.

441. Many, perhaps all, ciliated Infusoria at certain times undergo an *encysting process*, resembling the passage of Protophytes into the 'still' condition (§ 231), and apparently serving, like it, as a provision for their preservation under circumstances which do not permit the continuance of their ordinary vital activity. Previously to the formation of the cyst, the movements of the animalcule diminish in vigor, and gradually cease altogether; its form becomes more rounded; its oral aperture closes; and its cilia or other filamentous prolongations are either lost or retracted, as is well seen in *Vorticella* (Fig. 307, A). A new wreath of cilia, however, is developed near the base, and in this condition the animal detaches itself from its stem, and swims freely for a short time, soon passing, however, into the 'still' condition. The surface of the body then exudes a gelatinous excretion that hardens around it so as to form a complete coffin-like case, within which little of the original structure of the animal can be distinguished. Even after the completion of the cyst, however, the contained animalcule may often be observed to move freely within it, and may sometimes be caused to come forth from its prison by the mere application of warmth and moisture.

Fig. 307.

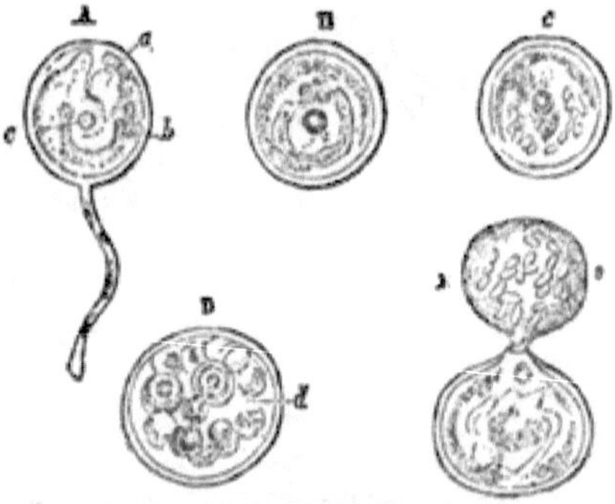

Encysting process in *Vorticella microstoma*:—A, full-grown individual in its encysted state; *a*, retracted oval circlet of cilia; *b*, nucleus; *c*, contractile vesicle; B, a cyst separated from its stalk;—c, the same more advanced, the nucleus broken-up into spore-like globules; D, the same more developed, the original body of the Vorticella, *d*, having become sacculated, and containing many clear spaces:—at E, one of the sacculations having burst through the enveloping cyst, a gelatinous mass, *e*, containing the gemmules, is discharged.

In the simplest form of the 'encysting process,' indeed, the animalcule seems to remain altogether quiescent through the whole period of its torpidity; so that, however long may be the duration of its imprisonment, it emerges without any essential change in its form or condition. But in other cases, this process seems to be subservient either to multiplication or to metamorphosis. For in *Vorticella*, the substance of the encysted body (B) appears to break up (C, D) into eight or nine segments, which, when set free by the bursting of the cyst, come forth as spontaneously moving spherules. Each of these soon increases in size, develops a ciliary wreath within which a mouth makes its appearance, and gradually assumes the form of the *Trichodina grandinella* of Ehrenberg. It then develops a posterior wreath of cilia, and multiplies by transverse fission; each half fixes itself by the end on which the mouth is situated, a short stem becomes developed, and the cilia-wreath disappears. A new mouth and cilia-wreath then form at the free extremity; and the growth of the stem completes the development into the true Vorticellan form.[1]—
In *Trichoda lynceus*, again, the 'encysting process' appears subservient to a like kind of metamorphosis; the form which emerges from the cyst

---

[1] Everts, "Untersuchungen an *Vorticella nebulifera*," quoted by Prof. Allman, *loc. cit.*

differing in many respects from that of the animalcule which became encysted. According to M. Jules Haime, by whom this history was very carefully studied,[1] the form to be considered as the larval one, is that shown in Fig. 308, A–E, which has been described by Prof. Ehrenberg under the name of *Oxytricha*. This possesses a long, narrow, flattened body, furnished with cilia along the greater part of both margins, and having also at its two extremities a set of larger and stronger hair-like filaments; and its mouth, which is an oblique slit on the right-hand side of its fore-part, has a fringe of minute cilia on each lip. Through this mouth large particles are not unfrequently swallowed, which are seen lying in the midst of the endosarc without any surrounding vesicle; and sometimes even an Animalcule of the same species, but in a different stage of its life, is seen in the interior of one of these voracious little devourers (B). In this phase of its existence, the *Trichoda* undergoes multiplication by transverse fission, after the ordinary mode (C, D); and it is usually one of the short-bodied 'doubles' (E) thus produced, that passes into the next phase. This phase consists in the assumption of the globu-

Fig. 308.

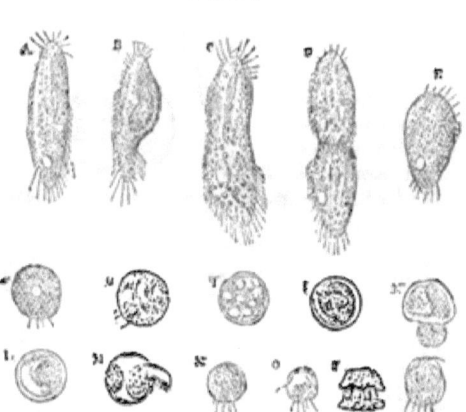

Metamorphoses of *Trichoda lynceus*:—A, larva (*Oxytricha*); B, a similar larva, after swallowing the animalcule represented at M; C, a very large individual on the point of undergoing fission; D, another in which the process has advanced further; E, one of the products of such fission; F, the same body become spherical and motionless; G, aspect of this sphere fifteen days afterwards; H, later condition of the same, showing the formation of the cyst; I, incipient separation between living substance and exuvial matter; K, partial discharge of the latter, with flattening of the sphere; L, more distinct formation of the contained animal; M, its escape from the cyst; N, its appearance some days afterwards; O, more advanced stage of the same; P, Q, perfect *Aspidiscus*, one as seen sideways, moving on its bristles, the other as seen from below (magnified twice as much as the preceding figures).

lar form, and the almost entire loss of the locomotive appendages (F); in the escape of successive portions of the granular sarcode, so that 'vacuoles' make their appearance (G); and in the formation of a gelatinous envelope or cyst, which, at first soft, afterwards acquires increased firmness (H). After remaining for some time in this condition, the contents of the cyst become clearly separated from their envelope; and a space appears on one side, in which ciliary movement can be distinguished (I). This space gradually extends all round, and a further discharge of granular matter takes place from the cyst, by which its form becomes altered

---

[1] "Annales des Sci. Nat.," Ser. 3, Tom. xix. (1853), p. 109.

(K); and the distinction between the newly-formed body to which the cilia belong, and the effete residue of the old, becomes more and more apparent (L). The former increases in size, whilst the latter diminishes; and at last the former makes its escape through an aperture in the wall of the cyst, a part of the latter still remaining within its cavity (M). The body thus discharged (N) does not differ much in appearance from that of the Oxytricha before its encystment (F), though of only about two-thirds its diameter; but it soon develops itself (O, P, Q) into an Animalcule very different from that in which it originated. First it becomes still smaller, by the discharge of a portion of its substance; numerous very stiff bristle-like organs are developed, on which the Animalcule creeps, as by legs, over solid surfaces; the external integument becomes more consolidated on its upper surface, so as to become a kind of carapace; and a mouth is formed by the opening of a slit on one side, in front of which is a single hair-like flagellum, which turns round and round with great rapidity, so as to describe a sort of an inverted cone, whereby a current is brought towards the mouth. This latter form has been described by Prof. Ehrenberg under the name of *Aspidisca*. It is very much smaller than the larva; the difference being, in fact, twice as great as that which exists between A and P, Q (Fig. 308), since the last two figures are drawn under a magnifying power double that employed for the preceding. How the *Aspidisca*-form in its turn gives origin to the *Oxytricha*-form, has not yet been made out.—A similar 'encysting process' has been observed to take place among several other forms of ciliated Infusoria; so that, considering the strong general resemblance in kind and degree of organization which prevails throughout the group, it does not seem unlikely that it may occur at some stage of the life of nearly all these Animalcules. And it is not improbably in the 'encysted' condition that their dispersion chiefly takes place, since they have been found to endure desiccation in this state, although in their ordinary condition of activity they cannot be dried-up without loss of life. When this circumstance is taken into account, in conjunction with the extraordinary rapidity of multiplication of these Animalcules, there seems no difficulty in accounting for the universality of their diffusion. It may be stated as a general fact, that wherever decaying Organic matter exists in a liquid state, and is exposed to air and warmth, it speedily becomes peopled with some or other of these minute inhabitants: and it may be fairly presumed that, as in the case of the Fungi, the dried cysts or germs of Infusoria are everywhere floating about in the air, ready to develop themselves wherever the appropriate conditions are presented; while all our knowledge of their history seems further to justify the belief, that (in some instances, at least) the same germs may develop themselves into a succession of forms so different, as to have been regarded as distinct specific or even generic types.

442. A very important advance was supposed to have been made in this direction by the asserted discovery of M. Balbiani[1] that a true process of *sexual generation* occurs among Infusoria; his observations having led him to the conclusion that male and female organs are combined in each individual of the numerous genera he has examined, but that the congress of two individuals is necessary for the impregnation of

---

[1] See his "Recherches sur les Phénomènes Sexuels des Infusoires," in Dr. Brown-Séquard's "Journal de la Physiologie," for 1861. An abstract of these researches is contained in the "Quart. Journ. of Microsc. Science," for July and October, 1862.

PLATE XIV

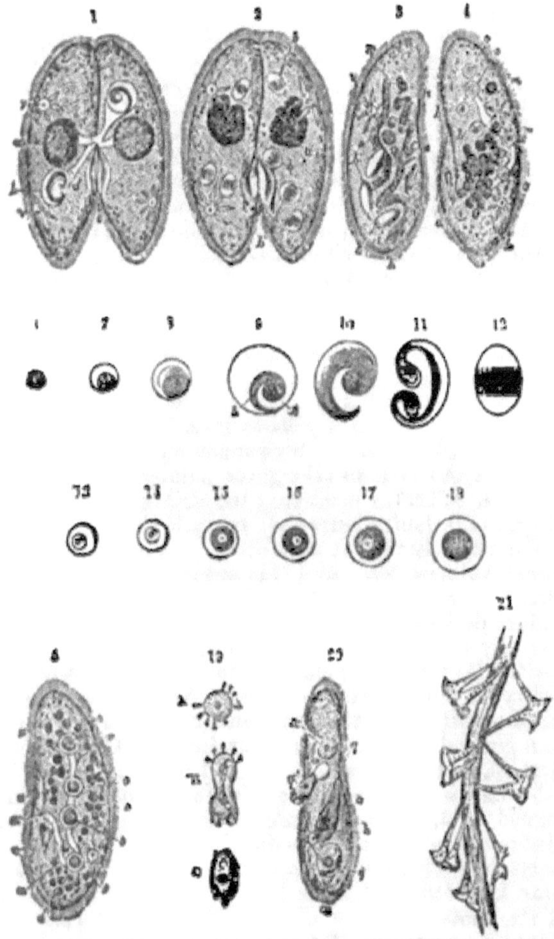

SEXUAL (?) REPRODUCTION OF INFUSORIA (after Balbiani).

Fig. 1. Conjugation of *Paramecium aurelia*; *a*, ovarium (nucleus); *b*, seminal capsule (nucleolus); *c*, oviducal canal; *d*, seminal canal; *e*, buccal fissure.
2. The same, more advanced; *a*, ovary, showing lobulated surface; *b*, *b*, secondary seminal capsules.
3. One of the individuals in a still more advanced state of conjugation, showing the ovary *a*, *a*, broken up into fragments connected by the tube *m*; *b*, *b*, seminal capsules; *v*, contractile vesicle.
4. *Paramecium*, ten hours after the conclusion of the conjugation; *a*, *a*, unchanged granular masses of the ovary, of which other portions have been developed into the ova, *o*, *o*, still contained within the connecting tube *m*; *b*, *b*, seminal capsules.
5. The same, three days after the completion of the conjugation.
6–12. Successive stages in the development of the seminal capsules.
13–18. Successive stages in the development of the ovules.
19. *Acinetæ* in different stages, A, B, C.
20. *Paramecium* containing three *Acineta*-parasites, *q*, *q*, *q*, lying in introverted pouches, of which the external openings are seen at *x*, *x*.
21. *Stentor* in conjugation.

the ova, those of each being fertilized by the spermatozoa of the other. He regards the 'nucleus' as an *ovarium* or aggregation of germs, whilst the 'nucleolus' is really a *testis* or aggregation of spermatozoids. The particular form and position which these organs present, and the nature of the changes which they undergo, vary in the several types of Infusoria; but as we have in the common *Paramecium aurelia* an example, which, although exceptional in some particulars, affords peculiar facilities for the observation of the process, and has been most completely studied by M. Balbiani, it is here selected for illustration.—This Animalcule, as is well known, multiplies itself with great rapidity (under favorable circumstances) by duplicative subdivision, which always takes place in the *transverse* direction; and the condition represented in Plate XIV., Figs. 1, 2, is not, as has been usually supposed, another form of the same process, but is really the sexual congress of two individuals previously distinct. When the period arrives at which the *Paramecia* are to propagate in this manner, they are seen assembling upon certain parts of the vessel, either towards the bottom or on the walls; and they are soon found coupled in pairs, closely adherent to each other, with their similar extremities turned in the same direction, and their two mouths closely applied to one another, but still continuing to move freely in the liquid, turning constantly round upon their axes. This conjugation lasts for five or six days, during which period very important changes take place in the condition of the reproductive organs. In order to distinguish these, the Animalcules should be slightly flattened by compression, and treated with acetic acid, which brings the reproductive apparatus into more distinct view, as shown in Figs. 1–5. In Fig. 1, each individual contains an ovarium a, which is shown to present in the first instance a smooth surface; and from this there proceeds an excretory canal or oviduct c, that opens externally at about the middle of the length of the body into the buccal fissure e. Each individual also contains a seminal capsule b, in which is seen lying a bundle of spermatozoids curved upon itself, and which communicates by an elongated neck with the orifice of the excretory canal. The successive stages by which the seminal capsule arrives at this condition, from that of a simple cell, whose granular contents resolve themselves (as it were) into a bundle of filaments, are shown in Figs. 6–10. In Fig. 2, the surface of the ovary a, is seen to present a lobulated appearance, which is occasioned by the commencement of its resolution into separate ova; while the seminal capsule is found to have undergone division into two or four secondary capsules b, b, each of which contains a bundle of spermatozoa now straightened out. This division takes place by the elongation of the capsule into the form represented in Fig. 11, and by the narrowing of the central portion whilst the extremities enlarge; the further multiplication being effected by the repetition of the same process of elongation and fission. In Fig. 3, which represents one of the individuals still in conjugation, the four seminal capsules b, b, are represented as thus elongated in preparation for another subdivision, whilst the ovary a, a, has begun, as it were, to unroll itself, and to break up into fragments which are connected by the tube m. It is in this condition that the object of the conjugation appears to be effected, by the passage of the seminal capsules of each individual, previously to their complete maturation, into the body of the other. In Fig. 4 is shown the condition of a *Paramecium* ten hours after the conclusion of the conjugation; the ovary has here completely broken up into separate granular masses, of which some *a, a,* remain unchanged, whilst others, *o, o, o, o,* either two, four, or eight in number, are converted into ovules that appear to be fertilized by the escape of the spermatozoa from the seminal capsules, these being now seen in process of withering. Finally, in fig. 5, which represents a *Paramecium* three days after the completion of the conjugation, are seen four complete ova, *o, o, o, o,* within the connecting tube *m, m ;* whilst the seminal capsules have now altogether disappeared. In figs. 13–18 are seen the successive stages of the development of the ovule, which seems at first (fig. 13) to consist of a germ-cell having within it a secondary cell containing minute granules, which is to become the 'vitelline vesicle.' This secondary cell augments in size, and becomes more and more opaque from the increase of its granular contents (figs. 14, 15, 16), forming the 'vitellus' or yolk; in the midst of which is seen the clear 'germinal vesicle,' which shows on its wall, as the ovule approaches maturity, the 'germinal spot' (fig. 17). The germinal vesicle is subsequently concealed (fig. 18) by the increase in the quantity and opacity of the vitelline granules. The fertilized ova seem to be expelled by the gradual shortening of the tube that contains them; and this shortening also brings together the scattered fragments of the granular substance of the original ovarium, so as to form a mass resembling that shown in fig. 1, *a,* by the evolution of which, after the same fashion, another brood of ova may be produced.

443. Now there can be no doubt as to the occurrence of 'conjugation' among Ciliated Infusoria; and this not only in the free-swimming, but also in the attached forms, as *Stentor* (Plate XIV., fig. 21). In *Vorticella,* according to several recent observers, what has been regarded as *gemmiparous* multiplication—the putting-forth of a bud from the base of the body—is really the conjugation of a small individual in the free-swimming stage with a fully-developed fixed individual, with whose body its own becomes fused. But it is doubtful whether such conjugation has any reference to the encysting process. According to Bütschli and Engelmann, the conjugating process results in the breaking up of the nucleus and (so-called) nucleolus of the conjugating individuals; these individuals separate again, and after the expulsion of the broken-up nuclear structures, the characteristic nucleus and nucleolus are reformed. The same excellent observers adduce strong grounds for distrusting Balbiani's assignment of sexual characters to the nucleus and nucleolus. For although a striation may be observed on the surface of the latter, no one has witnessed its subdivision into spermatozoidal filaments. And if embryos are really produced at the expense of the nucleus, what Balbiani described as sexual *ova* are really non-sexual *gemmules,* each consisting (like the zoospore of Protophytes) of a segment of the nucleus surrounded by an envelope of protoplasm.—There is still much uncertainty in regard to the embryonic forms of Ciliate Infusoria; some eminent observers asserting that the 'gemmule' in the first instance, besides forming a cilia-wreath, puts forth suctorial appendages (Plate XIV., fig. 19, A, B, C), by means of which it imbibes nourishment until the formation of its mouth permits it to obtain its supplies in the ordinary way; whilst others maintain these acinetiform bodies to be parasites, which even imbed themselves in the substance of the Infusoria they infest.[1]

---

[1] There can be no doubt that Stein was wrong in his original doctrine that the fully-developed *Acinetina* are only transition-stages in the development of *Vorticellina* and other Ciliated Infusoria. But the balance of evidence seems to the

444. It is obvious that no Classification of Infusoria can be of any permanent value, until it shall have been ascertained by the study of their entire life-history, what are to be accounted really distinct forms. And the differences between them, consisting chiefly in the shape of their bodies, the disposition of their cilia, the possession of other locomotive appendages, the position of the mouth, the presence of a distinct anal orifice, and the like, are matters of such trivial importance as compared with those leading features of their structure and physiology on which we have been dwelling, that it does not seem desirable to attempt in this place to give any detailed account of them. The life-history of the *ciliate Infusoria* is a subject pre-eminently worthy of the attention of Microscopists, who can scarcely be better employed than in tracing out the sequence of its phenomena, with the same care and assiduity as have been displayed by Messrs. Dallinger and Drysdale in the study of the *Monadina*.—"In pursuing our researches," say these excellent observers, "we have become practically convinced of what we have theoretically assumed—the absolute necessity for prolonged and patient observation of the same forms. Two observers, independently of each other, examining the same Monad, if their inquiries were not sufficiently prolonged, might, with the utmost truthfulness of interpretation, assert opposite modes of development. Competent optical means, careful interpretation, close observation, and *time*, are alone capable of solving the problem."

Section II.—Rotifera, or Wheel-Animalcules.

445. We now come to that higher group of Animalcules, which, in point of complexity of organization, is as far removed from the preceding, as Mosses are from the simplest Protophytes; the only point of real resemblance between the two groups, in fact, being the minuteness of size which is common to both, and which was long the obstacle to the recognition of the comparatively elevated character of the *Rotifera*, as it still is to the precise determination of certain points of their structure. Some of the Wheel-Animalcules are inhabitants of salt water only; but by far the larger proportion are found in collections of fresh water, and rather in such as are free from actively decomposing matter, than in those which contain organic substances in a putrescent state. Hence when they present themselves in Vegetable infusions, it is usually after that offensive condition which is favorable to the development of many of the Infusoria has passed-away; and they are consequently to be looked-for after the disappearance of many successions (it may be) of Animalcules of inferior organization. Rotifera are more abundantly developed in liquids which have been long and freely exposed to the open air, than in such as have been kept under shelter; certain kinds, for example, are to be met with in the little pools left after rain in the hollows of the lead with which the tops of houses are partly covered; and they are occasionally found in enormous numbers in cisterns which are not beneath roofs or otherwise covered over.[1] They are not, however, absolutely confined to collections of liquid: for there are a few species which can maintain their existence in damp earth; the common *Rotifer* is occasionally found in the interior of the leaf-cells of *Sphagnum* (§ 339); and at least two species of *Notommata* also are known to be para-

---

writer to be in favor of his later statement, that the bodies figured in Pl. XIV., fig. 19, are really Infusorian embryos, and not parasitic Acinetæ.

[1] See a remarkable instance of this in vol. i., p. 232, *note*.

sitic, the one in the large cells of *Vaucheria* (§ 219), and another in the sphere of *Volvox* (§ 236).—The Wheel-like organs from which the class derives its designation, are most characteristically seen in the common *Rotifer* (Fig. 310), where they consist of two disk-like lobes or projections of the body, whose margins are fringed with long cilia; and it is the uninterrupted succession of strokes given by these cilia, each row of which nearly returns (as it were) into itself, that gives rise by an optical illusion to the notion of 'wheels.' This arrangement, however, is by no means universal; in fact, it obtains in only a small proportion of the group; and by far the more general plan is that seen in Fig. 309, in

FIG. 309.   FIG. 310.

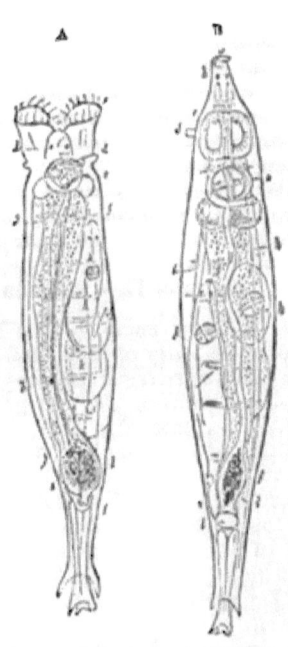

Brachionus pala.

Rotifer vulgaris, as seen at A with the wheels drawn-in, and at B with the wheels expanded:— a, mouth; b, eye-spots; c, wheels; d, calcar (antenna?); e, jaws and teeth; f, alimentary canal; g, glandular (?) mass enclosing it; h, longitudinal muscles; i, i, tubes of water-vascular system; k, young animal; l, cloaca.

which the cilia form one continuous line across the body, being disposed upon the sinuous edges of certain lobes or projections which are borne upon its anterior portion. Some of the chief departures from this plan will be noticed hereafter (§ 453).

446. The great transparence of the Rotifera permits their general structure to be easily recognized. They have usually an elongated form, similar on the two sides; but this rarely exhibits any traces of segmental division. The body is covered with a double envelope, both layers of which are extremely thin and flexible in some species, whilst in others the outer one seems to possess a horny consistence. In the former case the whole integument is drawn together in a wrinkled manner when the

body is shortened; in some of the latter the sheath has the form of a polype-cell, and the body lies loosely in it, the inner layer of the integument being separated from the outer by a considerable space (Fig. 312); whilst in others the envelope or *lorica* is tightly fitted to the body, and strongly resembles the horny casing of an Insect or the shell of a Crab, except that it is not jointed, and does not extend over the head and tail, which can be projected from the openings at its extremities, or completely drawn within it for protection (Fig. 313). In those Rotifera in which the flexibility of the body is not interfered with by the consolidation of the external integument, we usually find it capable of great variation in shape, the elongated **form** being occasionally exchanged for an almost globular one, as is seen especially when the animals are suffering from deficiency **of water;** whilst by alternating movements of contraction and extension, **they can** make their way over solid surfaces, after the manner of a Worm or a Leech, with considerable activity,—some even of the loricated species being rendered capable of this kind of progression by the contractility of the head and tail. All these, **too, can swim** readily through the water by the action of their cilia; **and there are** some species which are limited to the latter **mode of progression.** The greater number have an organ of attachment **at** the posterior **extremity** of the body, which is usually prolonged into **a tail,** by which they can affix themselves to any solid object; and this is their ordinary position, when keeping their 'wheels' in action for a supply of food or of water; they have no difficulty, however, in letting-go their hold and moving through the water in search of a new attachment, and may therefore be considered as perfectly free. The sessile species, in their adult stage, on the other hand, remain attached by the posterior extremity to the spot on which they have at first fixed themselves; and their cilia are consequently employed for no other purpose than that of creating currents in the surrounding water.

447. In considering the internal structure of Rotifera, we shall take as its type the arrangement which it presents in the *Rotifer vulgaris* (Fig. 310); and specify the principal variations exhibited elsewhere. The body of this animal, when fully extended, possesses greater length in proportion to its diameter than that of most others of its class; and the tail is composed of three joints or segments, which are capable of being drawn up, one within another, like the sliding tubes of a telescope, each having a pair of prongs or points at its extremity. Within the external integument of the body are seen a set of longitudinal muscular bands (h), which serve to draw the two extremities towards each other; and these are crossed by a set of transverse annular bands, which also are probably muscular, and serve to diminish the diameter of the body, and thus to increase its length. Between the wheels is a prominence bearing two red spots (b), and having the mouth (a) at its extremity; these red spots differ altogether from those common in Infusoria and Protophyta, each having a minute highly-refracting spherical lens set in red pigment, and being clearly a rudimentary eye; and the prominence that bears them may be considered, therefore, as a true head, notwithstanding that it is not clearly distinguishable from the body. This head also bears upon its under surface a projecting spur-like organ (d), which was thought by Prof. Ehrenberg to be a siphon for the admission of water to the cavity of the body for the purpose of respiration; this, however, is certainly not the case, the 'spur' being imperforate at its extremity; and there seems much **more** probability in the idea of Dujardin, that it represents the

antennæ or *palpi* of higher Articulata, the single organ being replaced in many Rotifera by a pair, of which each is furnished at its extremity with a brush-like tuft of hairs that can be retracted into the tube. The œsophagus, which is narrow in the *Rotifer*, but is dilated into a crop in *Stephanoceros* (Fig. 312) and in some other genera, leads to the masticating apparatus (Fig. 310, *e*), which in these animals is placed far behind the mouth, and in close proximity to the stomach.—The Masticating apparatus has been made the subject of attentive study by Mr. P. H. Gosse; who has given an elaborate account of the various types of form which it presents in the several subdivisions of the group.[1] The following description of one of the more complicated will serve our present purpose. The various movable parts are included in a muscular bulb, termed the *mastax* (Fig. 311, *a*), which intervenes between the buccal funnel (*m*) and the œsophagus (*p*). The mastax includes a pair of organs, which, from the resemblance of their action to that of hammers working on an anvil, may be called *mallei*, and a third, still more complex, termed the *incus*. Each malleus consists of two principal parts placed nearly at right angles to each other, the *manubrium* (*c*), and the *uncus* (*e*); these are articulated to one another by a sort of hinge-joint. The former, as its name imports, serves the purpose in some degree of a handle; and it is the latter which is the instrument for crushing and dividing the food. This is done by means of the finger-like processes with which it is furnished at the edge where it meets its fellow; these being five or six in number, set parallel to each other like the teeth of a comb. The incus also consists of distinct articulated portions, namely two stout *rami* (*a*) resting on what seems a slender footstalk (*h*)

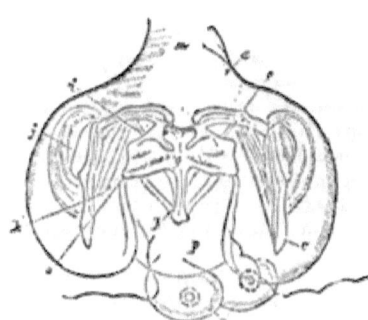

Fig. 311.

Masticating Apparatus of *Euchlanis deflexa*; —*a*, Mastax; *c*, manubrium, and *e*, uncus, of Malleus; *g*, rami, and *h*, fulcrum, of Incus; *i*, muscle connecting ramus and uncus; *j*, mucles passing from malleus to mastax; *k*, muscle connecting uncus and manubrium; *m*, buccal funnel; *n*, salivary glands; *p*, œsophagus.

termed the *fulcrum ;* when viewed laterally, however, the fulcrum is seen to be a thin plate, having the rami so jointed to one edge of it that they can open and close like a pair of shears. The uncus of each malleus falls into the concavity of its respective ramus, and is connected with it by a stout triangular muscle (*i*), which is seen passing from the hollow of the ramus to the under surface of the uncus. It is difficult to say with certainty what is the substance of which these firm structures are composed; it is not affected by solution of potass, but is instantly dissolved without effervescence by the mineral acids and by acetic acid. Besides the muscles already described, a thick band (*j*) embraces the upper and outer angle of the articulation of the malleus; and is inserted in the adjacent wall of the mastax; and a semi-crescentic band (*k*) is inserted by its broad end into the inferior and basal part of the uncus, and by its slender end into the middle of the inner side of the manubrium; the former of these

---

[1] "Philosophical Transactions," 1856, p. 419.

may be considered as an extensor, and the latter as a flexor, of the malleus. By these and other muscles which cannot be so clearly distinguished, the two unci are made to approach and recede by a perpendicular motion on the hinge-joint, so that their opposing faces come into contact, and their teeth bruise down the particles of food; but at the same time they are carried apart and approximated laterally by the movement of the free extremities of the manubria. The rami of the incus also open and shut with the working of the mallei: and by the conjoint action of the whole, the food is effectually comminuted in its passage downwards.[1]

448. The Alimentary Canal, which lies loose in the 'general cavity of the body,' is sometimes a simple tube, passing without enlargement or constriction from the masticating apparatus to the anal orifice at the posterior part of the body; whilst in other instances there is a marked distinction between the stomach and intestinal tube, the former being a large globular dilatation immediately below the jaws, whilst the latter is cylindrical and comparatively small. The alimentary canal of *Rotifer* (Fig. 310) most resembles the first of these types, but presents a dilatation (*l*) close to the anal orifice, which may be considered as a cloaca; that of *Brachionus* (Fig. 309) is rather formed upon the second. Connected with the alimentary canal are various glandular appendages, more or less developed; sometimes clustering round its walls as a mass of separate follicles, which seems to be the condition of the glandular investment (*g*) of the alimentary canal in *Rotifer*; in other cases having the form of cæcal tubuli. Some of these open into the stomach close to the termination of the œsophagus, and have been supposed to be salivary or pancreatic in their character, whilst others, which discharge their secretion into the intestinal tube, have been regarded, and probably with correctness, as the rudiment of a liver.—In the genus *Asplanchna* (Gosse), there is a wide departure from the ordinary *Rotifer* type; as the species belonging to it have neither intestine nor anus. The stomach consists of a large bag at the end of the gullet, about which, when the animals are quiet, the ovary is bent in a horseshoe form. The indigestible matters are ejected through the mouth. The curious absence of any digestive apparatus in the males of this group, will be presently noticed (§ 450).[2]

449. There does not appear to be any special Circulating apparatus in these animals; but the fluid which is contained in the perivisceral cavity is probably to be regarded as nutritive in its character; and its aeration is provided for by a peculiar apparatus, which seems to be a rudimentary form of the 'water-vascular system,' that attains a high development in the class of Worms. On either side of the body there is usually to be observed a long flexuous tube (Fig. 309), which extends from a contractile vessel common to both and opening into the cloaca (Fig. 310, *i, i*), towards the anterior region of the body, where it frequently subdivides into branches, one of which may arch over towards its opposite sides, and inosculate with a corresponding branch from its tube. Attached to each of these tubes are a number of peculiar organs (usually from two to eight on each side), in which a trembling movement is seen, very like that of a flickering flame; these appear to be pear-shaped sacs, attached by hollow stalks to the main tube, and each having a flagelliform cilium in

---

[1] See also the description of the mastax of *Melicerta ringens* and *Conochilus* by Mr. Bedwell in "Journ. of Roy. Micr. Soc.," Vol. i. (1878), p. 176.

[2] See Brightwell in "Ann. Nat. His.," Ser. 2, Vol. ii. (1848). p. 153; Dalrymple in "Philos. Trans." (1849), p. 339; and Gosse in "Ann. Nat. Hist.," Ser. 2, Vols. iii. (1848), p. 518; vi. (1850), p. 18; and viii. (1851), p. 198.

its interior, that is attached by one extremity to the interior of the sac, and vibrates with a quick undulatory motion in its cavity; and there can be little doubt that their function is to keep up a constant movement in the contents of the aquiferous tubes, whereby fresh water may be continually introduced from without for the aeration of the fluids of the body.[1] The Nervous system is represented by only a single ganglionic body (sometimes bilobed, however), which lies at one side of the œsophagus, in near proximity to the eye-spots, the spur-like organ, and the ciliated pit, and has also, in some Rotifers, an auditory vesicle attached to it. No nerve-trunks proceeding to the muscular bands have as yet been certainly distinguished.

450. The Reproduction of the Rotifera has not yet been completely elucidated. Although they were affirmed by Prof. Ehrenberg to be hermaphrodite, yet the existence of distinct sexes has been detected in so many genera (for the most part by Mr. Gosse[2]), that it may fairly be presumed to be the general fact. The male is inferior in size to the female; and sometimes differs so much in organization, that it would not be recognized as belonging to the same species, if the copulative act had not been witnessed. In all the cases yet known, as in the *Asplanchna* of which the separate male was the first discovered, there is an absolute and universal atrophy of the digestive system; neither mastax, jaws, œsophagus, stomach, nor intestines being discoverable in any male; no other organs, in fact, being fully developed, than those of generation. The male would appear, therefore, quite unfit to obtain aliment for itself; and its existence is probably a very brief one, being continued only so long as the store of nutriment supplied by the egg remains unexhausted. In the remarkable six-limbed Rotifer discovered by Dr. Hudson, and named by him *Pedalion mira*, the virgin female was found to lay female eggs during the greater part of the year, while male eggs, which are not found in the same individuals, "are half the size of the female ones, and are carried in clusters of often a score at a time." The males are very small in comparison with the females, and are very short-lived, sometimes dying within an hour. In *Rotifer*, however, as in a large proportion of the group, no males have yet been discovered, probably because they are produced only at certain times. The female organ consists of a single ovarian sac, which frequently occupies a large part of the cavity of the body, and opens at its lower end by a narrow orifice into the cloaca.— Although the number of eggs in these animals is so small, yet the rapidity with which the whole process of their development and maturation is accomplished, renders the multiplication of the race very rapid. The egg of the *Hydatina* is extruded from the cloaca within a few hours after the first rudiment of it is visible; and within twelve hours more the shell bursts, and the young animal comes forth. Three or four eggs being deposited at once, it was calculated by Prof. Ehrenberg that nearly *seventeen millions* may be produced within twenty-four days from a single individual. In *Rotifer* and several other genera, the development of the embryo takes-place whilst the egg is yet retained within the body of the parent (Fig. 310, *k*), and the young are extruded alive; whilst in

---

[1] See Prof. Huxley's account of these organs in his description of *Lacinularia socialis*, "Transact. of Microsc. Soc.," N.S., Vol. i. (1853), p. 1.

[2] "Philosophical Transactions," 1853, p. 313. See also Dr. Hudson in "Monthly Microsc. Journ.," Vol. xiii. (1875), p. 45.

[3] "Monthly Microsc. Journ.," Vol. viii. (1872), p. 209; and "Quart. Journ. Mic. Sci.," Vol. xii. (1872), p. 333.

some other instances the eggs, after their extrusion, remain attached to the posterior extremity of the body (Fig. 309), until the young are set free. The transparence of the egg-membrane, and also of the tissues, of the parent Rotifer, allows the process of development to be watched, even when the egg is retained within the body; and it is curious to observe, at a very early period, not merely the red eye spot of the embryo, but also a distinct ciliary movement. In general it would seem that whether the rupture of the egg-membrane takes place before or after the egg has left the body, the germinal mass within it is developed at once into the form of the young animal, which usually resembles that of its parent; no preliminary metamorphosis being gone through, nor any parts developed which are not to be permanent. In *Floscularia ornata*, however, the young leave the eggs in the shape of little maggots, from one end of which a tuft of cilia soon appears. The form changes in a few hours, the ciliated end becoming lobed, and the body rounded. The foot is developed later.[1]—In the curious *Notommata Werneckii*, which is found parasitic in the reproductive capsules of *Vaucheria* (§ 249), the young animal has the general organization of the free-swimming Rotifers, and leads a similarly active life; but when its eggs are becoming mature, it finds its way into one of these capsules and there undergoes a remarkable deformation, its characteristic organs disappearing, and its body becoming a large egg-sac, which seems to be nourished by absorption.[2]

451. Even in those species which usually hatch their eggs within their bodies, a different set of Ova is occasionally developed, which are furnished with a thick glutinous investment; these, which are extruded entire, and are laid one upon another, so as at last to form masses of considerable size in proportion to the bulk of the animals, seem not to be destined to come so early to maturity, but very probably remain dormant during the whole winter season, so as to produce a new brood in the spring. These 'winter-eggs' are inferred by Prof. Huxley, from the history of their development, to be really *gemmæ* produced by a non-sexual operation; while the bodies ordinarily known as ova, he considers to be true generative products. Prof. Cohn, however, states that he has ascertained, by direct experiment upon those species in which the sexes are distinct, that the bodies commonly termed 'ova' (Figs. 309, 310) are really *internal gemmæ*, since they are reproduced, through many successions, without any sexual process, just like the external gemmæ of *Hydra* (§ 515), or the internal gemmæ of *Entomostraca* (§ 609) and *Aphides* (§ 643); whilst the 'winter-eggs,' are only produced as the result of a true generative act.[3] By M. Balbiani, however, 't is affirmed (*loc. cit.*) that the 'winter-eggs,' like the ordinary eggs, are produced non-sexually; so that it would seem as if the intervention of the true generative act is only occasionally required for the continued propagation of these interesting creatures.

452. Certain Rotifera, among them the common Wheel-Animalcule, are remarkable for their tenacity of life, even when reduced to such a state of dryness that they will break in pieces when touched with the point of a needle (as the Author has himself ascertained); for they can be kept in this condition for any length of time, and will yet revive very speedily upon being moistened. Taking advantage of this fact, some

---

[1] See Mr. Slack's "Marvels of Pond Life," 2d Edit., p. 54.
[2] See Balbiani in "Journ. Roy. Microsc. Soc.," Vol. ii. (1879), p. 530.
[3] See his Memoir, 'Ueber die Fortpflanzung der Räderthiere,' in "Siebold and Kölliker's Zeitschrift," 1855.

Microscopists are in the habit of keeping by them stocks of desiccated Rotifers, which can be distributed in the condition of dry dusty powder. The desiccating process has been carried yet farther with the tribe of *Tardigrada* (§ 453, IV.); individuals of which have been kept in a vacuum for thirty days, with sulphuric acid and chloride of calcium, and yet have not lost their capability of revivification. These facts, taken in connection with the extraordinary rate of increase mentioned in the preceding paragraph, remove all difficulty in accounting for the extent of the difusion of these animals, and for their occurrence in incalculable numbers in situations where, a few days previously, none were known to exist. For their entire bodies may be wafted in a dry state by the atmosphere from place to place; and their return to a state of active life, after a desiccation of unlimited duration, may take place whenever they meet with the requisite conditions—moisture, warmth, and food. It is probable that the Ova are capable of sustaining treatment even more severe than the fully developed Animals can bear; and that the race is frequently continued by them when the latter have perished.—It is not requisite to suppose, however, that in any of the foregoing cases the desiccation is *complete;* for it appears that Wheel-Animalcules, in drying, exude a glutinous matter that forms a sort of impervious casing, which may keep-in the remaining fluid.[1] When acted on by heat as well as by drought, Rotifers and Tardigrades lose their vitality; yet the former have survived a gradual heating up to 200° Fahr.

453. The principles on which the various forms that belong to this Class should be systematically arranged, have not yet been satisfactorily determined. By Prof. Ehrenberg, the disposition of the ciliated lobes or wheel-organs, and the inclosure or non-inclosure of the body in a *lorica* or case, were taken as the basis of his classification; but as his ideas on both these points are inconsistent with the actual facts of organization, the arrangement founded upon them cannot be received. Another division of the class has been propounded by M. Dujardin, which is based on the several modes of life of the most characteristic forms. And in a third, more recently put forth by Prof. Leydig, the general configuration of the body, with the presence, absence, and conformation of the foot (or tail) are made to furnish the characters of the subordinate groups. Either of the two latter is certainly more *natural* than the first, as bringing together for the most part the forms which most agree in general organization, and separating those which differ; and we shall adopt that of M. Dujardin as most suitable to our present purpose.

I. The first group includes those that habitually live attached by the foot, which is prolonged into a pedicle; and it includes two families, the *Floscularians* and the *Melicertians,* the members of which are commonly found attached to the stems and leaves of aquatic plants, by a long pedicle or foot-stalk, bearing a somewhat bell-shaped body. In one of the most beautiful species, the *Stephanoceros Eichornii* (Fig. 312), this body has five long tentacles, beset with tufts of cilia, whilst the body is inclosed in a gelatinous cylindrical cell. At first sight, the tentacles of this Rotifer may seem to resemble those of the *Polyzoa;* but, if there are carefully illuminated, the filaments which beset them will be found to be much larger, to be arranged differently, and to exhibit only an occasional motion, not at all resembling the regular rhythmical vibrations of the

---

[1] See Davis in "Monthly Micros. Journ.," Vol. ix. (1863), p. 207; also Slack, at p. 241 of same volume.

cilia of Polyzoa.[1] In fact, they seem rather to deserve the designation of *setæ* (bristles); for "their action is spasmodic, it creates no vortex, and it is only by actual contact with these *setæ* that floating particles are whipped within the area inclosed by the lobes, where by the same whipping action they are twitched from point to point irregularly downwards, until they come within the range of a vortex that is due, not to any action of the *setæ*, but to a range of minute cilia in the funnel."[1] A careful comparison of *Stephanoceros* with other forms, shows that its tentacles are only extensions of the ciliated lobes which are common to all the members of these families; and the cylindrical 'cell' which envelops the body is formed by the gelatinous secretion from its surface, thrown-off in rings, the indications of which often remain as a series of constrictions. In respect of the length of the filaments projecting from its lobes, and the breadth of these expansions, *Floscularia* is still more aberrant.—The body of *Melicerta* is protected by a most curious cylindrical tube, composed of little rounded pellets agglutinated together; this is obviously an artificial construction, and the process by which it is built may be watched by any Microscopist who is fortunate enough to capture it.[2] Beneath a projection on its head, there is observed a small disk-like organ, in which, when the 'wheels' are at work, a movement is seen very much resembling that of a revolving ventilator. Towards this disk the greater proportion of the solid particles that may be drawn from the surrounding liquid into the vortex of the wheel-organs, are driven by their ciliary movement, a small part only being taken into the alimentary canal; and there they accumulate until the aggregation (probably cemented by a glutinous secretion furnished by the organ itself) acquires the size and form of one of the globular pellets of the case; the time ordinarily required being about three minutes. The head of the animal then bends itself down, the pellet-disk is applied to the edge of the tube, the newly-formed pellet is attached there, and, the head being lifted into its former position, the formation of a new pellet at once commences.—Another curious example of this family is presented by the *Conochilus volvox;* which is found in spherical clusters composed of a considerable number of individuals adherent by their tails, their bodies being arranged in a radiating manner, and the intervals between them being filled up by a gelatinous substance. There is not, however, any such organic connec-

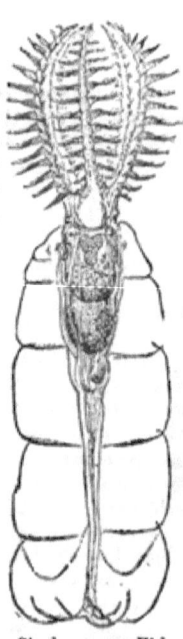

Fig. 312.

Stephanoceros Eichornii.

---

[1] In ordinary drawings, the filaments of the *Stephanoceros* are represented as short bristles; this is an error arising from bad instruments or defective illumination. It requires considerable skill to show these filaments, or those of the *Floscularia*, in their true length; but the beauty of the object is greatly increased when this is accomplished.

[2] See Mr. C. Cubitt's 'Observations on the Economy of Stephanoceros,' in 'Monthly Microsc. Journ.," Vol. iii. (1870), p. 242.

[3] See Gosse 'On the architectual instincts of *Melicerta ringens*,' in 'Trans. of Microsc. Soc.," Vol. iii. (1852), p. 58; also Bedwell in "Monthly Microsc. Journ.," Vol. xvi. (1877), p. 214; and Hudson in "Journ. Roy. Microsc. Soc.," Vol. ii. (1879), p. 1.

tion between them as exists in the *Ophrydium* (§ 440); and the uniting substance seems to be nothing else than the clear slimy secretion which probably all *Rotifera* exude from the surface of their bodies. It is into this that the eggs are extruded; and as they are hatched in it, the young produced from them remain to form part of the cluster; but, as its numbers increase, the cluster breaks up into two or more, which in their turn enlarge and then subdivide, so that a pond to whose bottom the 'winter eggs' of the year before have subsided, becomes alive with them in the early summer of the following year.'—The *Lacinularia socialis*, in like manner, forms transparent gelatinous-looking globular clusters, about 1-5th of an inch in diameter, which attach themselves to the leaves of aquatic plants.

II. The next of M. Dujardin's primary groups (ranged by him, however, as the third) consists of the ordinary *Rotifer* and its allies, which pass their lives in a state of alternation between the conditions of those attached by a pedicle, of those which habitually swim freely through the water, and of those which creep or crawl over hard surfaces.—As these have already been fully described, it is not requisite to dwell longer upon them.

III. The next group consists of those Rotifera which seldom or never attach themselves by the foot, but habitually swim freely through the water; and putting aside the peculiar aberrant form *Albertia* which has only been found as a parasite in the intestines of Worms, it may be divided into families, the *Brachionians* and the *Furcularians*. The former are for the most part distinguished by the short, broad, and flattened form of the body (Figs. 309, 313); which is, moreover, inclosed in a sort of cuirass formed by the consolidation of the external integument. This cuirass is often very beautifully marked on its surface, and may be prolonged into extensions of various forms, which are sometimes of very considerable length. The latter (corresponding almost exactly with the *Hydatineæ* of Prof. Ehrenberg) derived their name from the bifurcation of the foot into a sort of two-bladed forceps; their bodies are ovoidal or cylindrical, and are inclosed in a flexible integument, which is often seen to wrinkle itself into longitudinal and transverse folds at equidistant lines. To this family belongs the *Hydatina senta*, one of the largest of the Rotifera, which was employed by Prof. Ehrenberg as the chief subject of his examination of the internal structure of this group; as does also the *Asplanchna*, the curious condition of whose digestive apparatus has been already noticed (§ 448).

IV. The fourth of M. Dujardin's primary orders consists of the very curious tribe, first carefully investigated by M. Doyère, to which the name of *Tardigrada* has been given, on account of the slowness of their creeping movement. It seems now clear, however, that they have no near relationship to the true Rotifera; corresponding to them only in their minute size and simple structure. They are found in the same localities with the Rotifers, and, like them, can be revivified after desiccation (§ 452): but they have a vermiform body, divided transversely into five segments, of which one constitutes the head, whilst each of the others bears a pair of little fleshy protuberances, furnished with four curved hooks, and much resembling the pro-legs of a caterpillar. The head is entirely unpossessed of ciliated lobes; and the mouth, situated at the end of a sort of beak furnished with two longitudinal stylets, leads, through

---

[1] See Davis in "Monthly Microsc. Journ.," Vol. xvi. (1876), p. 1.

a muscular pharynx, into a wide alimentary canal, which gradually narrows to the anus. There are no special organs of circulation or respiration, but the nervous system is much more developed than in the Rotifera; a cerebral mass, bearing two eyes, giving origin to two longitudinal cords, on which are seated pairs of ganglia in connection with the members, as in Articulated animals generally. Their nearest affinities seem with the lowest forms of the *Arachnida*.

454. Nothwithstanding that all the best-informed Zoologists are now agreed in ranking the true *Rotifera* among *Articulated* animals, yet there is still a considerable discordance of opinion as to the precise part of that series in which they should stand. Prof. Leydig, who has devoted much attention to the study of the class, regards them as most allied to the *Crustacea*, and terms them 'Cilio-crustaceans;' and the curious Entomostracan-looking *Pedalion* of Dr. Hudson might seem a link with that group.[1] Prof. Huxley, on the other hand, has argued that they are more connected with the *Annelida*, through the resemblance which they bear to the early larval forms of that class (§ 595); while in their single bilobed nerve-ganglion and water-vascular system, they seem allied to *Planaria* (§ 593).[2]

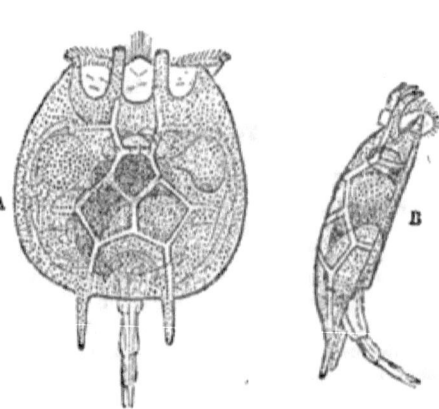

Fig. 313.

Noteus quadricornis: A, dorsal view; B, side view.

---

[1] See Prof. E. Ray Lankester's 'Remarks on *Pedalion*,' in "Quart. Journ. Microsc. Sci.," Vol. xii. (1878), p. 338.

[2] The following Treatises and Memoirs (in addition to those already referred to) contain valuable information in regard to the life-history of Animalcules and their principal forms:—Ehrenberg, "Die Infusionsthierchen," Berlin, 1838; Dujardin, "Histoire Naturelle des Zoophytes Infusoires," Paris, 1841; Pritchard, "History of Infusoria," 4th Ed., London, 1861 (a comprehensive repertory of information); Stein, "Der Organismus der Infusionsthiere," Leipzig, Erste Abtheilung, 1859, Zweite Abtheilung, 1867, Dritte Abtheilung, Hälfte i , 1878; Saville Kent's "Manual of the Infusoria," 1880-1; and Prof. Bütschli's *Protozoa* (1880, 1881) in the new edition of "Bronn's Thierreich."—For the RHIZOPODA and INFUSORIA specially, see Claparède and Lachmann, "Etudes sur les Infusoires et les Rhizopodes," Geneva, 1858-1861; Cohn, in "Siebold and Kölliker's Zeitschrift," 1851-4, and 1857; Lieberkühn, in "Müller's Archiv," 1856, and "Ann. of Nat. Hist.," 2d Ser., Vol. xviii., 1856; Engelmann, "Zur Naturgeschichte der Infusions-Thiere" (1862); and Prof. Butschli's "Studien über die Conjugation der Infusorien," etc., 1876.—For the ROTIFERA specially, see Leydig, in "Siebold and Kölliker's Zeitschrift," Bd. vi., 1854; Gosse on *Melicerta ringens*, in "Quart. Journ. of Microsc. Science," Vol. i. (1853), p. 1; Huxley on *Lacinularia socialis* in "Trans. act. of Microsc. Soc.," Ser. 2, Vol. i (1853), p. 1; and Cohn in "Siebold and Kölliker's Zeitschrift," Bde. vii., ix. (1856, 1858). Mr. Slack's "Marvels of Pond Life" (2d Edit., London, 1871) contains many interesting observations on the habits of Infusoria and Rotifera.

CHAPTER XII.

FORAMINIFERA AND RADIOLARIA.

455. RETURNING now to the lowest or *Rhizopod* type of Animal life (Chap. x)., we have to direct our attention to two very remarkable series of forms, almost exclusively Marine, under which that type manifests itself; all of them distinguished by *skeletons* so consolidated by Mineral deposit, as to retain their form and intimate structure long after the Animals to which they belonged have ceased to live, even for those undefined periods in which they have been imbedded as Fossils in strata of various geological ages. In the first of these groups, the *Foraminifera*, the skeleton usually consists of a *calcareous* many-chambered Shell, which closely invests the sarcode-body, and which, in a large proportion of the group, is perforated with numerous minute apertures; this shell, however, is sometimes replaced by a 'test,' formed of minute grains of sand cemented together; and there are a few cases (§ 397) in which the Animal has no other protection than a membranous envelope.—In the second group, the *Radiolaria*, the skeleton is always *siliceous*; and may be either composed of disconnected spicules, or may consist of a symmetrical open framework, or may have the form of a shell perforated by numerous apertures, which more or less completely incloses the body. —The *Foraminifera* probably take, and always have taken, the largest share of any Animal group in the maintenance of the solid carcareous portion of the Earth's crust; by separating from its solution in Ocean-water the Carbonate of Lime continually brought down by rivers from the land. The *Radiolaria* do the same, though in far less measure, for the Silex. And both extract from Sea-water the organic matter universally diffused through it, converting it into a form that serves for the nutrition of higher Marine animals.

SECTION I.—FORAMINIFERA.

456. The animals of this group belong to that *Reticularian* form of the Rhizopod type (§ 397), in which,—with a differentiation between the containing and the contained sarcodic substance which is involved in the formation of a definite investment,—a distinct *nucleus* (sometimes single, in other cases multiple) is probably always present.[1] The Shells of

---

[1] The *absence* of a nucleus was long supposed to be a characteristic of the animal of the *Foraminifera*; and its presence in *Gromia* (first detected by Dr. Wallich) was regarded as differentiating that type from the Foraminifera proper. But the researches of Hertwig and Lesser having established its presence in several true Foraminifera, and the Author's own observations on other forms having confirmed theirs, its general presence may be fairly assumed, until contradicted by more extended observation.

PLATE XV.

VARIOUS FORMS OF FORAMINIFERA (Original).

Fig. 1. *Cornuspira.*
    2. *Spiroloculina.*
    3. *Triloculina.*
    4. *Biloculina.*
    5. *Peneroplis.*
    6. *Orbiculina* (cyclical form).
    7. *Orbiculina* (young).
    8. *Orbiculina* (spiral form).
    9. *Lagena.*
  10. *Nodosaria.*

Fig. 11. *Cristellaria.*
  12. *Globigerina.*
  13. *Polymorphina.*
  14. *Textularia.*
  15. *Discorbina.*
  16. *Polystomella.*
  17. *Planorbulina.*
  18. *Rotalia.*
  19. *Nonionina.*

Foraminifera are, for the most part, *polythalamous* or many-chambered (Plate XV.); often so strongly resembling those of *Nautilus, Spirula,* and other Cephalopod Mollusks, that it is not surprising that the older Naturalists, to whom the structure of these animals was entirely unknown, ranked them under that Class. But independently of the entire difference in the character of the animal bodies by which the two kinds of shells are formed, there is a most important distinction between them in regard to the relation of the animal to the shell. For whilst, in the chambered shells of the *Nautilus* and other Cephalopods, the animal is a single individual tenanting only the last formed chamber, and withdrawing itself from each chamber in succession, as it adds to this another and larger one, the animal of a nautiloid Foraminifer has a *composite* body, consisting of a number (sometimes very large) of 'segments,' each repeating the rest, which continues to increase by *gemmation* or budding from the last-formed segment. And thus each of the chambers, however numerous they may be, is not only formed, but continues to be occupied, by its own segment; which is connected with the segments of earlier and later formation by a continuous 'stolon' (or creeping stem), that passes through apertures in the *septa* or partitions dividing the chambers.—From what we know of the semi-fluid condition of the sarcode-body in the Reticularian type (§ 397), there can be little doubt that there is an incessant circulatory change in the actual substance of each segment; so that the material taken-in as food by the segment nearest the surface or margin, is speedily diffused through the entire mass. The relation between these 'polythalamous' forms, therefore, and the *monothalamous* or single-chambered,—of which we have already had an example in *Gromia* (§ 397), and of which others will be presently described, —is simply that whereas any buds produced by the latter detach themselves to form separate individuals, those put forth by the former remain in continuity with the parent stock and with each other, so as to form a 'composite' Animal and a 'polythalamous' Shell.

457. According to the plan on which the gemmation takes place, will be the configuration of the shelly structure produced by the segmented body. Thus, if the bud should be put forth from the aperture of a *Lagena* (Plate XV., fig. 9) in the direction of the axis of its body, and a second shell should be formed around this bud in continuity with the first, and this process should be successively repeated, a straight rod-like shell would be produced (fig. 10), whose multiple chambers communicate with each other by the openings that originally constituted their mouths; the mouth of the last-formed chamber being the only aperture through which the sarcode-body, thus composed of a number of segments connected by a peduncle or 'stolon' of the same material, could now project itself or draw-in its food. The successive segments may be all of the same size, or nearly so, in which case the entire rod will approach the cylindrical form, or will resemble a line of beads; but it often happens that each segment is somewhat larger than the preceding (fig. 11), so that the composite shell has a conical form, the apex of the cone being the original segment, and its base the one last formed.—The method of growth now described is common to a large number of Foraminifera, chiefly belonging to the genus *Nodosarina;* but even in that genus we have every gradation between the *rectilineal* (fig. 10), and the *spiral* mode of growth (fig. 11); whilst in the genus *Peneroplis* (fig. 5) it is not at all uncommon for shells which commence in a spiral to exchange this in a more advanced stage for the rectilineal. When the

successive segments are added in a spiral direction, the character of the spire will depend in great degree upon the enlargement or non-enlargement of the successively-formed chambers; for sometimes it opens out very rapidly, every whorl being considerably broader than that which it surrounds, in consequence of the great excess of the size of each segment over that of its predecessor, as in *Peneroplis;* but more commonly there is so little difference between the successive segments, after the spire has made two or three turns, that the breadth of each whorl scarcely exceeds that of its predecessor, as is well seen in the section of the *Rotalia* represented in Fig. 330. An intermediate condition is presented by such a *Rotalia* as is shown in Fig. 314, which may be taken as a characteristic type of a very large and important group of Foraminifera, whose general features will be presently described. Again, a spiral may be either 'nau-

FIG. 314.

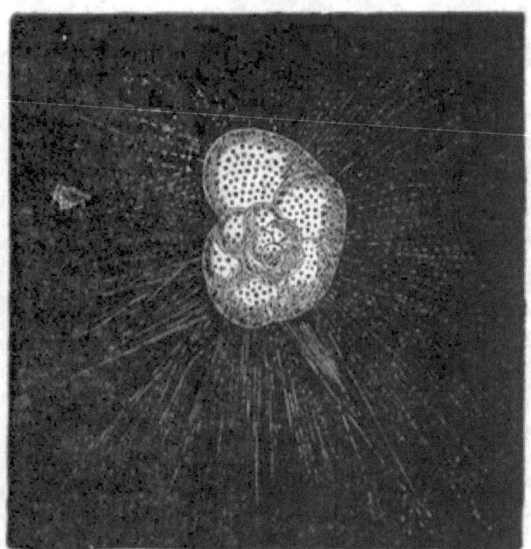

Rotalia ornata, with its pseudopodia extended.

tiloid' or 'turbinoid': the former designation being applied to that form in which the successive convolutions all lie in one plane (as they do in the Nautilus), so that the shell is 'equilateral' or similar on its two sides; whilst the latter is used to mark that form in which the spire passes obliquely round an axis, so that the shell becomes 'inequilateral,' having a more or less conical form, like that of a Snail or a Periwinkle, the first-formed chamber being at the apex. Of the former we have characteristic examples in *Polystomella* (Plate XV., fig. 16) and *Nonionina* (fig. 19); whilst of the latter we find a typical representation in *Rotalia Baccarii* (fig. 18). Further, we find among the shells whose increase takes place upon the spiral plan, a very marked difference as to the degree in which the earlier convolutions are invested and concealed by the latter. In the great *Rotaline* group, whose characteristic form is a turbinoid spiral, all the convolutions are usually visible, at least on one side (figs. 15, 17, 18), but among the *Nautiloid* tribes it more frequently happens that the last-formed whorl incloses the preceding to such an extent that they are scarcely, or not all, visible externally, as is the case in *Cristellaria* (fig. 11), *Polystomella* (fig. 16), and *Nonionina* (fig. 19).—The turbinoid spire may coil so rapidly round an elongated axis, that the number of chambers in each turn is very small; thus in *Globigerina* (fig. 12) there are usually only four; and in *Valvulina* the regular number is only three. Thus we are led to the *biserial* arrangement of the chambers which is characteristic of the *Textularian* group (fig. 14); in which we find the chambers arranged in two rows, each chamber communicating with that above and that below it on the opposite side, without any direct communication with the chamber of its own side, as will be understood by reference to Fig. 328, A, which shows a 'cast' of the sarcode body of the animal. On the other hand, we find in the nautiloid spire a tendency to pass (by a curious transitional form to be presently described, § 464) into the *cyclical* mode of growth; in which the original segment, instead of budding-forth on one side only, develops *gemmæ* all round, so that a ring of small chambers (or chamberlets) is formed around the primordial chamber, and this in its turn surrounds itself after the like fashion with another ring; and by successive repetitions of the same process the shell comes to have form of a disc made up of a great number of concentric rings, as we see in *Orbitolites* (Fig. 316) and in *Cycloclypeus* (Plate XVI., fig. 1).

458. These and other differences in the *plan of growth* were made by M. D'Orbigny the foundation of his Classification of this group, which, though at one time generally accepted, has now been abandoned by most of those who have occupied themselves in the study of the Foraminifera. For it has come to be generally admitted that 'plan of growth' is a character of very subordinate importance among the Foraminifera, so that any classification which is primarily based upon it must necessarily be altogether unnatural; those characters being of primary importance which have an immediate and direct relation to the Physiological condition of the Animal, and are thus indicative of the real affinities of the several groups which they serve to distinguish. The most important of these characters will now be noticed.[1]

459. Two very distinct types of Shell-structure prevail among ordinary Foraminifera—namely, the *porcellanous* and the *hyaline* or *vitreous*. The shell of the former, when viewed by reflected light, presents an opaque-white aspect which bears a strong resemblance to porcelain; but when thin natural or artificial laminæ of it are viewed by transmitted light, the opacity gives place to a rich brown or amber color, which in a few instances is tinged with crimson. No structure of any description can be detected in this kind of shell substance, which is apparently homogeneous throughout. Although the shells of this 'porcellanous' type often present the appearance of being perforated with foramina, yet this appearance is illusory, being due to a mere 'pitting' of the external surface, which, though often very deep, never extends through the whole thickness of the shell. Some kind of inequality of that surface, indeed, is extremely common in the shells of the 'porcellanous' Fora-

---

[1] This subject will be found amply discussed in the Author's "Introduction to the Study of the Foraminifera," published by the Ray Society; to which work he would refer such of his readers as may desire more detailed information in regard to it.

minifera; one of the most frequent forms of it being a regular alternation of ridges and furrows, such as is occasionally seen in *Miliola* (Plate xv., fig. 3), but which is an almost constant characteristic of *Peneroplis* (fig. 5). But no difference of texture accompanies either this or any other kind of inequality of surface; the raised and depressed portions being alike homogeneous.—In the shells of the *vitreous* or *hyaline* type, on the other hand, the proper shell-substance has an almost glassy transparence, which is shown by it alike in thin natural lamellæ, and in artificially prepared specimens of such as are thicker and older. It is usually colorless, even when (as in the case with many *Rotalinæ*) the substance of the animal is deeply colored; but in certain aberrant Rotalines the shell is commonly, like the animal body, of a rich crimson hue. All the shells of this type are beset more or less closely with *tubular perforations*, which pass directly, and (in general) without any subdivision, from one surface to the other. These tubuli are in some instances sufficiently coarse for their orifices to be distinguished with a low magnifying power, as 'punctations' on the surface of the shell, as is shown in Fig. 314; whilst in other cases they are so minute as only to be discernible in thin sections seen by transmitted light under a higher magnifying power, as is shown in Figs. 335, 336. When they are very numerous and closely set, the shell derives from their presence that kind of opacity which is characteristic of all minutely-tubular textures, whose tubuli are occupied either by air or by any substance having a refractive power different from that of the intertubular substance, however perfect may be the transparence of the latter. The straightness, parallelism, and isolation of these tubuli are well seen in vertical sections of the thick shells of the largest examples of the group, such as *Nummulina* (Fig. 335). It often happens, however, that certain parts of the shell are left unchannelled by these tubuli; and such are readily distinguished, even under a low magnifying power, by the readiness with which they allow transmitted light to pass through them, and by the peculiar vitreous lustre they exhibit when light is thrown obliquely on their surface. In shells formed upon this type, we frequently find that the surface presents either bands or spots which are so distinguished; the non-tubular *bands* usually marking the position of the septa, and being sometimes raised into ridges, though in other instances they are either level or somewhat depressed; whilst the non-tubular *spots* may occur on any part of the surface, and are most commonly raised into tubercles, which sometimes attain a size and number that give a very distinctive aspect to the shells that bear them.

460. Between the comparatively *coarse* perforations which are common in the *Rotaline* type, and the *minute* tubuli which are characteristic of the *Nummuline*, there is such a continuous gradation as indicates that their mode of formation, and probably their uses, are essentially the same. In the former, it has been demonstrated by actual observation that they allow the passage of pseudopodial extensions of the sarcode-body through every part of the external wall of the chambers occupied by it (Fig. 314); and there is nothing to oppose the idea that they answer the same purpose in the latter, since, minute as they are, their diameter is not too small to enable them to be traversed by the finest of the threads into which the branching pseudopodia of Foraminifera are known to subdivide themselves. Moreover, the close approximation of the tubuli in the most finely-perforated Nummulines, makes their collective area fully equal to that of the larger but more scattered pores of the most coarsely-perforated Rotalines. Hence it is obvious that the

tubulation or *non-tubulation* of Foraminiferal shells is the key to a very important Physiological difference between the Animal inhabitants of the two kinds respectively; for whilst every segment of the sarcode-body in the former case gives off pseudopodia, which pass at once into the surrounding medium, and contribute by their action to the nutrition of the segment from which they proceed, these pseudopodia are limited in the latter case to the *final* segment, issuing forth only through the aperture of the last chamber, so that all the nutrient material which they draw-in must be first received into the last segment, and be transmitted thence from one segment to another until it reaches the earliest. With this difference in the physiological condition of the Animal of these two types, is usually associated a further very important difference in the conformation of the Shell—viz., that whilst the aperture of communication between the **chambers, and** between the last chamber and the exterior, is usually very small in the 'vitreous' shells, serving merely to give passage to a slender *stolon* or thread of sarcode from which the succeeding segment may be budded-off, it is much wider in the 'porcellanous' shells, so as to give passage to a 'stolon' that may not only bud-off new segments, but may serve as the medium for transmitting nutrient material from the outer to the inner chambers.

461. Between the highest types of the *Porcellanous* and the *Vitreous* series respectively, which frequently bear a close resemblance to each other in *form*, there are certain other well-marked differences in *structure*, which clearly indicate their essential dissimilarity. Thus, for example, if we compare *Orbitolites* (Fig. 316) with *Cycloclypeus* (Plate XVI., fig. 1), we recognize the same plan of growth in each, the chamberlets being arranged in concentric rings around the primordial chamber; and to a superficial observer there would appear little difference between them. But a minuter examination shows that not only is the texture of the shell 'porcellanous' and non-tubular in *Orbitolites*, whilst it is 'vitreous' and minutely tubular in *Cycloclypeus;* but that the partitions between the chamberlets are *single* in the former, whilst they are *double* in the latter, each segment of the sarcode-body having its own proper shelly investment. Moreover, between these double partitions an additional deposit of calcareous substance is very commonly found, constituting what may be termed the *intermediate skeleton;* and this is traversed by a peculiar system of inosculating *canals*, which pass around the chamberlets in interspaces left between the two laminæ of their partitions, and which seem to convey through its substance extensions of the sarcode-body whose segments occupy the chamberlets. We occasionally find this 'intermediate skeleton' extending itself into peculiar *outgrowths*, which have no direct relation to the chambered shell; of this we have a very curious example in *Calcarina* (Plate XVI., fig. 3); and it is in these that we find the 'canal-system' attaining its greatest development. Its most regular distribution, however, is seen in *Polystomella* and in *Operculina;* and an account of it will be given in the description of those types.

462. PORCELLANEA.—Commencing, now, with the *Porcellanous* series, we shall briefly notice some of its most important forms, which are so related to each other as to constitute but the one family *Miliolida*. Its simplest **type** is presented by the *Cornuspira* (Plate XV., fig. 1) of our own coasts, found attached to Sea-weeds and Zoöphytes; this is a minute spiral shell, of which the interior forms a continuous tube not divided into chambers; the latter portion of the spire is often very much flattened-out, as in *Peneroplis* (fig. 5), so that the form of the mouth is

changed from a circle to a long narrow slit.—Among the commonest of the Foraminifera, and abounding near the shores of almost every sea, are some forms of the *Milioline* type, so named from the resemblance of some of their minute fossilized forms (of which enormous beds of limestone in the neighborhood of Paris are almost entirely composed) to millet-seeds. The peculiar mode of growth by which these are characterized, will be best understood by examining in the first instance the form which has been designated as *Spiroloculina* (Plate xv., fig. 2). This shell is a spiral, elongated in the direction of one of its diameters, and having in each turn a contraction at either end of that diameter, which partially divides each convolution into two chambers; the separation between the consecutive chambers is made more complete by a peculiar projection from the inner side of the cavity, known as the 'tongue' or 'valve,' which may be considered as an imperfect septum; of this a characteristic example is shown in the upper part of fig. 4. Now it is a very general habit in the Milioline type, for the chambers of the later convolutions to extend themselves over those of the earlier, so as to conceal them more or less completely; and this they very commonly do somewhat unequally, so that more of the earlier chambers are visible on one side than on the other. *Miliolæ* thus modified (fig. 3) have received the names of *Quinqueloculina* and *Triloculina* according to the number of chambers visible externally; but the extreme inconstancy which is found to mark such distinctions, when the comparison of specimens has been sufficiently extended, entirely destroys their value as differential characters. Sometimes the earlier convolutions are so completely concealed by the later, that only the two chambers of the last turn are visible externally; and in this type, which has been designated *Biloculina*, there is often such an increase in the breadth of the chambers as altogether changes the usual proportions of the shell, which has almost the shape of an egg when so placed that either the last or the penultimate chamber faces the observer (Plate xv., fig. 4). It is very common in Milioline shells for the external surface to present a 'pitting,' more or less deep, a ridge-and-furrow arrangement (fig. 3), or a honeycomb division; and these diversities have been used for the characterization of species. Not only, however, may every intermediate gradation be met-with between the most strongly marked forms, but it is not at all uncommon to find the surface smooth on some parts, whilst other parts of the surface in the same shell are deeply pitted or strongly ribbed or honeycombed; so that here again the inconstancy of these differences deprives them of all value as distinctive characters.

463. Reverting again to the primitive type presented in the simple spiral of *Cornuspira*, we find the most complete development of it in *Peneroplis* (Plate xv., fig. 5), a very beautiful form, which, although very rare on our own coasts, is one of the commonest of all Foraminifera in the shore-sands and shallow water dredgings of the warmer regions of every part of the globe. This is a nautiloid shell, of which the spire flattens itself out as it advances in growth; it is marked externally by a series of transverse bands, which indicate the position of the internal septa that divide the cavity into chambers; and these chambers communicate with each other by numerous minute pores traversing each of the septa, and giving passage to threads of sarcode that connect the segments of the body. At *a* is shown the 'septal plane' closing in the last-formed chamber, with its single row of pores through which the pseudopodial filaments extend themselves into the surrounding medium.

The surface of the shell, which has a peculiarly 'porcellanous' aspect, is marked by closely-set *striæ* that cross the spaces between the successive septal bands; these markings, however, do not indicate internal divisions, and are due to a surface-furrowing of the shelly walls of the chambers. This type passes into two very curious modifications; one having a spire which, instead of flattening itself out, remains turgid like that of a *Nautilus*, having only a single aperture, which sends out fissured extensions that subdivide like the branches of a tree, suggesting the name of *Dendritina;* the other having its spire continued in a rectilineal direction, so that the shell takes the form of a crosier, this being distinguished by the name of *Spirolina*. A careful examination of intermediate forms, however, has made it evident that these modifications, though ranked as of generic value by M. D'Orbigny, are merely *varietal;* a continuous gradation being found to exist from the elongated septal plane of Peneroplis, with its single row of isolated pores, to the arrow-shaped, oval, or even circular septal plane of Dendritina, with all its pores fused together (so to speak) into one dendritic aperture; and a like gradation being presented between the ordinary and the 'spiroline' forms into which both *Peneroplis* and *Dendritina* tend to elongate themselves.

464. From the ordinary nautiloid multilocular spiral, we now pass to a more complex and highly-developed form, which is restricted to tropical regions, but is there very abundant—that, namely, which has received the designation *Orbiculina* (Plate xv., figs. 6, 7, 8). The relation of this to the preceding will be best understood by an examination of its earlier stage of growth, represented in fig. 7; for here we see that the shell resembles that of Peneroplis in its general form, but that its principal chambers are divided by 'secondary septa' passing at right angles to the primary, into 'chamberlets' occupied by sub-segments of the sarcode-body. Each of these secondary septa is perforated by an aperture, so that a continuous gallery is formed, through which (as in Fig. 316) there passes a stolon that unites together all the sub-segments of each row. The chamberlets of successive rows alternate with one another in position; and the pores of the principal septa are so disposed, that each chamberlet of any row normally communicates with two chamberlets in each of the adjacent rows. The later turns of the spire very commonly grow completely over the earlier, and thus the central portion or 'umbilicus' comes to be protuberant, whilst the growing edge is thin. The spire also opens out at its growing margin, which tends to encircle the first-formed portion, and thus gives rise to the peculiar shape represented in fig. 8, which is the common *aduncal* type of this organism. But sometimes, even at an early age, the growing margin extends so far round on each side, that its two extremities meet on the opposite side of the original spire, which is thus completely inclosed by it; and its subsequent growth is no longer *spiral* but *cyclical*, a succession of *concentric rings* being added, one around the other, as shown in fig. 6. This change is extremely curious, as demonstrating the intimate relationship between the *spiral* and the *cyclical* plans of growth, which at first sight appear essentially distinct. In all but the youngest examples of *Orbiculina*, the septal plane presents more than a single row of pores, the number of rows increasing in the thickest specimens to six or eight. This increase is associated with a change in the form of the sub-segments of sarcode from little blocks to columns, and with a greater complexity in the general arrangement, such as will be more fully described here-

after in *Orbitolites* (§ 466). The largest existing examples of this type are far surpassed in size by those which make up a considerable part of a Tertiary Limestone on the Malabar coast of India, whose diameter reaches 7 or 8 lines.

465. A very curious modification of the same general plan is shown in *Alveolina*, a genus of which the largest existing forms (Fig. 315) are commonly about one third of an inch long, while far larger specimens

Fig. 315.

Alveolina Quoii:—*a, a*, septal plane, showing multiple pores.

are found in the Tertiary Limestones of Scinde. Here the spire turns round a very elongated axis, so that the shell has almost the form of a cylinder drawn to a point at each extremity. Its surface shows a series of longitudinal lines which mark the principal septa; and the bands that intervene between these are marked transversely by lines which show the subdivision of the principal chambers into 'chamberlets.' The chamberlets of each row are connected with each other, as in the preceding type, by a continuous gallery; and they communicate with those of the next row by a series of multiple pores in the principal septa, such as constitute the external orifices of the last-formed series, seen on its septal plane at *a, a*.

466. The highest development of that cyclical plan of growth which we have seen to be sometimes taken on by Orbiculina, is found in *Orbitolites*; a type which, long known as a very abundant fossil in the

Fig. 316.

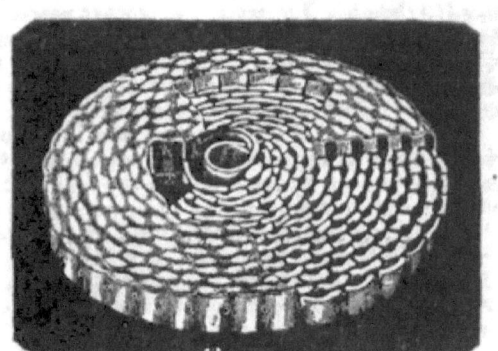

Simple disk of *Orbitolites complanatus*, laid open to show its interior structure:—*a*, central chamber; *b*, circumambient chamber, surrounded by concentric zones of chamberlets connected with each other by annular and radiating passages.

earlier Tertiaries of the Paris basin, has lately proved to be scarcely less abundant in certain parts of the existing Ocean. The largest recent specimens of it, sometimes attaining the size of a shilling, have hitherto been obtained only from the coast of New Holland, the Fijian reefs, and various other parts of the Polynesian Archipelago; but disks of comparatively minute size and simpler organization are to be found in almost all Foraminiferal sands and dredgings from the shores of the warmer regions of the globe, being especially abundant in those of some of the Philippine Islands, of the Red Sea, of the Mediterranean, and especially of the Ægean. When such disks are subjected to microscopic examination, they are found (if uninjured by abrasion) to present the structure represented in Fig. 316; where we see on the surface (by incident light) a number of rounded elevations, arranged in concentric zones around a sort of nucleus (which has been laid-open in the figure to show its internal structure); whilst at the margin we observe a row of rounded projections, with a single aperture or pore in each of the intervening depressions. In very thin disks the structure may often be brought into view by mounting them in Canada balsam and transmitting light through them; but in those which are too opaque to be thus seen-through, it is sufficient to rub-down one of the surfaces upon a stone, and then to mount the specimen in balsam. Each of the superficial elevations will then be found to be the roof or cover of an ovate cavity or 'chamberlet,' which communicates by means of a lateral passage with the chamberlet on either side of it in the same ring; so that each circular zone of chamberlets might be described as a continuous annular passage, dilated into cavities at intervals. On the other hand, each zone communicates with the zones that are internal and external to it, by means of passages in a radiating direction; these passages run, however, not from the chamberlets of the inner zone to those of the outer, but from the connecting passages of the former to the chamberlets of the latter; so that the chamberlets of each zone *alternate* in position with those of the zones internal and external to it. The radial passages from the outermost annulus make their way at once to the margin, where they terminate, forming the 'pores' which (as already mentioned) are to be seen on its exterior. The central nucleus, when rendered sufficiently transparent by the means just adverted-to, is found to consist of a 'primordial chamber' (*a*), usually somewhat pear-shaped, that communicates by a narrow passage with a much larger 'circumambient chamber' (*b*), which nearly surrounds it, and which sends off a variable number of radiating passages towards the chamberlets of the first zone, which forms a complete ring around the circumambient chamber.[1]

467. The idea of the nature of the living occupant of these cavities which might be suggested by the foregoing account of their arrangement, is fully borne-out by the results of the examination of the sarcode-body, which may be obtained by the maceration in dilute acid (so as to remove the shelly investment) of specimens of *Orbitolite* that have been gathered fresh and preserved in spirit. For this body is found to be composed (Fig. 317) of a multitude of segments of sarcode, presenting not the least trace of higher organization in any part, and connected together by 'stolons' of the like substance. The 'primordial' pear-shaped segment, *a*, is seen to have budded-off its 'circumambient' segment, *b*, by a narrow footstalk or stolon; and this circumambient segment, after passing almost entirely round the primordial, has budded-off three stolons, which swell into new sub-segments from which the first ring is formed. Scarcely any two specimens are precisely alike as to the mode in which the first ring

---

[1] Although the above may be considered the *typical* form of the Orbitolite, yet, in a very large proportion of specimens, the first few zones are not complete circles, the early growth having taken place from one side only; and there is a very beautiful variety in which this one-sidedness of increase imparts a distinctly *spiral* character to the early growth, which soon, however, gives place to the *cyclical*.—In the *Orbitolites tenuissimus* (Fig. 318) brought up from depths of 1,500 fathoms or more, the 'nucleus' is formed by three or four turns of a spiral closely resembling that of a *Cornuspira* (§ 462), with an interruption at every half-turn as in *Spiroloculina;* the growth afterwards becoming purely concentric.

originates from the 'circumambient segments;' for sometimes a score or more of radial passages extend themselves from every part of the margin of the latter (and this, as corresponding with the plan of growth afterwards followed) is probably the *typical* arrangement); whilst in other cases (as in the example before us) the number of these primary offsets is extremely small. Each zone is seen to consist of an assemblage of ovate sub-segments, whose height (which could not be shown in the figure) corresponds with the thickness of the disk; these sub-segments, which are all exactly similar and equal to one another, are connected by annular stolons; and each zone is connected with that on its exterior by radial extensions of those stolons passing-off between the sub-segments.

468. The radial extensions of the outermost zone issue-forth as pseudopodia from the marginal pores, searching-for and drawing in alimentary materials in the manner formerly described (§ 397); the whole of the soft body, which has no communication whatever with the exterior save through these marginal pores, being nourished by the transmission of the products of digestion from zone to zone, through similar bands of protoplasmic substance. In all cases in which the growth of the disk takes place with normal regularity, it is probable that a complete circular zone is added at once. Thus we find this simple type of organization giving origin to fabrics of by no means microscopic dimensions, in which, however, there is no other differentiation of parts than that concerned in the formation of the shell; every segment and every stolon (with the exception of the two forming the 'nucleus') being, so far as can be ascertained, a precise repetition of every other, and the segments of the nucleus differing from the rest in nothing else than their form. The equality of the endowments of the segments is shown by the fact—of which accident has repeatedly furnished proof—that a small portion of a disk, entirely separated from the remainder, will not only continue to live, but will so increase as to form a new disk (Fig. 318); the want of the 'nucleus' not appearing to be of the slightest consequence, from the time that active life is established in the outer zones.

FIG. 317.

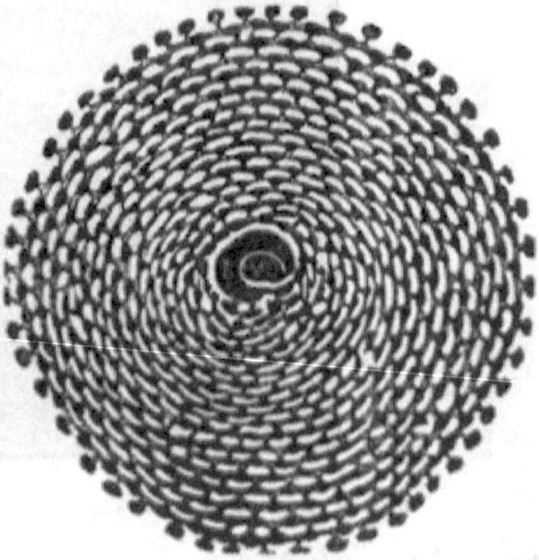

Composite Animal of Simple type of *Orbitolites complanatus;*—*a*, central mass of sarcode; *b*, circumambient segment, giving off peduncles, in which originate the concentric zones of sub-segments connected by annular bands.

469. One of the most curious features in the history of this type is its capacity for developing itself into a form which, whilst fundamentally the same as that previously described, is very much more complex. In all the larger specimens of *Orbitolite*, we observe that the marginal pores, instead of constituting but a single row, form many rows one above another, and besides this, the chamberlets of the two surfaces, instead of being rounded or ovate in form, are usually oblong and straight-sided, their long diameters lying in a radial direction,

like those of the cyclical type of *Orbiculina* (Plate XV., fig. 6). When a vertical section is made through such a disk, it is found that these oblong chambers constitute two *superficial* layers, between which are interposed *columnar* chambers of a rounded form; and these last are connected to-

Fig. 318.

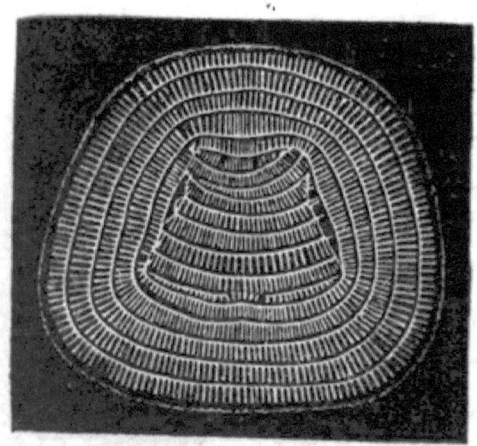

Disk of *Orbitolites tenuissimus*, formed round fragment of previous disk.

gether by a complex series of passages, the arrangement of which will be best understood from the examination of a part of the sarcode-body that occupies them (Fig. 319). For the oblong superficial chambers are occupied by sub-segments of sarcode, $c\,c$, $d\,d$, lying side by side, so as to form part of an annulus, but each of them disconnected from its neighbors, and communicating only by a double footstalk with the two annular 'stolons,' $a\,a'$, $b\,b'$ which obviously correspond with the single stolon of 'simple' type (Fig. 317). These indirectly connect together not merely all the superficial chamberlets of each zone, but also the columnar sub-segments of the intermediate layer; for these columns ($e\,e$, $e'\,e'$) terminate above and below in the annular stolons, sometimes passing directly from one to the other, but sometimes going out of their direct course to coalesce with another column. The columns of the successive zones (two sets of which are shown in the figure) communicate with each other by threads of sarcode, in such a manner that (as in the simple type) each column is thus brought into connection with two columns of the zone next interior, to which it alternates in position.

Fig. 319.

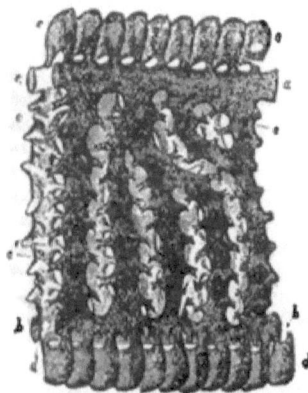

Portion of Animal of Complex type of *Orbitolites complanatus*:—$a\,a'$, $b\,b'$, the upper and lower rings of two concentric zones; $c\,c$, the upper layer of superficial sub-segments, and $d\,d$, the lower layer, connected with the annular bands of both zones; $e\,e$ and $e'\,e'$, vertical sub-segments of the two zones.

Similar threads, passing off from the outermost zone, through the multiple ranges of marginal pores, would doubtless act as pseudopodia.

470. Now this plan of growth is so different from that previously described, that there would at first seem ample ground for separating the *simple* and the *complex* types as distinct species. But the test furnished by the examination of *a large number of specimens*, which ought never to be passed-by when it can possibly be appealed to, furnishes these very singular results :—1st. That the two forms must be considered as specifically identical; since there is not only a gradational passage from one to the other, but they are often combined in the same individual, the *inner* and first-formed portion of a large disk frequently presenting the simple type, whilst the *outer* and later-formed part has developed itself upon the complex:—2d. That although the last-mentioned circumstance would naturally suggest that the change from the one plan to another may be simply a feature of advancing age, yet this cannot be the case; since, although the complex sometimes evolves itself even from the very first (the 'nucleus,' though resembling that of the simple form, sending out two or more tiers of radiating threads), more frequently the simple prevails for an indefinite number of zones, and then changes itself in the course of a few zones into the complex.—No department of Natural History could furnish more striking instances than are afforded by the different forms presented by the Foraminiferal types now described, of the wide *range of variation* that may occur within the limits of one and the same species; and the Microscopist needs to be specially put on his guard as to this point, in respect to the lower types of Animal as to those of Vegetable life, since the determination of form seems to be far less precise among such than it is in the higher types.

471. In what manner the reproduction of *Orbitolites* is accomplished, we can as yet do little more than guess; but from appearances sometimes presented by the sarcode-body, it seems reasonable to infer that *gemmules*, corresponding with the zoöspores of Protophytes (§ 244), are occasionally formed by the breaking-up of the sarcode into globular masses; and that these, escaping through the marginal pores, are sent forth to develop themselves into new fabrics. Of the mode wherein that sexual operation is performed, however, in which alone true Generation consists, nothing whatever is known.

472. ARENACEA.—In certain forms of the preceding family, and especially in the genus *Miliola*, we not unfrequently find the shells encrusted with particles of sand, which are imbedded in the proper shell-substance. This incrustation, however, must be looked on as (so to speak) accidental; since we find shells that are in every other respect of the same type, altogether free from it. A similar accidental incrustation presents itself among certain 'vitreous' and perforate shells; but there, too, it is on usually a basis of true shell, and the sandy incrustation is often entirely absent. There is, however, a group of Foraminifera in which the true shell is constantly and entirely *replaced* by a sandy envelope, which is distinguished as a 'test;' the arenaceous particles being held together only by a cement exuded by the animal. It is not a little curious that the forms of these arenaceous 'tests' should represent those of many different types among both the 'porcellanous' and the 'vitreous' series; whilst yet they graduate into one another in such a manner, as to indicate that all the members of this 'arenaceous' group are closely related to each other, so as to form a series of their own. And it is further remarkable, that while the Deep-sea dredgings recently carried

down to depths of from 1,000 to 2,500 fathoms, have brought up few forms of either 'porcellanous' or 'vitreous' Foraminifera that were not previously known, they have added greatly to our knowledge of the 'arenaceous' types, the number and variety of which far exceed all previous conception. These have not yet been systematically described; but the following notice of a few of the more remarkable, will give some idea of the interest attaching to this portion of the new *Fauna* which has been brought to light by Deep-sea exploration.

473. In the midst of the sandy mud which formed the bottom where the warm area of the 'Globigerina-mud' (§ 480) abutted on that over which a glacial stream flowed, there were found a number of little pellets, varying in size from a large pin's head to that of a large pea, formed of an aggregation of sand-grains, minute Foraminifers, etc., held together by a tenacious protoplasmic substance. On tearing these open, the whole interior was found to have the same composition; and no trace of any structural arrangement could be discovered in their mass. Hence they might be supposed to be mere accidental agglomerations, were it not for their conformity to the 'monerozoic' type previously described (§ 393); for just as a simple 'moner,' by a differentiation of its homogeneous sarcode, becomes an *Amœba*, so would one of these uniform blendings of sand and sarcode, by a separation of its two components,—the sand forming the investing 'test,' and the sarcode occupying its interior,—become the arenaceous *Astrorhiza*. This type, which abounds on the sea-bed in certain localities, presents remarkable variations of form: being sometimes globular, sometimes stellate, sometimes cervicorn. But the same general arrangement prevails throughout; the cavity being occupied by a dark-green sarcode, whilst the 'test' is composed of loosely aggregated sand-grains not held together by any recognizable cement, and has *no definite orifice*, so that the pseudopodia must issue from interstices between the sand-grains, which spaces are probably occupied during life with living protoplasm, that continues to hold together the sand-grains after death. These are by no means microscopic forms; the 'stellate' varieties ranging to 0.3 or even 0.4 inch in diameter, and the 'cervicorn' to nearly 0.5 inch in length.[1]

474. The purely *Arenaceous* Foraminifera are arranged by Mr. H. B. Brady[2] (by whom they have been specially studied) under two Families: the first of which, *Astrorhizida*, includes with the preceding a number of coarse sandy forms, usually of considerable size, and essentially monothalamous, though sometimes imperfectly chambered by constrictions at intervals. Some of the more interesting examples of this family will now be noticed; beginning with the *Saccamina* (Sars), which is a remarkably regular type, composed of coarse sand-grains firmly cemented together in a globular form, so as to form a wall nearly smooth on the outer, though rough on the inner surface, with a projecting neck surrounding a circular mouth (Fig. 319,* *a*, *b*, *c*,). This type, which occurs in extraordinary abundance in certain localities (as the entrance of the Christiania-fjord), is of peculiar interest from the fact that it has been discovered in a fossil state by Mr. H. B. Brady, in a clay seam between two layers of Carboniferous Limestone. Its size is that of very minute seeds.—In striking contrast to the preceding is another single-chambered

---

[1] See the description and figures of this type given by the Author in "Quart. Journ. Microsc. Sci.," Vol. xvi (1876), p. 221.
[2] See his "Notes" in "Quart Journ. of Microsc. Soc.," N.S., Vol. xix. (1879), p. 20; and Vol. xxi. (1881), p. 31.

type, distinguished by the whiteness of its 'test,' to which the Author has given the name of *Pilulina*, from its resemblance to a homœopathic 'globule' (Fig. 319,* *d*, *e*). The form of this is a very regular sphere; and its orifice, instead of being circular and surrounded by a neck, is a slit or fissure with slightly raised lips, and having a somewhat S-shaped curvature. It is by the structure of its 'test,' however, that it is especially distinguished; for this is composed of the finest ends of sponge-spicules, very regularly 'laid' so as to form a kind of felt, through the substance of which very fine sand-grains are dispersed. This 'felt' is somewhat flexible, and its components do not seem to be united by any kind of cement, as it is not affected by being boiled in strong nitric acid; its tenacity, therefore, seems entirely due to the wonderful manner in which the separate siliceous fibres are 'laid.'—It is not a little curious that

Fig. 319*

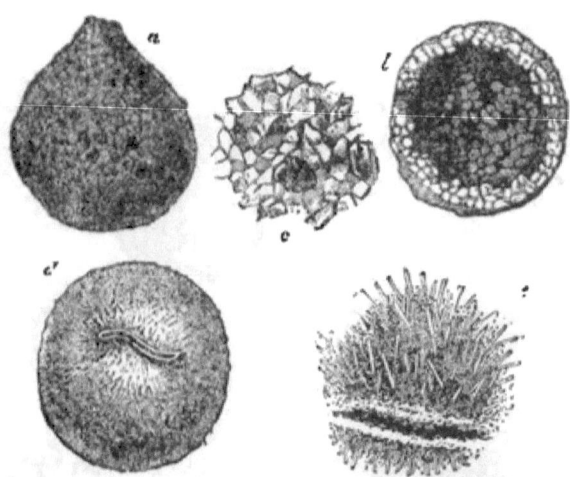

Arenaceous Foraminifera:—*a*, *Saccamina sphaerica;* *b*, the same laid open; *c*, portion of the test enlarged to show its component sand-grains:—*d*, *Pilulina Jeffreysii;* *e*, portion of the test enlarged, showing the arrangement of the sponge-spicules.

these two forms should present themselves in the same dredging; and that there should be no perceptible difference in the character of their sarcode-bodies, which, as in the preceding case, have a dark-green hue.— The *Marsipella elongata* (Fig. 320, *d*), on the other hand, is somewhat fusiform in shape, and has its two extremities elongated into tubes, with a circular orifice at the end of each. The materials of the 'tests' differ remarkably according to the nature of the bottom whereon they live. When they come up with 'Globigerina mud,' in which sponge-spicules abound, whilst sand-grains are scarce, they are almost entirely made up of the former, which are 'laid' in a sort of lattice-work, the interspaces of which are filled up by fine sand-grains; but when they are brought up from a bottom on which sand predominates, the larger part of the 'test' is made up of sand-grains and minute Foraminifera, with here and there a sponge-spicule (Fig. 320, *d*, *f*). In each case, however, the tubular extensions (one of which sometimes forms a sort of proboscis, *e*, nearly equalling the body itself in length) are entirely made up of sponge-spicules laid side by side with extraordinary regularity.—The genus *Rhabdammina* (Sars) resembles Saccamina in the structure of its 'test,' which is composed of sand-grains very firmly cemented together; but the grains are of smaller size, and they are so disposed as to present a smooth surface internally, though the exterior is rough. What is most remarkable about this, is the geometrical regularity of its form, which is typically *triradiate* (Fig. 321, *c*), the rays diverging at equal angles from the central cavity, and each being a tube (*d*) with an orifice at its extremity. Not unfrequently, however, it is *quadri-radiate*, the rays diverging at right angles; and occasionally a fifth ray presents itself, its radiation, however, being on a different plane. The three rays are normally of equal length; but one of them is sometimes shorter than the other two; and when this is the case, the angle between the long rays increases at the expense of the other two, so that the long rays lie more nearly in a straight line. Sometimes the place of the third ray is indicated only by a litttle knob: and then the two long rays have very nearly the same direction.

FIG. 320.

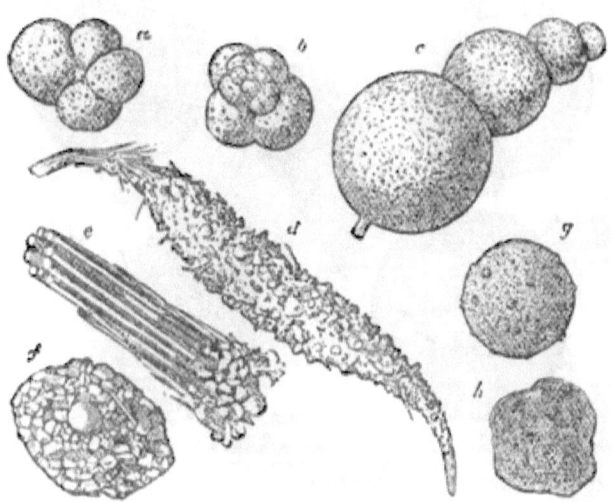

Arenaceous Foraminifera:—*a, b*, upper and lower aspects of *Halophragmium globigeriniforme*; *c, Hormosina globulifera; d, Marsipella elongata; e*, terminal portion, and *f*, middle portion of the same, enlarged; *g, Thurammina papillata; h*, portion of its inner surface enlarged.

We are thus led to forms in which there is no vestige of a third ray, but merely a single straight tube, with an orifice at each end; and the length of this, which often exceeds half an inch, taken in connection with the abundance in which it presents itself in dredgings in which the triradiate forms are rare, seems to preclude the idea that these long single rods are broken rays of the latter.—It is undoubtedly in this group that we are to place the genus *Haliphysema;* which, from constructing its 'test' entirely of sponge-spicules, and even including these in its pseudopodial expansions, has been ranked as a Sponge, although observation of it in its living state leaves no doubt whatever of its Rhizopodal character.[1]

[1] See Saville Kent in "Ann. of Nat. Hist.," Ser. 5, Vol. ii. (1878); Prof. R. Lankester in "Quart. Journ. Microsc. Sci.," Vol. xix. (1868), p. 476; and Prof. Möbius's "Foraminifera von Mauritius."

FORAMINIFERA AND RADIOLARIA. 81

475. *Lituolida.*—The type of this family, which is named after it, is a large, sandy, many-chambered fossil form occurring in the Chalk, to which the name *Lituola* was given by Lamarck, from its resemblance in shape to a crozier. A great variety of recent forms, mostly obtained by deep-sea dredging, are now included in it; as bearing a more or less close resemblance to it and to each other in their chambered structure, and in the arrangement of the sand-grains of which their tests are formed.— These grains are, for the most part, finer than those of which the tests of the preceding family are constructed, and are set (so to speak) more artistically; and a considerable quantity of a cement exuded by the animal is employed in uniting them. This is often mixed up with sandy particles of extreme fineness, to form a sort of 'plaster' with which the exterior of the test is smoothed off, so as to present quite a polished surface.—It is remarkable that the cement contains a considerable quantity of oxide of iron, which imparts a ferruginous hue to the 'tests' in which it is largely employed. The forms of the *Lituoline* 'tests' often simulate in a very curious way those of the simpler types of the *Vitreous* series.

Fig. 321.

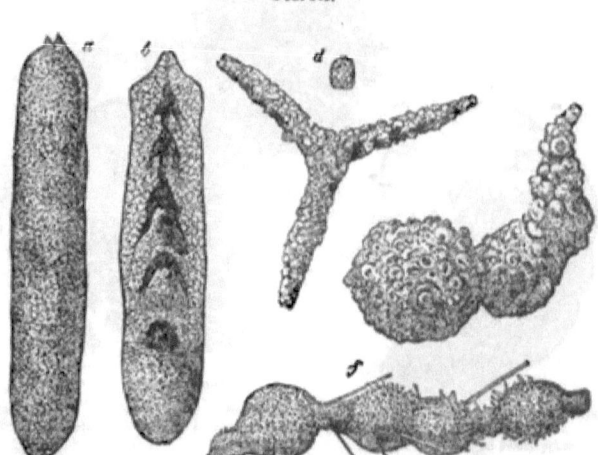

Arenaceous Foraminifera:—*a, b,* Exterior and sectional views of *Reophax rudis*; *c, Rhabdammina abyssorum*; *d,* cross section of one of its arms; *e, Reophax scorpiurus*; *f, Hormosina Carpenteri.*

Thus, the long, spirally coiled undivided sandy tube of *Ammodiscus* is the isomorph of *Spirillina* (§ 479). In the genus *Halophragmium* (Fig. 320 *a, b*), we have a singular imitation of the *Globigerine* type; and in *Thurammina papillata* (Fig. 320, *g*) a not less remarkable imitation of the *Orbuline.* This last is specially noteworthy for the admirable manner in which its component sand-grains are set together; these being small and very uniform in size; and being disposed in such a manner as to present a smooth surface both inside and out (Fig. 320, *h*), whilst there are at intervals nipple-shaped protuberances, in every one of which there is a rounded orifice. A like perfection of finish is seen in the test of *Hormosina globulifera* (Fig. 320, *c*), which is composed of a succession of globular chambers rapidly increasing in size, each having a narrow tubular neck with a rounded orifice, which is received into the next segment. In other species of the same genus, there is a nearer approach to

6

the ordinary nodosarine type, their tests being sometimes constructed with the regularity characteristic of the shells of the true *Nodosaria* (Plate XV., fig. 10); whilst in other cases the chambers are less regularly disposed (Fig. 321, *f*), having rather the character of bead-like enlargements of a tube, whilst their walls show a less exact selection of material, sponge-spicules being worked-in with the sand-grains, so as to give them a hirsute aspect. A greater rudeness of structure shows itself in the nodosarine forms of the genus *Reophax;* in which not only are the sand-grains of the test very coarse, but small Foraminifera are often worked-up with them (Fig. 321, *e*). A straight, many-chambered form of the same genus (Fig. 321, *a, b*) is remarkable for the peculiar finish of the neck of each segment; for whilst the test generally is composed of sand-grains as loosely aggregated as those of which the test of *Astrorhiza* is made up, the grains that form the neck are firmly united by ferruginous cement, forming a very smooth wall to the tubular orifice.

FIG. 322.

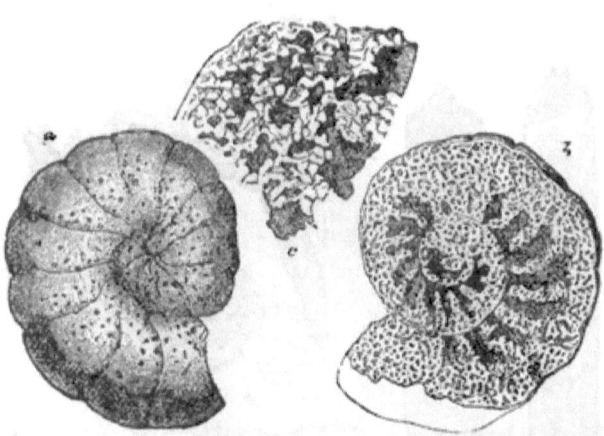

Cyclammina cancellata:—showing at *a*, its external aspect; *b*, its internal structure; *c*, a portion of its outer wall more highly magnified, showing the sand-grains of which L. is built up, and the passages excavated in its substance.

476. The highest development of the 'Arenaceous' type at the present time is found in the forms that imitate the very regular *nautiloid* shells, both of the 'porcellaneous' and the 'vitreous' series; and the most remarkable of these is the *Cyclammina cancellata* (Fig. 322), which has been brought up in considerable abundance from depths ranging downwards to 1,900 fathoms, the largest examples being found within 700 fathoms. The test (Fig. 322, *a*) is composed of aggregated sand-grains firmly cemented together and smoothed over externally with 'plaster,' in which large glistening sand-grains are sometimes set at regular intervals, as if for ornament. On laying open the spire, it is found to be very regularly divided into chambers by partitions formed of cemented sand-grains (*b*); a communication between those chambers being left by a fissure at the inner margin of the spire, as in *Operculina* (Plate XVI., fig. 2). One of the most curious features in the structure of this type, is the extension of the cavity of each chamber into passages excavated in its thick external wall; each passage being surrounded by a very regular arrangement of sand-grains, as shown at *c*. It not unfrequently happens

that the outer layer of the test is worn-away, and the ends of the passages then show themselves as pores upon its surface; this appearance, however, is abnormal, the passages simply running from the chamber-cavity into the thickness of its wall, and having (so long as this is complete) no external opening. This 'labyrinthic' structure is of great interest, from its relation not only to the similar structure of the large Fossil examples of the same type, but also to that which is presented in other gigantic Fossil arenaceous forms to be presently described.

477. Although some of the Nautiloid *Lituolæ* are among the largest of existing Foraminifera, having a diameter of 0.3 inch, they are mere dwarfs in comparison with two gigantic Fossil forms, whose structure has been elucidated by Mr. H. B. Brady and the Author.[1] Geologists,

Fig. 323.

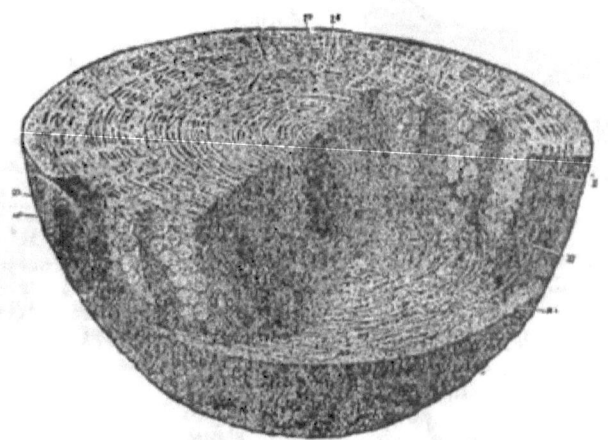

General view of the internal structure of *Parkeria*:—In the horizontal section, l^1, l^2, l^3, l^4, mark the four thick layers; in the vertical sections, A marks the internal surface of a layer separated by concentric fracture; B, the appearance presented by a similar fracture passing through the radiating processes; c, the result of a tangential section passing through the cancellated substance of a lamella; D, the appearance presented by the external surface of a lamella separated by a concentric fracture which has passed through the radial processes; E, the aspect of a section taken in a radial direction, so as to cross the solid lamellæ and their intervening spaces; c^1, c^2, c^3, c^4, successive chambers of nucleus.

who have worked over the Greensand of Cambridgeshire have long been familiar with solid spherical bodies which there present themselves not unfrequently, varying in size from that of a pistol-bullet to that of a small cricket-ball; and whilst some regarded them as Mineral concretions, others were led by certain appearances presented by their surfaces, to suppose them to be fossilized Sponges. A specimen having been fortunately discovered, however, in which the original structure had remained unconsolidated by mineral infiltration, it was submitted by Prof. Morris to the Author, who was at once led by his examination of it to recognize it as a member of the Arenaceous group of Foraminifera, to which he gave the designation *Parkeria*, in compliment to his valued friend and coadjutor, Mr. W. K. Parker. A section of the sphere taken through its centre (Fig. 323) presents an aspect very much resembling that of an

---

[1] See their 'Description of *Parkeria* and *Loftusia*,' in "Philosophical Transactions," 1869, p. 721.

Orbitolite (§ 466), a series of chamberlets being concentrically arranged round a 'nucleus'; and as the same appearance is presented, whatever be the direction of the section, it becomes apparent that these chamberlets, instead of being arranged in successive *rings* on a single plane, so as to form a disk, are grouped in concentric *spheres*, each completely investing that which preceded it in date of formation. The outer wall of each chamberlet is itself penetrated by extensions of the cavity into its substance, as in the *Cyclammina* last described; and these passages are separated by partitions very regularly built up of sand-grains, which also close-in their extremities, as is shown in Fig. 324. The concentric spheres are occasionally separated by walls of more than ordinary thickness; and such a wall is seen in Fig. 323 to close-in the last formed series of chamberlets. But these walls have the same 'labyrinthic' structure as the thinner ones; and an examination of numerous specimens shows that they are not formed at any regular intervals. The 'nucleus' is always composed of a single series of chambers arranged end to end, sometimes in a straight line, as in Fig. 323, c^1, c^2, c^3, c^4, sometimes forming a spiral, and in one instance returning upon itself. But the outermost chamber enlarges, and extends itself over the whole 'nucleus,' very much as the 'circumambient' chamber of the Orbitolite extends itself round the primordial chamber (§ 466); and radial prolongations given off from this in every direction form the first investing sphere, round which the entire series of concentric spheres are successively formed. Of the sand of which this remarkable fabric is constructed, about 60 per cent consists of phosphate of lime, and nearly the whole remainder of carbonate of lime.—Another large Fossil arenaceous type, constructed upon the same general plan, but growing spirally round an elongated axis after the manner of *Alveolina* (Fig. 315), and attaining a length of three inches, has been described by Mr. H. B. Brady (*loc. cit.*), under the name *Loftusia*, after its discoverer, the late Mr. W. K. Loftus, who brought it from the Turko-Persian frontier, where he found it imbedded in "a blue marly limestone" probably of early Tertiary age.

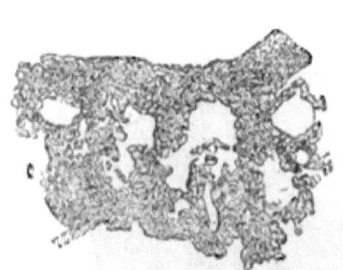

Fig. 324.

Portion of one of the lamellæ of *Parkeria*, showing the sand-grains of which it is built up, and the passages extending into its substance.

478. There is nothing, it seems to the Author, more wonderful in Nature, than the building-up of these elaborate and symmetrical structures by mere 'jelly-specks,' presenting no trace whatever of that definite 'organization' which we are accustomed to regard as necessary to the manifestations of Conscious Life. Suppose a Human mason to be put down by the side of a pile of stones of various shapes and sizes, and to be told to build a dome of these, smooth on both surfaces, without using more than the least possible quantity of a very tenacious but very costly cement in holding the stones together. If he accomplished this well, he would receive credit for great intelligence and skill. Yet this is exactly what these little 'jelly-specks' do on a most minute scale; the 'tests' they construct, when highly magnified, bearing comparison with the most skilful masonry of Man. From *the same sandy bottom*, one species picks

up the *coarser* quartz-grains, unites them together with a ferruginous cement secreted from its own substance, and thus constructs a flask-shaped 'test' having a short neck and a single large orifice. Another picks up the *finer* grains, and puts them together with the same cement into perfectly spherical 'tests' of the most extraordinary finish, perforated with numerous small pores, disposed at pretty regular intervals. Another selects the *minutest* sand-grains and the terminal portions of sponge-spicules, and works these up together—apparently with no cement at all, but by the mere 'laying' of the spicules—into perfect white spheres, like homœopathic globules, each having a single fissured orifice. And another, which makes a straight many-chambered 'test,' the conical mouth of each chamber projecting into the cavity of the next, while forming the walls of its chambers of ordinary sand-grains rather loosely held together, shapes the conical mouths of the successive chambers by firmly cementing to each other the quartz-grains which border it.—To give these actions the vague designation 'instinctive,' does not in the least help us to account for them; since what we want is, to discover the *mechanism* by which they are worked out; and it is most difficult to conceive how so artificial a selection can be made by creatures so simple.

470. VITREA.—Returning now to the Foraminifera which form true *shells* by the calcification of the superficial layer of their sarcode-bodies, we shall take a similar general survey of the *vitreous* series, whose shells are perforated by multitudes of minute foramina (Fig. 314). Thus, *Spirillina* has a minute, spirally convoluted, undivided tube, resembling that of Cornuspira (Plate xv., fig. 1), but having its wall somewhat coarsely perforated by numerous apertures for the admission of pseudopodia. The 'monothalamous' forms of this growth mostly belong to the Family *Lagenida;* which also contains a series of transition-forms leading up gradationally to the 'polythalamous' Nautiloid type. In *Lagena* (Plate xv., fig. 9) the mouth is narrowed and prolonged into a tubular neck, giving to the shell the form of a microscopic flask; this neck terminates in an everted lip, which is marked with radiating furrows.—A mouth of this kind is a distinctive character of a large group of many-chambered shells, of which each single chamber bears a more or less close resemblance to the simple Lagena, and of which, like it, the external surface generally presents some kind of ornamentation, which may have the form either of longitudinal ribs or of pointed tubercles. Thus the shell of *Nodosaria* (fig. 10) is obviously made up of a succession of lageniform chambers, the neck of each being received into the cavity of that which succeeds it; whilst in *Cristellaria* (fig. 11) we have a similar succession of chambers, presenting the characteristic radiate aperture, and often longitudinally ribbed, disposed in a nautiloid spiral. Between *Nodosaria* and *Cristellaria,* moreover, there is such a gradational series of connecting forms, as shows that no essential difference exists between these two types, which must be combined into one genus, *Nodosarina;* and it is a fact of no little interest, that these varietal forms, of which many are to be met-with on our own shores, but which are more abundant on those of the Mediterranean and especially of the Adriatic, can be traced backwards in Geological time even as far as the New Red Sandstone period.—In another genus, *Polymorphina,* we find the shell to be made up of lageniform chambers arranged in a double series, alternating with each other on the two sides of a rectilinear axis (fig. 13); here, again, the forms of the individual chambers, and the mode in which they are set one upon another, vary in such a manner as to give rise to very marked differences

in the general configuration of the shell, which are indicated by the name it bears.

480. *Globigerinida.*—Returing once again to the simple 'monothalamous' condition, we have in *Orbulina*—a minute spherical shell that presents itself in greater or less abundance in Deep-sea dredgings from almost every region of the globe—a globular chamber with porous walls, and a simple circular aperture that is frequently replaced by a number of large pores scattered throughout the wall of the sphere. It is maintained by some that *Orbulina* is really a detached generative segment of *Globigerina*, with which it is generally found associated.—The shell of *Globigerina* consists of an assemblage of nearly spherical chambers (Fig. 325), having coarsely porous walls, and cohering externally into a more or less regular turbinoid spire, each turn of which consists of four chambers progressively increasing in size. These chambers, whose total number seldom exceeds sixteen, do not communicate directly with each other, but open separately into a common 'vestibule' which occupies the centre of the under side of the spire.—This type has recently attracted great attention, from the extraordinary abundance in which it occurs at great depths over large areas of the Ocean-bottom. Thus its minute shells

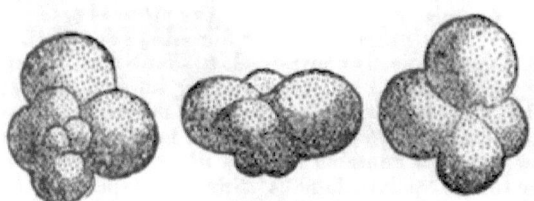

Fig. 325.

Globigerina bulloides, as seen in three positions.

have been found to constitute no less than 97 per cent of the 'ooze' brought up from depths of from 1260 to 2000 fathoms in the middle of the northern parts of the Atlantic Ocean. The surface-layer of this ooze, the thickness of which is entirely unknown, consists of Globigerinæ whose chambers are occupied by the sarcodic bodies of the animals, and which may therefore be presumed to be living on the bottom; whilst its deeper layers are almost entirely composed of dead and disintegrating shells of the same type. The younger shells, consisting of from eight to twelve chambers, are thin and smooth; but the older shells are thicker, their surface is raised into ridges that form a hexagonal areolation round the pores (Fig. 326, A); and this thickening is shown by examination of thin sections of the shell (B) to be produced by an exogenous deposit around the original chamber-wall (corresponding with the 'intermediate skeleton' of the more complex types), which sometimes contains little flask-shaped cavities filled with sarcode—as was first pointed-out by Dr. Wallich. But the sweeping of the upper waters of the Ocean by the 'tow-net' (§ 217), which was systematically carried-on during the voyage of the 'Challenger,' brought into prominence the fact that these waters in all but the coldest seas are inhabited by *floating* Globigerinæ, whose shells are beset with multitudes of delicate calcareous spines, which extend themselves radially from the angles at which the ridges meet, to a length equal to four or five times the diameter of the shell (Fig. 327). Among the basis of these spines, the sarcodic substance of the body exudes through the pores of the shell, forming a flocculent fringe around it; and this extends itself on each of the spines, creeping up one side to its extremity, and passing down the other, with the peculiar flowing movement already described (§ 395). The whole of this sarcodic extension is at once retracted if the cell which holds the Globigerina receives a sudden shock, or a drop of any irritating fluid is added to the water it contains.—It is maintained by Sir Wyville Thomson that the bottom-deposit is formed by the continual 'raining-down' of the Globigerinæ of the upper waters, which (he affirms) only *live* at or near the surface, and which, when they die, lose their spines and subside. But it has been shown by the careful comparison made by Mr. H. B. Brady between the surface-gatherings and the bottom-deposits of the same areas, that the two are often so marked, as to forbid the idea that the latter are solely derived from the former.[1] For not only are there several specific types

Fig. 376.

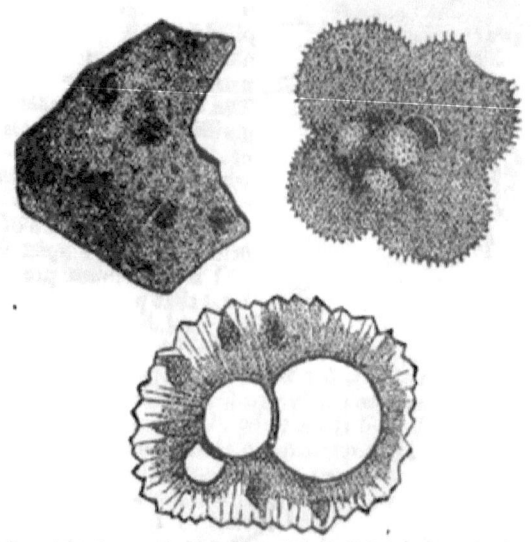

Globigerina, from Atlantic ooze showing thickening of shell by exogenous deposit:—A, entire shell, showing areolated ridges of surface; B, portion of shell more highly magnified, showing orifices of tubuli and large cavities filled with sarcode; C, section of shell showing exogenous deposit upon original chamber-wall, which is raised into ridges with tubuli between them, and includes sarcodic cavities.

found in each, which do not present themselves in the other, but, as a rule, the shells of the types common to both are larger and thicker in the latter than they are in the former. This evidence strongly supports the conclusion originally drawn by the Author from his own examination of the Globigerina-ooze, that the shells forming its surface-layer must *live on the bottom*, being incapable of floating in consequence of their weight; and that if they have passed the earlier part of their lives in the upper waters, they drop down as soon as the calcareous deposit continually exuding from the body of each animal, instead of being employed in the formation of new chambers, is applied to the thickening of those

---

[1] "Quart. Journ. Microsc. Sci.," Vol. xix. (1879), p. 295.

previously formed.—That many types of Foraminifera *pass their whole lives* at depths of at least 2000 fathoms, is proved, in regard to those forming Calcareous shells, by their attachment to stones, corals, etc.; and in the case of the Arenaceous types, by the fact that they can only procure *on the bottom* the sand of which their 'tests' are made up.

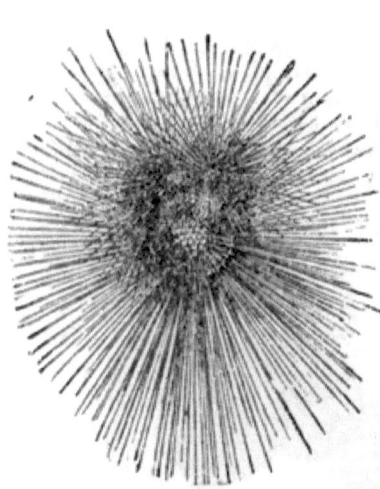

Fig. 327.

Globigerina, as captured by tow-net, floating at or near surface.

481. A very remarkable type has recently been discovered, adherent to shells and corals brought from tropical seas, to which the name *Carpenteria* has been given; this may be regarded as a highly developed form of Globigerina, its first-formed portion having all the essential characters of that genus. It grows attached by the apex of its spire; and its later chambers increase rapidly in size, and are piled on the earlier in such a manner as to form a depressed cone with an irregular spreading base. The essential character of Globigerina—the separate orifice of each of its chambers—is here retained with a curious modification; for the central vestibule, into which they all open, forms a sort of vent whose orifice is at the apex of the cone, and is sometimes prolonged into a tube that proceeds from it; and the external wall of this cone is so marked-out by septal bands, that it comes to bear a strong resemblance to a minute *Balanus* (acorn-shell), for which this type was at first mistaken. The principal chambers are partly divided into chamberlets by incomplete partitions, as we shall find them to be in *Eozoön* (§ 494). The presence of sponge-spicules in large quantity in the chambers of many of the best-preserved examples of this type, was for some time a source of perplexity; but this is now explained by the interesting observations made by Prof. Möbius[1] on a large branching and spreading form of *Carpenteria*, which he recently met-with on a reef near Mauritius, and to which he has given the name of *C. raphidodendron*. For the pseudopodia of this Rhizopod have the habit, like those of *Haliphysema* (§ 474), of taking into themselves sponge-spicules, which they draw into the chambers, so that they become incorporated with the sarcode-body.

482. A less aberrant modification of the Globigerine type, however, is presented in the two great series which may be designated (after the leading forms in each) as the *Textularian* and the *Rotalian*. For notwithstanding the marked difference in their respective plans of growth, the characters of the individual chambers are the same; their walls being coarsely-porous, and their apertures being oval, semi-oval, or crescent-shaped, sometimes merely fissured. In *Textularia* (Plate xv., fig. 14) the chambers are arranged biserially along a straight axis, the position of those on the two sides of it being alternate, and each chamber opening

---

[1] See his "Foraminifera von Mauritius," Plates v., vi.

into those above and below it on the opposite side by a narrow fissure; as is well shown in such 'internal casts' (Fig. 328, A) as exhibit the forms and connections of the segments of sarcode by which the chambers were occupied during life. In the genus *Bulimina* the chambers are so arranged as to form a spire like that of a Bulimus, and the aperture is a curved fissure whose direction is nearly transverse to that of the fissure of Textularia; but in this, as in the preceding type, there is an extraordinary variety in the disposition of the chambers. In both, moreover, the shell is often covered by a sandy incrustation, so that its perforations are completely hidden, and can only be made visible by the removal of the adherent crust. And so many cases are now known, in which the shell of *Textularinæ* is entirely replaced by a sandy test, that some Systematists prefer to range this group among the *Arenacea*.

483. In the *Rotalian* series, the chambers are disposed in a turbinoid spire, opening one into another by an aperture situated on the lower and inner side of the spire, as shown in Plate XV., fig. 18; the forms and connections of the segments of their sarcode-bodies being shown in such 'internal casts' as are represented in Fig. 328, B. One of the lowest and simplest forms of this type is that very common one now distin-

FIG. 328. FIG. 329.

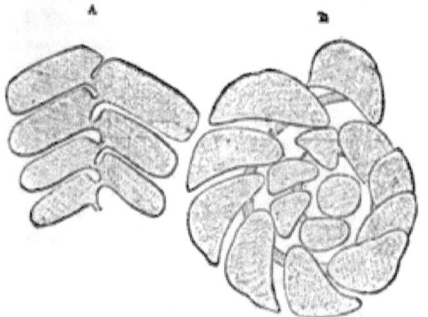

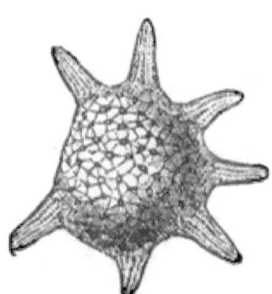

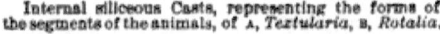

Internal siliceous Casts, representing the forms of the segments of the animals, of A, *Textularia*, B, *Rotalia*.

Tinoporus baculatus.

guished as *Discorbina*, of which a characteristic example is represented in Plate XV., fig. 15. The early form of *Planorbulina* is a rotaline spire, very much resembling that of Discorbina; but this afterwards gives place to a cyclical plan of growth (fig. 17); and in those most developed forms of this type which occur in warmer seas, the earlier chambers are completely overgrown by the latter, which are often piled-up in an irregular 'acervuline' manner, spreading over the surfaces of shells, or clustering round the stems of zoöphytes.—In the genus *Tinoporus* there is a more regular growth of this kind, the chambers being piled successively on the two sides of the original median plane, and those of adjacent piles communicating with each other obliquely (like those of Textularia) by large apertures, whilst they communicate with those directly above and below by the ordinary pores of the shell. The simple or smooth form of this genus presents great diversities of shape, with great constancy, in its internal structure; being sometimes spherical, sometimes resembling a minute sugar-loaf, and sometimes being irregularly flattened-out. A peculiar form of this type (Fig. 329), in which the walls of the piles are thickened at their meeting-angles into solid columns that appear on the surface as tubercles, and are sometimes prolonged into spinous outgrowths that radiate from the central mass, is of very common occurrence in shore-sands and shallow-water dredgings on some parts of the Australian coast and among the Polynesian islands.—To the simple form of this genus we are probably to refer a large part of the fossils of the Cretaceous and early Tertiary period, that have been described under the name *Orbitolina*, some of which attain a very large size. Globular *Orbitolinæ*, which appear to have been artificially perforated and strung as beads, are not unfrequently found associated with the "flint-implements" of gravel-beds.—Another very curious modification of the Rotaline type is presented by *Polytrema*, which so much resembles a Zoöphyte as to have been taken for a minute Millepore; but which is made up of an aggregation of 'globigerine' chambers communicating with each other like those of Tinoporus, and differs from that genus in nothing else than its erect and usually branching manner of growth, and the freer communication between its chambers. This, again, is of special interest in relation to *Eozoön;* showing that an indefinite zoöphytic mode of growth is perfectly compatible with truly Foraminiferal structure.

484. In *Rotalia*, properly so called, we find a marked advance towards the highest type of Foraminiferal structure; the partitions that divide the chambers being composed of two laminæ, and spaces being left between them which give passage to a system of canals, whose general distribution is shown in Fig. 330. The proper walls of the chambers, moreover, are thickened by an extraneous deposit or 'intermediate skeleton,' which sometimes forms radiating outgrowths; but this peculiarity of conformation is carried much further in the genus *Calcarina*, which has been so designated from its resemblance to a spur-rowel (Plate XVI., fig. 3). The solid club-shaped appendages with which the shell is provided, entirely belonging to the 'intermediate skeleton' *b*, which is quite independent of the chambered structure *a;* and this body is nourished by a set of canals containing prolongations of the sarcode-body, which not only furrow the surface of these appendages, but are seen to traverse their interior when this is laid open by section, as shown at *c*. In no other recent Foraminifer does the 'canal system' attain a like development; and its distribution in this minute shell, which has been made out by careful microscopic study, affords a valuable clue to its meaning in the gigantic fossil organism *Eozoön Canadense* (§ 494). The resemblance which *Calcarina* bears to the radiate forms of *Tinoporus* (Fig. 329), which are often found with them in the same dredgings, is frequently extremely striking; and in their early growth the two can scarcely be distinguished, since both commence in a 'rotaline' spire with radiating appendages; but whilst the successive chambers of Calcarina continue to be added on the same plane, those of Tinoporus are heaped-up in less regular piles.

485. Certain beds of Carboniferous Limestone in Russia are entirely made up, like the more modern Nummulitic Limestone (§ 489), of an aggregation of the remains of a peculiar type of Foraminifera, to which the name *Fusulina* (indicative of its fusiform or spindle-shape) has been given (Fig. 331). In general aspect and plan of growth it so much resembles *Alveolina*, that its relationship to that type would scarcely be questioned by the superficial observer. But when its mouth is examined, it is found to consist of a single slit in the middle of the lip; and the

interior, instead of being minutely divided into chamberlets, is found to
consist of a regular series of simple chambers; while from each of these
proceeds a pair of elongated extensions, which correspond to the 'alar
prolongations' of other spirally growing Foraminifera (§ 486), but which,
instead of wrapping round the preceding whorls, are prolonged in the
direction of the axis of the spire,
those of each whorl projecting beyond those of the preceding, so that
the shell is elongated with every increase in its diameter. Thus it appears that in its general plan of
growth, *Fusulina* bears much the
same relation to a symmetrical Rotaline or Nummuline shell, that
Alveolina bears to *Orbiculina;* and
this view of its affinities is fully confirmed by the Author's microscopic
examination of the structure of its
shell. For although the Fusulina-
Limestone of Russia has undergone
a degree of metamorphism, which
so far obscures the tubularity of its
component shells, as to prevent him
from confidently affirming it, yet
the appearances he could distinguish were decidedly in its favor.
And having since received specimens from the Upper Coal Mea-
sures of Iowa, U. S., which are in a much more perfect state of preservation, he is able to state with certainty, not only that *Fusulina* is
tubular, but that its tubulation is of the large coarse nature that marks
its affinity rather to the *Rotaline* than to the *Nummuline* series.—This
type is of peculiar interest, as having long been regarded as the oldest
form of Foraminifera which was known to have occurred in sufficient
abundance to form Rocks by the aggregation of its individuals. It will
be presently shown, however, that in point both of antiquity and of
importance, it is far surpassed by another (§ 493).

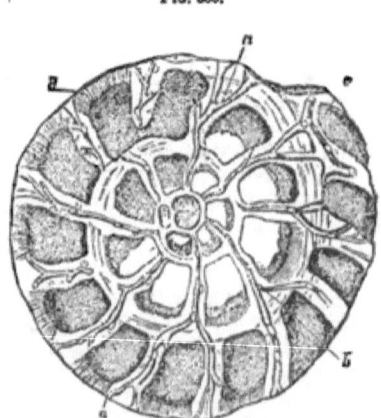

FIG. 330.

Section of *Rotalia Schroetteriana* near its base and parallel to it; showing *a, a*, the radiating interseptal canals; *b*, their internal bifurcations; *c*, a transverse branch; *d*, tubular wall of the chambers.

486. *Nummulinida.*—All the most elaborately constructed, and the
greater part of the largest, of the 'vitreous' Foraminifera belong to the
group of which the well-known *Nummulite* may be taken as the representative. Various plans of growth prevail in the family; but its distinguishing characters consist in the completeness of the wall that surrounds
each segment of the body (the septa being double instead of single as
elsewhere), the density and fine porosity of the shell-substance, and the
presence of an 'intermediate skeleton,' with a 'canal-system' for its
nutrition. It is true that these characters are also exhibited in the highest of the Rotaline series (§ 484), whilst they are deficient in the genus
Amphistegina, which connects the Nummuline series with the Rotaline;
but the occurrence of such modifications in their border-forms is common
to other truly Natural groups. With the exception of Amphistegina, all
the genera of this family are symmetrical in form; the spire being nautiloid in such as follow that plan of growth, whilst in those which follow
the cyclical plan there is a constant equality on the two sides of the
median plane: but in Amphistegina there is a reversion to the rotalian

type in the turbinoid form of its spire, as in the characters already specified, although its general conformity to the Nummuline type is such as to leave no reasonable doubt as to its title to be placed in this family. Notwithstanding the want of symmetry of its spire, its accords with *Operculina* and *Nummulina* in having its chambers extended by 'alar prolongations' over each surface of the previous whorl; but on the under side these prolongations are almost entirely cut off from the principal chambers, and are so displaced as apparently to alternate with them in position; so that M. D'Orbigny, supposing them to constitute a distinct series of chambers, described its plan of growth as a biserial spiral, and made this the character of a separate Order.[1]

487. The existing *Nummulinida* are almost entirely restricted to tropical climates; but a beautiful little form, the *Polystomella crispa* (Plate xv., fig. 16), the representative of a genus that presents the most regular and complete development of the 'canal system' anywhere to be met with, is

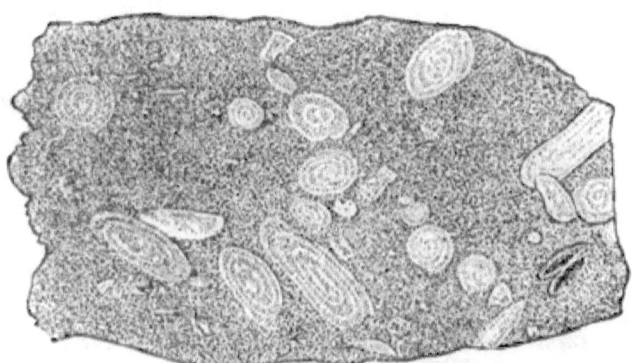

Section of *Fusulina*-Limestone.

common on our own coasts. The peculiar surface-marking shown in the figure consists in a strongly marked ridge and furrow plication of the shelly wall of each segment along its posterior margin; the furrows being sometimes so deep as to resemble fissures opening into the cavity of the chamber beneath. No such openings, however, exist; the only communication which the sarcode-body of any segment has with the exterior, being either through the fine tubuli of its shelly walls, or through the row of pores that are seen in front view along the inner margin of the septal plane, collectively representing a fissured aperture divided by minute bridges of shell. The meaning the plication of the shelly wall comes to be understood, when we examine the conformation of the segments of the sarcode-body, which may be seen in the common *Polystomella crispa* by dissolving away the shell of fresh specimens by the action of dilute acid, but which may be better studied in such internal casts (Fig. 332) of the sarcode-body and canal-system of the large *P. craticulata* of the Australian coast, as may sometimes be obtained by the same means from

[1] For an account of this curious modification of the Nummuline plan of growth, the real nature of which was first elucidated by Messrs. Parker and Rupert Jones, see the Author's 'Introduction to the Study of the Foraminifera' (published by the Ray Society).

dead shells which have undergone infiltration with ferruginous silicates.[1] Here we see that the segments of the sarcode-body are smooth along their anterior edge b, b', but that along their posterior edge, a, they are prolonged backwards into a set of 'retral processes;' and these processes lie under the ridges of the shell, whilst the shelly wall dips down into the spaces between them, so as to form the furrows seen on the surface. The connections of the segments by stolons, c, c', passing through the pores at the inner margin of each septum, are also admirably displayed in such 'casts.' But what they serve most beautifully to demonstrate is the canal-system, of which the distribution is here most remarkably complete and symmetrical. At d, d,' d,' are seen three turns of a spiral canal which passes along one end of all the segments of the like number of convolutions, whilst a corresponding canal is found on the side which is the figure is undermost; these two spires are connected by a set of meridional canals, e, e', e^2, which pass down between the two layers of the septa that divide the segments; whilst from each of these there passes off towards the surface a set of pairs of diverging branches, f, f',

Fig. 392.

Internal Cast of *Polystomella craticulata*:—a, retral processes, proceeding from the posterior margin of one of the segments; b, b', smooth anterior margin of the same segment; c, c', stolons connecting successive segments, and uniting themselves with the diverging branches of the meridional canals; d, d', d^2, three turns of one of the spiral canals; e, e', e^2, three of the meridional canals; f, f', f^2, their diverging branches.

f^2, which open upon the surface along the two sides of each septal band, the external openings of those on its anterior margin being in the furrows between the retral processes of the next segment. These canals appear to be occupied in the living state by prolongations of the sarcode-body; and the diverging branches of those of each convolution unite themselves, when this is inclosed by another convolution, with the stolon-processes

---

[1] It was by Prof. Ehrenberg that the existence of such 'casts' in the Green Sands of various Geological periods (from the Silurian to the Tertiary) was first pointed out, in his Memoir 'Ueber den Grünsand und seine Einläuterung des organischen Lebens,' in "Abhandlungen den Königl. Akad. der Wissenschaften," Berlin, 1855. It was soon afterwards shown by the late Prof. Bailey ("Quart. Journ. Microsc. Sci.," Vol. v., 1857, p. 83) that the like infiltration occasionally takes place in recent Foraminifera, enabling similar 'casts' to be obtained from them by the solution of their shells in dilute acid; the Author, as well as Messrs. Parker and Rupert Jones, soon afterwards obtained most beautiful and complete internal casts from recent Foraminifera brought from various localities; and a large collection of green sands yielding similar casts was made in the 'Challenger.'

connecting the successive segments of the latter, as seen at c'. There can be little doubt that this remarkable development of the canal-system has reference to the unusual amount of shell-substance which is deposited as an 'intermediate skeleton' upon the layer that forms the proper walls of the chambers, and which fills up with a solid 'boss' what would otherwise be the depression at the umbilicus of the spire. The substance of this 'boss' is traversed by a set of straight canals, which pass directly from the spiral canal beneath, towards the external surface, where they open in little pits, as is shown in Pl. xv., fig. 16; the umbilical boss in *P. crispa*, however, being much smaller in proportion than it is in *P. craticulata*. There is a group of Foraminifera to which the term *Nonionina* is properly applicable, that is probably to be considered as a sub-genus of Polystomella; agreeing with it in its general conformation, and especially in the distribution of its canal system; but differing in its aperture, which is here a single fissure at the inner edge of the septal plane (Plate xv., fig. 19), and in the absence of the 'retral processes' of the segments of the sarcode-body, the external walls of the chambers being smooth. This form constitutes a transition to the ordinary Nummuline type, of which Polystomella is a more aberrant modification.

488. The Nummuline type is most characteristically represented at the present time by the genus *Operculina;* which is so intimately united to the true *Nummulite* by intermediate forms, that it is not easy to separate the two, notwithstanding that their typical examples are widely dissimilar. The former genus (Plate XVI., fig. 2) is represented on our own coast by very small and feeble forms; but it attains a much higher development in Tropical seas, where its diameter sometimes reaches 1-4th of an inch. The shell is a flattened nautiloid spire, the breadth of whose earlier convolutions increases in a regular progression, but of which the last convolution (in full-grown specimens) usually flattens itself out like that of Peneroplis, so as to be very much broader than the preceding. The external walls of the chambers, arching over the spaces between the septa, are seen at b, b; and these are bounded at the outer edge of each convolution by a peculiar band a, termed the 'marginal cord.' This cord, instead of being perforated by minute tubuli like those which pass from the inner to the outer surface of the chamber-walls without division or inosculation (Fig. 335), is traversed by a system of comparatively large inosculating passages seen in cross section at a'; and these form part of the canal-system to be presently described. The principal cavities of the chambers are seen at c, c; while the 'alar prolongations' of those cavities over the surface of the preceding whorl are shown at c', c'. The chambers are separated by the septa, d, d, d, formed of two laminæ of shell, one belonging to each chamber, and having spaces between them in which lie the 'interseptal canals,' whose general distribution is seen in the septa marked e, e, and whose smaller branches are seen irregularly divided in the septa d', d', whilst in the septum d'' one of the principal trunks is laid open through its whole length. At the approach of each septum to the marginal cord of the preceding, is seen the narrow fissure which constitutes the principal aperture of communication between the chambers; in most of the septa, however, there are also some isolated pores (to which the lines point that radiate from e, e) varying both in number and position. The interseptal canals of each septum take their departure at its inner extremity from a pair of spiral canals, of which one passes along each side of the marginal cord; and they communicate at their outer extremity with the canal-system of the

PLATE XVI.

Fig. 1.

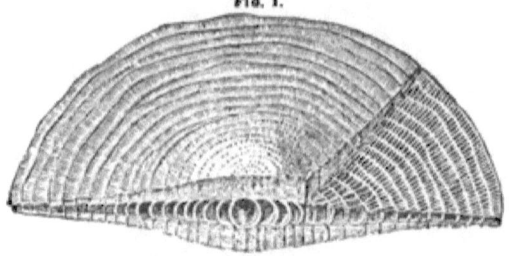

Fig. 2.

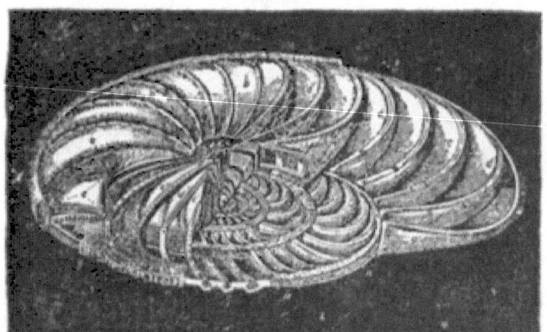

Fig. 3.

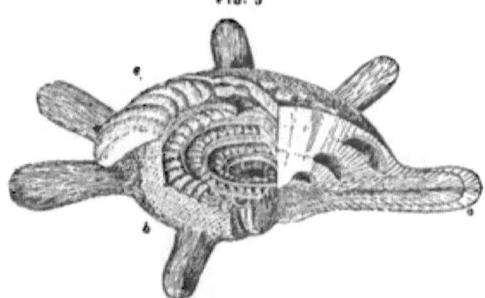

VARIOUS FORMS OF FORAMINIFERA (Original).

Fig. 1. *Cycloclypeus*, showing external surface, and vertical and horizontal sections.

2. *Operculina*, laid open to show its internal structure ; *a*, marginal cord, seen in cross section at *a'* ; *b, b*, external walls of the chambers; *c, c*, cavities of the chambers; *c' c'*, their alar prolongations; *d, d*, septa, divided at *d' d'* and at *d''*, so as to lay open the interseptal canals, the general distribution of which is seen in the septa *e, e* ; the lines radiating from *e, e*, point to the secondary pores; *g, g*, non-tubular columns.

3. *Calcarina*, laid open to show its internal structure :—*a*, chambered portion ; *b*, intermediate skeleton; *c*, one of the radiating prolongations proceeding from it, with extensions of the canal-system.

'marginal cord,' as shown in Fig. 337. The external walls of the chambers are composed of the same finely-tubular shell-substance that forms them in the Nummulite; but, as in that genus, not only are the septa themselves composed of vitreous non-tubular substance, but that which lies over them, continuing them to the surface of the shell, has the same character; showing itself externally in the form sometimes of continuous ridges, sometimes of rows of tubercles, which mark the position of the septa beneath. These non-tubular plates or columns are often traversed by branches of the canal-system, as seen at *g*, *q*. Similar columns of non-tubular substance, of which the summits show themselves as tubercles on the surface, are not unfrequently seen between the septal bands, giving a variation to the surface-marking, which, taken in conjunction with variations in general conformation, might be fairly held sufficient to characterize distinct species, were it not that on *a comparison of a great number of specimens*, these variations are found to be so gradational, that no distinct line of demarcation can be drawn between the individuals which present them.

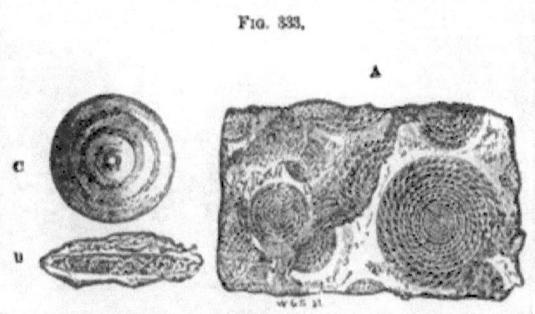

FIG. 333.

A, piece of *Nummulitic Limestone* from Pyrenees, showing Nummulites laid open by fracture through median plane; B, vertical section of *Nummulite*; C, *Orbitoides*.

489. The Genus *Nummulina*, of which the fossil forms are commonly known as *Nummulites*, though represented at the present time by small and comparatively infrequent examples, was formerly developed to a vast extent; the Nummulitic Limestone chiefly made-up by the aggregation of its remains (the material of which the Pyramids are built) forming a band, often 1,800 miles in breadth and frequently of enormous thickness, that may be traced from the Atlantic shores of Europe and Africa, through Western Asia to Northern India and China, and likewise over vast areas of North America (Fig. 333). The diameter of a large proportion of fossil Nummulites ranges between half an inch and an inch; but there are some whose diameter does not exceed 1-16th of an inch, whilst others attain the gigantic diameter of $4\frac{1}{2}$ inches. Their typical form is that of a double-convex lens; but sometimes it much more nearly approaches the globular shape, whilst in other cases it is very much flattened; and great differences exist in this respect among individuals of what must be accounted one and the same species. Although there are some Nummulites which closely approximate *Operculinæ* in their mode of growth, yet the typical forms of this genus present certain well-marked distinctive peculiarities. Each convolution is so completely invested by that which succeeds it, and the external wall or spiral lamina of the new convolution is so completely separated from that of the convolution it

incloses by the 'alar prolongations' of its own chambers (the peculiar arrangement of which will be presently described), that the spire is scarcely if at all visible on the external surface. It is brought into view, however, by splitting the Nummulite through the median plane, which may often be accomplished simply by striking it on one edge with a hammer, the opposite edge being placed on a firm support; or, if this method should not succeed, by heating it in the flame of a spirit-lamp, and then throwing it into cold water or striking it edgeways. Nummulites usually show many more turns, and a more gradual rate of increase in the breadth of the spire, than Foraminifera generally; this will be apparent from an examination of the vertical section shown in Fig. 334, which is taken from one of the commonest and most characteristic fossil examples of the genus, and which shows no fewer than ten convolutions in a fragment that does not nearly extend to the centre of the spire. This section also shows the complete inclosure of the older convolutions by the newer, and the interposition of the alar prolongations of the chambers between the successive layers of the spiral lamina. These prolongations are variously arranged in different examples of the genus: thus in some, as *N. distans*, they keep their own separate course, all tending radially towards the centre; in others, as *N. lævigata*, their partitions inosculate with each other, so as to divide the space intervening between each layer and the next into an irregular network, presenting in vertical section the appearance shown in Fig. 334; whilst in *N. garansensis* they are broken up into a number of chamberlets, having little or no direct communication with each other.

FIG. 334.

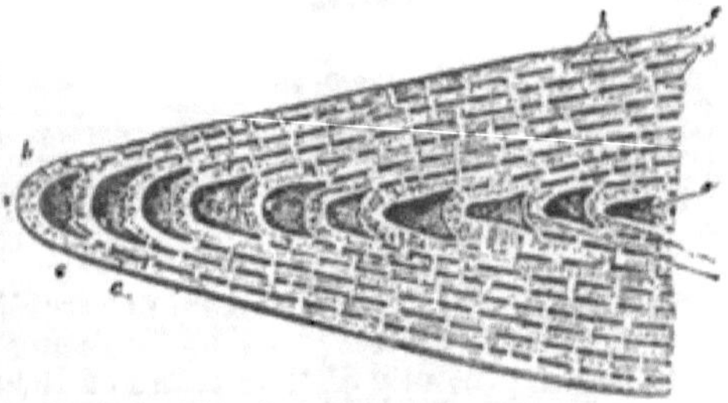

Vertical section of portion of *Nummulina lævigata*;— *a*, margin of external whorl; *b*, one of the outer row of chambers; *c, c*, whorl invested by *a*; *d*, one of the chambers of the fourth whorl from the margin; *e, e'*, marginal portions of the inclosed whorls; *f*, investing portions of outer whorl; *g, g*, spaces left between the investing portion of successive whorls; *h, h*, sections of the partitions dividing these.

490. Notwithstanding that the inner chambers are thus so deeply buried in the mass of investing whorls, yet there is evidence that the segments of sarcode which they contained were not cut off from communication with the exterior, but that they may have retained their vitality to the last. The shell itself is almost everywhere minutely porous, being penetrated by parallel tubuli which pass directly from one surface to the other. These tubes are shown, as divided lengthways by a vertical section, in Fig. 335, *a, a;* whilst the appearance they present when cut across in a horizontal section is shown in Fig. 336, the transparent shell-substance *a, a, a*, being closely dotted with minute punctations which mark their orifices. In that portion of the shell, however, which forms the margin of each whorl (Fig. 335, *b, b*), the tubes are larger, and diverge from each other at greater intervals; and it is shown by horizontal sections that they communicate freely with each other laterally, so as to form a network such as is seen at *b, b*, Fig. 337. At certain other points, *d, d, d* (Fig. 335), the shell-substance is not perforated by tubes, but is peculiarly dense in its texture, forming solid pillars, which seem to strengthen the other parts; and in Nummulites whose

7

surfaces have been much exposed to attrition, it commonly happens that the pillars of the superficial layer, being harder than the ordinary shell-substance, and being consequently less worn down, are left as prominences, the presence of which has often been accounted (but erroneously) as a specific character. The successive chambers of the same whorl communicate with each other by a passage left between the inner edge of the partition that separates them, and the 'marginal cord' of the preceding whorl; this passage is sometimes a single large broad aperture, but is more commonly formed by the more or less complete coalescence of several separate perforations, as is seen in Fig. 334, *b*. There is also, as in Operculina, a variable number of isolated pores in most of the septa, forming a secondary means of communication between the chambers.—The Canal-system of *Nummulina* seems to be distributed upon essentially the same plan as in *Operculina;* its passages, however, are usually more or less obscured by fossilizing material. A careful examination will generally disclose traces of them in the middle of the parti-

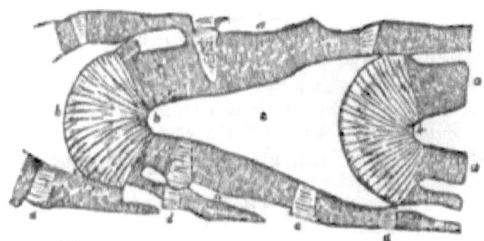

Fig. 335.

Portion of a thin Section of *Nummulina lævigata*, taken in the direction of the preceding, highly magnified to show the minute structure of the shell:—*a, a*, portions of the ordinary shell-substance traversed by parallel tubuli; *b, b*, portions forming the marginal cord, traversed by diverging and larger tubuli; *c*, one of the chambers laid open; *d, d, d*, pillars of solid substance not perforated by tubuli.

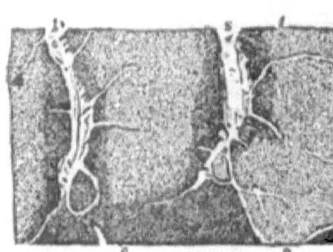

Fig. 336.

Portion of Horizontal Section of Nummulite, showing the structure of the walls and of the septa of the chambers:—*a, a, a*, portion of the wall covering three chambers, the punctations of which are the orifices of tubuli; *b, b*, septa between these chambers, containing canals which send out lateral branches, *c, c*, entering the chambers by larger orifices, one of which is seen at *d*.

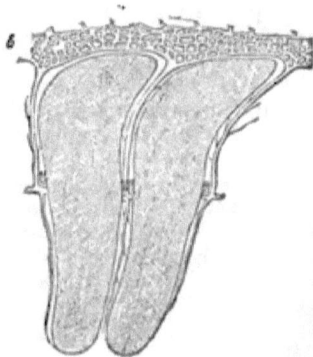

Fig. 337.

Internal cast of two of the chambers, *a, a*, of *Nummulina striata*, with the network of Canals, *b, b*, in the marginal cord, communicating with canals passing between the chambers.

tions that divide the chambers (Fig. 336, *b, b*), while from these may be seen to proceed the lateral branches (*c, c*), which, after burrowing (so to speak) in the walls of the chambers, enter them by large orifices (*d*). These 'interseptal' canals, and their communication with the inosculating system of passages excavated in the marginal cord, are extremely well seen in the 'internal cast' represented in Fig. 337.

491. A very interesting modification of the Nummuline type is presented in the genus *Heterostegina* (Fig. 338), which bears a very strong resemblance to *Orbiculina* in its plan of growth, whilst in every other respect it is essentially different. If the principal chambers of an *Operculina* were divided into chamberlets by secondary partitions in a direction transverse to that of the principal septa, it would be converted into a *Heterostegina*; just as a *Peneroplis* would be converted by the like subdivision into an *Orbiculina* (§ 464). Moreover, we see in Heterostegina, as in Orbiculina, a great tendency to the opening-out of the spire with the advance of age; so that the apertural margin extends round a large part of the shell, which thus tends to become discoidal. And it is not a little

FIG. 338.

FIG. 339.

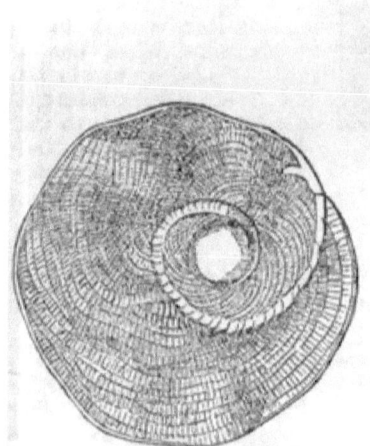
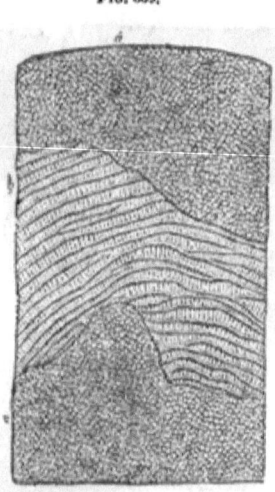

Heterostegina.

Section of *Orbitoides Fortisii*, parallel to the surface; traversing at *a, a*, the superficial layer, and at *b, b*, the median layer.

curious that we have in this series another form, *Cycloclypeus*, which bears exactly the same relation to Heterostegina, that Orbitolites does to Orbiculina; in being constructed upon the *cyclical* plan from the commencement, its chamberlets being arranged in rings around a central chamber (Plate XVI., fig. 1). This remarkable genus, at present only known by specimens dredged up from considerable depths off the coast of Borneo, is the largest of existing Foraminifera; some specimens of its discs in the British Museum having a diameter of 2¼ inches. Notwithstanding the difference of its plan of growth, it so precisely accords with the Nummuline type in every character which essentially distinguishes the genus, that there cannot be a doubt of the intimacy of their relationship. It will be seen from the examination of that portion of the figure which shows *Cycloclypeus* in vertical section, that the solid layers of shell by which the chambered portion is inclosed are so much thicker, and consist of so many more lamellæ, in the central portion of the disk, than they do nearer its edge, that new lamellæ must be progressively added to the surfaces of the disk, concurrently with the addition of new rings of chamberlets to its margin. These lamellæ, however, are closely applied one to the other, without any intervening spaces; and they are all traversed by columns of non-tubular substance, which spring from the septal bands, and gradually increase in diameter with their approach to the surface, from which they project in the central portion of the disk as glistening tubercles.

492. The Nummulitic Limestone of certain localities (as the South-west of France, North-eastern India, etc.) contains a vast abundance of discoidal bodies termed *Orbitoides* (Fig. 333, c), which are so similar to Nummulites as to have been taken for them, but which bear a much closer resemblance to Cycloclypeus. These are only known in the fossil state; and their structure can only be ascertained by the examination of sections thin enough to be translucent. When one of these disks (which vary in size, in different species, from that of a four-penny piece to that of half-a-crown) is rubbed-down so as to display its internal organization, two different kinds of structure are usually seen in it; one being composed of chamberlets of very definite form, quadrangular in some species,

Fig. 340.

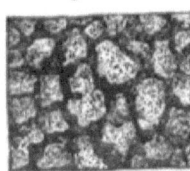 

Portions of the Section of *Orbitoides Fortisii* shown in Fig. 339, more highly magnified;—a superficial layer; b, median layer.

Fig. 341.

Vertical Sections of *Orbitoides Fortisii*, showing the large central chamber at a, and the median layer surrounding it, covered above and below by the superficial layer.

circular in others, arranged with a general but not constant regularity in concentric circles (Figs. 339, 340, b, b); the other, less transparent, being formed of minuter chamberlets which have no such constancy of form, but which might almost be taken for the pieces of a dissected map (a, a). In the upper and lower walls of these last, minute punctuations may be observed, which seem to be the orifices of the connecting tubes whereby they are perforated. The relations of these two kinds of structure to each other are made evident by the examination of a vertical section (Fig. 341): which shows that the portion b, Figs. 339, 340, forms the median plane, its concentric circles of chamberlets being arranged round a large central chamber, as in *Cycloclypeus;* whilst the chamberlets of the portion a are irregularly superposed one upon the other, so as to form sev-

Fig. 342.

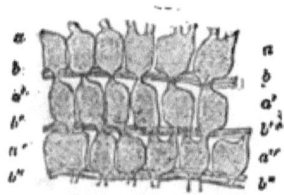

Internal Cast of portion of median plane of *Orbitoides Fortisii*, showing at a a, a' a', a'' a'', six chambers of each of three zones, with their mutual communications; and at b b, b' b', b'' b'', portions of three annular canals.

eral layers which are most numerous towards the centre of the disk, and thin-away gradually towards its margin. The disposition and connections of the chamberlets of the median layer in *Orbitoides* seem to correspond very closely with those which have been already described as prevailing in *Cycloclypeus;* the most satisfactory indications to this effect being furnished by the siliceous 'internal casts' to be met with in certain Green Sands, which afford a model of the sarcode-body of the animal. In such a fragment (Fig. 342) we recognize the chamberlets of three successive zones, a, a', a'', each of which seems normally to communicate by one or two passages with the chamberlets of the zone internal and external to its own; whilst between the chamberlets of the same zone there seems to be no direct connection. They are brought into rela-

Fig. 343.

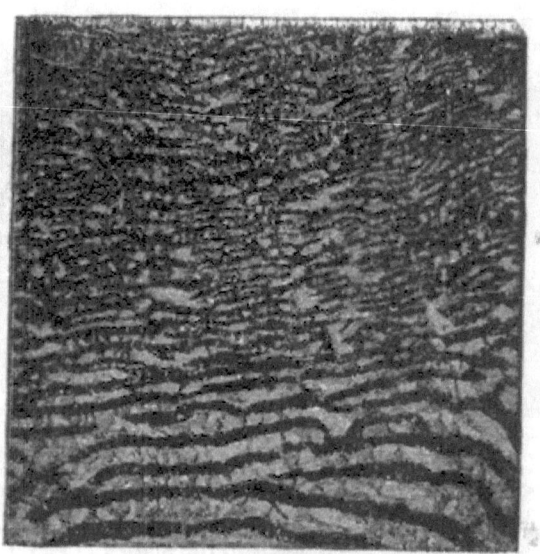

Vertical Section of *Eozoön Canadense*, showing alternation of Calcareous (light) and Serpentinous (dark) lamellae.

tion, however, by means of annular canals, which seem to represent the spiral canals of the Nummulite, and of which the 'internal casts' are seen at $b\ b,\ b'\ b',\ b''\ b''$.

493. A most remarkable Fossil, referable to the Foraminiferal type, has been recently discovered in strata much older than the very earliest that were previously known to contain Organic remains; and the determination of its real character may be regarded as one of the most interesting results of Microscopic research. This fossil, which has received the name *Eozoön Canadense* (Fig. 343), is found in beds of Serpentine Limestone that occur near the base of the *Laurentian* formation[1] of

---

[1] This *Laurentian formation* was first identified as a regular series of stratified rocks, underlying the equivalents not merely of the Silurian, but also of the Upper and Lower Cambrian systems of this country, by Sir William Logan, the former able Director of the Geological Survey of Canada.

Canada, which has its parallel in Europe in the 'fundamental gneiss' of Bohemia and Bavaria, and in the very earliest stratified rocks of Scandinavia and Scotland. These beds are found in many parts to contain masses of considerable size, but usually of indeterminate form, disposed after the manner of an ancient Coral Reef, and consisting of alternating layers—frequently numbering from 50 to 100—of Carbonate of Lime and Serpentine (silicate of magnesia). The regularity of this alternation, and the fact that it presents itself also between other Calcareous and Siliceous minerals, having led to a suspicion that it had its origin in Organic structure, thin sections of well-preserved specimens were submitted to microscopic examination by Dr. Dawson of Montreal, who at once recognized its Foraminiferal nature:[1] the *calcareous* layers presenting the characteristic appearances of true *shell*, so disposed as to form an irregularly chambered structure, and frequently traversed by systems of ramifying canals corresponding to those of *Calcarina* (§ 484); whilst the *serpentinous* or other siliceous layers were regarded by him as having been formed by the infiltration of silicates in solution into the cavities originally occupied by the sarcode-body of the animal,—a process of whose occurrence at various Geological periods, and also at the present time, abundant evidence has already been adduced. Having himself taken up the investigation (at the instance of Sir William Logan), the Author was not only able to confirm Dr. Dawson's conclusions, but to adduce new and important evidence in support of them.[2] Although this determination has been called in question, on the ground that some resemblance to the supposed organic structure of Eozoön is presented by bodies of purely Mineral origin,[3] yet, as it has been accepted not only by most of those whose knowledge of Foraminiferal structure gives weight to their judgment (among whom the late Prof. Max Schulze may be specially named), but also by Geologists who have specially studied the Micro-mineralogical structure of the older Metamorphic rocks,[4] the Author feels justified in here describing *Eozoön* as he believes it to have existed when it originally extended itself as an animal growth over vast areas of the sea-bottom in the Laurentian epoch.

---

[1] This recognition was due, as Dr. Dawson has explicitly stated in his original Memoir ("Quarterly Journal of the Geological Society," Vol xxi., p. 54), to his acquaintance not merely with the Author's previous researches on the minute structure of the *Foraminifera*, but with the special characters presented by thin sections of *Calcarina* which had been transmitted to him by the Author. Dr. D. has given an excellent account of the Geological and Mineralogical relations of *Eozoön*, as well of its Organic structure, in a small book entitled "The Dawn of Life."

[2] For a fuller account of the results of the Author's own study of *Eozoön*, and of the basis on which the above reconstruction is founded, see his Papers, in "Quart. Journ. of Geol. Soc.," Vol. xxi., p. 59. and Vol. xxii., p. 219, and in the "Intellectual Observer," Vol. vii. (1865), p. 278; and his 'Further Researches,' in "Ann. of Nat. Hist.," June, 1874.

[3] See the Memoirs of Profs. King and Rowney, in "Quart. Journ. of Geol. Soc.," Vol. xxii., p. 185; and "Ann. of Nat. Hist.," May, 1874.

[4] Among these the Author is permitted to mention Prof. Geikie, of Edinburgh, who has thus studied the older rocks of Scotland, and Prof. Bonney, of Cambridge and London, who has made a like study of the Cornish and other Serpentines. By both these eminent authorities he is assured that they have met with no purely Mineral structure in the least resembling *Eozoön*, either in its regular alternation of Calcareous and Serpentinous lamellæ, or in the dendritic extensions of the latter into the former; and while they accept as entirely satisfactory the doctrine of its Organic origin maintained by the Author, they find themselves unable to conceive of any Inorganic agency by which such a structure could have been produced.

494. Whilst essentially belonging to the *Nummuline* group, in virtue of the fine tubulation of the shelly layers forming the 'proper wall' of its chambers, *Eozoön* is related to various types of recent Foraminifera in its other characters. For in its indeterminate zoöphytic mode of growth, it agrees with *Polytrema* (§ 483); in the incomplete separation of its chambers, it has its parallel in *Carpenteria* (§ 481); whilst in the high development of its 'intermediate skeleton' and of the 'canal-system' by which this is formed and nourished, it finds its nearest representative in *Calcarina* (§ 484). Its calcareous layers were so superposed, one upon another, as to include between them a succession of 'storeys' of *chambers* (Plate XVII., fig. 1, A', A', A², A³); the chambers of each 'storey' usually opening one into another, as at *a, a*, like apartments *en suite;* but being occasionally divided by complete *septa*, as at *b. b*. These septa are traversed by passages of communication between the chambers which they separate; resembling those which, in existing types, are occupied by *stolons* connecting together the segments of the sarcode-body. Each layer of shell consists of two finely-tubulated or 'nummuline' lamellæ, B, B, which form the boundaries of the chambers beneath and above, serving (so to speak) as the *ceiling* of the former, and as the *floor* of the latter; and of an intervening deposit of homogeneous shell-substance C, C, which constitutes the 'intermediate skeleton.' The tubuli of this 'nummuline layer' (Fig. 344) are usually filled-up (as in the Nummulites of the 'nummulitic limestone') by mineral infiltration, so as in

Fig. 344.

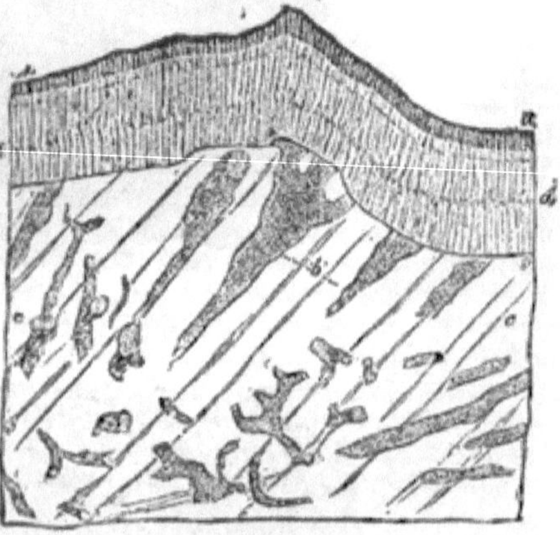

Vertical Section of a portion of one of the Calcareous lamellæ of *Eozoön Canadense:*—*a a*, Nummuline layer, perforated by parallel tubuli, which show a flexure along the line *a' a'*; beneath this is seen the intermediate skeleton, *c, c*, traversed by the large canals, *b, b*, and by oblique cleavage planes, which extend also into the Nummuline layer.

transparent sections to present a fibrous appearance; but it fortunately happens that through their having in some cases escaped infiltration, the tubulation is as distinct as it is even in recent Nummuline shells (Fig. 344), bearing a singular resemblance in its occasional waviness to that of the Crab's claw (§ 613). The thickness of this interposed layer varies considerably in different parts of the same mass; being in general greatest near its base, and progressively diminishing towards its upper surface. The 'intermediate skeleton' is occasionally traversed by large passages (D), which seem to establish a connection between the successive layers of chambers; and it is penetrated by arborescent systems of canals (E, E), which are often distributed both so extensively and so minutely through its substance, as to leave very little of it without a branch. These canals take their origin, not directly from the chambers, but from irregular *lacunæ* or interspaces between the outside of the proper chamber-walls and the 'intermediate skeleton,' exactly as in *Calcarina* (§ 484); the extensions of the sarcode-body which occupied them having apparently

104 THE MICROSCOPE AND ITS REVELATIONS.

PLATE XVII.

Fig. 1.

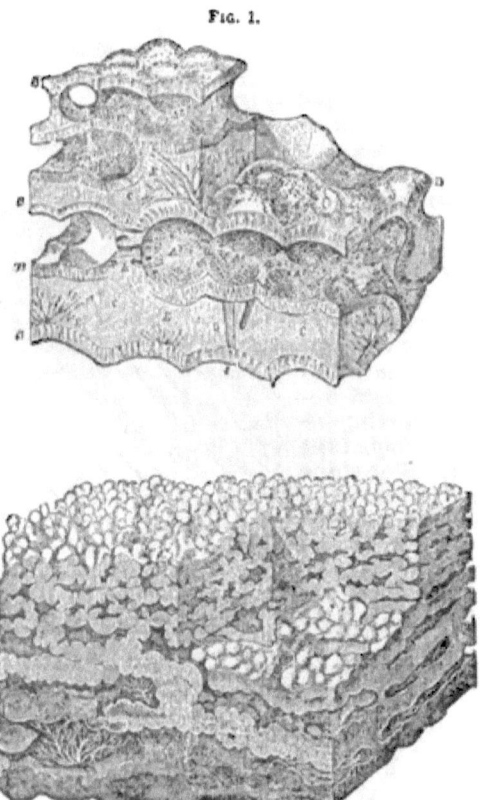

STRUCTURE OF EOZOÖN CANADENSE (Original).

Fig. 1. Portion of its **calcareous Shell, as it** would appear if the Serpentine that fills its chambers were dissolved away:—A^1, A^1, chambers of lower story, opening into each other at a, a, but occasionally separated by a septum b, b; A^2, A^2, chambers of upper story; B, B, proper walls of the chambers, formed of a finely-tubular or nummuline substance; c, c, intermediate skeleton, occasionally traversed by large stolon-passages, D, connecting the chambers of different stories, and penetrated by the arborescent systems of canals E, E, E.

2. Decalcified portion **showing the Serpentinous** *internal cast* of the chambers, canals, and tubuli of the original; presenting an exact model of the animal substance which originally filled them.

been formed by the coalescence of the pseudopodial filaments that passed through the tubulated lamellæ.

495. In the fossilized condition in which *Eozoön* is most commonly found, not only the cavities of the chambers, but the canal-systems to their smallest ramifications, are filled up by the siliceous infiltration which has taken the place of the original sarcode-body, as in the cases already cited (§ 487 *note*); and thus when a piece of this fossil is subjected to the action of dilute acid, by which its calcareous portion is dissolved-away, we obtain an *internal cast* of its chambers and canal-system (Plate XVII., fig. 2), which, though altogether dissimilar in *arrangement*, is essentially analogous in *character* to the 'internal casts' represented in Figs. 328, 332. This cast presents us, therefore, with a *model* in hard Serpentine of the soft sarcode-body which originally occupied the chambers, and extended itself into the ramifying canals, of the calcareous shell; and, like that of *Polystomella* (§ 487), it affords an even more satisfactory elucidation of the relations of these parts, than we could have gained from the study of the living organism. We see that each of the layers of serpentine, forming the lower part of such a specimen, is made up of a number of coherent segments, which have only undergone a partial separation; these appear to have extended themselves horizontally without any definite limit; but have here and there developed new segments in a vertical direction, so as to give origin to new layers. In the spaces between these successive layers, which were originally occupied by the calcareous shell, we see the 'internal casts' of the branching canal-system; which give us the exact models of the extensions of the sarcode-body that originally passed into them.— But this is not all. In specimens in which the nummuline layer constituting the 'proper wall' of the chambers was originally well preserved, and in which the decalcifying process has been carefully managed (so as not, by too rapid an evolution of carbonic acid gas, to disturb the arrangement of the serpentinous residuum), that layer is represented by a thin white film covering the exposed surfaces of the segments; the superficial aspect of which, as well as its sectional view, are shown in fig. 2. And when this layer is examined with a sufficient magnifying power, it is found to consist of extremely minute needle-like fibres of Serpentine, which sometimes stand upright, parallel, and almost in contact with each other, like the fibres of asbestos (so that the film which they form has been termed the 'asbestiform layer'), but which are frequently grouped in converging brush-like bundles, so as to be very close to each other in certain spots at the surface of the film, whilst widely separated in others. Now these fibres, which are less than 1-10,000th of an inch in diameter, are the 'internal casts' of the tubuli of the Nummuline layer (a precise parallel to them being presented in the 'internal cast' of a *recent* Amphistegina in the Author's possession); and their arrangement presents all the varieties which have been mentioned (§ 488) as existing in the shells of *Operculina*.—Thus these delicate and beautiful siliceous fibres represent those *pseudopodial* threads of sarcode, which originally traversed the minutely-tubular walls of the chambers; and *a precise model* of the most ancient animal of which we have any knowledge, notwithstanding the extreme softness and tenuity of its substance, is thus presented to us, with a completeness that is scarcely even approached in any later fossil.

496. In the upper part of the 'decalcified' specimen shown in Plate XVII., fig. 2, it is to be observed that the segments are confusedly heaped

together, instead of being regularly arranged in layers; the *lamellated* mode of growth having given place to the *acervuline*. This change is by no means uncommon among Foraminifera; an irregular piling-together of the chambers being frequently met-with in the later growth of types, whose earlier increase takes place upon some much more definite plan. After what fashion the *earliest* development of *Eozöon* took place, we have at present no knowledge whatever; but in a *young* specimen which has been recently discovered, it is obvious that each successive 'storey' of chambers was limited by the closing-in of the shelly layer at its edges, so as to give to the entire fabric a definite form closely resembling that of a straightened *Peneroplis* (Plate XV., fig. b). Thus it is obvious that the chief peculiarity of *Eozöon* lay in its capacity for *indefinite extension*; so that the product of a single germ might attain a size comparable to that of a massive Coral.—Now this, it will be observed, is simply due to the fact that its increase by gemmation takes place *continuously*; the new segments successively budded-off remaining in connection with the original stock, instead of detaching themselves from it as in Foraminifera generally. Thus the little *Globigerina* forms a shell of which the number of chambers does not usually seem to increase beyond *sixteen*, any additional segments detaching themselves so as to form separate shells; but by the repetition of this multiplication, the sea-bottom of large areas of the Atlantic Ocean at the present time has come to be covered with accumulations of *Globigerinæ*, which, if fossilized, would form beds of Limestone not less massive than those which have had their origin in the growth of *Eozöon*.—The difference between the two modes of increase may be compared to the difference between a Plant and a Tree. For in the Plant the individual organism never attains any considerable size, its extension by gemmation being limited; though the aggregation of individuals produced by the detachment of its buds (as in a Potato field) may give rise to a mass of vegetation as great as that formed in the largest Tree by the continuous putting forth of new buds.

497. It has been hitherto only in the Laurentian Serpentine-Limestone of Canada, that *Eozöon* has presented itself in such a state of preservation as fully to justify the assumption of its Organic nature. But from the greater or less resemblance which is presented to this by Serpentine-Limestones occurring in various localities, among strata that seem the Geological equivalents of the Canadian Laurentians, it seems a justifiable conclusion that this type was very generally diffused in the earlier ages of the Earth's history; and that it had a large (and probably the chief) share in the production of the most ancient Calcareous strata, separating Carbonate of Lime from its solution in Ocean-water, in the same manner as do the Polypes by whose growth Coral-reefs and islands are being upraised at the present time.

An elaborate work, "Der Bau des Eozöon Canadense" (1878) has been recently published by Prof. Möbius of Kiel, in which the structure of *Eozöon* is compared with that of various types of Foraminifera, and, as it differs from that of every one of them, is affirmed not to be organic at all, but purely Mineral. Upon this the Author would remark, that if the validity of this mode of reasoning be admitted, *any* Fossil whose structure does not correspond with that of some existing type, is to be similarly rejected. Thus, the *Stromatopora* of Silurian and Devonian rocks, which some Palæontologists regard as a Coral, others as Polyzoary, others as a Calcareous Sponge, and others as Foraminifer, would not be a fossil at all, because it differs from every known living form. Yet the suggestion that it is of Mineral origin would be scouted as absurd by every Palæontologist. Again, it is urged by Prof. Möbius that as the supposed canal-system of *Eozöon* has not the constancy and regularity of distribution which it presents in existing Foraminifera, it must be accounted a Mineral infiltration. To this the Author would reply:—(1) That a prolonged and careful study of this 'canal-system,' in a great variety of modes, with an amount of material at his disposal many times greater than Prof. Möbius could command, has satisfied him that in well-preserved specimens the canal-system, so far from being vague and indefinite, has a very regular plan of distribution;—(2) That this plan does **not** differ more from the arrangements characteristic of the several types of existing *Foraminifera*, than these differ from each other, its *general* conformity to them being such as to satisfy Prof. Max Schultze (one of the ablest Foraminiferalists of his time) of its Foraminiferal character:—and (3) that not only does the distribution of the canal-system of *Eozōon* differ in certain essential features from every form of Mineral infiltration hitherto brought to light, but that *canal-systems in no respect differing from each other in distribution are occupied by different minerals*,—a fact which seems conclusively to point to their *pre-existence* in the Calcareous layers, and the *subsequent* penetration of these minerals into the passages previously occupied by sarcode,—precisely as has happened in those 'internal casts' of existing **Foraminifera** (§ 497) which Prof. Möbius altogether ignores.

The argument for the Foraminiferal nature of *Eozōon* is essentially a *cumulative* one, resting on a number of *independent probabilities*, no one of which, taken separately, has the cogency of a *proof;* yet the accordance of them all with that hypothesis has an almost demonstrative value, no other hypothesis accounting at once for the whole assemblage of facts.—As it is the Author's intention to set forth this in the best and completest form he can devise, at the earliest possible period, he would beg for *a suspension of judgment* on the part of those who have credited Prof. Möbius with having completely settled the question; the small amount of evidence contained in his Memoir bearing no comparison to that of an opposite bearing of which the Author is in possession.

498. *Collection and Selection of Foraminifera.*—Many of the Foraminifera attach themselves in the living state to Sea-weeds, Zoöphytes, etc.; and they should, therefore, be carefully looked-for on such bodies, especially when it is desired to observe their internal organization and their habits of life. They are often to be collected in much larger numbers, however, from the sand or mud dredged-up from the sea-bottom, or even from that taken from between the tide-marks. In a paper containing some valuable hints on this subject,[1] Mr. Legg mentions that, in walking over the Small-mouth Sand, which is situated on the north-side of Portland Bay, he observed the sand to be distinctly marked with white ridges, many yards in length, running parallel with the edge of the water; and upon examining portions of these, he found Foraminifera in considerable abundance. One of the most fertile sources of supply that our own coasts afford, is the *ooze* of the Oyster-beds, in which large numbers of living specimens will be found; the variety of specific forms, however, is usually not very great. In separating these bodies from the particles of sand, mud, etc., with which they are mixed, various methods may be adopted, in order to shorten the tedious labor of picking them out, one by one, under the Simple Microscope; and the choice to be made among these will mainly depend upon the condition of the Foraminifera, the importance (or otherwise) of obtaining them alive, and the nature of the substances with which they are mingled.—Thus, if it be desired to obtain *living* specimens from the oyster-ooze, for the examination of their soft parts, or for preservation in an Aquarium, much time will be saved by stirring the mud (which should be taken from the surface only of the deposit) in a jar with water, and then allowing it to stand for a few moments; for the finer particles will remain diffused though the liquid, while the coarser will subside; and as the Foraminifera (in the present case) will be among the *heavier*, they will be found

---

[1] "Transaction of Microscopical Society," 2d Series, Vol. ii. (1854), p. 19.

at the bottom of the vessel with comparatively little extraneous matter, after this operation has been repeated two or three times. It would always be well to examine the first deposit let fall by the water that has been poured-away; as this may contain the smaller and lighter forms of Foraminifera.—But supposing that it be only desired to obtain the *dead* shells from a mass of sand brought-up by the dredge, a very different method should be adopted. The whole mass should be exposed for some hours to the heat of an oven, and be turned-over several times, until it is found to have been thoroughly dried throughout; and then, after being allowed to cool, it should be stirred in a large vessel of water. The chambers of their shells being now occupied by air alone (for the bodies of such as were alive will have shrunk-up almost to nothing), the Foraminifera will be the *lightest* portion of the mass; and they will be found floating on the water, while the particles of sand, etc., subside. Another method, devised by Mr. Legg, consists in taking advantage of the relative sizes of different kinds of Foraminifera and of the substances that accompany them. This, which is especially applicable to the sand and rubbish obtainable from Sponges (which may be got in large quantity from the sponge-merchants), consists in sifting the whole aggregate through successive sieves of wire-gauze, commencing with one of 10 wires to the inch, which will separate large extraneous particles, and proceeding to those of 20, 40, 70, and 100 wires to the inch, each (especially that of 70) retaining a much larger proportion of Foraminiferal shells than of the accompanying particles; so that a large portion of the extraneous matter being thus got rid of, the final selection becomes comparatively easy.—Certain forms of Foraminifera are found attached to Shells, especially bivalves (such as the *Chamaceæ*) with foliated surfaces; and a careful examination of those of tropical seas, when brought home 'in the rough,' is almost sure to yield most valuable results.—The **final** selection of specimens for mounting should always be made under **some** appropriate form of Single Microscope (§§ 43–48); a fine camel-**hair** pencil, with the point wetted between the lips, being the instrument which may be most conveniently and safely employed, even for the most delicate specimens. In mounting Foraminifera as Microscopic objects, the method to be adopted must entirely depend upon whether they are to be viewed by *transmitted* or by *reflected* light. In the former case it should be mounted in Canada balsam (§ 210); the various precautions to prevent the retention of air-bubbles, which have been already described, being carefully observed. In the latter no plan is so simple, easy, and effectual, as the attaching them with a little gum to wooden slides (Fig. 124). They should be fixed in various positions, so as to present all the different aspects of the shell, particular care being taken that its mouth is clearly displayed; and this may often be most readily managed by attaching the specimens *sideways* to the wall of the circular depression of the slide. Or the specimens may be attached to disks fitted for being held in Morris's Disk-holder (Fig. 95); whilst for the examination of specimens in every variety of position, Mr. R. Beck's Disk-holder (Fig. 94) will be found extremely convenient. Where, as will often happen, the several individuals differ considerably from one another, special care should be taken to arrange them in *series* illustrative of their range of variation and of the mutual connections of even the most diverse forms.— For the display of the internal structure of Foraminifera, it will often be necessary to make extremely thin sections, in the manner already described (§§ 192–194); and much time will be saved by attaching a number of specimens to the glass slide at once, and by grinding them down together (§ 192, *note*). For the preparation of sections, however, of the extreme thinness that is often required, those which have been thus reduced should be transferred to separate slides, and finished-off each one by itself.

RADIOLARIA.

499. It has been shown that one series of forms belonging to the *Rhizopod* type is characterized by the *radiating* arrangement of their rod-like *pseudopodia* (§ 399), suggesting the designation *Heliozoa* or 'sun-animalcules;' and that even among those fresh-water forms that do not depart widely from the common *Actinophrys* (Fig. 285), there are some whose bodies are inclosed in a complete siliceous skeleton. Now just as the

Fossil *Radiolaria* from Barbadoes.—*a*, Podocyrtis mitra; *b*, Rhabdolithus sceptrum; *c*, Lychnocanium falciferum; *d*, Eucyrtidium tubulus; *e*, Flustrella concentrica; *f*, Lychnocanium lucerna; *g*, Eucyrtidium elegans; *h*, Dictyospyris clathrus; *i*, Eucyrtidium Mongolfieri; *k*, Stephanolithis spinescens; *l*, S. nodosa; *m*, Lithocyclia ocellus; *n*, Cephalolithis sylvina; *o*, Podocyrtis cothurnata; *p*, Rhabdolituus pipa.

Reticularian type of Rhizopod life culminates in the marine calcareous-shelled *Foraminifera*, so does the Heliozoic type seem to culminate in the marine *Radiolaria;* which, living for the most part near the surface of the ocean, form *siliceous* skeletons (often of marvellous symmetry and beauty), that fall to the bottom on the death of the animals that produced them, and may remain unchanged, like those of the Diatoms, through unlimited periods of time. Some of these skeletons, mingled with those of Diatoms, had been detected by Prof. Ehrenberg in the midst of various deposits of Foraminiferal origin, such as the Calcareous Tertiaries of Sicily and Greece, and of Oran in Africa; and he established for them the group of *Polycystina*, to which he was able also to refer a beautiful series of forms making-up nearly the whole of a siliceous sandstone prevailing through an extensive district in the island of Barbadoes (Fig. 345). Nothing, however, was known of the nature of the animals that formed them, until they were discovered and studied in the living state by Prof. J. Müller;[1] who established the group of *Radiolaria*, including therein, with the *Polycystina* of Ehrenberg, the *Acanthometrina* (§ 505) first recognized by himself, and the *Thalassicolla* (§ 506) which had been discovered by Prof. Huxley. Not long afterwards appeared the magnificent and 'epoch-making' work of Prof. Haeckel;[2] and since that time much has been added by various observers to our knowledge of this group, which still remains, however, very imperfect. For the following general account of its characters, the Author is indebted to the valuable summary of "Recent Researches in regard to the Radiolaria" lately given by Prof. Mivart.[3]

500. Each individual Radiolarian consists of two portions of colored or colorless sarcode: one portion nucleated and central; the other portion peripheral, and almost always containing certain yellow corpuscles. These two portions are separated by a chitinous membrane called the *capsule;* but this is so porous as to allow of their free communication with each other. The yellow corpuscles seem to be true 'cells;' having a regular membranous wall, with protoplasmic contents (including starch-granules), and distinct nuclei; and multiplying themselves by subdivision. But there is considerable doubt whether they are really parts of the animal body, as they have been found in vigorous life when the rest of the animal is dead and decaying; and they are regarded by Cienkowski as parasites. The *pseudopodia* radiate in all directions (Plate XVIII., figs. 3, 4) from the deeper portion of the extra-capsular sarcode; they have generally much persistency of direction, and very little flexibility; in some species (but not ordinarily) they branch and anastomose; while in others they are inclosed in hollow rods that form part of the siliceous skeleton, and issue forth from the extremities of these. A flow of granules takes place along them; and the mode in which they obtain food-particles (consisting of Diatoms and other minute Algæ, marine Infusoria, etc.), and draw them into the sarcode-bodies of the Radiolarians, appears to correspond entirely with their action in *Actinophrys* and other Heliozoa (§ 399).

501. In most *Radiolaria*, skeletal structures are developed in the sarcode-body, either inside or outside the capsule, or in both positions; sometimes in the form of investing networks having more or less of a *spheroidal* form (Plate XIX., figs. 1, 2), or of *radiating* spines (fig. 3), or of combinations of these (figs. 4, 5). But in many cases the skeleton consists only of a few scattered spicules; and this is especially the case in certain large composite forms or 'colonies' (Fig. 350) which may consist of as many as a thousand zooids, aggregated together in various forms, discoidal, cylindrical, spheroidal, chain-like, or even necklace-like. The 'colonies' seem to be produced, like the multiple segments of the bodies of Foraminifera (§ 456), by the non-sexual multiplication of a primordial zooid; but whether this multiplication takes place by fission, or by the budding-off of portions of the sarcode-body, has not yet been clearly

---

[1] 'Ueber die *Thalassicollen, Polycystinen* und *Acanthometren* des Mittelmeeres,' in "Abhandlungen der Königl. Akad. der Wissensch. zu Berlin," 1858, and separately published; also 'Ueber die im Hafen von Messina beobachteten *Polycystinen*,' in the "Monatsberichte" of the Berlin Academy for 1855, pp. 671–676.
[2] "Die *Radiolarien* (Rhizopoda Radiaria)," Berlin, 1862.
[3] "Journal of the Linnæan Society," Vol. xiv. (Zool.), p. 136.

PLATE XVIII.

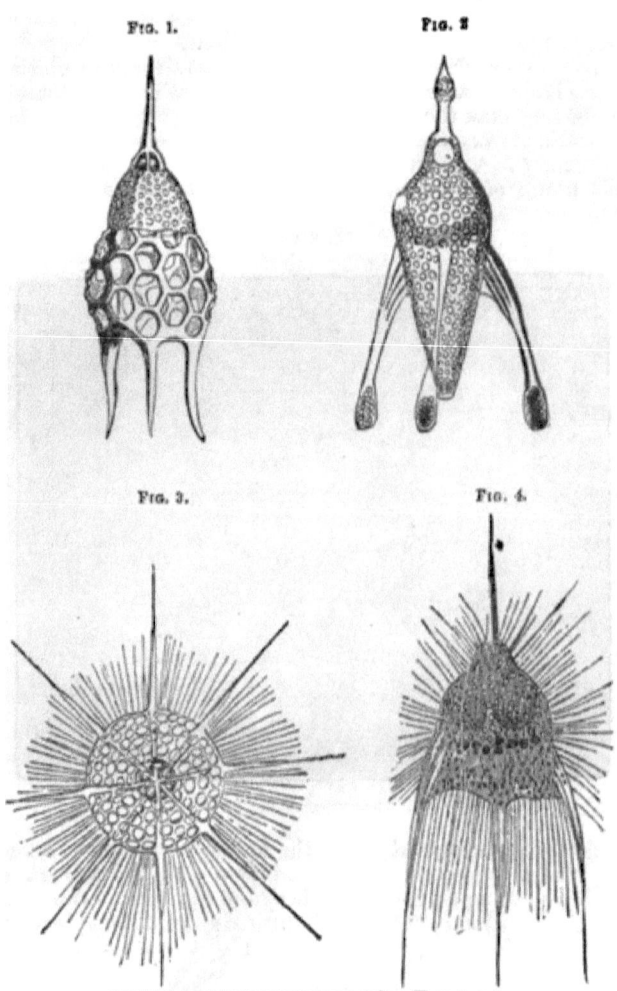

VARIOUS FORMS OF POLYCYSTINA (after Ehrenberg).

Fig. 1. *Podoyrtis Schomburgkii.*
2. *Rhopalocanium ornatum.*
3. *Haliomma hystrix.*
4. *Pterocanium, with animal.*

made-out. The emission of flagellated zoospores, very similar to those of *Clathrulina* (Fig. 288), has been observed in many Radiolarians; but of the mode in which they are produced, and of their subsequent history, very little is at present known.—Until the structure and life-history of the animals of this very interesting type shall have been more fully elucidated, no satisfactory classification of them can be framed; and nothing more will be here attempted than to indicate some of the principal forms under which the Radiolarian type presents itself.

502. *Discida.*—Among the beautiful siliceous structures which are met with in the Radiolarian sandstone of Barbadoes (Fig. 345) there is none more interesting than the skeleton of *Astromma* (Fig. 346); in which we have a remarkable example of the *range of variation* that is compatible with conformity to a general plan of structure. As in other forms of Haeckel's group of *Discida,* there is in this skeleton a combination of

Fig. 346

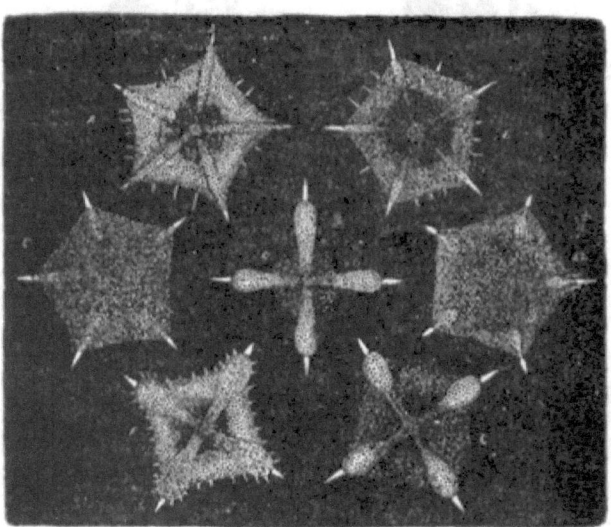

Varietal modifications of *Astromma.*

radial and of circumferential parts; the former consisting of solid spoke-like rods, whilst the latter is composed of a siliceous network more or less completely filling up the spaces between the rays. The radial part of the skeleton predominates in the beautiful 4-rayed example represented at D, having the form of a Maltese cross; whilst in F and G it still shows itself very conspicuously, though the spaces between the rays are in great part filled up by the circumferential network. In the 5-rayed specimens A and B, on the other hand, the radial portion is much less developed, whilst the circumferential becomes more discoidal. And in c and E, while the circumferential network forms a pentagonal disk, the radial portion is represented only by solid projections at its angles. The transition between the extreme forms is found to be so gradual when a number of specimens are compared, that no lines of specific distinction can be drawn between them; and the difference in the *number* of rays is probably of no more account in these low forms of Animal life, than it is in the discoidal Diatoms (§ 290).—Other discoidal forms, showing a like combination of radial and circumferential parts are represented in Figs. 347 and 348, and also in Fig. 345, *e, m.*

503. *Entosphærida.*—In this group the siliceous shell is spheroidal, and is formed *within* the capsule; and it is not traversed by radii, although prolongations of the shell often extend themselves radially outwards, as in *Cladococcus* (Plate XIX., fig. 5). Sometimes the central sphere is inclosed in two, three, or even more concentric spheres connected by radii, as in the beautiful *Actinomma* (Plate XIX., fig. 2); reminding us of the wonderful concentric spheres carved in ivory by the Chinese.—One of the most common examples of this group is the *Haliomma Humboldtii* (Fig. 349), in which the shell is double.

504. *Polycystina.*—This name, which originally included the preceding group, is now restricted to those which have the shell formed *outside* the capsule. This shell may, as in the preceding, be a simple sphere composed of an open siliceous network, as in *Ethmosphæra* (Plate XIX., fig. 1); or it may consist of two or three concentric spheres connected by radii; or, again, it may put forth radial outgrowths, which

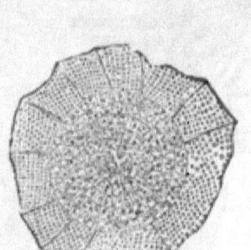

Fig. 347.

Perichlamydium prætextum.

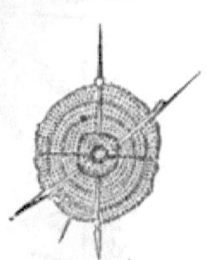

Fig. 348.

Stylodyctya gracilis.

sometimes extend themselves to several times the diameter of the shell, and ramify more or less minutely, as in *Arachnosphæra* (Plate XIX., fig. 4). But more frequently the shell opens-out at one pole into a form more or less bell-like, as in *Podocyrtis* (Plate XVIII., fig. 1, and Fig. 345, *a, o*), *Rhopalocanium* (Plate XVIII., fig. 2), and *Pterocanium* (Plate XVIII., fig. 4); or it may be elongated into a somewhat cylindrical form, one pole remaining closed, while the other is more or less contracted, as in *Eucyrtidium* (Fig. 345, *d, g, i*).—The transition between these forms again, proves to be as gradational, when many specimens are compared,[1] as it is among *Foraminifera* (§ 488).

505. *Acanthometrina.*—In this group the animal is not inclosed within a shell, but is furnished with a very regular skeleton composed of elongated spines, which radiate in all directions from a common centre (Plate XIX., fig. 3). The soft sarcode-body is spherical in form, and occupies the spaces left between the bases of these spines, which are sometimes partly inclosed (as in the species represented) by transverse projec-

---

[1] The general Plan of structure of the *Polycystina*, and the signification of their immense variety of forms, were ably discussed by Dr. Wallich, in the "Trans. of the Microsc. Soc.," N.S., Vol. xii. (1865), p. 75.

PLATE XIX.

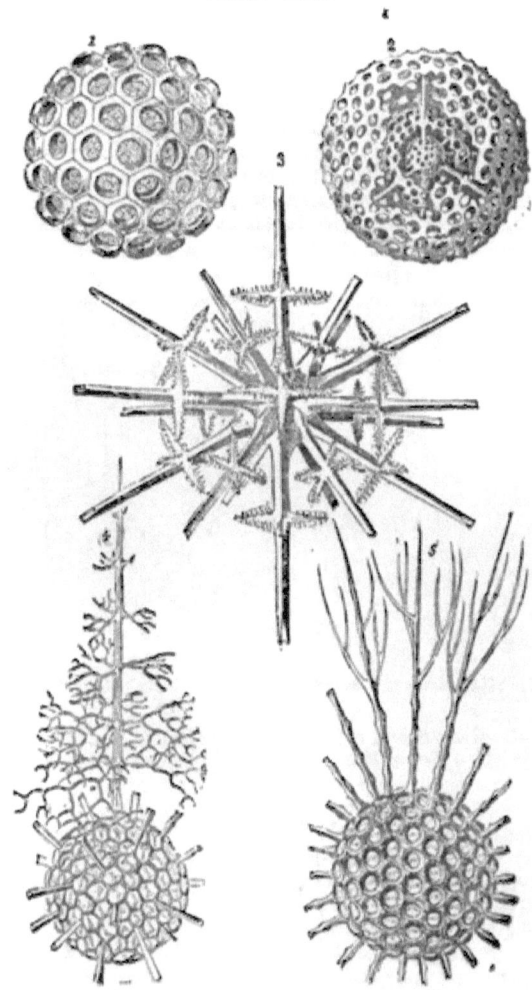

VARIOUS FORMS OF RADIOLARIA (after Hæckel).

Fig. 1. *Ethmosphæra siphonophora.*
2. *Actinomma inerme.*
3. *Acanthometra xiphicantha.*
4. *Arachnosphæra obligacantha.*
5. *Cladococcus viminalis.*

tions. The 'capsule' is pierced by the pseudopodia, whose convergence may be traced from without inwards, after passing through it; and it is itself enveloped in a layer of less tenacious protoplasm, resembling that of which the pseudopodia are composed. One species, the *Acanthometra echinoides*, which presents itself to the naked eye as a crimson-red point, the diameter of the central part of its body being about 6-1000ths of an inch, is very common on some parts of the coast of Norway, especially during the prevalence of westerly winds; and the Author has himself met with it abundantly near Shetland, in the floating brown masses termed *madre* by the fishermen (who believe them to furnish food to the herring), which consist mainly of this Acanthometra mingled with Entomostraca.

506. *Collozoa.*—To this group belong these remarkable composite forms, which, exhibiting the characteristic Radiolarian type in their individual zooids, are aggregated into masses in which the skeleton is repre-

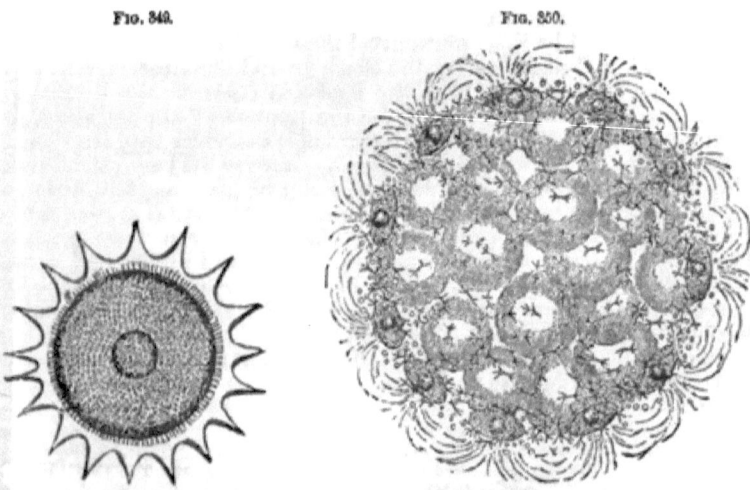

Fig. 349. Fig. 350.

Haliomma Humboldtii.   *Sphærozoum ovodimare.*

sented only by scattered spicules, as in *Sphærozoum* (Fig. 350) and *Thalassicolla.*—These 'sea-jellies,' which so abound in the seas of warm latitudes as to be among the commonest objects collected by the Tow-net, are small gelatinous rounded bodies, of very variable size and shape, but usually either globular or discoidal. Externally they are invested by a layer of condensed sarcode, which sends forth pseudopodial extensions that commonly stand out like rays, but sometimes inosculate with each other so as to form network. Towards the inner surface of this coat are scattered a great number of oval bodies resembling cells, having a tolerably distinct membraniform wall and a conspicuous round central nucleus. Each of these bodies appears to be without any direct connection with the rest; but it serves as a centre round which a number of minute yellowish-green vesicles are disposed. Each of these groups is protected by a siliceous skeleton, which sometimes consists of separate spicules (as in Fig. 350), but which may be a thin perforated sphere, like that of certain Polycystina, sometimes extending itself into radial prolongations.

The internal portion of each mass is composed of an aggregation of large vesicle-like bodies, imbedded in a softer sarcodic substance.[1]

507. From the researches made during the 'Challenger' expedition, it appears that the *Radiolaria* are very widely diffused through the waters of the ocean, some forms being more abundant in tropical and others in temperature seas; and that they live not only at or near the surface, but also at considerable depths. Their siliceous skeletons accumulate in some localities (in which the calcareous remains of Foraminifera are wanting) to such an extent as to form a 'Radiolarian ooze;' and it is obvious that the elevation of such a deposit into dry land would form a bed of siliceous sandstone resembling the well-known Barbadoes rock, which is said to attain a thickness of 1100 feet, or a similar rock of yet greater thickness in the Nicobar Islands.—Few Microscopic objects are more beautiful than an assemblage of the most remarkable forms of the Barbadian *Polycystina* (Fig. 345), especially when seen brightly illuminated upon a black ground; since (for the reason formerly explained, § 103) their solid forms then become much more apparent than they are when these objects are examined by light transmitted through them. And when they are mounted in Canada-balsam, the Black-ground illumination, either by the Webster-condenser (§ 100), the Spot-lens (104), or the Paraboloid (§ 105), is much to be preferred for the purpose of display, although minute details of structure can be better made out when they are viewed as transparent objects with higher powers. Many of the more solid forms, when exposed to a high temperature on a slip of platinum foil, undergo a change in aspect which renders them peculiarly beautiful as opaque objects; their glassy transparence giving place to an enamel-like opacity. They may then be mounted on a black ground, and illuminated either with a Side-condenser, or with the Parabolic Speculum (§ 114).—No class of objects is more suitable than these to the Binocular Microscope; its stereoscopic projection causing them to be presented to the mind's eye in complete relief, so as to bring-out with the most marvellous and beautiful effect all their delicate sculpture.[2]

---

[1] See Prof. Huxley (to whom we owe our first knowledge of these forms) in "Ann. Nat. Hist.," Ser. 2, Vol. viii. (1851), p. 433; also Prof. Müller, of Berlin, in "Quart. Journ. Microsc. Sci.," Vol. iv. (1856), p. 72, and in his Treatise "Ueber die Thalassicollen, Polycystinen, und Acanthometren des Mittelmeeres;' and the magnificent work of Prof. Haeckel, "Die Radiolarien."—Great additions to our knowledge of this group may be expected from the collections made in the 'Challenger' expedition.

[2] For a fuller description of the Fossil forms of this group, see Prof. Ehrenberg's Memoirs in the "Monatsberichte" of the Berlin Academy for 1846, 1847, and 1850; also his 'Microgeologie,' 1854; and "Ann. of Nat. Hist.," Vol. xx. (1847).—The best method of separating the *Polycystina* from the Barbadoes sandstone is described by Mr. Furlong in the "Quart. Journ. of Microsc. Sci.," N. S., Vol. i. (1861), p. 64.

CHAPTER XIII.

SPONGES AND ZOOPHYTES.

I. SPONGES.

508. THE determination of the real character of the animals of this Class has been entirely effected by the Microscopic examination of their minute structure; for until this came to be properly understood, not only was the general nature of these organisms entirely misapprehended, but they were regarded by many naturalists as having no certain claim to a place in the Animal Kingdom. It may now be unhesitatingly affirmed that a Sponge is essentially an aggregate of Protozoic units, of which some correspond in every particular to the collared *Flagellata* (Fig. 295), whilst others resemble *Amœbæ* (Fig. 289),—the two conditions being probably only different stages of the same life-history. These units are held together by a continuous sarcode-body, which clothes the skeletal framework that represents our usual idea of a Sponge. In the simpler forms of sponges, however, this framework is altogether absent; in others it is represented only by calcareous or siliceous 'spicules,' which are dispersed through the sarcodic substance (Fig. 352, B); in others, again, the skeleton is a keratose (horny) network, which may be entirely destitute (as in our ordinary Sponge) of any mineral support, but which is often strengthened by calcareous or siliceous spicules (Fig. 352, A); whilst in what may be regarded as the highest types of the group, the siliceous component of the skeleton increases, and the keratose diminishes, until the skeleton consists of a beautiful siliceous network resembling spun-glass (§ 511). But whatever may be the condition of the *skeleton*, that of the body that clothes it remains essentially the same; and the peculiarity that chiefly distinguishes the Sponge-colony from the plant-like colonies of the Flagellate Infusoria (Fig. 296), is that whilst the latter extend themselves *outwards* by repeated ramification, sending their zooid-bearing branches to meet the water they inhabit, the surface of the former extends itself *inwards*, forming a system of passages and cavities lined by these and the amœboid zooids, through which a current of water is drawn-in to meet them by the action of the flagella. The minute pores (Fig. 351, *b, b*) with which the surface *a, a*, of the living Sponge is beset, lead to incurrent passages that open into chambers lying beneath it (*c, c*); and it is especially on the walls of these 'ampullaceous sacs,' that the *flagellate* zooids present themselves. The water drawn-in by their agency is driven outwards through a system of excurrent canals, which, uniting into larger trunks, proceed to the *oscula* or projecting vents *d*, from each of which, during the active life of the Sponge, a stream of water, carrying out excrementitious matter, is continually

issuing. The in-current brings into the chambers both food-material and oxygen; and from the manner in which colored particles experimentally diffused through the water wherein a Sponge is living, are received into its sarcodic substance, it seems clear that the nutrition of the entire fabric is the resultant of the feeding action of the separate amœboid and flagellate units, each of which takes-in, after its kind, the food-particles brought by the current of water, and imparts the product of its digestion to them to the general sarcodic mass.[1]

509. The continuous sarcode-substance or 'cyloblastema' that clothes the skeleton of the Sponge and constitutes its living body, includes great numbers of 'cytodes (§ 392), in various stages of development; which, like isolated *Amœbæ*, are constantly undergoing changes in form and position. Their long slender pseudopodia, radiating towards those of their neighbors, often unite together to form a complex network; and it seems to be by their agency, that the continual contractions and expansions of the oscula are produced, which are very characteristic of the living Sponge. It would seem, indeed, as if they combined in themselves the functions of nerve and muscle-elements, which are differentiated in the higher forms of animal life. Any one of these amœboids, again, detached from the mass, may lay the foundation of a new 'colony.' In the aggregate mass produced by its continuous segmentation, certain globular clusters are distinguishable, each having a cavity in its interior; and the amœboids that form the wall of this cavity become metamorphosed into collared flagellate zooids whose flagella project into it. Thus is formed one of the characteristic 'ampullaceous sacs;' which, at first closed, afterwards communicates with the exterior, on the one hand, by an incurrent passage, and on the other with the excurrent canal-system leading to the oscula.—Besides this reproduction by 'micro-spores,' there is another form of non-sexual reproduction by 'macro-spores;' which are clusters of amœboids encysted in firm capsules, frequently strengthened on their exterior by a layer of spicules of very peculiar form. These 'seed-like bodies,' which answer to the encysted states of many protophytes, are met with in the substance of the sponge, chiefly in winter; and after being set free through the oscula, they give exit to their contained amœboids, each of which may found a new colony.—A

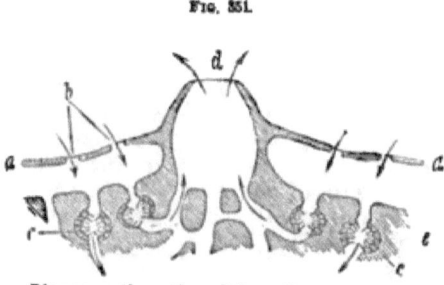

Fig. 351.

Diagrammatic section of *Spongilla*:—*a, a*, superficial layer; *b*, inhalant apertures; *c, c*, flagellated chambers; *d*, exhalant oscule; *e*, deeper substance of the sponge.

---

[1] This view of the nature and living action of *Sponges*, originally suggested by Dujardin, was definitely put forth by the late Prof. H. James-Clark, as the result of an admirable series of researches on Sponges and Flagellate Infusoria, in the Transactions of the Boston Society of Natural History for 1868, reproduced in the "Ann. Nat. Hist." for the same year. See also his Memoir on *Spongilla* in "Amer. Journ. Sci.," 1871, pp. 426–436; reproduced in 'Monthly Microsc. Journ.,' Vol. vii., (1872), p. 104.- His observations have been since fully confirmed by Messrs. Carter and Saville Kent; who have published a succession of Papers in the "Annals of Natural History," the general conclusions of which are embodied in Chap. v. of Mr. S. Kent's "Manual of the Infusoria."

true process of sexual generation, moreover, is said to take place in Sponges; certain of the amœboids, like certain cells of *Volvox* (§ 240), becoming 'sperm-cells,' and developing spermatozoa by the metamorphosis of their nuclei; while others become 'germ-cells,' developing themselves by segmentation (when fertilized) into the bodies known as 'ciliated gemmules,' which are set free from the walls of the canals, swim forth from the vents, and for a time move actively through the water. According to Prof. Haeckel, the fertilized germ-cells are to be regarded as true *ova*, and the products of their segmentation as *morulæ*, which, by invagination (§ 391), become *gastrulæ;* and he argues that the whole system of canals and ampullaceous sacs is really, like the system of canals in the Sponge-like *Alcyonium* (§ 529), an extension of the primitive gastric cavity; the *oscula* of Sponges being the undeveloped representatives of the *polypes* of the Zoöphyte.—As it is doubtful, however, whether the supposed Sponge-spermatozoa are anything else than ordinary flagellated monads, and as the development of the supposed ovum by no means conforms to the ordinary *gastræa* type, the question

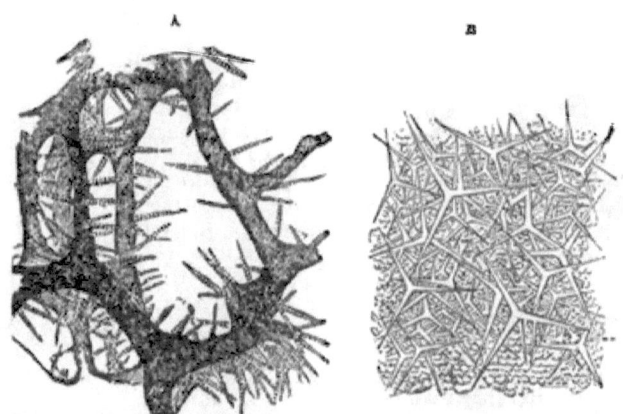

Fig. 32.

A. Portion of *Halichondria* (?) from Madagascar, with siliceous spicules projecting from the keratose network.
B. Triradiate spicules of *Grantia compressa*, lying in the midst of its cytoblastema.

whether Sponges are strictly *Protozoa*, or are to be regarded as constituting the lowest form of the *Metazoic* type, must be considered (in the Author's opinion) as still an open one.[1]

510. The arrangement of the keratose reticulation in the Sponges with which we are most familiar, may be best made out by cutting thin slices of a piece of Sponge submitted to firm compression, and viewing these slices, mounted upon a dark ground, with a low magnifying power, under incident light. Such sections, thus illuminated, are not merely striking objects; but serve to show, very characteristically, the general disposition of the larger canals and of the smaller pores with which they communicate. In the ordinary Sponge, the fibrous skeleton is almost entirely destitute of spicules; the absence of which, in fact, is one im-

---

[1] See Chap. v. of Mr. Saville Kent's "Manual of the Infusoria," and Chap. v. of Mr. Balfour's "Comparative Embryology," as well as Prof. Haeckel's important work on the Calcareous Sponges.

portant condition of that flexibility and compressibility on which its uses depend. When spicules exist in connection with such a skeleton, they are usually either altogether imbedded in the fibres, or are implanted into them at their bases, as shown in Fig. 352, A. But smaller and simpler Sponges, such as *Grantia*, have no horny skeleton; and their spicules are imbedded in the general substance of the body (Fig. 351, B).—Sponge-spicules are much more frequently Siliceous than Calcareous; and the variety of forms presented by the siliceous spicules is much greater than that which we find in the comparatively small division in which they are composed of carbonate of lime. The long needle-like spicules (Fig. 353), which are extremely abundant in several Sponges, lying close together in bundles, are sometimes straight, sometimes slightly curved; they are sometimes pointed at both ends, sometimes at one only; one or both ends may be furnished with a head like that of a pin, or may carry three or more diverging points which sometimes curve back so as to form hooks (Fig. 488, H). When the spicules project from the horny framework, they are usually somewhat conical in form, and their surface is often beset with little spines, arranged at regular intervals, giving them a jointed appearance (Fig. 352, A). Sponge-spicules frequently occur, however, under forms very different from the preceding; some being short and many-branched, and the branches being themselves very commonly stunted into mere tubercles (some examples of which type are presented in Fig. 488, A, C); whilst others are stellate, having a central body with conical spines projecting from it in all directions (as at D of the same figure). Great varieties present themselves in the stellate form, according to the relative predominance of the body and of the rays:

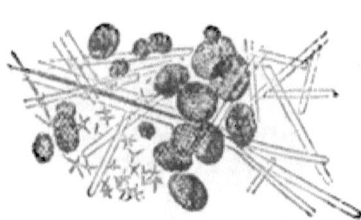

Siliceous Spicules of *Pachymatisma*.

in those represented in Fig. 353, the rays, though very numerous, are extremely short; in other instances the rays are much longer, and scarcely any central nucleus can be said to exist. The varieties in the form of Sponge-spicules are, in fact, almost endless; and a single Sponge often presents two or more (as shown in Fig. 353), the *stellate* spicules usually occurring either in the interspaces between the elongated kinds, or in the external crust.[1] The spicules of Sponges cannot be considered, like the *raphides* of Plants (§ 359), simply as deposits of Mineral matter in a crystalline state; for the forms of many of them are such as no mere crystallization can produce; they generally (at least in the earlier stage of their formation) possess internal cavities, which contain organic matter; and the calcareous spicules, whose mineral matter can be readily dissolved away by an acid, are found to have a distinct animal basis. Hence it seems probable that each spicule was originally a segment of sarcode, which has

---

[1] A minute account of the various forms of spicules contained in Sponges is given by Mr. Bowerbank in his First Memoir 'On the Anatomy and Physiology of the Spongiadæ,' in ' Philos. Transact.,'' 1858, pp. 279–332; and in his " Monograph of the British Spongiadæ" published by the Ray Society.—The Calcareous Sponges have been made by Prof. Haeckel the subject of an elaborate Monograph, "Die Kalkschwämme," Berlin, 1872.

undergone either calcification or silification; and by the self-shaping power of which, the form of the spicule is mainly determined.

511. There is an extremely interesting group of Sponges, in which the horny skeleton is entirely replaced by a *siliceous* framework of great firmness and of singular beauty of construction. This framework may be regarded as fundamentally consisting of an arrangement of *six-rayed* spicules, the extensions of which come to be, as it were, soldered to one another; and hence the group is distinguished as *hexiradiate*. Of this type the beautiful *Euplectella* of the Manilla Seas—which was for a long time one of the greatest of zoological rarities, but which now, under the name of 'Venus's flower-basket,' is a common ornament of our drawing-rooms—is one of the most characteristic examples. Another example is presented by the *Holtenia Carpenteri*, of which four specimens, dredged up from a depth of 530 fathoms between the Faroe Islands and the North of Scotland, were among the most valuable of the 'treasures of the deep' obtained during the first Deep-sea Exploration (1868), carried on by Sir Wyville Thomson and the Author. This is a turnip-shaped body, with a cavity in its interior, the circular mouth of which is surrounded with a fringe of elongated siliceous spicules; whilst from its base there hangs a sort of beard of siliceous threads, that extend themselves, sometimes to a length of several feet, into the Atlantic mud (§ 480) on which these bodies are found. The framework is much more massive than that of *Euplectella*, but it is not so exclusively mineral; for if it be boiled in nitric acid it is resolved into separate spicules, these being not soldered together by siliceous continuity, but held together by animal matter. Besides the regular hexiradiate spicules, there is a remarkable variety of other forms, which have been fully described and figured by Sir Wyville Thomson.[1] One of the greatest features of interest in this *Holtenia*, is its singular resemblance to the *Ventriculites* of the Cretaceous formation (§ 699). Subsequent investigations have shown that it is very widely diffused, and that it is only one of several Deep-sea forms, including several of singularly beautiful structure, which are the existing representatives of the old Ventriculite type. One of these was previously known, from being occasionally cast up on the shores of Barbadoes after a storm. This *Dictyocalyx pumiceous* has the shape of a mushroom, the diameter of its disk sometimes ranging to a foot. A small portion of its reticulated skeleton is a singularly beautiful object, when viewed with incident light under a low magnifying power.

512. With the exception of the genus *Spongilla*, all known Sponges are marine; but they differ very much in habit of growth. For whilst some can only be obtained by dredging at considerable depth, others live near the surface, whilst others attach themselves to the surfaces of rocks, shells, etc., between the tide-marks. The various species of *Grantia*, in which, of all the marine Sponges, the flagellate zooids can most readily be observed, belong to this last category. They have a peculiarly simple structure, each being a sort of bag whose wall is so thin that no system of canals is required; the water absorbed by the outer surface passing directly towards the inner, and being expelled by the mouth of the bag. The flagella may be plainly distinguished with a 1-8th inch objective on some of the cells of the gelatinous substance scraped from the interior of the bag; or they may be seen *in situ*, by making very thin transverse

---

[1] See his elaborate Memoir in "Philos. Transact.," 1870; and his "Depths of the Sea" (1872), p. 71.

sections of the substance of the sponge. It is by such sections alone that the internal structure of Sponges, and the relation of their spicular and horny skeletons to their fleshy substance, can be demonstated. They are best made by the imbedding process (§§ 189, 190).—In order to obtain the *spicules* in an isolated condition, the animal matter must be got rid of, either by incineration, or by chemical reagents. The latter method is preferable, as it is difficult to free the mineral residue from carbonaceous particles by heat alone. If (as is commonly the case) the spicules are *siliceous*, the Sponge may be treated with strong nitric or nitromuriatic acid, until its animal substance is dissolved away; if, on the other hand, they be *calcareous*, a strong solution of potass may be employed instead of the acid. The operation is more rapidly accomplished by the aid of heat; but if the saving of time be not of importance, it is preferable on several accounts to dispense with it. The spicules, when obtained in a separate state, should be mounted in Canada balsam.— Sponge-tissue may often be distinctly recognized in sections of Agate, Chalcedony, and other siliceous concretions, as will be more fully stated hereafter (§ 699).

ZOOPHYTES.

513. **Under** the general designation *Zoophytes* it will be still convenient to group those animals which form composite skeletons or 'polyparies' of a more or less plant-like character; associating with them the *Acalephs*, which are now known to be the 'sexual zooids,' of Polypes (§ 518); but excluding the *Polyzoa* (Chap. XV.) on account of their truly Molluscoid structure, notwithstanding their Zoophytic forms and habits of life. The animals belonging to this group may be considered as formed upon the primitive *gastrula* type (§ 391): their gastric cavity (though sometimes extending itself almost indefinitely) being lined by the original *endoderm*, and their surface being covered by the original *ectoderm;* and these two lamellæ not being separated by the interposition of any body-cavity or *cœlom*. It is a fact of great interest, that although the product of the development of a *morula* is here a distinctly individualized Polype, in which several mutually dependent parts make up a single organic whole, yet that these parts still retain much of their independent Protozoic life; which is manifested in two very remarkable modes. In the first place, the digestive sac is observed to be lined by a layer of amœboid cells, which send out pseudopodial prolongations into its cavity, by whose agency (it may be pretty certainly affirmed) the nutrient material is first introduced into the body-substance. This was first noticed by Prof. Allman in the beautiful Hydroid polype *Myriothela;*[1] the like has been since shown by Mr. Jeffery Parker to be true of the ordinary *Hydra;*[2] and Prof. E. Ray Lankester has made the same observation upon the curious little Medusa lately found in a *fresh-water* tank.[3] (It may be mentioned in this connection, that Metschnikoff has seen the cells which line the alimentary canal of the lower Planarian worms gorging themselves with colored food-particles, exactly in the manner of *Amœbœ*.)—The second 'survival' of Protozoic independence is shown in the extraordinary power possessed by *Hydra*, *Actinia*, etc., to reproduce the entire organism from a mere fragment

---

[1] "Philos. Transact.," 1875, p. 552.
[2] "Proceed. of Roy. Soc.," Vol. xxx. (1880), p. 61.
[3] "Quart. Journ. Microsc. Sci.," N.S., Vol. xx. (1880), p. 371.

(§ 515).—This great division includes the two principal groups, the HYDROZOA and the ACTINOZOA; the former comprehending the *Polypes*, and the latter the *Anemonies*. In the Hydrozoa there is no separation between the digestive cavity and the external body-wall; and the reproductive organs are external. In the Actinozoa the wall of the digestive sac is separated from the external body-wall by an intervening space, which communicates with it, and must be regarded as an extension of it; and this is subdivided into chambers by a series of vertical partitions, to which the reproductive organs are attached.—As most of the Hydrozoa or Hydroid Polypes are essentially Microscopic animals, they need to be described with some minuteness; whilst in regard to the Actinozoa those points only will be dwelt-on, which are of special interest to the Microscopist.

514. HYDROZOA.—The type of this group is the *Hydra* or fresh-water polype, a very common inhabitant of pools and ditches, where it is most commonly to be found attached to the leaves or stems of aquatic plants, floating pieces of stick, etc. Two species are common in this country, the *H. viridis* or green Polype, and the *H. vulgaris*, which is usually orange-brown, but sometimes yellowish or red (its color being liable to some variation according to the nature of the food on which it has been subsisting); a third less common species, the *H. fusca*, is distinguished from both the preceding by the length of its tentacles, which in the former are scarcely as long as the body, whilst in the latter they are, when fully extended, many times longer (Fig. 354). The body of the Hydra consists of a simple bag or sac, which may be regarded as a stomach, and is capable of varying its shape and dimensions in a very remarkable degree; sometimes extending itself in a straight line so as to form a long narrow cylinder, at other times being seen (when empty) as a minute contracted globe, whilst, if distended with food, it may present the form of an inverted flask or bottle, or even of a button. At the upper end of this sac is a central opening, the mouth; and this is surrounded by a circle of tentacles or 'arms,' usually from six to ten in number, which are arranged with great regularity around the orifice. The body is prolonged at its lower end into a narrow base, which is furnished with a suctorial disk; and the Hydra usually attaches itself by this while it allows its tendril-like tentacles to float freely in the water. The wall of the body is composed of cells imbedded in sarcode-substance; and between its two layers there is a space chiefly occupied by undifferentiated sarcode, having many 'vacuoles' or 'lacunæ' (which often seem to communicate with one another) excavated in its substance. The arms are made-up of the same materials as the body; but their surface is beset with little wart-like prominences, which, when carefully examined, are found to be composed of clusters of 'thread-cells,' having a single large cell with a long spiculum in the centre of each. The structure of these thread-cells or 'urticating organs' will be described hereafter (§ 528); at present it will be enough to point-out that this apparatus, repeated many times on each tentacle, is doubtless intended to give to the organ a great prehensile power; the minute filaments forming a rough surface adapted to prevent the object from readily slipping out of the grasp of the arm, whilst the central spicule or 'dart' is projected into its substance, probably conveying into it a poisonous fluid secreted by a vesicle at its base. The latter inference is founded upon the oft-repeated observation, that if the living prey seized by the tentacles have a body destitute of hard integument, as is the case with the minute aquatic Worms which consti-

tute a large part of its aliment, this speedily dies, even though, instead of being swallowed, it escapes from their grasp; whilst, on the other hand, minute Entomostraca, Insects, and other animals with hard envelopes, may escape without injury, even after having been detained for some time in the polype's embrace. The contractility of the tentacles (the interior of which is traversed by a canal that communicates with the cavity of the stomach) is very remarkable, especially in the *Hydra fusca;* whose arms, when extended in search of prey, are not less than seven or eight inches in length; whilst they are sometimes so contracted, when the stomach is filled with food, as to appear only like little tubercles

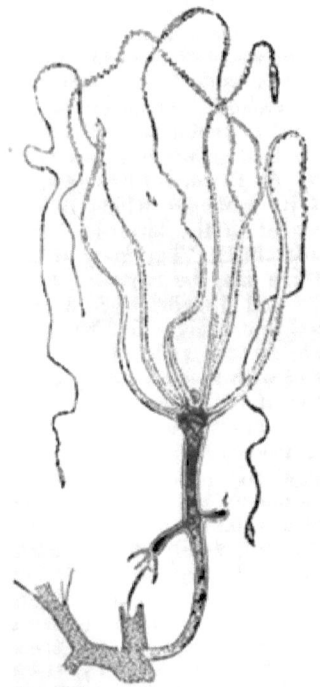

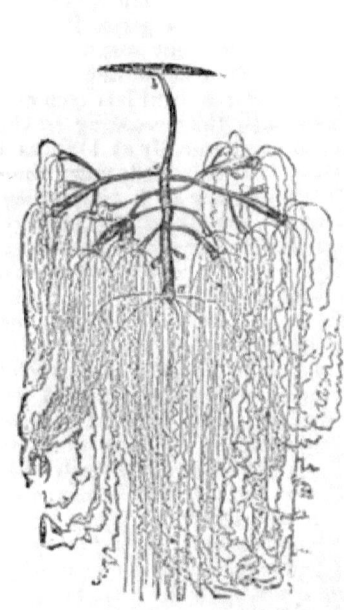

Hydra fusca, with a young bud at *b,* and a more advanced bud at *c.*

Hydra fusca in gemmation; *a,* mouth; *b,* base; *c,* origin of one of the buds.

around its entrance. By means of these instruments the Hydra is enabled to draw its support from animals whose activity, as compared with its own slight powers of locomotion, might have been supposed to remove them altogether from its reach; for when, in its movements through the water, a minute Worm or a Water-flea happens to touch one of the tentacles of the Polype, spread-out as these are in readiness for prey, it is immediately seized by this, other arms are soon coiled around it, and the unfortunate victim is speedily conveyed to the stomach, within which it may frequently be seen to continue moving for some little time. Soon, however, its struggles cease, and its outline is obscured by a turbid film, which gradually thickens, so that at last its form is wholly lost. The soft parts are soon completely dissolved, and the harder indigestible portions are rejected through the mouth. A second orifice has been observed at the lower extremity of the stomach; but this would not seem to be properly regarded as anal, since it is not used for the discharge of such exuviæ; it is probably rather to be considered as representing, in the Hydra, the entrance to that ramifying cavity, which, in the *Compound Hydrozoa*, brings into mutual connection the lower extremities of the stomachs of all the individual polypes (Plate xx.).

515. The ordinary mode of reproduction in this animal is by a 'gemmation' resembling that of Plants. Little bud-like processes (Fig. 354, *b, c*) developed from its external surface gradually come to resemble the parent in character, and to possess a digestive sac, mouth, and tentacles; for a long time, however, their cavity is connected with that of the parent, but at last the communication is cut-off by the closure of the canal of the foot-stalk, and the young polype quits its attachment and goes in quest of its own maintenance. A second generation of buds is sometimes observed on the young polype before quitting its parent; and as many as *nineteen* young *Hydræ* in different stages of development have been seen connected with a single original stock (Fig. 355). This process takes place most rapidly under the influence of warmth and abundant food; it is usually suspended in winter, but may be made to continue by keeping the polypes in a warm situation and well supplied with food. Another very curious endowment seems to depend on the same condition—the extraordinary power which one portion possesses of reproducing the rest. Into whatever number of parts a *Hydra* may be divided, each may retain its vitality, and give origin to a new and entire fabric; so that *thirty* or *forty* individuals may be formed by the section of one.—The *Hydra* also propagates itself, however, by a truly sexual process; the fecundating apparatus, or vesicle producing 'sperm-cells,' and the ovum (containing the 'germ-cell,' imbedded in a store of nutriment adapted for its early development) being both evolved in the substance of the walls of the stomach —the male apparatus forming a conical projection just beneath the arms, while the female ovary, or portion of the body-substance in which the ovum is generated, has the form of a knob protruding from the middle of its length. It would appear that sometimes one individual Hydra develops only the male cysts or sperm-cells, while another develops only the female cysts or ovisacs; but the general rule seems to be that the same individual forms both organs. The fertilization of the ova, however, cannot take-place until after the rupture of the spermatic cyst and of the ovisac, by which the contents of both are set entirely free from the body of the parent.—The autumn is the chief time for the development of the sexual organs; but they also present themselves in the earlier part of the year, chiefly between April and July. According to Ecker, the eggs of *H. viridis* produced early in the season, run their course in the summer of the same year; while those produced in the autumn, pass the winter without change. When the ovum is nearly ripe for fecundation, the ovary bursts its ectodermal covering, and remains attached by a kind of pedicle. It seems to be at this stage that the act of fecundation occurs; a very strong elastic shell or capsule then forms round the ovum, the surface of which is in some cases studded with spine-like points, in others tuberculated, the divisions between the tubercles being polygonal. The ovum finally drops from its pedicle, and attaches itself by means of a mucous secretion, till the hatching of the young Hydra, which comes

forth provided with four rudimentary tentacles like buds.—The Hydra possesses the power of free locomotion, being able to remove from the spot to which it has attached itself, to any other that may be more suitable to its wants; its changes of place, however, seem rather to be performed under the influence of *light*, towards which the Hydra seeks to move itself, than with reference to the search after food.[1]

516. The *Compound Hydroids* may be likened to a *Hydra* whose gemmæ, instead of becoming detached, remain permanently connected with the parent; and as these in their turn may develop gemmæ from their own bodies, a structure of more or less arborescent character, termed a *polypary*, may be produced. The form which this will present, and the relation of the component polypes to each other, will depend upon the mode in which the gemmation takes-place: in all instances, however, the entire cluster is produced by continuous growth from a single individual; and the stomachs of the several polypes are united by tubes, which proceed from the base of each, along the stalk and branches, to communicate with the cavity of the central stem. Whatever may be the form taken by the stem and branches constituting the polypary of a Hydroid colony, they will be found to be, or to contain, fleshy tubes having two distinct layers; the inner (endoderm) having nutritive functions; the outer (ectoderm) usually secreting a hard cortical layer, and thus giving rise to fabrics of various forms. Between these a muscular coat is sometimes noticed. The fleshy tube, whether single or compound, is called a *cœnosarc;* and through it the nutrient matter circulates. The 'zooids,' or individual members of the colony, are of two kinds: one, the *polypite*, or *alimentary* zooid, resembling the Hydra in essential structure, and more or less in aspect; the other, *gonozooid*, or *sexual* zooid, developed at certain seasons only, in buds of particular shape.

517. The simplest division of the Hydroida is that adopted by Mr. Hincks,[2] who groups them under the sub-order *Athecata* and *Thecata*, the latter being again divided into the *Thecaphora* and the *Gymnochroa*. In the first, neither the 'polypites' nor the sexual zooids bear true protective cases; in the second, the polypites are lodged in cells, or, as Mr. Hincks prefers to call them, *calycles*, many of which resemble exquisitely formed crystal cups, variously ornamented, and sometimes furnished with lids or opercula; in the third, which contains the Hydras, there is no polypary, and the reproductive zooids (gonozooids) are always fixed and developed in the body-walls. According to Mr. Hincks, the two sexes are sometimes borne on the same colony, but more commonly the zoophyte is diœcious. The cases, however, are much less rare than has been supposed, in which both male and female are mingled on the same shoots. The sexual zooids either remain attached, and discharge their contents at maturity, or become free and enter upon an independent existence. The free forms nearly always take the shape of *Medusæ* (jelly fish), swimming by rhythmical contractions of their bell or umbrella. The digestive cavity is in the handle (manubrium) of the bell; and the generative elements (sperm-cells or ova) are developed either between the membranes of the manubrium, or in special sacs in the canals radiating

---

[1] A very full account of the structure and development of *Hydra* has recently been published by Kleinenberg; of whose admirable Monograph a summary is given by Prof. Allman, with valuable remarks of his own, in "Quart. Journ. Microsc. Sci.," N.S., Vol. xiv. (1874), p. 1. See also the important Paper by Mr. Jeffery Parker already cited.

[2] "History of British Hydroid Zoophytes," 1868.

from it. The ova, when fertilized by the spermatozoa, undergo 'segmentation' according to the ordinary type (§ 581), the whole yolk-mass subdividing successively into 2, 4, 8, 16, 32 or more parts, until a 'mulberry mass' is formed; this then begins to elongate itself, its surface being at first smooth, and showing a transparent margin, but afterwards becoming clothed with cilia, by whose agency these little *planulæ*, closely resembling ciliated Infusoria, first move about within the capsule, and then swim forth freely when liberated by the opening of its mouth. At this period the embryo can be made out to consist of an outer and an inner layer of cells, with a hollow interior; after some little time the cilia disappear, and one extremity becomes expanded into a kind of disk by which it attaches itself to some fixed object; a mouth is formed, and tentacles sprout forth around it; and the body increases in length and thickness, so as gradually to acquire the likeness of one of the parent polypes, after which the 'polypary' characteristic of the genus is gradually evolved by the successive development of polype-buds from the first-formed polype and its subsequent offsets.—The Medusæ of these polypes (Fig. 358) belong to the division called 'naked-eye,' on account of the (supposed) eye-spots usually seen surrounding the margin of the bell at the base of the tentacles.

518. A characteristic example of this production of Medusa-like 'gonozooids' is presented by the form termed *Syncoryne Sarsii* (Fig. 356) belonging to the sub-order *Athecata*. At A is shown the alimentary zooid, or polypite, with its tentacles, and at B the successive stages *a*, *b*, *c*, of the sexual zooids, or medusa-buds. When sufficiently developed, the medusa swims away, and as it grows to maturity enlarges its manubrium, so that it hangs below the bell. The Medusæ of the genus *Syncoryne* (as now restricted) have the form named *Sarsia* in honor of the Swedish naturalist Sars. Their normal character is that of free swimmers; but Agassiz ascertained that in some cases, towards the end of the breeding season, the sexual zooids remain fixed, and mature their products while attached to the zoophyte.' This condition of the sexual zooids is very common amongst the Hydroida; and various intermediate stages may be traced in different genera, between the mode in which the gonozooids are produced in the common Hydra, as already described, and that of Syncoryne. In *Tubularia* the gonozooids, though permanently attached, are furnished with swimming bells, having four tubercles representing marginal tentacles. A common and interesting species *Tubularia indivisa* receives its specific name from the infrequency with which branches are given off from the stem, these for the most part standing erect and parallel, like the stalks of corn, upon the base to which they are attached. This beautiful Zoophyte, which sometimes grows between the tide-marks, but is more abundantly obtained by dredging in deep water, often attains a size which renders it scarcely a microscopic object; its stems being sometimes no less than a foot in height and a line in diameter. Several curious phenomena, however, are brought into view by Microscopic examination. The Polype-stomach is connected with the cavity of the stem by a circular opening, which is surrounded by a sphincter; and an alternate movement of dilatation and contraction takes place in it, fluid being apparently forced up from below, and then expelled again, after which the sphincter closes in preparation for a recurrence of the operation; this, as observed by Mr. Lister, being repeated at intervals of

---

[1] Hincks, *op. cit.*, p. 49.

PLATE XX.

CAMPANULARIA GELATINOSA (after Van Beneden).

A, Upper part of the stem and branches, of the natural size.
B, Small portion enlarged, showing the structure of the animal; *a*, terminal branch bearing polypes; *b*, polype-bud partially developed; " horny cell containing the expanded polype *d; e*, ovarian capsule, containing medusiform gemmæ in various stages of development; *f*, fleshy substance extending through the stem and branches, and connecting the different polype-cells and ovarian capsules; *g*, annular constrictions at the base of the branches.

eighty seconds. Besides the foregoing movement, a regular flow of fluid carrying with it solid particles of various sizes, may be observed along the whole length of the stem, passing in a somewhat spiral direction.—It is worthy of mention here, that when a Tubularia is kept in confinement, the polype-heads almost always drop off after a few days, but are soon renewed again by a new growth from the stem beneath; and this exuviation and regeneration may take place many times in the same individual.[1]

519. It is in the Families *Campanularida* and *Sertularida* (whose polyparies are commonly known as 'corallines'), that the horny branch-

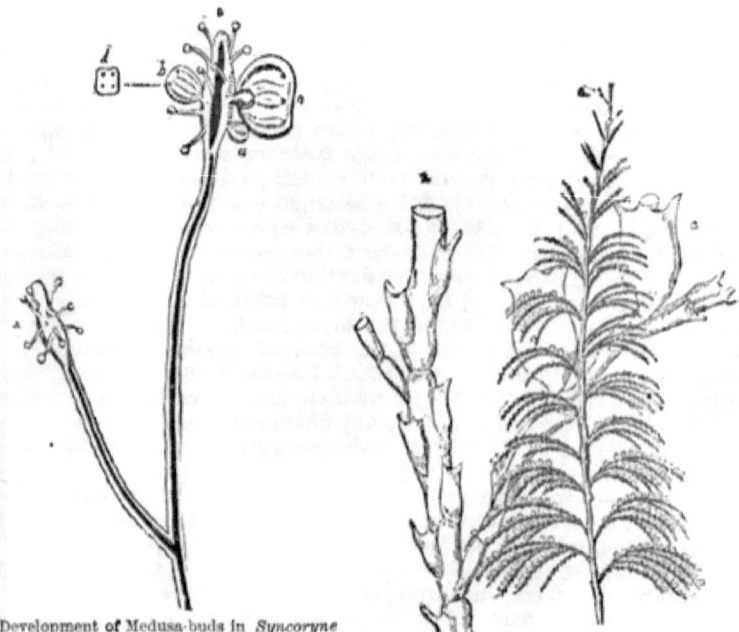

Fig. 356.

Development of Medusa-buds in *Syncoryne Sarsii*:—A, an ordinary polype, with its club-shaped body covered with tentacles:—B, a polype putting forth Medusoid gemmæ; a, a very young bud; b, a bud more advanced, the quadrangular form of which, with the four nuclei whence the cirrhi afterwards spring, is shown at d; c, a bud still more advanced.

Fig. 357.

Sertularia cupressina: A, natural size; B, portion magnified.

ing fabric attains its completest development; not only affording an investment to the stem, but forming cups or cells for the protection of the polypites, as well as capsules for the reproductive gonozooids. Both these families thus belong to the Sub-order *Thecata*. In the *Campanularida* the polype-cells are campanulate or bell-shaped, and are borne at the extremities of ringed stalks (Plate xx., c); in the *Sertularida*, on the other hand, the polype-cells lie along the stem and branches, at-

---

[1] The British *Tubularida* form the subject of a most complete and beautiful Monograph by Prof. Allman, published by the Ray Society.

tached either to one side only, or to both sides (Fig. 357). In both, the general structure of the individual polypes (Plate xx., *d*) closely corresponds with that of the *Hydra;* and the mode in which they obtain their food is essentially the same. Of the products of digestion, however, a portion finds its way down into the tubular stem, for the nourishment of the general fabric; and very much the same kind of circulatory movement can be seen in *Campanularia* as in Tubularia, the circulation being most vigorous in the neighborhood of growing parts. It is from the 'cœnosarc' (*f*) contained in the stem and branches, that new polype-buds (*b*) are evolved; these carry before them (so to speak) a portion of the horny integument, which at first completely invests the bud; but as the latter acquires the organization of a polype, the case thins away at its most prominent part, and an opening is formed through which the young polype protrudes itself.

520. The origin of the reproductive capsules or 'gonothecæ' (*e*) is exactly similar; but their destination is very different. Within them are evolved, by a budding process, the generative organs of the Zoophyte; and these in the *Campanularida* may either develop themselves into the form of independent Medusoids, which completely detach themselves from the stock that bore them, make their way out of the capsule, and swim-forth freely, to mature their sexual products (some developing sperm-cells, and others ova), and give origin to a new generation of polypes; or, in cases in which the Medusoid structure is less distinctly pronounced, may not completely detach themselves, but (like the flower-buds of a Plant) expand one after another at the mouth of the capsule, withering and dropping-off after they have matured their generative products. In the *Sertularida*, on the other hand, the Medusan conformation is wanting, as the gonozooids are always fixed; the reproductive cells (Fig. 357, *a*), which were shown by Prof. Edward Forbes to be really metamorphosed branches, developing in their interior certain bodies which were formerly supposed to be ova, but which are now known to be 'medusoids' reduced to their most rudimentary condition. Within these are developed,—in separate gonothecæ, sometimes perhaps on distinct polyparies,—spermatozoa and ova; and the latter are fertilized by the entrance of the former whilst still contained within their capsules. The fertilized ova, whether produced in free or in attached medusoids, develop themselves in the first instance into ciliated 'gemmules,' which soon evolve themselves into true polypes, from every one of which a new composite polypary may spring.

521. There are few parts of our coast which will not supply some or other of the beautiful and interesting forms of Zoophytic life which have been thus briefly noticed, without any more trouble in searching for them than that of examining the surfaces of rocks, stones, sea-weeds, and dead shells between the tide-marks. Many of them habitually live in that situation; and others are frequently cast-up by the waves from the deeper waters, especially after a storm. Many kinds, however, can only be obtained by means of the dredge. For observing them during their living state, no means is so convenient as the Zoophyte-trough (§ 124).—In mounting Compound Hydrozoa, as well as Polyzoa, it will be found of great advantage to place the specimens alive in the cells they are permanently to occupy, and to then add Osmic acid drop by drop to the sea-water; this has the effect of causing the protrusion of the animals, and of rendering their tentacles rigid. The liquid may be withdrawn, and replaced by Goadby's solution, Dean's Gelatine, Glycerine

jelly, weak Spirit, diluted Glycerine, a mixture of Spirit and Glycerine with Sea-water or any other menstruum, by means of the Syringe; and it is well to mount specimens in several different menstrua, marking the nature and strength of each, as some forms are better preserved by one and some by another.' The size of the cell must of course be proportioned to that of the object; and if it be desired to mount such a specimen as may serve for a characteristic illustration of the mode of growth of the species it represents, the large shallow cells, whose walls are made by cementing four strips of glass to the plate that forms the bottom (§ 174), will generally be found preferable.—The horny polyparies of the *Sertularida*, when mounted in Canada balsam, are beautiful objects for the Polariscope; but in order to prepare them successfully, some nicety of management is required. The following are the outlines of the method recommended by Dr. Golding Bird, who very successfully practised it:—The specimens selected, which should not exceed two inches in length, are first to be submitted, while immersed in water of 120°, to the vacuum of an air-pump. The ebullition which will take-place within the cavities, will have the effect of freeing the polyparies from dead polypes and other animal matter; and this cleansing process should be repeated several times. The specimens are then to be dried, by first draining them for a few seconds on bibulous paper, and then by submitting them to the vacuum of an air-pump, within a thick earthenware ointment-pot fitted with a cover, which has been previously heated to about 200°; by this means the specimens are very quickly and completely dried, the water being evaporated so quickly that the cells and tubes hardly collapse or wrinkle. The specimens are then placed in camphine, and again subjected to the exhausting process, for the displacement of the air by that liquid; and when they have been thoroughly saturated, they should be mounted in Canada balsam in the usual mode. When thus prepared, they become very beautiful transparent objects for low magnifying powers; and they present a gorgeous display of colors when examined by Polarized light, with the interposition of a plate Selenite, the effect being much enhanced by the use of Black-ground illumination.

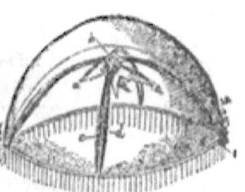

FIG. 358.

522. No result of Microscopic research was more unexpected than the discovery of the close relationship subsisting between the Hydroid *Zoophytes* and the Medusoid *Acalephæ* (or 'jelly-fish'). We now know that the small free-swimming Medusoids belonging to the 'naked-eyed' group, of which *Thaumantias* (Fig. 358) may be taken as a representative, are really to be considered as the detached sexual apparatus of the Zoophytes from which they have been budded-off, endowed with independent organs of nutrition and locomotion, whereby they become capable of maintaining their own existence and of developing their sexual products. The general conformation of these organs will be understood from the accompanying figure. Many of this group are very beautiful objects for Microscopic examination, being small enough to be viewed

Thaumantias pilosella one of the 'naked-eyed' Medusæ;—*a a*, oral tentacles; *b*, stomach; *c*, gastro-vascular canals, having the ovaries, *d d*, on either side, and terminating in the marginal canal, *e e*.

---

[1] See Mr. J. W. Morris in "Quart. Journ. of Microsc. Science," N.S., Vol. ii. (1862), p. 116.

entire in the Zoophyte-trough. There are few parts of the coast on which they may not be found, especially on a calm warm day, by skimming the surface of the sea with the Tow-net (§ 217); and they are capable of being stained and preserved in cells, after being hardened by osmic acid.

523. The history of the large and highly-developed *Medusæ* or ACALEPHÆ which are commonly known as 'jelly-fish,' is essentially similar; for their progeny have been ascertained to develop themselves in the first instance under the Polype-form, and to lead a life which in all essential respects is zoophytic; their development into Medusæ taking place only in the closing phase of their existence, and then rather by gemmation from the original polype, than by a metamorphosis of its own fabric. The huge *Rhizostoma* found commonly swimming round our coasts, and the beautiful *Chrysaora* remarkable for its long 'furbelows' which act as organs of prehension, are Oceanic Acalephs developed from very small polypites, which fix themselves by a basal cup or disk. The embryo emerges from the cavity of its parent, within which the first stages of its development have taken place, in the condition of a ciliated 'gemmule,' of rather oblong form, very closely resembling an Infusory Animalcule, but destitute of a mouth. One end soon contracts and attaches itself, however, so as to form a foot; the other enlarges and opens to form a mouth, four tubercles sprouting around it, which grow into tentacles; whilst the central cells melt-down to form the cavity of the stomach. Thus a Hydra-like polype is formed, which soon acquires many additional tentacles; and this, according to the observations of Sir J. G. Dalyell on the *Hydra tuba*, which is the polype-stage of the *Chrysaora*, leads in every important particular the life of a Hydra; propagates like it by repeated gemmation, so that whole colonies are formed as offsets from a single stock; and can be multiplied like it by artificial division, each segment developing itself into a perfect Hydra. There seems to be no definite limit to its continuance in this state, or to its power of giving origin to new polype-buds; but when the time comes for the development of its sexual gonozooids, the polype quits its original condition of a minute bell with slender tentacles (Fig. 359, c, a), assumes a cylindrical form, and elongates itself considerably; a constriction or indentation is then seen around it, just below the ring which encircles the mouth and gives origin to the tentacles; and similar constrictions are soon repeated round the lower parts of the cylinder, so as to give to the whole body somewhat the appearance of a rouleau of coins (Fig. 359, A); a sort of fleshy bulb, *a*,

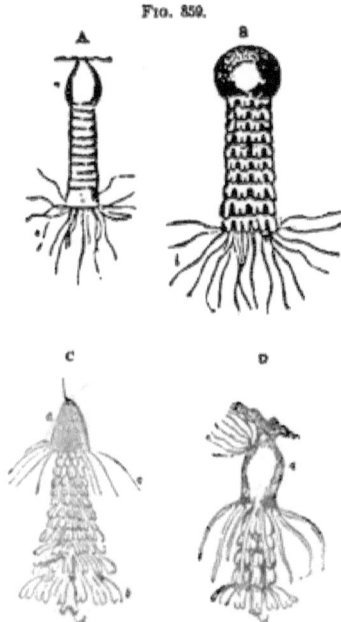

FIG. 859.

Successive stages A, B, C, D, of development of *Chrysaora*:—*a*, elongated and constricted Polype-body; *b*, its original circle of tentacles; *c*, its secondary circle of tentacles; *d*, proboscis of most advanced Medusa-disk; *e*, polype-bud from side of polype-body.

somewhat of the form of the original polype, being still left at the attached extremity. The number of circles is indefinite, and all are not formed at once, new constrictions appearing below, after the upper portions have been detached; as many as 30 or even 40 have thus been produced in one specimen. The constrictions then gradually deepen, so as to divide the cylinder into a pile of saucer-like bodies; the division being most complete above, and the upper disks usually presenting some increase in diameter; and whilst this is taking place, the edges of the disks become divided into lobes (B), each lobe soon presenting the cleft with the supposed rudimentary eye at the bottom of it, which is to be plainly seen in the detached Medusæ (Fig. 360, c). Up to this period, the tentacles of the original polype surmount the highest of the disks; but before the detachment of the topmost disk, this circle disappears, and a new one is developed at the summit of the bulb which remains at the base of the pile (c, c). At last the topmost and largest disk begins to exhibit a sort of convulsive struggle; it becomes detached, and swims freely away; and the same series of changes takes-place from above downwards, until the whole pile of disks is detached and converted into free-swimming Medusæ. But the original polypoid body still remains, and may return to its original polype-like mode of gemmation (D, e); becoming the progenitor of a new colony, every member of which may in its turn bud-off a pile of Medusa-disks.

Fig. 360.

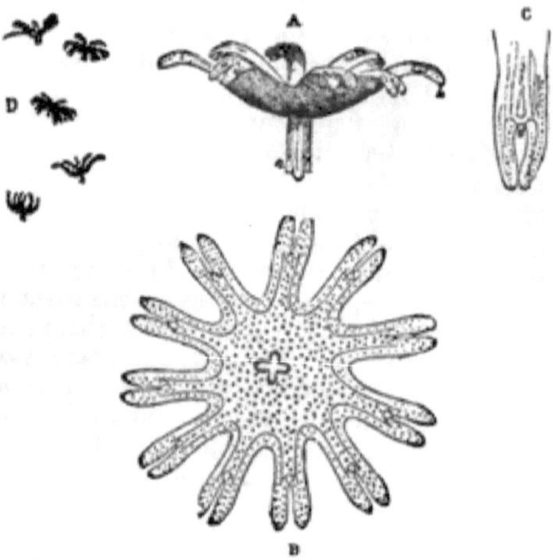

Development of *Chrysaora* from *Hydra tuba*.—A, detached individual viewed sideways, and enlarged, showing the proboscis a, and b the bifid lobes; B, individual seen from above, showing the bifid lobes of the margin, and the quadrilateral mouth; c, one of the bifid lobes still more enlarged, showing the rudimentary eye (?) at the bottom of the cleft; D, group of young Medusæ, as seen swimming in the water, of the natural size.

524. The bodies thus detached have all the essential characters of the adult *Medusæ*. Each consists of an umbrella-like disk, divided at its edge into a variable number of lobes, usually eight; and of a stomach, which occupies a considerable proportion of the disk, and projects downwards in the form of a proboscis, in the centre of which is the quadrangular mouth (Fig. 360, A, B). As the animal advances towards maturity, the intervals between the segments of the border of the disk gradually fill-up, so that the divisions are obliterated; tubular prolongations of the stomach extend themselves over the disk; and from its border there sprout forth tendril-like filaments, which hang down like a fringe around its margin. From the four angles of the mouth, which, even in the youngest detached animal, admits of being greatly extended and protruded, prolongations are put forth, which form the four large tentacles of the adult. The young Medusæ are very voracious, and grow rapidly, so as to attain a very large size. The *Cyaneæ* and *Chrysaoræ*, which are

common all round our coasts, often have a diameter of from 6 to 15 inches; while the *Rhizostoma* sometimes reaches a diameter of from two to three feet. The quantity of solid matter, however, which their fabrics contain is extremely small. It is not until adult age has been attained, that the generative organs make their appearance, in four chambers disposed around the stomach, which are occupied by plaited membranous ribands containing sperm-cells in the male and ova in the female, and the embryoes evolved from the latter, when they have been fertilized by the agency of the former, repeat the extraordinary cycle of phenomena which has been now described, developing themselves in the first instance into Hydroid Polypes, from which Medusoids are subsequently budded-off.

525. This cycle of phenomena is one of those to which the term 'alternation of generations' was applied by Steenstrup,[1] who brought together under this designation a number of cases in which generation A does not produce a form resembling itself, but a different form, B; whilst generation B gives origin to a form which does not resemble itself, but returns to the form A, from which B itself sprang. It was early pointed out, however, by the Author,[2] that the term 'alternation of generations' does not appropriately represent the facts either of this case, or of any of the other cases grouped under the same category: the real fact being that the two organisms, A and B, constitute two stages in the life-history of *one generation;* and the production of one form from the other being in only one instance by a truly *generative* or sexual act, whilst in the other it is by a process of *gemmation* or budding. Thus the *Medusæ* of both orders (the 'naked-eyed' and the 'covered-eyed' of Forbes) are detached flower-buds, so to speak, of the Hydroid Zoophytes which bud them off; the Zoophytic phase of life being the most conspicuous in such *Thecata* as *Campanularida* and *Sertularida*, whose Medusa-buds are of small size and simple conformation, and not unfrequently do not detach themselves as independent organisms; whilst the Medusan phase of life is the most conspicuous in the ordinary Acalephs, their Zoophytic stage being passed in such obscurity as only to be detected by careful research.—The Author's views on this subject, which were at first strongly contested by Prof. E. Forbes, and other eminent Zoologists, have now come to be generally adopted.

526. ACTINOZOA.—Of this group, the common Sea-Anemonies may be taken as types; constituting, with their allies, the order *Zoantharia*, or Helianthoid polypes, which have numerous tentacles disposed in several rows. Next to them come the *Alcyonaria*, consisting of those whose polypes, having only six or eight broad short tentacles, present a star-like aspect when expanded; as is the case with various composite Sponge-like bodies, unpossessed of any hard skeleton, which inhabit our own shores, and also with the Red Coral and the *Tubipora* of warmer seas, which have a stony skeleton that is internal in the first case and external in the second, as also with the Sea-pens, and the *Gorgoniæ* or Sea-fans. A third order, *Rugosa*, consists of fossil Corals, whose stony polyparies are intermediate in character between those of the two preceding. And lastly, the *Ctenophora*, free swimming gelatinous animals, many of which are beautiful objects for the Microscope, are by most Zoologists ranked with the Actinozoa.

---

[1] See his Treatise on "The Alternation of Generations," published by the Ray Society.
[2] "Brit. and For. Med.-Chir. Review," Vol. i. (1848), p. 192, *et seq.*

527. Of the *Zoantharia*, the common *Actinia* or 'sea anemone' may be taken as the type; the individual polypites of all the composite fabrics included in the group being constructed upon the same model. In by far the larger proportion of these Zoophytes, the bases of the polypites, as well as the soft flesh that connects together the members of aggregate masses, are consolidated by calcareous deposit into stony Corals; and the surfaces of these are beset with 'cells,' usually of a nearly circular form, each having numerous vertical plates or *lamellæ* radiating from its centre towards its circumference, which are formed by the consolidation of the lower portions of the radiating partitions that divide the space intervening between the stomach and the general integument of the animal into separate chambers. This arrangement is seen on a large scale in the *Fungia* or 'mushroom-coral' of tropical seas, which is the stony base of a solitary Anemone-like animal; on a far smaller scale, it is seen in the little *Caryophyllia*, a like solitary Anemone of our own coasts, which is scarcely distinguishable from an Actinia by any other character than the presence of this disk, and also on the surface of many of those stony corals known as 'madrepores;' whilst in some of these the individual polype-cells are so small, that the lamellated arrangement can only be made-out when they are considerably magnified. Portions of the surface of such Corals, or sections taken at a small depth, are very beautiful objects for low powers, the former being viewed by reflected, and the latter by transmitted light. And thin sections of various fossil Corals of this group are very striking objects for the lower powers of the Oxyhydrogen Microscope.

528. The chief point of interest to the Microscopist, however, in the structure of these animals, lies in the extraordinary abundance and high development of those 'filiferous capsules,' or 'thread-cells,' the presence of which on the tentacles of the Hydroid polypes has been already noticed (§ 514), and which are also to be found, sometimes sparingly, sometimes very abundantly, in the tentacles surrounding the mouth of the Medusæ, as well as on other parts of their bodies. If a tentacle of any of the Sea-anemonies so abundant on our coasts (the smaller and more transparent kinds being selected in preference) be cut-off, and be subjected to gentle pressure between the two glasses of the Aquatic-box or the Compressorium, multitudes of little dart-like organs will be seen to project themselves from its surface near its tip; and if the pressure be gradually augmented, many additional darts will every moment come into view. Not only do these organs present different forms in different species, but even in one and the same individual very strongly marked diversities are shown, of which a few examples are given in Fig. 361. At A, B, C, D, is shown the appearance of the 'filiferous capsules,' whilst as yet the thread lies coiled-up in their interior; and at E, F, G, H, are seen a few of the most striking forms which they exhibit when the thread or dart has started-forth. These thread-cells are found not merely in the tentacles and other parts of the external integument of Actinozoa, but also in the long filaments which lie in coils within the chambers that surround the stomach, in contact with the sexual organs which are attached to the lamellæ dividing the chambers. The latter sometimes contain 'sperm-cells' and sometimes ova, the two sexes being here divided, not united in the same individual.—What can be the office of the filiferous filaments thus contained in the interior of the body, it is difficult to guess-at. They are often found to protrude from rents in the external tegument, when any violence has been used in detaching the

animal from its base; and when there is no external rupture, they are often forced through the wall of the stomach into its cavity, and may be seen hanging out of the mouth. The largest of these capsules, in their unprotected state, are about 1-300th of an inch in length; while the thread or dart, in *Corynactis Allmanni*, when fully extended, is not less than 1-8th of an inch, or thirty-seven times the length of its capsule.[1]

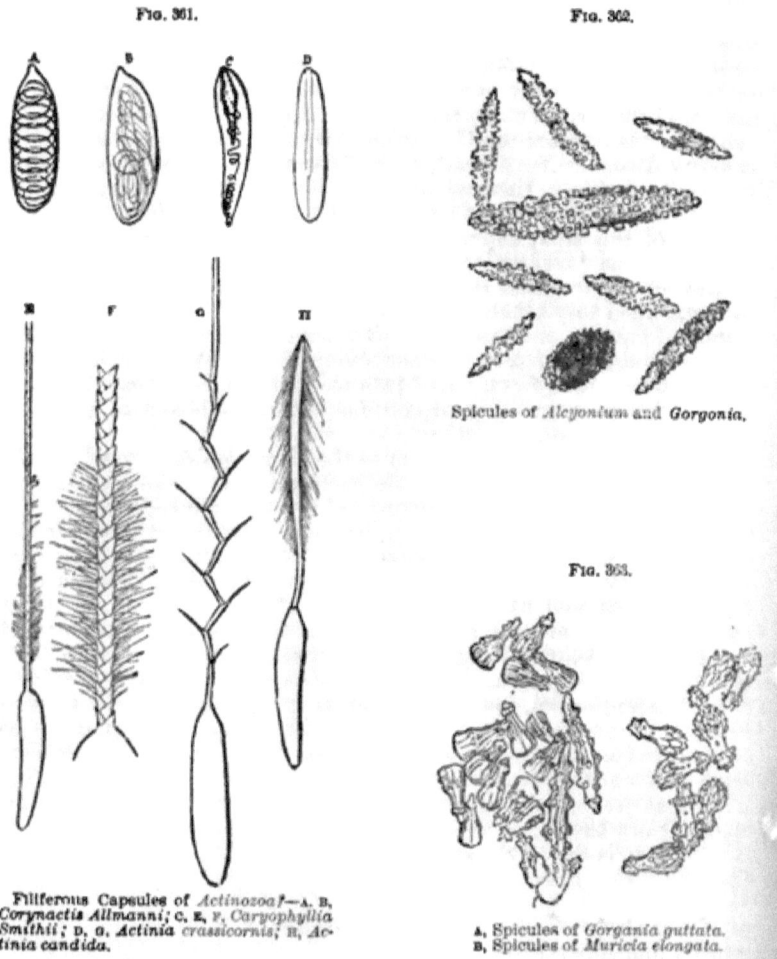

Fig. 361.

Fig. 362.

Spicules of *Alcyonium* and *Gorgonia*.

Fig. 363.

Filiferous Capsules of *Actinozoa?*—A, B, *Corynactis Allmanni*; C, E, F, *Caryophyllia Smithii*; D, G, *Actinia crassicornis*; H, *Actinia candida*.

A, Spicules of *Gorgonia guttata*.
B, Spicules of *Muricia elongata*.

529. Of the *Alcyonaria* a characteristic example is found in the *Alcyonium digitatum* of our coasts; a lobed sponge-like mass, covered with a tough skin; which is commonly known under the name of 'deadman's toes,' or by the more elegant name of 'mermaids' fingers.' When

---

[1] See Mr. Gosse's "Naturalist's Rambles on the Devonshire Coast," and Prof. Möbius 'Ueber den Bau, etc., der Nesselkapseln einiger Polypen und Quallen,' in "Abhandl. Naturw. Vereins zu Hamburg," Band v., 1866.

a specimen of this is first torn from the rock to which it has attached itself, it contracts into an unshapely mass, whose surface presents nothing but a series of slight depressions arranged with a certain regularity. But after being immersed for a little time in a jar of sea-water, the mass swells-out again, and from every one of these depressions an eight-armed polype is protruded, "which resembles a flower of exquisite beauty and perfect symmetry. In specimens recently taken, each of the petal-like tentacula is seen with a hand-glass to be furnished with a row of delicately-slender *pinnæ* or filaments, fringing each margin, and arching onwards; and with a higher power, these pinnæ are seen to be roughened throughout their whole length, with numerous prickly rings. After a day's captivity, however, the petals shrink up into short, thick, unshapely masses, rudely notched at their edges" (Gosse). When a mass of this sort is cut-into it is found to be channelled-out somewhat like a Sponge, by ramifying canals; the vents of which open into the stomachal cavities of the polypes, which are thus brought into free communication with each other,—a character that especially distinguishes this Order. A movement of fluid is kept-up within these canals (as may be distinctly seen through their transparent bodies) by means of cilia lining the internal surfaces of the polypes; but no cilia can be discerned on their external surfaces. The tissue of this spongy polypidom is strengthened throughout, like that of Sponges (§ 510), with mineral spicules (always, however, calcareous), which are remarkable for the elegance of their forms; these are disposed with great regularity around the bases of the polypes, and even extend part of their length upwards on their bodies. In the *Gorgonia*, or sea-fan, whilst the central part of the polypidom is consolidated into a horny axis, the soft flesh which clothes this axis is so full of tuberculated spicules, especially in its outer layer, that, when this dries-up they form a thick yellowish or reddish incrustation upon the horney stem; this crust is, however, so friable, that it may be easily rubbed down between the fingers, and when examined with the Microscope, it is found to consists of spicules of different shapes and sizes, more or less resembling those shown in Figs. 362, 363, sometimes colorless, but sometimes of a beautiful crimson, yellow, or purple. These spicules are best seen by Black-ground illumination, especially when viewed by the Binocular Microscope. They are, of course, to be separated from the animal substance in the same manner as the calcareous spicules of Sponges (§ 512); and they should be mounted, like them, in Canada balsam. The spicules always possess an organic basis; as is proved by the fact, that when their lime is dissolved by dilute acid, a gelatinous-looking residuum is left, which preserves the form of the spicule.

530. The *Ctenophora*, or 'comb-bearers,' are so named from the comb-like arrangement of the rows of tiny 'paddles,' by the movement of which the bodies of these animals are propelled. A very beautiful and not uncommon representative of this order is furnished by the *Cydippe pileus* (Fig. 364), very commonly known as the *Beroë*, which designation, however, properly appertains to another animal (Fig. 365) of the same grade of organization. The body of *Cydippe* is a nearly-globular mass of soft jelly, usually about 3-8ths of an inch in diameter; and it may be observed, even with the naked eye, to be marked by eight bright bands, which proceed from pole to pole like meridian lines. These bands are seen with the Microscope to be formed of rows of flattened filaments, far larger than ordinary cilia, but lashing the water in the same manner; they sometimes act quite independently of one

another, so as to give to the body every variety of motion, but sometimes work altogether. If the sun light should fall upon them when they are in activity, they display very beautiful iridescent colors. In addition to these 'paddles,' the *Cydippe* is furnished with a pair of long tendril-like filaments, arising from the bottom of a pair of cavities in the posterior part of the body, and furnished with lateral branches (A); within these cavities they may lie doubled-up, so as not to be visible externally; and when they are ejected, which often happens quite suddenly, the main filaments first come-forth, and the lateral tendrils subsequently uncoil themselves, to be drawn-in again and packed-up within the cavities with almost equal suddenness. The mouth of the animal, situated at one of the poles, leads first to a quadrifid cavity bounded by four folds which seem to represent the oral proboscis of the ordinary Medusæ (Fig. 359); and this leads to the true stomach, which passes towards the opposite pole, near to which it bifurcates, its branches passing towards the polar

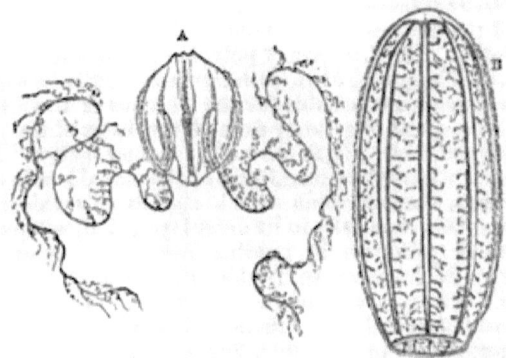

Fig. 364.   Fig. 365.

Cydippe pileus, with its tentacles extended.   *Beroë Forskalii*, showing the tubular prolongations of the stomach.

surface on either side of a little body which has every appearance of being a nervous ganglion, and which is surmounted externally by a fringe-like apparatus that seems essentially to consist of sensory tentacles.[1] From the cavity of the stomach, tubular prolongations pass-off beneath the ciliated bands, very much as in the true *Beroë* (B); these may easily be injected with colored liquids, by the introduction of the extremity of a fine-pointed glass-syringe (Fig. 106) into the mouth. The liveliness of this little creature, which may sometimes be collected

---

[1] It is commonly stated that the two branches of the alimentary canal open on the surface by two pores situated in the hollow of the fringe, one on either side of the nervous ganglion. The Author, however, has not been able to satisfy himself of the existence of such excretory pores in the ordinary *Cydippe* or *Beroë*, although he has repeatedly injected their whole alimentary canal and its extensions, and has attentively watched the currents produced by ciliary action in the interior of the bifurcating prolongations, which currents always appear to him to **return** as from cæcal extremities. He is himself inclined to believe that this arrangement has reference solely to the nutrition of the nervous ganglion and tentacular apparatus, which lies imbedded (so to speak) in the bifurcation of the alimentary canal, so as to be able to draw its supply of nutriment direct from that cavity.

in large quantities at once by the Stick-net, renders it a most beautiful subject for observation when due scope is given to its movements; but for the sake of Microscopic examination, it is of course necessary to confine these.—Various species of true *Beroë*, some of them even attaining the size of a small lemon, are occasionally to be met with on our coasts; in all of which the movements of the body are effected by the like agency of paddles arranged in meridional bands. These are splendidly luminous in the dark, and the luminosity is retained even by fragments of their bodies, being augmented by agitation of the water containing them. — All the *Ctenophora* are reproduced from eggs, and are already quite advanced in their development by the time they are hatched. Long before they escape, indeed, they swim about with great activity within the walls of their diminutive prison; their rows of locomotive paddles early attaining a large size, although the long flexile tentacles of *Cydippe* are then only short stumpy protuberances. Through the embryonic forms of the two groups, Prof. Alex. Agassiz considers the *Ctenophora* as related to *Echinodermata*.

Those who may desire to acquire a more systematic and detailed acquaintance with the Zoophyte-group, may be especially referred to the following Treatises and Memoirs, in addition to those already cited, and to the various recent systematic Treatises on Zoology:—Dr. Johnston's "History of British Zoophytes," Prof. Milne-Edwards's "Recherches sur les Polypes," and his "Histoire des Corallaires" (in the 'Suites à Buffon'), Paris, 1857, Prof. Van Beneden 'Sur les Tubulaires,' and 'Sur les Campanulaires,' in "Mém. de l'Acad. Roy. de Bruxelles," Tom. xvii., and his "Recherches sur l'Hist. Nat. des Polypes qui fréquentent les Côtes de Belgique," *Op. cit.*, Tom. xxxvi., Sir J. G. Dalyell's 'Rare and Remarkable Animals of Scotland," Vol. i., Trembley's "Mém. pour servir à l'histoire d'un genre de Polype d'Eau douce," M. Hollard's 'Monographie du Genre *Actinia*,' in "Ann. des Sci. Nat." Ser. 3, Tom. xv., Prof. Max Schultze, 'On the Male Reproductive Organs of *Campanularia geniculata*,' in "Quart. Journ. of Microsc. Sci.," Vol. iii. (1855), p. 59, Prof. Agassiz's beautiful Monograph on American Medusæ, forming the third volume of his "Contributions to the Natural History of the United States of America," Mr. Hincks's "British Hydroid Zoophytes," Prof. Allman's admirable Memoirs on *Cordylophora* and *Myriothela* in the Philos. Transact. for 1853 and 1875, Prof. J. R. Greene's "Manual of the Sub-Kingdom *Cœlenterata*," which contains a Bibliography very complete to the date of its publication, and the articles 'Actinozoa,' 'Ctenophora,' and 'Hydrozoa,' in the Supplement to the Natural History Division of the "English Cyclopædia." The *Ctenophora* are specially treated of in Vol. iii. of Prof. Agassiz's "Contributions to the Natural History of the United States." See also Prof. Alex. Agassiz "Seaside Studies in Natural History," and his "Illustrated Catalogue of the Museum of Comparative Anatomy at Harvard College," Prof. James-Clark in "American Journal of Science," Ser. 2, Vol. xxxv., p. 348, Dr. D. Macdonald in "Transact. Roy. Soc. Edinb.," Vol. xxiii., p 515, Mr. H. N. Moseley 'On the Structure of a species of *Millepora*,' in "Philos. Trans.," 1877, p. 117, and 'On the Structure of the *Stylasteridæ*,' Ibid., 1878, p. 425; and on the '*Acalephæ*,' Prof. Haeckel's " Beiträge zur Naturgeschichte der Hydromedusen," the masterly work of the brothers Hertwig, "Das Nervensystem und die Sinnesorgane der Medusen," 1878, and the Memoir of Prof. Schäfer 'On the Nervous System of *Aurelia aurita*,' in "Philos. Trans.," 1878, p. 563.

CHAPTER XIV.

ECHINODERMATA.

531. As we ascend the scale of Animal life, we meet with such a rapid advance in complexity of structure, that it is no longer possible to acquaint one's-self with any organism by Microscopic examination of it as a whole; and the dissection or analysis which becomes necessary, in order that each separate part may be studied in detail, belongs rather to the Comparative Anatomist than to the ordinary Microscopist. This is especially the case with the *Echinus* (Sea-Urchin), *Asterias* (Star-fish), and other members of the class Echinodermata; even a general account of whose complex organization would be quite foreign to the purpose of this work. Yet there are certain parts of their structure which furnish Microscopic objects of such beauty and interest that they cannot by any means be passed by; while the study of their Embryonic forms, which can be prosecuted by any Sea-side observer, brings into view an order of facts of the highest scientific interest.

532. It is in the structure of that Calcareous Skeleton which probably exists under some form in every member of this class, that the ordinary Microscopist finds most to interest him. This attains its highest development in the *Echinida;* in which it forms a box-like shell or 'test,' composed of numerous polygonal plates jointed to each other with great exactness, and beset on its external surface with 'spines,' which may have the form of prickles of no great length, or may be stout club-shaped bodies, or, again, may be very long and slender rods. The intimate structure of the shell is everywhere the same; for it is composed of a *network*, which consists of Carbonate of Lime with a very small quantity of animal matter as a basis, and which extends in every direction (*i.e.*, in thickness as well as in length and breadth), its *areolæ* or interspaces freely communicating with each other (Figs. 366, 367). These 'areolæ,' and the solid structure which surrounds them, may bear an extremely variable proportion one to the other; so that in two masses of equal size, the one or the other may greatly predominate; and the texture may have either a remarkable lightness and porosity, if the network be a very open one like that of Fig. 366, or may possess a considerable degree of compactness, if the solid portion be strengthened. Generally speaking, the different layers of this network, which are connected together by pillars that pass from one to the other in a direction perpendicular to their plane, are so arranged that the perforations in one shall correspond to the intermediate solid structure in the next; and their transparence is such that when we are examining a section thin enough to contain only two or three such layers, it is easy, by properly focussing the Microscope, to bring either one of them into distinct view. From this very simple but very beautiful arrangement, it comes to pass that the plates of which the entire 'test' is made-up possess a very consider-

able degree of strength, notwithstanding that their porousness is such, that if a portion of a fractured edge, or any other part from which the investing membrane has been removed, be laid upon fluid of almost any description, this will be rapidly sucked-up into its substance.—A very beautiful example of the same kind of calcareous skeleton, having a more regular conformation, is furnished by the disk or 'rosette' which is contained in the tip of every one of the tubular suckers put forth by the

FIG. 366.

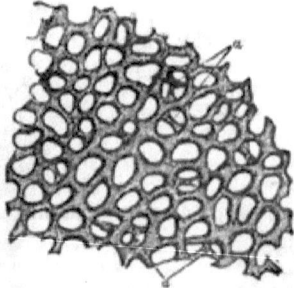

FIG. 367.

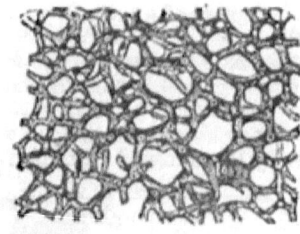

Section of Shell of *Echinus*, showing the calcareous network of which it is composed:—*a a*, portions of a deeper layer.

Transverse Section of central portion of Spine of *Acrocladia*, showing its more open network.

living Echinus from the 'ambulacral pores' that are seen in the rows of smaller plates interposed between the larger spine-bearing plates of its box-like shell. If the entire disk be cut-off, and be mounted when dry in Canada balsam, the calcareous rosette may be seen sufficiently well; but its beautiful structure is better made-out when the animal membrane that incloses it has been got-rid of by boiling in a solution of caustic potass; and the appearance of one of the five segments of which it is composed, when thus prepared, is shown in Fig. 368.

533. The most beautiful display of this reticulated structure however, is shown in the structure of the 'spines' of *Echinus*, *Cidaris*, etc.; in which it is combined with solid ribs or pillars, disposed in such a manner as to increase the strength of these organs; a regular and elaborate pattern being formed by their intermixture, which shows considerable variety in different species.—When we make a thin transverse section (Plate II., fig. 1) of almost any spine belonging to the genus *Echinus* (the small spines of our British species, however, being exceptional in this respect) or its immediate allies, we see it to be made up of a number of concentric layers, arranged in a manner that strongly reminds us of the concentric rings of an Exogenous tree (Fig. 254). The number of these layers is extremely variable; depending not merely upon the age

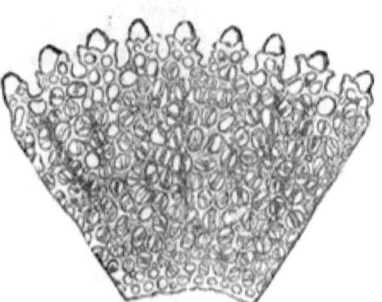

One of the segments of the calcareous skeleton of an Ambulacral Disk of *Echinus*.

of the spine, but (as will presently appear) upon the part of its length from which the section happens to be taken. The centre is usually occupied by a very open network (Fig. 367); and this is bounded by a row of transparent spaces (like those at *a, a', b b', c c,'* etc., Fig. 369), which on a cursory inspection might be supposed to be void, but are found on examination to be the sections of solid ribs or pillars, which run in the direction of the length of the spines, and form the exterior of every layer. Their solidity becomes very obvious, when we either examine a section of a spine whose substance is pervaded (as often happens) with a coloring matter of some depth, or when we look at a very thin section by black-ground illumination. Around the innermost circle of these solid pillars there is another layer of the calcareous network, which again is surrounded by another circle of solid pillars; and this arrangement may be repeated many times, as shown in Fig. 369, the outermost row of pillars forming the projecting ribs that are commonly to be distinguished on the surface of the spine. Around the cup-shaped base of the spine is a membrane which is continuous with that covering the surface of the shell, and serves not merely to hold-down the cup upon the tubercle over which it works, but also by its contractility to move the spine in any required direction.

FIG. 369.

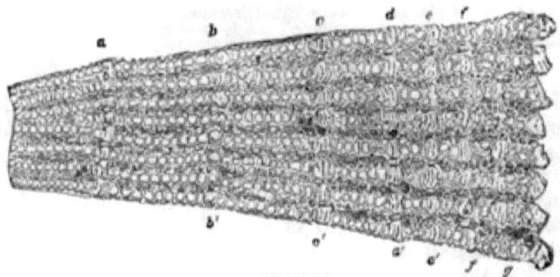

Portion of transverse section of Spine of *Acrocladia mammillata*.

This membrane is probably continued onwards over the whole surface of the spine, although it cannot be clearly traced to any distance from the base, and the new formations may be presumed to take place in its substance. Each new formation completely ensheathes the old; not merely surrounding the part previously formed, but also projecting considerably beyond it; and thus it happens that the number of layers shown in a transverse section will depend in part upon the place of that section. For if it cross near the base, it will traverse every one of the successive layers from the very commencement; whilst if it cross near the apex, it will traverse only the single layer of the last growth, notwithstanding that, in the club-shaped spines, this terminal portion may be of considerably larger diameter than the basal; and in any intermediate part of the spine, so many layers will be traversed, as have been formed since the spine first attained that length. The basal portion of the spine is enveloped in a reticulation of a very close texture, without concentric layers; forming the cup or socket which works over the tubercle of the shell.

534. Their combination of elegance of pattern with richness of coloring, renders well-prepared specimens of these Spines among the most beautiful objects that the Microscopist can anywhere meet with.

The large spines of the various species of the genus *Acrocladia* furnish sections most remarkable for size and elaborateness, as well as for depth of color (in which last point, however, the deep purple spines of *Echinus lividus* are pre-eminent); but for exquisite neatness of pattern, there are no spines that can approach those of *Echinometra heteropora* (Plate II., fig. 1) and *E. lucunter*. The spines of *Heliocidaris variolaris* are also remarkable for their beauty.—No succession of concentric layers is seen in the spines of the British *Echini*, probably because (according to the opinion of the late Sir J. G. Dalyell) these spines are cast off and renewed every year; each new formation thus going to make an entire spine, instead of making an addition to that previously existing.—Most curious indications are sometimes afforded by sections of Echinus-spines, of an extraordinary power of reparation inherent in these bodies. For irregularities are often seen in the transverse sections, which can be accounted for in no other way than by supposing the spines to have received an injury when the irregular part was at the exterior, and to have had its loss of substance supplied by the growth of new tissue, over which the subsequent layers have been formed as usual. And sometimes

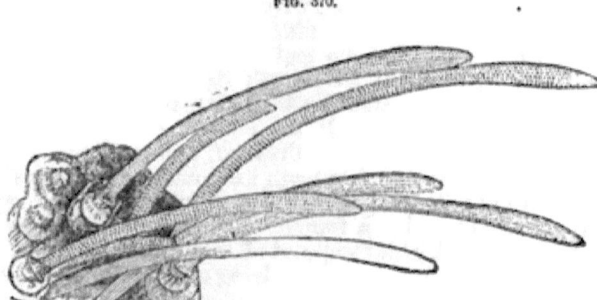

Fig. 370.

Spines of *Spatangus*.

a peculiar ring may be seen upon the surface of a spine, which indicates the place of a complete fracture; all beyond it being a new growth, whose unconformableness to the older or basal portion is clearly shown by a longitudinal section.[1]—The spines of *Cidaris* present a marked departure from the plan of structure exhibited in Echinus; for not only are they destitute of concentric layers, but the calcareous network which forms their principal substance is encased in a solid calcareous sheath perforated with tubules, which seems to take the place of the separate pillars of the Echini. This is usually found to close in the spine at its tip also; and thus it would appear that the entire spine must be formed at once, since no addition could be made either to its length or to its diameter, save on the outside of the sheath, where it is never to be found. The sheath itself often rises up in prominent points or ridges on the surface of these spines; thus giving them a character by which they may be distinguished from those of Echini.—The slender, almost filamentary spines of *Spatangus* (Fig. 370), and the innumerable minute

---

[1] See the Author's description of such Reparations in the "Monthly Microscopical Journal," Vol. iii. (1870), p. 225.

hair-like processes attached to the shell of *Clypeaster*, are composed of the like regularly-reticulated substance; and these are very beautiful objects for the lower powers of the Microscope, when laid upon a black ground and examined by reflected light without any further preparation. —It is interesting also to find that the same structure presents itself in the curious *Pedicellariæ* (forceps-like bodies mounted on long stalks), which are found on the surface of many Echinida, and the nature of which was formerly a source of much perplexity to Naturalists, some having maintained that they are parasites, whilst others considered them as proper appendages of the Echinus itself. The complete conformity which exists between the structure of their skeleton and that of the animal to which they are attached, removes all doubt of their being truly appendages to it, as observation of their actions in the living state would indicate.

535. Another example of the same structure is found in the peculiar framework of plates which surrounds the interior of the oral orifice of the shell, and which includes the five *teeth* that may often be seen projecting externally through that orifice; the whole forming what is known as the 'lantern of Aristotle.' The texture of the plates or jaws resembles that of the shell in every respect, save that the network is more open; but that of the teeth differs from it so widely, as to have been likened to that of the bone and dentine of Vertebrate animals. The careful investigations of Mr. James Salter,[1] however, have fully demonstrated that the appearances which have suggested this comparison are to be otherwise explained; the plan of structure of the *tooth* being essentially the same as that of the *shell*, although greatly modified in its working-out. The complete tooth has somewhat the form of that of the front tooth of a Rodent; save that its concave side is strengthened by a projecting 'keel,' so that a transverse section of the tooth presents the form of a $\perp$. This keel is composed of cylindrical rods of carbonate of lime, having club-shaped extremities lying obliquely to the axis of the tooth (Fig. 371, A, *d*); these rods do not adhere very firmly together, so that it is difficult to keep them in their places in making sections of the part. The convex surface of the tooth (*c, c, c*) is covered with a firmer layer, which has received the name of 'enamel;' this is composed of shorter rods, also obliquely arranged, but having a much more intimate mutual adhesion than we find among the rods of the keel. The principal part of the substance of the tooth (A, *b*) is made-up of what may be called the 'primary plates;' these are triangular plates of calcareous shell-substance, arranged in two series (as shown at B), and constituting a sort of framework with which the other parts to be presently described become connected. These plates may be seen by examining the growing base of an adult tooth that has been preserved with its attached soft parts in alcohol, or (which is preferable) by examining the base of the tooth of a fresh specimen, the minuter the better. The lengthening of the tooth below, as it is worn-away above, is mainly affected by the successive addition of new 'primary plates.' To the outer edge of the primary plates, at some little distance from the base, we find attached a set of lappet-like appendages, which are formed of similar plates of calcareous shell-substance, and are denominated by Mr. Salter 'secondary plates.' Another set of appendages termed 'flabelli-

---

[1] See his Memoir 'On the Structure and Growth of the Tooth of Echinus,' in "Philos. Transact." for 1861, p. 387.

form processes' is added at some little distance from the growing base; these consist of elaborate reticulations of calcareous fibres, ending in fan-shaped extremities. And at a point still further from the base, we find the different components of the tooth connected together by 'soldering particles,' which are minute calcareous disks interposed between the previously-formed structures; and it is by the increased development of this connective substance, that the intervening spaces are narrowed into the semblance of tubuli like those of bone or dentine. Thus a vertical section of the tooth comes to present an appearance very like that of the *bone* of a Vertebrate animal, with its lacunæ, canaliculi, and lamellæ; but in a transverse section the body of the tooth bears a stronger resemblance to *dentine;* whilst the keel and enamel-layer more resemble an oblique section of *Pinna* than any other form of shell-structure.

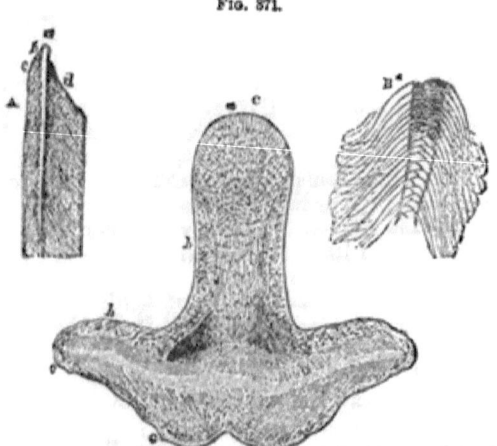

Fig. 371.

Structure of the Tooth of *Echinus:*—A, vertical section, showing the form of the apex of the tooth as produced by wear, and retained by the relative hardness of its elementary parts; *a*, the clear condensed axis; *b*, the body formed of plates; *c*, the so-called enamel; *d*, the keel:—B, commencing growth of the tooth, as seen at its base, showing its two systems of plates; the dark appearance in the central portion of the upper part is produced by the incipient reticulations of the flabelliform processes:—c, transverse section of the tooth, showing at *a* the ridge of the keel, at *b* its lateral portion, resembling the shell in texture; at *c, c,* the enamel.

536. The calcareous plates which form the less compact skeletons of the *Asteriada* ('star-fish' and their allies), and of the *Ophiurida* ('sand-stars' and 'brittle-stars'), have the same texture as those of the shell of Echinus. And this presents itself, too, in the spines or prickles of their surface, when these (as in the great *Goniaster equestris*) are large enough to be furnished with a calcareous framework, and are not mere projections of the horny integument. An example of this kind, furnished by the *Astrophyton* (better known as the *Euryale*), is represented in Fig. 372. The spines with which the arms of the species of *Ophiocoma* ('brittle-star') are beset, are often remarkable for their beauty of conformation; those of *O. rosula*, one of the most common kinds, might serve (as Prof. E. Forbes justly remarked), in point of lightness and beauty, as models for the spire of a cathedral. These are seen to the greatest advantage when mounted in Canada balsam, and viewed by the Binocular Microscope with black-ground illumination. —It is interesting to remark that the minute tooth of *Ophiocoma*

clearly exhibits, with scarcely any preparation, that gradational transition between the ordinary reticular structure of the shell and the peculiar substance of the tooth, which, in the adult tooth of the *Echinus*, can only be traced by making sections of it near its base. The tooth of *Ophiocoma* may be mounted in balsam as a transparent object, with scarcely any grinding down; and it is then seen that the basal portion of the tooth is formed upon the open reticular plan characteristic of the 'shell,' whilst this is so modified in the older portion by subsequent addition, that the upper part of the tooth has a bone-like character.

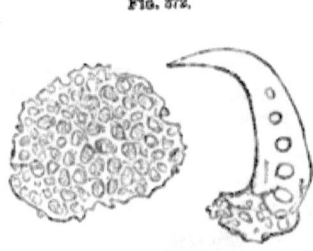

Fig. 372.

Calcareous plate and claw of *Astrophyton* (Euryale).

537. The calcareous skeleton is very highly developed in the *Crinoidea*; their stems and branches being made-up of a calcareous network closely resembling that of the shell of the Echinus. This is extremely well seen, not only in the recent *Pentacrinus Caput Medusæ*, a somewhat rare animal of West Indian seas, but also in a large proportion of the fossil Crinoids, whose remains are so abundant in many of the older Geological formations; for notwithstanding that these bodies have been penetrated in the act of fossilization by a Mineral infiltration, which seems to have substituted itself for the original fabric (a regularly-crystalline cleavage being commonly found to exist in the fossil stems of Encrinites, etc., as in the fossil spines of Echinida), yet their organic structure is often most perfectly preserved.[1] In the circular stems of *Encrinites*, the texture of the calcareous network is uniform, or nearly so, throughout; but in the pentangular *Pentacrini*, a certain figure or pattern is formed by variations of texture in different parts of the transverse section.[2]

538. The minute structure of the Shells, Spines, and other solid parts of the skeleton of Echinodermata can only be displayed by thin sections made upon the general plan already described (§§ 192–195). But their peculiar texture requires that certain precautions should be taken; in the first place, in order to prevent the section from breaking whilst being reduced so the desirable thickness; and in the second, to prevent the interspaces of the network from being clogged by the particles abraded in the reducing process.—A section of the Shell, Spine, or other portion of the skeleton should first be cut with a fine saw, and be rubbed on a flat file until it is about as thin as ordinary card, after which it should be smoothed on one side by friction with water on a Water-of-Ayr stone. It should then, after careful washing, be dried, first on white blotting-paper, afterwards by exposure for some time to a gentle heat, so

---

[1] The calcareous skeleton even of living Echinoderms has a crystalline aggregation, as is very obvious in the more solid spines of *Echinometræ*, etc.; for it is difficult, in sawing these across, to avoid their tendency to *cleavage* in the oblique plane of calcite. And the Author is informed by Mr. Sorby, that the calcareous deposit which fills up the areolæ of the fossilized skeleton has always the same crystalline system with the skeleton itself, as is shown not merely by the uniformity of their cleavage, but by their similar action on Polarized light.

[2] See Figs. 74–76 of the Author's Memoir on "Shell Structure" in the Report of the British Association for 1847.

that no water may be retained in the interstices of the network, which would oppose the complete penetration of the Canada balsam. Next, it is to be attached to a glass-slip by balsam hardened in the usual manner; but particular care should be taken, first, that the balsam be brought to exactly the right degree of hardness, and second, that there be enough not merely to attach the specimen of the glass, but also to saturate its substance throughout. The right degree of hardness is that at which the balsam can be with difficulty indented by the thumb-nail; if it be made harder than this, it is apt to chip-off the glass in grinding, so that the specimen also breaks away; and if it be softer, it holds the abraded particles, so that the openings of the network become clogged with them. If, when rubbed-down nearly to the required thinness, the section appears to be uniform and satisfactory throughout, the reduction may be completed without displacing it; but if (as often happens) some inequality in thickness should be observable, or some minute air bubbles should show themselves between the glass and the under surface, it is desirable to loosen the specimen by the application of just enough heat to melt the balsam (special care being taken to avoid the production of fresh air-bubbles), and to turn it over so as to attach the side last polished to the glass, taking care to remove or to break with the needle-point any air-bubbles that there may be in the balsam covering the part of the glass on which it is laid. The surface now brought uppermost is then to be very carefully ground down; special care being taken to keep its thickness uniform through every part (which may even be better judged-of by the touch than by the eye), and to carry the reducing process far enough, without carrying it too far. Until practice shall have enabled the operator to judge of this by passing his finger over the specimen, he must have continual recourse to the Microscope during the latter stages of his work; and he should bear constantly in mind, that, as the specimen will become much more transparent when mounted in balsam and covered with glass, than it is when the ground surface is exposed, he need not carry his reducing process so far as to produce at once the entire transparence he aims at, the attempt to accomplish which would involve the risk of the destruction of the specimen. In 'mounting' the specimen, liquid balsam should be employed, and only a very gentle heat (not sufficient to produce air-bubbles, or to loossen the specimen from the glass) should be applied; and if, after it has been mounted, the section should be found too thick, it will be easy to remove the glass cover and to reduce it further, care being taken to harden to the proper degree the balsam which has been newly laid-on.

539. If a number of sections are to be prepared at once (which it is often useful to do for the sake of economy of time, or in order to compare sections taken from different parts of the same spine), this may be most readily accomplished by laying them down, when cut-off by the saw, without any preliminary preparation save the blowing of the calcareous dust from their surfaces, upon a thick slip of glass well covered with hardened balsam; a large proportion of its surface may thus be occupied by the sections attached to it, the chief precaution required being that all the sections come into equally close contact with it. Their surfaces may then be brought to an exact level, by rubbing them down, first upon a flat piece of grit (which is very suitable for the rough grinding of such sections), and then upon a large Water-of-Ayr stone whose surface is 'true.' When this level has been attained, the ground surface is to be well washed and dried, and some balsam previously hardened is

to be spread over it, so as to be sucked-in by the sections, a moderate heat being at the same time applied to the glass slide; and when this has been increased sufficiently to loosen the sections without overheating the balsam, the sections are to be turned-over, one by one, so that the *ground* surfaces are now to be attached to the glass slip, special care being taken to press them all into close contact with it. They are then to be very carefully rubbed-down, until they are nearly reduced to the required thinness; and if, on examining them from time to time, their thinness should be found to be uniform throughout, the reduction of the entire set **may be** completed at once; and when it has been carried sufficiently **far, the** sections, loosened by warmth, are to be taken-up on a camel-hair **brush** dipped in turpentine, and transferred to separate slips of glass **whereon** some liquid balsam has been previously laid, in which they are to be mounted in the usual manner. It more frequently happens, however, that, notwithstanding every care, the sections, when ground in a number together, are not of uniform thickness, owing to some of them being underlaid by a thicker stratum of balsam than others; and it is **then** necessary to transfer them to separate slips before the reducing process is completed, **attaching** them with hardened balsam, and finishing each section **separately.**

540. A very curious *internal* skeleton, **formed of** detached plates or spicules, is found in many members of this **class;** often forming an in-

FIG. 373.

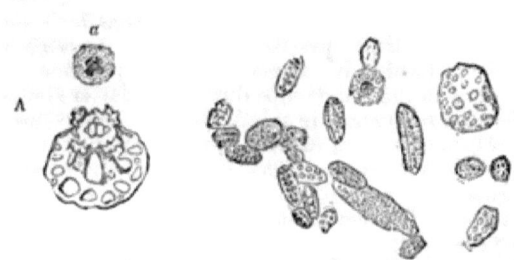

Calcareous plates in skin of *Holothuria*.

vestment like a coat of mail to some of the viscera, especially the ovaries. The forms of these plates and spicules are generally so diverse, even in closely-allied species, as to afford very good differential characters.—This subject is one that has been as yet but very little studied, Mr. Stewart being the only Microscopist who has given much attention to it;[1] but it is well worthy of much more extended research.

541. It now remains for us to notice the curious and often very beautiful structures, which represent, in the order *Holothurida*, the solid calcareous skeleton of the others already noticed. All the animals belonging to this Order are distinguished by the flexibility and absence of firmness of their envelopes; and excepting in the case of certain species which have a set of calcareous plates, supporting teeth, disposed around the mouth, very much as in the Echinida, we do not find among them any representation that is apparent to the unassisted eye, of that skeleton which constitutes so distinctive a feature of the class generally. But a microscopic examination of their integument at once brings to view the existence of great numbers of minute isolated plates, every one

---

[1] See his Memoir in the "Linnæan Transactions," Vol. **xxv.**, p. 365.

of them presenting the characteristic reticulated structure, which are set with greater or less closeness in the substance of the skin. Various forms of the plates which thus present themselves in *Holothuria* are shown in Fig. 373; and at A is seen an oblique view of the kind marked *a*, more highly magnified, showing the very peculiar manner wherein one part is superposed on the other, which is not at all brought into view when it is merely seen through in the ordinary manner.—In the *Synapta*, one of the long-bodied forms of this order, which abounds in the Adriatic Sea, and of which two species (the *S. digitata* and *S. inhærens*) occasionally occur upon our own coasts,[1] the calcareous plates of the integument have the regular form shown at A, Fig. 374; and each

FIG. 374.

Calcareous Skeleton of *Synapta*:—A, plate imbedded in Skin; B, the same, with its anchor-like spine attached; C, anchor-like spine separated.

of these carries the curious anchor-like appendage, C, which is articulated to it by the notched piece at the foot, in the manner shown (in side view) at B. The anchor-like appendages project from the surface of the skin, and may be considered as representing the spines of Echinida. —Nearly allied to the Synapta is the *Chirodota*, the integument of which is entirely destitute of 'anchors,' but is furnished with very remarkable wheel-like plates; those represented in Fig. 375 are found in the skin of *Chirodota violacea*, a species inhabiting the Mediterranean. These 'wheels' are objects of singular beauty and delicacy, being especially remarkable for the very minute notching (scarcely to be discerned in the figures without the aid of a magnifying-glass) which is traceable around the inner margin of their 'tires'—There can be scarcely any reasonable doubt that every member of this Order has some kind of calcareous skeleton, disposed in a manner conformable to the examples now cited; and it would be very valuable to determine how far the marked peculiarities by which they are respectively distinguished, are characteristic of genera and species. The plates may be obtained separately by the usual method of treating the skin with a solution of potass; and they should be mounted in Canada balsam. But their position in the skin can only be ascertained by making sections of the integument, both vertical and parallel to its surface; and these sections, when dry, are most advantageously mounted in the same medium, by which their transparence is

FIG. 375.

Wheel-like plates from Skin of *Chirodota violacea*.

---

[1] See Woodward in "Proceedings of Zoological Society," July 13, 1858.

greatly increased. All the objects of this class are most beautifully displayed by the Black-ground illumination; and their solid forms are seen with increased effect under the Binocular. The Black-ground illumination applied to *very thin* sections of Echinus spines brings out some effects of marvellous beauty; and even in these the solid form of the network connecting the pillars is better seen with the Binocular than it can be with the ordinary Microscope.[1]

542. *Echinoderm-Larvæ.*—We have now to notice that most remarkable set of objects furnished to the Microscopic inquirer by the *larval* states of this class; for our knowledge of which we are chiefly indebted to the painstaking and widely-extended investigations of Prof. J. Müller. All that our limits permit is a notice of two of the most curious forms of these larvæ, by way of sample of the wonderful phenomena which his researches brought to light, and to which the attention of Microscopists who have the opportunity of studying them should be the more assiduously directed; as even the most delicate of these organisms have been found capable of such perfect preservation, as to admit of being studied, when mounted as preparations, even better than when alive (§ 545, *a*). The peculiar feature by which the early history of the Echinoderms generally seems to be distinguished is this,—that the embryonic mass of cells is converted, not into a larva which subsequently attains the adult form by a process of metamorphosis, but into a peculiar 'zooid' or *pseudembryo*, which seems to exist for no other purpose than to give origin to the Echinoderm by a kind of internal gemmation, and to carry it to a distance by its active locomotive powers, so as to prevent the spots inhabited by the respective species from being overcrowded by the accumulation of their progeny. The larval zooids are formed upon a type quite different from that which characterizes the adults; for instead of a *radial* symmetry, they exhibit a *bilateral*, the two sides being precisely alike, and each having a ciiliated fringe along the greater part or the whole of its length. The two fringes are united by a superior and an inferior transverse ciliated band : and between these two the mouth of the zooid is always situated. Further, although the adult Star-fish and Sand-stars have usually neither intestinal tube nor anal orifice, their larval zooids, like those of other Echinoderms, always possess both. The external forms of these larvæ, however, vary in a most remarkable degree, owing to the unequal evolution of their different parts ; and there is also a considerable diversity in the several Orders, as to the proportion of the fabric of the larva which enters into the composition of the adult form. In the fully developed Star-fish and Sea-urchin, the only part retained is a portion of the stomach and intestine, which is pinched-off, so to speak, from that of the larval zooid.

543. One of the most remarkable forms of Echinoderm-larvæ is that which has received the name of *Bipinnaria* (Fig. 376), from the symmetrical arrangement of its natatory organs. The mouth (*a*), which opens in the middle of a transverse furrow, leads through an œsophagus *a'* to a large stomach, around which the body of a Star-fish is developing itself; and on one side of this mouth are observed the intestinal tube and anus (*b*). On either side of the anterior portion of the body are six or more

---

[1] It may be here pointed out that the reticulated appearance is sometimes deceptive; what seems to be *solid* network being in many instances a *hollow* network of passages channelled-out in a solid calcareous substance. Between these two conditions, in which the relation beween the solid frame-work and the intervening space is completely reversed, there is every intermediate gradation.

narrow fin-like appendages, which are fringed with cilia; and the posterior part of the body is prolonged into a sort of pedicle, bilobed towards its extremity, which also is covered with cilia. The organization of this larva seems completed, and its movements through the water become very active, before the mass at its anterior extremity presents anything of the aspect of the Star-fish; in this respect corresponding with the movements of the *pluteus* of the Echinida (§ 545). The temporary mouth of the larva does not remain as the permanent mouth of the Star-fish; for the œsophagus of the latter enters on what is to become the dorsal side of its body, and the true mouth is subsequently formed by the thinning-away of the integument on its ventral surface. The young Star-fish is separated from the Bipinnarian larva by the forcible contractions of the connecting stalk, as soon as the calcareous consolidation of its integument has taken-place and its true mouth has been formed, but long before it has attained the adult condition; and as its ulterior development has not hitherto been observed in any instance, it is not yet known what are the species in which this mode of evolution prevails. The larval zooid continues active for several days after its detachment; and it is possible, though perhaps scarcely probable, that it may develop another Asteroid by a repetition of this process of gemmation.

Fig. 376.

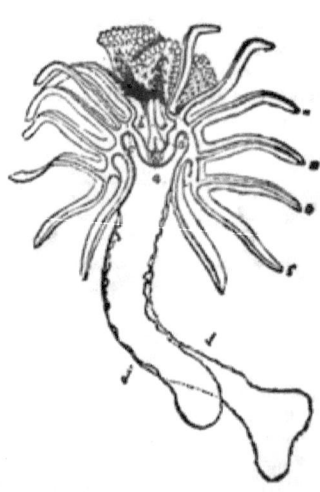

Bipinnaria asterigera, or Larva of Star-fish:—a, mouth; a', œsophagus; b, intestinal tube and anal orifice; c, furrow in which the mouth is situated; d d', bilobed peduncle; 1, 2, 3, 4, 5, 6, 7, ciliated arms.

544. In the Bipinnaria, as in other larval zooids of the Asteriada, there is no internal calcareous frame-work; such a frame-work, however, is found in the larvæ of the *Echinida* and *Ophiurida*, of which the form delineated in Fig. 377 is an example. The embryo issues from the ovum as soon as it has attained, by repeated 'segmentation' of the yolk (§ 581), the condition of the 'mulberry-mass;' and the superficial cells of this are covered with cilia, by whose agency it swims freely through the water. So rapid are the early processes of development, that no more than from twelve to twenty-four hours intervene between fecundation and the emersion of the embryo; the division into two, four, or even eight segments taking-place within three hours after impregnation. Within a few hours after its emersion, the embryo changes from the spherical into a sub-pyramidal form with a flattened base; and in the centre of this base is a depression, which gradually deepens, so as to form a mouth that communicates with a cavity in the interior of the body, which is surrounded by a portion of the yolk-mass that has returned to the liquid granular state. Subsequently a short intestinal tube is found, with an anal orifice opening on one side of the body. The pyramid is at first triangular, but it afterwards becomes quadrangular; and the angles are greatly prolonged round the mouth (or base), whilst the apex of the pyramid is sometimes much extended in the opposite direction, but is sometimes rounded off into a kind of dome (Fig. 377, A). All parts of

this curious body, and especially its most projecting portions, are strengthened by a frame-work of thread-like calcareous rods (*e*). In this condition the embryo swims freely through the water, being propelled by the action of the cilia, which clothe the four angels of the pyramid and its projecting arms, and which are sometimes thickly set upon two or four projecting lobes (*f*); and it has received the designation of *pluteus*. The mouth is usually surrounded by a sort of proboscis, the angles of which are prolonged into four slender processes (*g, g, g, g*), shorter than the four outer legs, but furnished with a similar calcareous frame-work.

545. The first indication of the production of the young Echinus from its 'pluteus,' is given by the formation of a circular disk (Fig. 377, A, c), on one side of the central stomach (*b*); and this disk soon presents five prominent tubercles (B), which subsequently become elongated into tubular cirrhi. The disk gradually extends itself over the stomach, and between its cirrhi the rudiments of spines are seen to protrude (C); these, with the cirrhi, increase in length, so as to project against the envelope

FIG. 377.

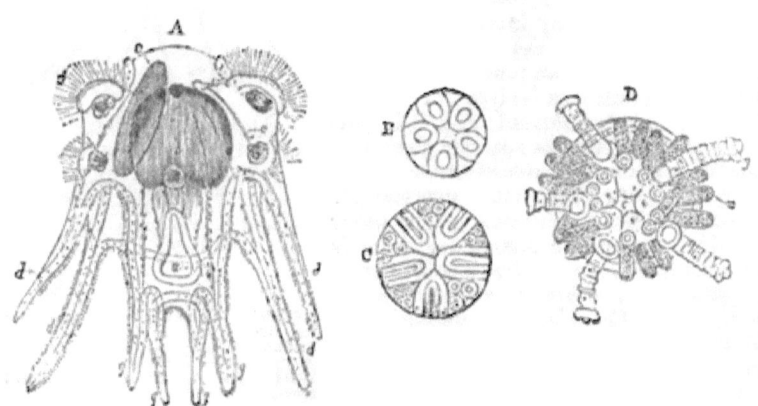

Embryonic development of *Echinus*:—A, *Pluteus larva* at the time of the first appearance of the disk; *a*, mouth in the midst of the four-pronged proboscis; *b*, stomach; *c*, Echinoid disk; *d, d, d, d*, four arms of the pluteus-body; *e*, calcareous framework; *f*, ciliated lobes; *g, g, g, g*, ciliated processes of the proboscis:—B, Disk with the first indication of the cirrhi:—C, Disk with the origin of the spines between the cirrhi;—D, more advanced disk, with the cirrhi, *g*, and spines, *x*, projecting considerably from the surface. (N.B—In B, C, and D, the Pluteus is not represented, its parts having undergone no change, save in becoming relatively smaller.)

of the pluteus, and to push themselves through it; whilst, at the same time, the original angular appendages of the pluteus diminish in size, the ciliary movement becomes less active, being superseded by the action of the cirrhi and spines, and the mouth of the pluteus closes-up. By the time that the disk has grown over half of the gastric sphere, very little of the pluteus remains, except some of the slender calcareous rods; and the number of cirrhi and spines rapidly increases. The calcareous framework of the shell at first consists, like that of the Star-fishes, of a series of isolated networks developed between the cirrhi; and upon these rest the first-formed spines (D). But they gradually become more consolidated, and extend themselves over the granular mass, so as to form the series of plates constituting the shell. The mouth of the Echinus (which is altogether distinct from that of the pluteus) is formed at that side of the granular mass over which the shell is last extended; and the first indication of it consists in the appearance of the five calcareous concretions, which are the summits of the five portions of the frame-work of jaws and teeth that surround it. All traces of the original pluteus are now lost; and the larva, which now presents the general aspect of an Echinoid animal, gradual augments in size, multiplies the number of its plates, cirrhi, and spines, evolves itself into its particular generic and specific type, and undergoes various changes of internal structure, tending to the development of the complete organism.

a. An excellent summary of the developmental history of the several Echinoderm-types, with references to the principal Memoirs which treat of it, will be found in Chap. xx. of Mr. Balfour's "Comparative Embryology."—In collecting the free-swimming larvæ of Echinodermata, the Stick-net should be carefully employed in the manner already described (§ 217); and the search for them is of course most likely to be successful in those localities in which the adult forms of the respective species abound, and on warm calm days, in which they seem to come to the surface in the greatest numbers. The following mode of preparing and mounting them has been kindly communicated to the Author by Mr. Percy Sladen:—"For killing and preserving Echinoderm zooids, I have come to prefer either Osmic acid or the Picro-sulphuric mixture of Kleinenberg (§ 199, *e*) of one-third strength. The latter, of course, destroys all calcareous structures, but the soft parts are preserved in a wonderful manner. If the diluted Kleinenberg's mixture is used, let the zooids remain in it for one or two hours; then wash them *thoroughly* in 70 per cent Spirit until all trace of acid is removed; then stain; then again wash in 70 per cent Spirit, transfer them to 90 per cent Spirit for some hours, and lastly to absolute Alcohol. Transfer them from this to Oil of Cloves; and finally mount in Canada balsam in the usual manner.—If Osmic acid be used, place three or four of the living zooids in a watch-glass of sea-water, and add a drop of the 1 per cent solution. They should not remain even in this weak solution for more than a minute; and should then be thoroughly washed in a superabundance of 35 per cent Spirit, to prevent the deposit of crystals of salt consequent on the action of the osmic acid. Then transfer the specimens to 70 per cent Spirit; and proceed as in the other case.

546. One of the most interesting to the Microscopist of all Echinodermata is the *Antedon*[1] (more generally known as *Comatula*), or 'feather-star' (Fig. 378), which is the commonest existing representative of the great fossil series *Crinoidea*, or 'lily star,' that were among the most abundant types of this class in the earlier epochs of the world's history. Like these, the *young* of Antedon is attached by a stalk to a fixed base, as shown in Fig. 379; but when it has arrived at a certain stage of development, it drops off from this like a fruit from its stalk; and the animal is thenceforth free to move through the ocean-water it inhabits. It can swim with considerable activity; but it exerts this power chiefly to gain a suitable place for attaching itself by means of the jointed prehensile cirrhi put forth from the under side of the central disk (Fig. 378); so that, notwithstanding its locomotive power, it is nearly as stationary in its free adult condition, as it is in its earlier Pentacrinoid stage. The *pentacrinoid larva*[2]—first discovered by Mr. J. V. Thomp-

---

[1] The Author has found himself obliged, by the accepted rules of Zoological nomenclature, to adopt the designation *Antedon*, instead of the much better known and very appropriate name given to this type of Lamarck. See his 'Researches on the Structure, Physiology, and Development of *Antedon rosaceus*,' Part. I., in "Philos. Transact.," 1866, p. 671.

[2] The Pentacrinoid larvæ of *Antedon* have been found abundantly (attached to Sea-weeds and Zöophytes) at Millport on the Clyde, and in Lamlash Bay, Arran; in Kirkwall Bay, Orkney; in Lough Strangford, near Belfast, and in the Bay of Cork; and at Ilfracombe, and in Salcombe Bay, Devon.

son of Cork, in 1823, but originally supposed by him to be a permanently attached Crinoid—forms a most beautiful object for the lower powers of the Microscope, when well preserved in fluid, and viewed by a strong incident light (Plate XXI., fig. 3); and a series of specimens in different stages of development, shows most curious modifications in the form and arangement of the various component pieces of its calcareous skeleton. In its earliest stage (Fig. 379, A), the body is inclosed in a calyx composed of two circles of plates; namely, five *basals*, forming a sort of pyramid whose apex points downward, and is attached to the highest joint of the stem: and five *orals* superposed on these, forming when closed a like pyramid whose apex points upwards, but usually separating to give passage to the tentacles, of which a circlet surrounds the mouth. In this condition there is no rudiment of arms. In the more advanced stage shown at B, the arms have begun to make their

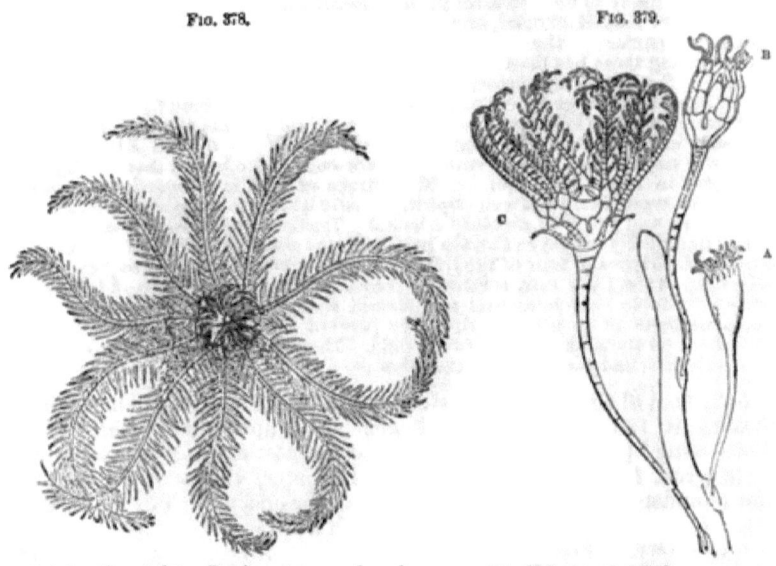

Fig. 378.

Antedon (Comatula) or Feather-star, seen from its under side.

Fig. 379.

Crinoid Larva of *Antedon*:—A, B, C, successive stages of development.

appearance; and the skeleton, when carefully examined, is found to consist of the following pieces, as shown in Plate XXI., fig. 1:—b, b, the circlet of *basals* supported on the top of the stem: r^1 the circlet of *first radials*, now interposed between the basals and the orals, and alternating with both; between two of these is interposed the single *anal* plate, a; whilst they support the *second* and the *third radials* (r^2, r^3), from the latter of which the bifurcating arms spring; finally, between the second radials we see the five *orals*, lifted from the basals on which they originally rested, by the interposition of the first radials. In the more advanced stage shown in Fig. 379, C, and on a larger scale in Plate XXI., figs. 2, 3, we find the highest joint of the stem beginning to enlarge, to form the *centro-dorsal* plate (fig. 2, $c\ d$), from which are beginning to spring the dorsal cirrhi ($c\ i\ r$), that serve to anchor the animal when it drops from the stem; this supports the *basals* (b), on which rest the

ECHINODERMATA. 155

PLATE XXI.

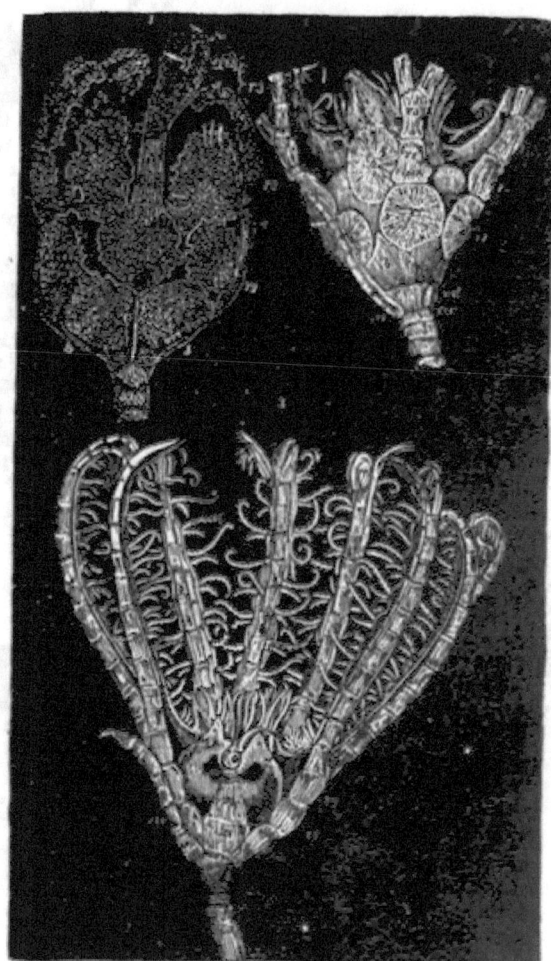

PENTACRINOID LARVA OF ANTEDON (Original).

Fig. 1. Skeleton of early Pentacrinoid, under Black-ground illumination, showing its component plates:—b, b, basals, articulated below to the highest point of the stem; r^1, r^1, first radials, between two of which is seen the single anal plate, a; r^2, second radials; r^3, third radials, giving off the bifurcating arms at their summit; o, o, orals.

2, 3. Back and front views of a more advanced Pentacrinoid, as seen by incident light, one of the pair of arms being cut away in Fig. 3, in order to bring the mouth and its surrounding parts into view:—b, b, basals, r^1, r^2, r^3, first, second, and third radials; a, anal, now carried upwards by the projection of the vent v; o, o, orals; cir, dorsal cirrhi, developed from the highest joint of the stem.

first radials (r^1); whilst the *anal* plate (a) is now lifted nearly to the level of the *second radials* (r^2), by the development of the anal funnel or vent (v) to which it is attached. The *oral* plates are not at first apparent, as they no longer occupy their first position; but on being carefully looked-for, they are found still to form a circlet around the mouth (fig. 3, o, o), not having undergone any increase in size, whilst the visceral disk and the calyx in which it is lodged have greatly extended. These *oral plates* finally disappear by absorption; while the *basals* are at first concealed by the great enlargement of the centro-dorsal (which finally extends so far as to conceal the first radials also); and at last undergo metamorphosis into a beautiful 'rosette,' which lies between the cavity of the centro-dorsal and that of the calyx.—In common with other members of its Class, the *Antedon* is represented in its earliest phase of development by a free-swimming 'larval zooid' or *pseudembryo*, which was first observed by Busch, and has been since carefully studied by Pro. Wyville Thomson[1] and Goette.[2] This zooid has an elongated egg-like form, and is furnished with transverse bands of cilia, and with a mouth and anus of its own. After a time, however, rudiments of the calcareous plates forming the stem and calyx begin to show themselves in its interior; a disk is then formed at the posterior extremity, by which it attaches itself to a Sea-weed (very commonly *Laminaria*), Zoophyte, or Polyzoary; the calyx, containing the true stomach, with its central mouth surrounded by tentacles, is gradually evolved; and the sarcodic substance of the pseudembryo, by which this calyx and the rudimentary stem were originally invested, gradually shrinks, until the young Pentacrinoid presents itself in its charateristic form and proportions.[3]

---

[1] 'On the Development of *Antedon rosaceus*' in "Philos. Transact." for 1865, p. 513.

[2] "Archiv f. Mikrosk. Anat.," Bd. xii., p. 583.

[3] The general results of the Author's own later studies of this most interesting type (the key to the life-history of the entire Geological succession of *Crinoidea*) are embodied in a notice communicated to the "Proceedings of the Royal Society," for 1876, p. 211, and in a subsequent note, p. 451. Of the further contributions recently made to our knowledge of it, the Memoir of Dr. H. Ludwig 'Zur Anatomie der Crinoideen' (Leipzig, 1877), forming part of his "Morphologische Studien an Echinodermen," is the most important.

CHAPTER XV.

POLYZOA AND TUNICATA.

547. At the lower extremity of the great series of Molluscous animals, we find two very remarkable groups, whose mode of life has much in common with Zoophytes, whilst their type of structure is conformable in essential particulars to that of the true Mollusks. These animals are for the most part microscopic in their dimensions; and as some members of both these groups are found on almost every coast, and are most interesting objects for anatomical examination, as well as for observation in the living state, a brief general account of them will be here appropriate.

548. POLYZOA.—The group which is known under this name to British naturalists (corresponding with that which by Continental Zoologists is designated *Bryozoa*) was formerly ranked as an order of Zoophytes; and it has been entirely by Microscopic study that its comparatively high organization has been ascertained.—The animals of the Polyzoa, in consequence of their universal tendency to multiplication by gemmation, are seldom or never found solitary, but form clusters or colonies of various kinds, and as each is inclosed in either a horny or a calcareous sheath or 'cell,' a composite structure is formed, closely corresponding with the 'polypidom' of a Zoophyte, which has been appropriately designated the *polyzoary*. The individual cells of the polyzoary are sometimes only connected with each other by their common relation to a creeping stem or *stolon*, as in *Laguncula* (Plate XXII.); but more frequently they bud-forth directly, one from another, and extend themselves in different directions over plane surfaces, as is the case with *Flustræ*, *Lepraliæ*, etc. (Fig. 380); whilst not unfrequently the polyzoary develops itself into an arborescent structure (Fig. 381), which may even present somewhat of the density and massiveness of the Stony Corals. Each individual, designated as a *polypide* or polype-like animal, is composed externally of a sort of sac, of which the outer or tegumentary layer is either simply membranous, or is horny, or in some instances calcified, so as to form the cell; this investing sac is lined by a more delicate membrane, which closes its orifice, and which then becomes continuous with the wall of the alimentary canal; this lies freely in the visceral sac, floating (as it were) in the liquid which it contains.

549. The principal features in the structure of this group will be best understood from the examination of a characteristic example, such as the *Laguncula repens;* which is shown in the state of expansion at A, Plate XXII., and in the state of contraction at B and C. The mouth is surrounded by a circle of tubular tentacles, which are clothed by vibratile cilia; these tentacles, in the species we are considering, vary from ten to twelve in number, but in some other instances they are more numerous.

By the ciliary investment of the tentacles, the Polyzoa are at once distinguishable from those Hydroid polypes to which they bear a superficial resemblance, and with which they were at one time confounded; and accordingly, whilst still ranked among Zoophytes, they were characterized as *ciliobrachiate*. The tentacula are seated upon an annular disk, which is termed the *lophophore*, and which forms the roof of the visceral or perigastric cavity; and this cavity extends itself into the interior of the tentacula, through perforations in the lophophore, as is shown at D, Plate XXII., representing a portion of the tentacular circle on a larger scale, *a a* being the tentacula, *b b* their internal canals, *c* the muscles of the tentacula, *d* the lophophore, and *e* its retractile muscles. The mouth situated in the centre of the lophophore, as shown at A, leads to a funnel-shaped cavity or pharynx, *b*, which is separated from the œsophagus, *d*, by a valve at *c;* and this œsophagus opens into the stomach *e*, which occupies a considerable part of the visceral cavity. (In the *Bowerbankia* and some other Polyzoa, a muscular stomach or gizzard for the trituration of the food intervenes between the œsophagus and the true digestive stomach). The walls of the stomach, *h*, have considerable thickness; and they are beset with minute follicles, which seem to have the character of a rudimentary liver. This, however, is more obvious in some other members of the group.

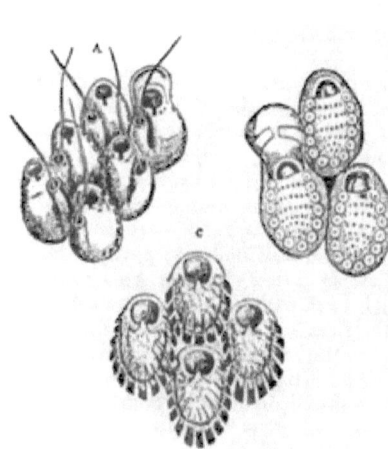

FIG. 380.

Cells of *Lepraliæ*.—A, *L. Hyndmanni;* B, *L. figularis;* C, *L. verrucosa.*

The stomach is lined, especially at its upper part, with vibratile cilia, as seen at *e, g;* and by the action of these the food is kept in a state of constant agitation during the digestive process. From the upper part of the stomach, which is (as it were) doubled upon itself, the intestine *i* opens, by a pyloric orifice, *f*, which is furnished with a regular valve; within the intestine are seen at *k* particles of excrementitious matter, which are discharged by the anal orifice at *l*. No special circulating apparatus here exists; but the liquid which fills the cavity that surrounds the viscera contains the nutritive matter which has been prepared by the digestive operation, and which has transuded through the walls of the alimentary canal; a few corpuscles of irregular size are seen to float in it. The visceral sacs of the different polypides put forth from the same stem appear to communicate with each other. No other respiratory organs exist than the tentacula; into whose cavity the nutritive fluid is probably sent from the perivisceral cavity, for aeration by the current of water that is continually flowing over them.

550. The production of *gemmæ* or buds may take place either from the bodies of the polypides themselves, which is what always happens when the cells are in mutual apposition; or from the connecting stem or 'stolon,' where the cells are distinct one from the other as in *Laguncula*. In the latter case there is first seen a bud-like protuberance of the horny

PLATE XXII.

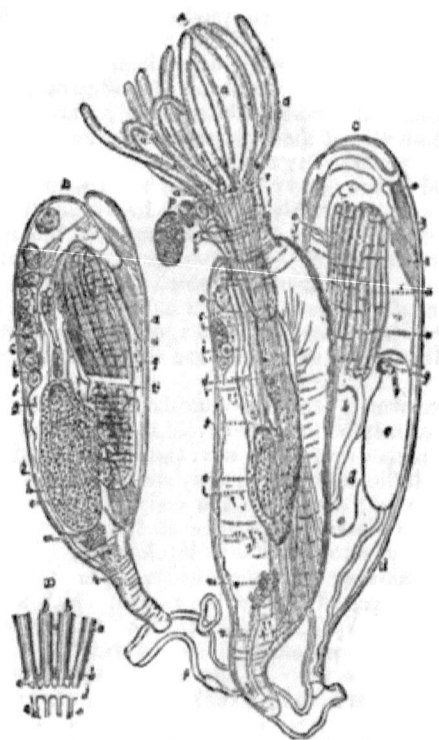

STRUCTURE OF LAGUNCULA REPENS (after Van Beneden).

A, Polypide expanded; B, polypide retracted; c, another view of the same, with the visceral apparatus in outline, that the manner in which it is doubled on itself, with the tentacular crown and muscular system, may be more distinctly seen:—*a, a,* tentacula ; *b,* pharynx ; *c,* pharyngeal valve ; *d,* œsophagus; *e,* stomach ; *f,* its pyloric orifice ; *g,* cilia on its inner surface ; *h,* biliary follicles lodged in its wall ; *i,* intestine; *k,* particles of excrementitious matter; *l,* anal orifice; *m,* testis: *n,* ovary; *o,* ova lying loose in the peri-visceral cavity; *p,* outlet for their discharge; *q,* spermatozoa in the perivisceral cavity; *r, s, t, u, v, w, x,* muscles.

D, Portion of the Lophophore more enlarged:—*a, a,* tentacula; *b, b,* their internal canals; *c,* their muscles; *d,* lophophore; *e,* its retractor muscles.

external integument, into which the soft membranous lining prolongs itself; the cavity thus formed, however, is not to become (as in Hydra and its allies) the stomach of the new zooid; but it constitutes the chamber surrounding the digestive viscera, which organs have their origin in a thickening of the lining membrane, that projects from one side of the cavity into its interior, and gradually shapes itself into the alimentary canal with its tentacular appendages. Of the production of gemmæ from the polypides themselves, the best examples are furnished by the *Flustræ* and their allies. From a single cell of the Flustra, five such buds may be sent-off, which develop themselves into new polypides around it; and these, in their turn, produce buds from their unattached margins, so as rapidly to augment the number of cells. To this extension there seems no definite limit; and it often happens that the cells in the central portion of the leaf like expansion of a Flustra are devoid of contents and have lost their vitality, whilst the edges are in a state of active growth.—Independently of their propagation by gemmation, the Polyzoa have a true sexual generation; the sexes, however, being usually, if not invariably, united in the same polypides. The sperm-cells are developed in a glandular body, the testis m, which lies beneath the base of the stomach; when mature they rupture, and set free the spermatozoa q q, swim freely in the liquid of the visceral cavity. The ova, on the other hand, are formed in an ovarium n, which is lodged in the membrane lining the tegumentary sheath near its outlet; the ova, having escaped from this into the visceral cavity, as at o, are fertilized by the spermatozoa which they there meet with; and are finally discharged by an outlet at p, beneath the tentacular circle.

551. These creatures possess a considerable number of muscles, by which their bodies may be projected from their sheaths, or drawn within them; of these muscles, r, s, t, u, v, w, x, the direction and points of attachment sufficiently indicate the uses; they are for the most part *retractors*, serving to draw-in and double up the body, to fold-together the circle of tentacula, and to close the aperture of the sheath, when the animal has been completely withdrawn into its interior. The *projection* and *expansion* of the animal, on the contrary, appear to be chiefly accomplished by a general pressure upon the sheath, which will tend to force-out all that can be expelled from it. The tentacles themselves are furnished with distinct muscular fibres, by which their separate movements seem to be produced. At the base of the tentacular circle, just above the anal orifice, is a small body (seen at A, a), which is a nervous ganglion; as yet no branches have been distinctly seen to be connected with it in this species; but its character is less doubtful in some other Polyzoa.—Besides the independent movements of the individual polypides, other movements may be observed, which are performed by so many of them simultaneously, as to indicate the existence of some connecting agency; and such connecting agency, it is affirmed by Dr. Fritz Müller,[1] is furnished by what he terms a 'colonial-nervous system.' In a *Serialaria* having a branching polyzoary that spreads itself on sea-weeds over a space of three or four inches, he states that a nervous ganglion may be distinguished at the origin of each branch, and another ganglion at the origin of each polypide-bud; all these ganglia being connected together, not merely by principal trunks, but also by plexuses of nerve-

---

[1] See his Memoir in "Wiegmann's Archiv," 1860, p. 311; translated in "Quart. Journ. of Microsc. Science," New Ser., Vol. i. (1861), p. 300.

fibres, which may be distinctly made-out with the aid of Chromic acid in the cylindrical joints of the polyzoary. His views, however, have not been universally accepted; some observers still maintaining that what he regards as nerve-fibres are only connective tissue.

552. Of all the Polyzoa of our own coasts, the *Flustræ* or 'sea-mats' are the most common; these present flat expanded surfaces, resembling in form those of many sea-weeds (for which they are often mistaken), but exhibiting, when viewed with even a low magnifying power, a most beautiful network, which at once indicates their real character. The cells are arranged on both sides; and it was calculated by Dr. Grant, that as a single square inch of an ordinary Flustra contains 1800 such cells, and as an average specimen presents about 10 square inches of surface, it will consist of no fewer than 18,000 polypides. The want of transparence in the cell-wall, however, and the infrequency with which the animal projects its body far beyond the mouth of the cell, render the Polyzoa of this genus less favorable subjects for microscopic examination than are those of the *Bowerbankia*, a Polyzoon with a trailing stem and separated cells like those of Laguncula, which is very commonly found clustering around the base of masses of Flustræ. It was in this that many of the details of the organization of the interesting group we are considering were first studied by Dr. A. Farre, who discovered it in 1837, and subjected it to a far more minute examination than any Polyzoon had previously received;[1] and it is one of the best adapted of all the marine forms yet known, for the display of the beauties and wonders of this type of organization.—The *Halodactylus* (formerly called Alcyonidium), however, is one of the most remarkable of all the marine forms for the comparatively large size of the tentacular crowns; these, when expanded, being very distinctly visible to the naked eye, and presenting a spectacle of the greatest beauty when viewed under a sufficient magnifying power. The polyzoary of this genus has a spongy aspect and texture, very much resembling that of certain Alcyonian Zoophytes (§ 529), for which it might readily be mistaken when its contained animals are all withdrawn into their cells; when these are expanded, however, the aspect of the two is altogether different, as the minute plumose tufts which then issue from the surface of the Halodactylus, making it look as if it were covered with the most delicate downy film, are in striking contrast with the larger, solid-looking polypes of Alcyonium. The opacity of the polyzoary of the Halodactylus renders it quite unsuitable for the examination of anything more than the tentacular crown and the œsophagus which it surmounts; the stomach and the remainder of the visceral apparatus being always retained within the cell. It furnishes, however, a most beautiful object for the Binocular Microscope, when mounted with all its polypides expanded, in the manner described in § 521.—Several of the fresh-water Polyzoa are peculiarly interesting subjects for Microscopic examination; alike on account of the remarkable distinctness with which the various parts of their organization may be seen, and the very beautiful manner in which their ciliated tentacula are arranged upon a deeply-crescentic or horseshoe-shaped *lophophore*. By this peculiarity the fresh-water Polyzoa are separated as a distinct sub-class from the marine, the former being designated as *Hippocrepia* (horseshoe-like), while the latter are termed

---

[1] See his Memoir 'On the Minute Structure of some of the higher forms of Polypi,' in the "Philosophical Transactions" for 1837, p. 387.

Infundibulata (funnel-like). The cells of the *Hippocrepia* are for the most part lodged in a sort of gelatinous substratum, which spreads over the leaves of aquatic plants, sometimes forming masses of considerable size; but in the very curious and beautiful *Cristatella*, the polyzoary is unattached, so as to be capable of moving freely through the water.[1]

553. The *Infundibulata* or Marine Polyzoa, constituting by far the most numerous division of the class, are divided into four Orders, as follows:—I. *Cheilostomata*, in which the mouth of the cell is *sub-terminal*, or not quite at its extremity (Fig. 380), is somewhat crescentic in form, and is furnished with a movable (generally membranous) lip, which closes it when the animal retreats. This includes a large part of the species that most abound on our own coasts, notwithstanding their wide differences in form and habit. Thus the polyzoaries of some (as *Flustra*) are horny and flexible, whilst those of others (as *Eschara* and *Retepora*) are so penetrated with calcareous matter as to be quite rigid; some grow as independent plant-like structures (as *Bugula* and *Gemellaria*), whilst others, having a like arborescent form, creep over the surfaces of rocks or stones (as *Hippothoa*); and others, again, have their cells in close apposition, and form crusts which possess no definite figure (as is the case with *Lepralia* and *Membranipora*).—II. The second order, *Cyclostomata*, consists of those Polyzoa which have the mouth at the *termination* of tubular calcareous cells, without any movable appendage or lip (Fig. 381). This includes a comparatively small number of genera, of which *Crisia* and *Tubulipora* contain the largest proportion of the species that occur on our own coasts.—III. The distinguishing character of the third order, *Ctenosomata*, is derived from the presence of a comb-like circular fringe of bristles, connected by a delicate membrane, around the mouth of the cell, when the animal is projected from it; this fringe being drawn in when the animal is retracted. The Polyzoaries of this group are very various in character, the cells being sometimes horny and separate (as in *Laguncula* and *Bowerbankia*), sometimes fleshy and coalescent (as in *Halodactylus*).—IV. In the fourth order, *Pedicellineæ*, which includes only a single genus, *Pedicellina*, the lophophore is produced upwards on the back of the tentacles, uniting them at their base in a sort of muscular calyx, and giving to the animal when expanded somewhat the form of an inverted bell, like that of *Vorticella* (Fig. 305).—As the Polyzoa altogether resemble Hydroid Zoophytes in their habits, and are found in the same localities, it is not requisite to add anything to what has already been said (§ 521), respecting the collection, examination, and mounting of this very interesting class of objects.[2]

554. A large proportion of the Polyzoa of the first Order are furnished with very peculiar motile appendages, which are of two kinds, *avicularia* and *vibracula*. The avicularia or 'bird's-head processes,' so named from the striking resemblance they present to the head and jaws of a bird (Fig. 381, B), are generally 'sessile' upon the angles or

---

[1] See Prof. Allman's beautiful "Monograph on the British Fresh-water Polyzoa," published by the Ray Society, 1857.

[2] For a more detailed account of the Structure and Classification of the Marine Polyzoa, see Prof. Van Beneden's 'Recherches sur les Bryozoaires de la Côte d'Ostende,' in "Mém. de l'Acad. Roy. de Bruxelles," tom. xvii.; Mr. G. Busk's "Catalogue of the Marine Polyzoa in the Collection of the British Museum;" Mr. Hincks's "British Marine Polyzoa," 1880; and Nitsche, 'Beiträge zur Kenntniss der Bryozoën, in Zeitschrift f. wiss. Zool.," Bde. xx., xxi., xxiv.

margins of the cells, that is, are attached at once to them without the intervention of a stalk, as at A, being either ' projecting ' or 'immersed;' but in the genera *Bugula* and *Bicellaria*, where they are present at all, they are 'pedunculate,' or mounted on footstalks (B). Under one form or the other, they are wanting in but few of the genera belonging to this order; and their presence or absence furnishes valuable characters for the discrimination of species. Each avicularium has two 'mandibles,' of which one is fixed, like the upper jaw of a bird, the other movable, like its lower jaw; the latter is opened and closed by two sets of muscles which are seen in the interior of the 'head;' and between them is a peculiar body, furnished with a pencil of bristles, which is probably a tactile organ, being brought forwards when the mouth is open, so that the bristles project beyond it, and being drawn back when the mandible closes. The avicularia keep up a continual snapping action during the life of the polyzoary; and they may often be observed

Fig. 381.

A. Portion of *Cellularia ciliata*, enlarged; B, one of the 'bird's-head' processes of *Bugula avicularia*, more highly magnified, and seen in the act of grasping another.

to lay hold of minute Worms or other bodies, sometimes even closing upon the beaks of adjacent organs of the same kind, as shown at B. In the pedunculate forms, besides the snapping action, there is a continual rhythmical nodding of the head upon the stalk; and few spectacles are more curious than a portion of the polyzoary of *Bugula avicularia* (a very common British species) in a state of active vitality, when viewed under a power sufficiently low to allow a number of these bodies to be in sight at once. It is still very doubtful what is their precise function in the economy of the animal; whether it is to retain within the reach of the ciliary current, bodies that may serve as food; or whether it is, like the Pedicellariæ of Echini (§ 534), to remove extraneous particles that may be in contact with the surface of the polyzoary. The latter would seem to be the function of the *vibracula*, which are long bristle-shaped organs (Fig. 380, A), each one springing at its base out of a sort of cup that contains muscles by which it is kept in almost constant motion, sweeping slowly and carefully over the surface of the polyzoary, and removing what might be injurious to the delicate inhabitants of the cells when their tentacles are protruded. Out of 191 species of Cheilostomatous Polyzoa described by Mr. Busk, no fewer than 126 are furnished either with Avicularia, or with Vibracula, or with both these organs.[1]

555. TUNICATA.—The Tunicated Mollusca are so named from the in-

---

[1] See Mr. G. Busk's 'Remarks on the Structure and Function of the Avicularian and Vibracular Organs of Polyzoa,' in "Transact. of Microsc. Soc.," Ser. 2, Vol. ii. (1854), p. 26.

closure of their bodies in a 'tunic,' which is sometimes leathery or even cartilaginous in its texture, and which very commonly includes calcareous spicules, whose forms are often very beautiful. They present a strong resemblance to the Polyzoa, not merely in their general plan of conformation, but also in their tendency to produce composite structures by gemmation; they are differentiated from them, however, by the absence of the ciliated tentacles which form so conspicuous a feature in the external aspect of the Polyzoa, by the presence of a distinct circulating apparatus, and by their peculiar respiratory apparatus, which may be regarded as a dilatation of their pharynx. In their habits, too, they are for the most part very inactive, exhibiting scarcely anything comparable to those rapid movements of expansion and retraction which it is so interesting to watch among the Polyzoa; whilst, with the exception of the *Salpidæ* and other floating species which are chiefly found in seas warmer than those that surround our coast, and the curious *Appendicularia* to be presently noticed (§ 560), they are rooted to one spot during all but the earliest period of their lives. The larger forms of the *Ascidian* group, which constitutes the bulk of the class, are always solitary; either not propagating by gemmation at all, or, if this process does take place, the gemmæ being detached before they have advanced far in their development.—Although of special importance to the Comparative Anatomist and the Zoologist, this group does not afford much to interest the ordinary Microscopist, except in the peculiar actions of its respiratory and circulatory apparatus. In common with the composite forms of the group, the solitary Ascidians have a large branchial sac, with fissured walls, resembling that shown in Figs. 382 and 384; into this sac water is admitted by the oral orifice, and a large proportion of it is caused to pass through the fissures, by the agency of the cilia with which they are fringed, into a surrounding chamber, whence it is expelled through the anal orifice. This action may be distinctly watched through the external walls in the smaller and more transparent species; and not even the ciliary action of the tentacles of the Polyzoa affords a more beautiful spectacle. It is peculiarly remarkable in one species that occurs on our own coasts, the *Ascidia parallelogramma*,[1] in which the wall of the branchial sac is divided into a number of areolæ, each of them shaped into a shallow funnel; and round one of these funnels each branchial fissure makes two or three turns of a spiral. When the cilia of all these spiral fissures are in active movement at once, the effect is most singular.—Another most remarkable phenomenon presented throughout the group, and well seen in the solitary Ascidian just referred-to, is the *alternation* in the direction of the Circulation. The heart, which lies at the bottom of the branchial sac, is composed of two chambers imperfectly divided from each other; one of these is connected with the principal trunk leading to the body, and the other with that leading to the branchial sac. At one time it will be seen that the blood flows *from* the respiratory apparatus to the cavity of the heart in which its trunk terminates, which then contracts so as to drive it into the other cavity, which in its turn contracts and propels it through the systemic trunk *to* the body at large; but after this course has been maintained for a time, the heart ceases to pulsate for a moment or two, and the course is reversed, the blood flowing into the heart *from* the body generally, and being propelled *to* the branchial sac. After this

---

[1] See Alder in "Ann. of Nat. Hist.," 3d Ser., Vol. xi. (1863), p. 157; and Hancock in "Journ. of Linn. Soc.," Vol. ix., p. 333.

reversed course has continued for some time, another pause occurs, and the first course is resumed. The length of time intervening between the changes does not seem by any means constant. It is usually stated at from half-a-minute to two minutes in the composite forms; but in the solitary *Ascidia parallelogramma* (a species very common in Lamlash Bay, Arran), the Author has repeatedly observed an interval of from five to fifteen minutes, and in some instances he has seen the circulation go-on for half-an-hour, or even longer, without change,—always, however, reversing at last.

556. The *Compound Ascidians* are very commonly found adherent to Sea-weeds, Zoophytes, and stones between the tide-marks; and they present objects of great interest to the Microscopist, since the small size and transparence of their bodies when they are detached from the mass in which they are imbedded, not only enables their structure to be clearly discerned without dissection, but allows many of their living actions to be watched. Of these we have a characteristic example in *Amoroucium proliferum;* of which the form of the composite mass and the anatomy of a single individual are displayed in Fig. 382. Its clusters appear almost completely inanimate, exhibiting no very obvious movements when irritated; but if they be placed when fresh in sea-water, a slight pouting of the orifices will soon be perceptible, and a constant and energetic series of currents will be found to enter by one set and to be ejected by the other, indicating that all the machinery of active life is going-on within these apathetic bodies. In the tribe of *Polyclinians* to which this genus belongs, the body is elongated, and may be divided into three regions, the thorax (A) which is chiefly occupied by the respiratory sac, the abdomen (B) which contains the digestive apparatus, and the post-abdomen (C) in which the heart and generative organs are lodged. At the summit of the thorax is seen the oral orifice *c*, which leads to the branchial sac *e*; this is perforated by an immense number of slits, which allow part of the water to pass into the space between the branchial sac and the muscular mantle, where it is especially collected in the thoracic sinus *f*. At *k* is

Fig. 382.

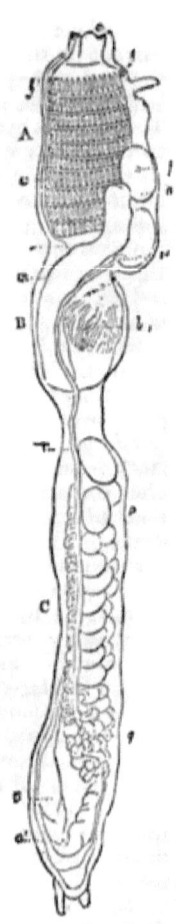

Compound mass of *Amoroucium proliferum* with the anatomy of a single zooid:—A, thorax; B, abdomen; C, post-abdomen;—*c*, oral orifice; *e*, branchial sac; *f*, thoracic sinus; *i*, anal orifice; *i'*, projection overhanging it; *j*, nervous ganglion; *k*, œsophagus; *l*, stomach surrounded by biliary tubuli; *m*, intestine; *n*, termination of intestine in cloaca; *o*, heart; *o'*, pericardium; *p*, ovarium; *p'*, egg ready to escape; *q*, testis; *r*, spermatic canal; *r'*, termination of this canal in the cloaca.

seen the œsophagus, which is continuous with the lower part of the pharyngeal cavity; this leads to the stomach *l*, which is surrounded by biliary follicles; and from this passes-off the intestine *m*, which terminates at *n* in the cloaca, or common vent. A current of water is continually drawn-in through the mouth by the action of the cilia of the branchial sac and of the alimentary canal; a part of this current passes through the fissures of the branchial sac into the thoracic sinus, and thence into the cloaca; whilst another portion, entering the stomach by an aperture at the bottom of the pharyngeal sac, passes through the alimentary canal, giving up any nutritive materials it may contain, and carrying away with it any excrementitious matter to be discharged; and this having met the respiratory current in the cloaca, the two mingled currents pass forth together by the anal orifice *i*. The long post-abdomen is principally occupied by the large ovarium, *p*, which contain ova in various stages of development. These, when matured and set-free, find their way into the cloaca; where two large ova are seen (one marked *p'*, and the other immediately below it) waiting for expulsion. In this position they receive the fertilizing influence from the testis, *q*, which discharges its products by the long spermatic canal, *r*, that opens into the cloaca, *r'*. At the very bottom of the post-abdomen we find the heart, *o*, inclosed in its pericardium, *o'*.— In the group we are now considering, a number of such animals are imbedded together in a sort of gelatinous mass, and covered with an integument common to them all; the composition of this gelatinous substance is remarkable as including Cellulose, which generally ranks as a Vegetable product. The mode in which new individuals are developed in this mass, is by the extension of *stolons* or creeping stems from the bases of those previously existing; and from each of these stolons several buds may be put-forth, every one of which may evolve itself into the likeness of the stock from which it proceeded, and may in its turn increase and multiply after the same fashion. A communication between the circulating systems of the different individuals is kept-up, through their connecting stems, during the whole of life; and thus their relationship to each other is somewhat like that of the several polypes on the polypidom of a *Campanularia* (§ 519).

557. In the family of *Didemnians* the post-abdomen is absent, the heart and generative apparatus being placed by the side of the intestine in the abdominal portion of the body. The zooids are frequently arranged in star-shaped clusters, their anal orifices being all directed towards a common vent which occupies the centre.—This shortening is still more remarkable, however, in the family of *Boctryllians*, whose beautiful stellate gelatinous incrustations are extremely common upon Sea-weeds and submerged rocks (Fig. 383). The anatomy of these animals is very similar to that of the *Amoroucium* already described; with this exception, that the body exhibits no distinction of cavities, all the organs being brought together in one, which must be considered as thoracic. In this respect there is an evident approximation towards the solitary species.[1]

558. This approximation is still closer, however, in the 'social' Ascidians, or *Clavellinidæ;* in which the general plan of structure is nearly the same, but the zooids are simply connected by their stolons

---

[1] For more special information respecting the *Compound Ascidians*, see especially the admirable Monograph of Prof. Milne-Edwards on that group; Mr. Lister's Memoir 'On the Structure and Functions of Tubular and Cellular Polypi, and of Ascidiæ,' in the "Philos. Transact.," 1834; and the Art. *Tunicata*, by Prof. T. Rupert Jones, in the "Cyclopædia of Anatomy and Physiology."

(Fig. 384), instead of being included in a common investment; so that their relation to each other is very nearly the same as that of the polypides of *Laguncula* (§ 549), the chief difference being that a regular circulation takes place through the stolon in the one case, such as has no existence in the other. A better opportunity of studying the living actions of the Ascidians can scarcely be found, than that which is afforded by the genus *Perophora*, first discovered by Mr. Lister; which occurs not unfrequently on the south coast of England and in the Irish Sea, living attached to Sea-weeds, and looking like an assemblage of minute globules of jelly, dotted with orange and brown, and linked by a silvery winding thread. The isolation of the body of each zooid from that of its fellows, and the extreme transparence of its tunics, not only enable the movements of the fluid within the body to be distinctly discerned, but also allow the action of the cilia that border the slits of the respiratory sac to be clearly made out. This sac is perforated with four rows of narrow oval openings, through which a portion of the water that enters its oral orifice (*g*) escapes into the space between the sac and the mantle, and is

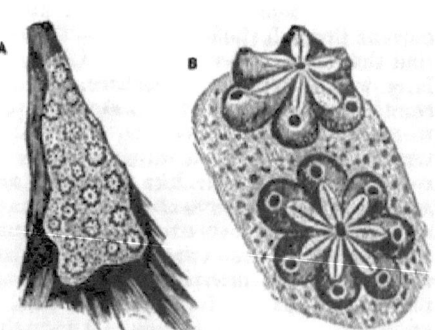

Botryllus violaceus:—A, cluster on the surface of a Fucus:—B, portion of the same enlarged.

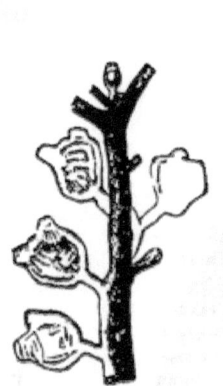
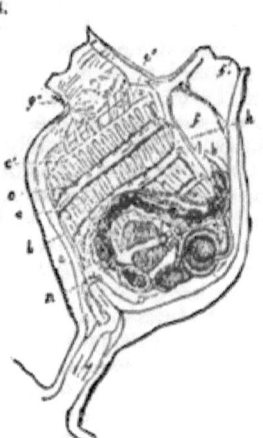

A, Group of *Perophora* (enlarged), growing from a common stalk:—B, single *Perophora*; *a*, test; *b*, inner sac; *c*, branchial sac, attached to the inner sac along the line *c' c'*; *e e*, finger-like processes projecting inwards; *f*, cavity between test and internal coat; *f'*, anal orifice or funnel; *g*, oral orifice; *g'*, oral tentacula; *h*, downward stream of food; *h'*, œsophagus; *i*, stomach; *k*, vent; *l*, ovary (?); *n*, vessels connecting the circulation in the body with that in the stalk.

thus discharged immediately by the anal funnel (*f*). Whatever little particles, animate or inanimate, the current of water brings, flow into the sac, unless stopped at its entrance by the tentacles (g'), which do not appear fastidious. The particles which are admitted usually lodge somewhere on the sides of the sac, and then travel horizontally until they arrive at that part of it down which the current proceeds to the entrance of the stomach (i), which is situated at the bottom of the sac. Minute animals are often swallowed alive, and have been observed darting about in the cavity for some days, without any apparent injury either to themselves or to the creature which incloses them. In general, however, particles which are unsuited for reception into the stomach are rejected by the sudden contraction of the mantle (or muscular tunic), the vent being at the same time closed, so that they are forced out by a powerful current through the oral orifice.—The curious alternation of the circulation that is characteristic of the Class generally (§ 555), may be particularly well studied in *Perophora*. The creeping-stalk (Fig. 384) that connects the individuals of any group, contains two distinct canals, which send off branches into each peduncle. One of these branches terminates in the heart, which is nothing more than a contractile dilatation of the principal trunk; this trunk subdivides into vessels (or rather *sinuses*, which are mere channels not having proper walls of their own), of which some ramify over the respiratory sac, branching off at each of the passages between the oval slits, whilst others are first distributed to the stomach and intestine, and to the soft surface of the mantle. All these reunite so as to form a trunk, which passes to the peduncle and constitutes the returning branch. Although the circulation in the different bodies is brought into connection by the common stem, yet that of each is independent of the rest, continuing when the current through its own footstalk is interrupted by a ligature; and the stream which returns from the branchial sac and the viscera is then poured into the posterior part of the heart, instead of entering the peduncle.

559. The *development* of the Ascidians, the early stages of which are observable whilst the ova are still within the cloaca of the parent, presents some phenomena of much interest to the Microscopist. After the ordinary repeated segmentation of the yolk, whereby a 'mulberry mass' is produced (§ 531), a sort of ring is seen encircling its central portion; but this soon shows itself as a tapering tail-like prolongation from one side of the yolk, which gradually becomes more and more detached from it, save at the part from which it springs. Either whilst the egg is still within the cloaca, or soon after it has escaped from the vent, its envelope bursts, and the larva escapes; and in this condition it presents very much the appearance of a tadpole, the tail being straightened out, and propelling the body freely through the water by its lateral strokes. The centre of the body is occupied by a mass of liquid yolk; and this is continued into the interior of three prolongations which extend themselves from the opposite extremity, each terminating in a sort of sucker. After swimming about for some hours with an active wriggling movement, the larva attaches itself to some solid body by means of one of these suckers; if disturbed from its position, it at first swims about as before; but it soon completely loses its activity, and becomes permanently attached; and important changes manifest themselves in its interior. The prolongations of the central yolk-substance into the anterior processes and tail are gradually drawn back, so that the whole of it is concentrated into one mass; and the tail, now consisting only of the gelatinous envelope, is either detached entire from the body by the contraction of the connect-

ing portion, or withers, and is thrown off gradually in shreds. The shaping of the internal organs out of the yolk-mass takes place very rapidly, so that by the end of the second day of the sedentary state the outlines of the branchial sac and of the stomach and intestine may be traced; no external orifices, however, being as yet visible. The pulsation of the heart is first seen on the third day, and the formation of the branchial and anal orifices takes-place on the fourth; after which the ciliary currents are immediately established through the branchial sac and alimentary canal.—The embryonic development of other Ascidians, solitary as well as composite, takes-place on a plan essentially the same as the foregoing, a free tadpole-like larva being always produced in the first instance.[1]

560. This larval condition is represented in a very curious adult free-swimming form, termed *Appendicularia*, which is frequently to be taken with the Tow-net on our own coasts. The animal has an oval or flask-like body, which in large specimens attains the length of one-fifth of an inch, but which is often not more than one-fourth or one-fifth of that size. It is furnished with a tail-like appendage three or four times its own length, broad, flattened, and rounded at its extremity; and by the powerful vibrations of this appendage it is propelled rapidly through the water. The structure of the body differs greatly from that of the Ascidians, its plan being much simpler; in particular, the pharyngeal sac is entirely destitute of ciliated branchial fissures opening into a surrounding cavity; but two canals, one on either side of the entrance to the stomach, are prolonged from it to the external surface; and by the action of the long cilia with which these are furnished, in conjunction with the cilia of the branchial sac, a current of water is maintained through its cavity. From the observations of Prof. Huxley, however, it appears that the direction of this current is by no means constant; since, although it usually enters by the mouth and passes out by the ciliated canals, it sometimes enters by the latter and passes out by the former. The caudal appendage has a central axis, above and below which is a riband-like layer of muscular fibres; a nervous cord, studded at intervals with minute ganglia, may be traced along its whole length.—By Mertens, one of the early observers of this animal, it was said to be furnished with a peculiar gelatinous envelope or *Haus* (house), very easily detached from the body, and capable of being re-formed after having been lost. Notwithstanding the great numbers of specimens which have been studied by Müller, Huxley, Leuckart, and Gegenbaur, neither of these excellent observers has met with this appendage; but it has been since seen by Prof. Allman, who describes it as an egg-shaped gelatinous mass, in which the body is imbedded, the tail alone being free; whilst from either side of the central plane there radiates a kind of double fan, which seems to be

---

[1] The study of the development of *Ascidians* has derived a new interest and importance from the discovery made by Kowalevsky in 1857, that their free-swimming larvæ present a most striking parallelism to Vertebrate embryoes, in exhibiting the beginnings of a spinal marrow and a spinal column; thus bridging over the gulf that was supposed to separate them from Invertebrata, and (when taken in connection with the curious Ascidian affinities of *Amphyoxus*, the lowest Vertebrate at present known) affording strong reason to believe in the derivation of the Vertebrate and Tunicate types from a common original. See his Memoir 'Entwickelungsgeschichte der einfachen Ascidien,' in "Mém. St. Pétersb. Acad. Sci.," Tom. x., 1867, and the abstract of it in "Quart. Journ. Microsc. Sci.," Vol. x., N.S. (1870), p. 59; also Prof. Haeckel's "History of Creation," Vol. ii., pp. 152, 200.

formed by a semicircular membranous lamina folded upon itself. It is surmised by Prof. Allman, with much probability, that this curious appendage is 'nidamental,' having reference to the development and protection of the young; but on this point further observations are much needed; and any Microscopist, who may meet with *Appendicularia* furnished with its 'house,' should do all he **can to** determine its structure and its relations to the body of the **animal**.[1]

---

[1] For details in respect to the structure of *Appendicularia*, see Huxley, in "Philos. Transact." for 1851, and in "Quart. Journ. of Microsc. Science," Vol. iv. (1856), p. 181; also Allman in the same Journal, Vol. vii. (1859), p. 86; Gegenbaur in Siebold and Kölliker's "Zeitschrift," Bd. vi (1855), p. 406; Leuckart's "Zoologische Untersuchungen," Heft ii., 1854; and Fol's 'Etudes sur les Appendiculaires' in "Archiv. Zool. Expérim.," Tom. i, (1872), p. 57.—For the *Tunicata* generally, see Prof. T. Rupert Jones, in Vol. iv. of the "Cyclop. of Anatomy and Physiology;" Mr. Alder's 'Observations on the British Tunicata,' in "Ann. of Nat. Hist.," Ser. 4, Vol. xi. (1863), p. 153; and Mr. Hancock's Memoir 'On the Anatomy and Physiology of the Tunicata,' in the "Journal of the Linnæan Society," Vol. ix., p. 309.

CHAPTER XVI.

MOLLUSCOUS ANIMALS GENERALLY.

561. THE various forms of 'Shell-fish,' with their 'naked' or shelless allies, furnish a great abundance of objects of interest to the Microscopist; of which, however, the greater part may be grouped under three heads:—namely, (1) the structure of the *shell*, which is most interesting in the CONCHIFERA and BRACHIOPODA, in both of which classes the shells are 'bivalve,' while the animals differ from each other essentially in general plan of structure; (2) the structure of the *tongue* or *palate* of the GASTEROPODA, most of which have 'univalve' shells, others, however, being 'naked;' (3) the *developmental history* of the embryo, for the study of which certain of the Gasteropods present the greatest facilities.—These three subjects, therefore, will be first treated of systematically; and a few miscellaneous facts of interest will be subjoined.

562. *Shells of Mollusca.*—These investments were formerly regarded as mere inorganic exudations, composed of calcareous particles, cemented together by animal glue; Microscopic examination, however, has shown that they possess a definite *structure*, and that this structure presents certain very remarkable variations in some of the groups of which the Molluscous series is composed.—We shall first describe that which may be regarded as the characteristic structure of the ordinary Bivalves; taking as a type the group of *Margaritaceæ*, which includes the *Avicula* or 'pearl-oyster' and its allies, the common *Pinna* ranking amongst the latter. In all these shells we readily distinguish the existence of two distinct layers; an *external*, of a brownish-yellow color; and an *internal*, which has a pearly or 'nacreous' aspect, and is commonly of a lighter hue.

563. The structure of the *outer* layer may be conveniently studied in the shell of *Pinna*, in which it commonly projects beyond the inner, and there often forms laminæ sufficiently thin and transparent to exhibit its general characters without any artificial reduction. If a small portion of such a lamina be examined with a low magnifying power by transmitted light, each of its surfaces will present very much the appearance of a honeycomb; whilst its broken edge exhibits an aspect which is evidently fibrous to the eye, but which, when examined under the Microscope with reflected light, resembles that of an assemblage of segments of basaltic columns (Fig. 488, P). This outer layer is thus seen to be composed of a vast number of *prisms*, having a tolerably uniform size, and usually presenting an approach to the hexagonal shape. These are arranged perpendicularly (or nearly so) to the surface of the lamina of the shell; so that its thickness is formed by their length, and its two surfaces by their extremities. A more satisfactory view of these prisms is obtained by grinding-down a lamina until it possesses a high degree of

transparence; the prisms being then seen (Fig. 385) to be themselves composed of a very homogeneous substance, but to be separated by definite and strongly marked lines of division. When such a lamina is submitted to the action of dilute acid, so as to dissolve away the carbonate of lime, a tolerably firm and consistent membrane is left, which exhibits the prismatic structure just as perfectly as did the original shell (Fig. 386); its hexagonal divisions bearing a strong resemblance to the walls of the cells of the pith or bark of a Plant. By making a section of the shell perpendicularly to its surface, we obtain a view of the prisms cut in the direction of their length (Fig. 387); and they are frequently seen to be marked by

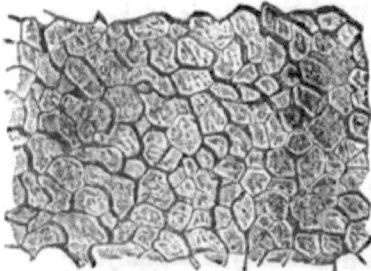

Fig. 385.

Section of shell of *Pinna*, taken transversely to the directions of its prisms.

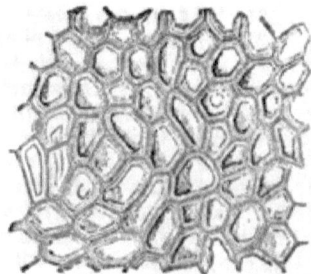

Fig. 386.

Membranous basis of the same.

delicate transverse striæ (Fig. 388), closely resembling those observable on the prisms of the enamel of teeth, to which this kind of shell structure may be considered as bearing a very close resemblance, except as regards the mineralizing ingredient. If a similar section be decalcified by dilute acid, the membranous residuum will exhibit the same resem-

Fig. 387.

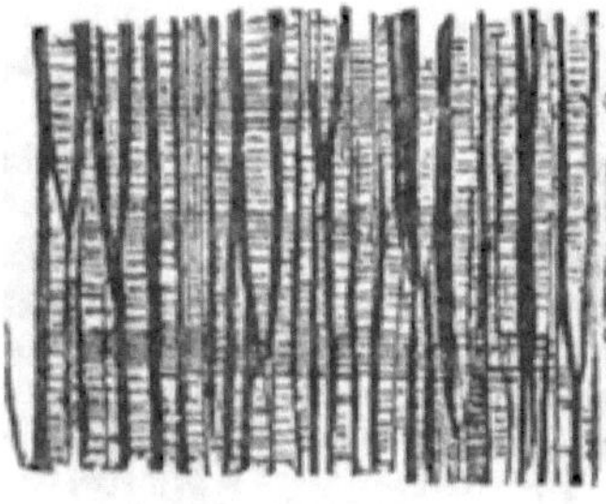

Section of the shell of *Pinna*, in the direction of its prisms.

Fig. 388.

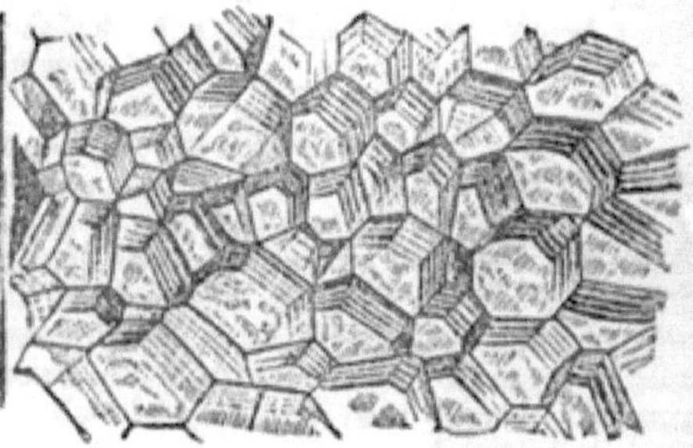

Oblique Section of Prismatic Shell-substance.

blance to the walls of prismatic cells viewed longitudinally, and will be seen to be more or less regularly marked by the transverse striæ just alluded to. It sometimes happens in recent, but still more commonly in fossil shells, that the decay of the animal membrane leaves the contained prisms without any connecting medium; as they are then quite isolated, they can be readily detached one from another; and each one may be observed to be marked by the like striations, which, when a sufficiently high magnifying power is used, are seen to be minute grooves, apparently resulting from a thickening of the intermediate wall in those situations. These appearances seem best accounted-for, by supposing that each is lengthened by successive additions at its base, the lines of junction of which correspond with the transverse striation; and this view corresponds well with the fact, that the shell-membrane not unfrequently shows a tendency to split into thin laminæ along the lines of striation; whilst we occasionally meet with an excessively thin natural lamina lying between the thicker prismatic layers, with one of which it would have probably coalesced, but for some accidental cause which preserved its distinctness. That the prisms are not formed in their entire length at once, but that they are progressively lengthened and consolidated at their lower extremities, would appear also from the fact that where the shell presents a deep color (as in *Pinna nigrina*), this color is usually disposed in distinct strata, the outer portion of each layer being the part most deeply tinged, whilst the inner extremities of the prisms are almost colorless.

564. This 'prismatic' arrangement of the carbonate of lime in the shells of *Pinna* and its allies, has been long familiar to Conchologists, and regarded by them as the result of crystallization. When it was first more minutely investigated by Mr. Bowerbank[1] and the Author,[2] and was shown to be connected with a similar arrangement in the membranous residuum left after the decalcification of the shell-substance by acid, Microscopists generally[3] agreed to regard it as a 'calcified epidermis;' the long prismatic cells being supposed to be formed by the coalescence of the epidermic cells in piles, and giving their shape to the deposit of carbonate of lime formed within them. The progress of inquiry, however, has led to an important modification of this interpretation; the Author being now disposed to agree with Prof. Huxley[4] in the belief that the entire thickness of the shell is formed as an *excretion* from the surface of the epidermis, and that the horny layer which in ordinary shells forms their external envelope or 'periostracum,'[5] being here thrown out at the same time with the calcifying material, is converted into the likeness of a cellular membrane by the pressure of the prisms that are formed by crystallization at regular distances in the midst of it. —The peculiar conditions under which calcareous concretions form themselves in an organic matrix, have been carefully studied by Mr. Rainey and Dr. W. M. Ord; of whose researches some account will be given hereafter (§ 711).

565. The *internal* layer of the shells of the *Margaritaceæ* and some other families has a 'nacreous' or iridescent lustre, which depends (as Sir D. Brewster has shown[6]) upon the striation of its surface with a

---

[1] 'On the Structure of the Shells of Molluscous and Conchiferous Animals,' in "Transact. of Microsc. Society," 1st Ser. (1844), Vol. i., p. 123.

[2] 'On the Microscopic Structure of Shells,' in "Reports of British Association" for 1844 and 1847.

[3] See Mr. Quekett's "Histological Catalogue of the College of Surgeons' Museum," and his "Lectures on Histology," Vol. ii.

[4] See his article 'Tegumentary Organs,' in "Cyclopædia of Anatomy and Physiology," Supplementary Volume, pp. 489-492.

[5] The *Periostracum* is the yellowish-brown membrane covering the surface of many shells, which is often (but erroneously) termed their *epidermis*.

[6] "Philosophical Transactions," 1814, p. 397.—The late Mr. Barton (of the Mint) succeeded in producing an artificial iridescence on metallic buttons, by

series of grooved lines, which usually run nearly parallel to each other (Fig. 389). As these lines are not obliterated by any amount of polishing, it is obvious that their presence depends upon something peculiar in the texture of this substance, and not upon any mere superficial arrangement. When a piece of the nacre (commonly known as 'mother-of-pearl') of the *Avicula* or 'pearl-oyster' is carefully examined, it becomes evident that the lines are produced by the cropping-out of laminæ of shell situated more or less obliquely to the plane of the surface. The greater the *dip* of these laminæ, the closer will their edges be; whilst the less the angle which they make with the surface, the wider will be the interval between the lines. When the section passes for any distance in the plane of a lamina, no lines will present themselves on that space. And thus the appearance of a section of nacre is such as to have been aptly compared by Sir J. Herschel to the surface of a smoothed dealboard, in which the woody layers are cut perpendicularly to their surface in one part, and nearly in their plane in another. Sir D. Brewster (*loc. cit.*) appears to have supposed that nacre consists of a multitude of layers of carbonate of lime alternating with animal membrane; and that the presence of the grooved lines on the most highly-polished surface is due to the wearing away of the edges of the animal laminæ, whilst those of the hard calcareous laminæ stand out. If each line upon the nacreous surface, however, indicates a distinct layer of shell-substance, a very thin section of 'mother-of-pearl' ought to contain many hundred laminæ, in accordance with the number of lines upon its surface; these being frequently no more than 1-7500th of an inch apart. But when the nacre is treated with dilute acid so as to dissolve its calcareous portion, no such repetition of membranous layers is to be found; on the contrary, if the piece of nacre be the product of one act of shell-formation, there is but a single layer of membrane. This layer, however, is found to present a more or less folded or plaited arrangement; and the lineation of the nacreous surface may perhaps be thus accounted for.—A similar arrangement is found in *pearls;* which are rounded concretions projecting from the inner surface of the shell of *Avicula*, and possessing a nacreous structure corresponding to that of 'mother-of-pearl.' Such concretions are found in many other shells, especially the fresh-water mussels, *Unio* and *Anodon;* but these are usually less remarkable for their pearly lustre; and, when formed at the edge of the valves, they may be partly or even entirely made-up of the prismatic substance of the external layer, and may be consequently altogether destitute of the pearly character.

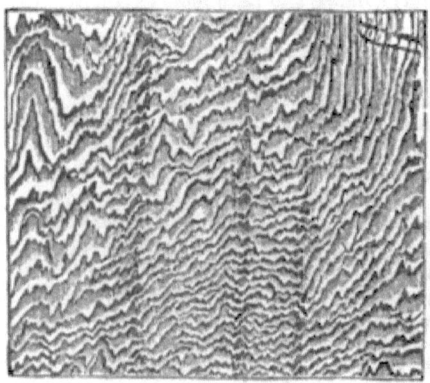

Fig. 389.

Section of nacreous lining of Shell of *Avicula margaritacea* (Pearl-oyster).

---

drawing closely-approximating lines with a diamond-point upon the surface of the steel die by which they were struck.

566. In all the genera of the *Margaritaceæ*, we find the external layer of the shell prismatic, and of considerable thickness; the internal layer being nacreous. But it is only in the shells of a few families of Bivalves, that the combination of organic with mineral components is seen in the same distinct form; and these families are for the most part nearly allied to *Pinna*. In the *Unionidæ* (or 'fresh-water mussels'), nearly the whole thickness of the shell is made-up of the internal or 'nacreous' layer; but a uniform stratum of prismatic substance is always found between the nacre and the periostracum, really constituting the inner layer of the latter, the outer being simply horny.—In the *Ostraceæ* (or oyster tribe) also, the greater part of the thickness of the shell is composed of a 'sub-nacreous' substance (§ 568) representing the inner layer of the shells of Margaritaceæ, its successively-formed laminæ, however, having very little adhesion to each other; and every one of these laminæ is bordered at its free edge by a layer of the prismatic substance, distinguished by its brownish-yellow color. In these and some other cases, a distinct membranous residuum is left after the decalcification of the prismatic layer by dilute acid; and this is most tenacious and substantial, where (as in the *Margaritaceæ*) there is no proper periostracum. Generally speaking, a thin prismatic layer may be detected upon the external surface of Bivalve shells, where this has been protected by a periostracum, or has been prevented in any other manner from undergoing abrasion; thus it is found pretty generally in *Chama*, *Trigonia*, and *Solen*, and occasionally in *Anomia* and *Pecten*.

567. In many other instances, however, nothing like a cellular structure can be distinctly seen in the delicate membrane left after decalcification; and in such cases the animal basis bears but a very small proportion to the calcareous substance, and the shell is usually extremely hard. This hardness appears to depend upon the mineral arrangement of the carbonate of lime; for whilst in the *prismatic* and ordinary *nacreous* layer this has the crystalline condition of *calcite*, it can be shown in the hard shell of *Pholas* to have the arrangement of *arragonite;* the difference between the two being made evident by Polarized light. A very curious appearance is presented by a section of the large hinge-tooth of *Mya arenaria* (Fig. 390), in which the carbonate of lime seems to be deposited in nodules that possess a crystalline structure resembling that of the mineral termed *Wavellite*. Approaches to this curious arrangement are seen in many other shells.

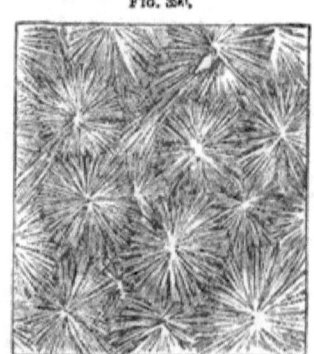

Section of hinge-tooth, of *Mya arenaria*.

568. There are several Bivalve shells which almost entirely consist of what may be termed a *sub-nacreous* substance; their polished surfaces being marked by lines, but these lines being destitute of that regularity of arrangement which is necessary to produce the iridescent lustre. This is the case, for example, with most of the *Pectinidæ* (or scallop tribe), also with some of the *Mytilaceæ* (or mussel tribe), and with the common Oyster. In the internal layer of by far the greater number of Bivalve

shells, however, there is not the least approach to the nacreous aspect; nor is there anything that can be described as definite structure;[1] and the residuum left after its decalcification is usually a structureless 'basement-membrane.'

569. The ordinary account of the mode of growth of the shells of Bivalve Mollusca,—that they are progressively enlarged by the deposition of new laminæ, each of which is in contact with the internal surface of the preceding, and extends beyond it,—does not express the whole truth; for it takes no account of the fact that most shells are composed of two layers of very different texture, and does not specify whether *both* these layers are thus formed by the entire surface of the 'mantle' whenever the shell has to be extended, or whether only *one* is produced. An examination of Fig. 391 will clearly show the mode in which the operation is effected. This figure represents a section of one of the valves of *Unio occidens*, taken perpendicularly to its surface, and passing from the margin or lip (at the left hand of the figure) towards the hinge (which would be at some distance beyond the right). This section brings into view the two substances of which the shell is composed; traversing the

FIG. 391.

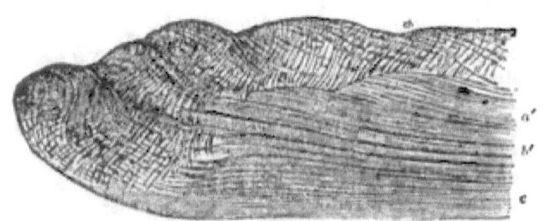

Vertical section of the lip of one of the valves of the shell of *Unio:*—*a*, *b*, *c*, successive formations of the outer prismatic layer; *a'*, *b'*, *c'*, the same of the inner nacreous layer.

outer or prismatic layer in the direction of the length of its prisms, and passing through the nacreous lining in such a manner as to bring into view its numerous laminæ, separated by the lines $a\,a'$, $b\,b'$, $c\,c'$, etc. These lines evidently indicate the successive formations of this layer; and it may be easily shown by tracing them towards the hinge on the one side and towards the margin on the other, that at every enlargement of the shell its whole interior is lined by a new nacreous lamina in immediate contact with that which preceded it. The number of such laminæ, therefore, in the oldest part of the shell, indicates the number of enlargements which it has undergone. The outer or prismatic layer of the growing shell, on the other hand, is only formed where the new structure projects beyond the margin of the old; and thus we do not find one layer of it overlapping another, except at the lines of junction of two distinct formations. When the shell has attained its full dimensions, however, new laminæ of both layers still continue to be added, and thus the lip becomes thickened by successive formations of prismatic structure, each being applied to the inner surface of the preceding, instead of to its free margin. —A like arrangement may be well seen in the *Oyster;* with this differ-

---

[1] For an explanation of the real nature of what was formerly described by the Author as 'tubular' Shell-substance, see § 316.

ence, that the successive layers have but a comparatively slight adhesion to each other.

570. The shells of *Terebratulæ*, however, and of most other *Brachiopods*, are distinguished by peculiarities of structure which differentiate them from all others. When thin sections of them are microscopically examined, they exhibit the appearance of long flattened prisms (Fig. 392, A, b), which are arranged with such obliquity that their rounded extremities crop-out upon the inner surface of the shell in an imbricated (tile-like) manner (a). All true *Terebratulidæ*, both recent and fossil, exhibit another very remarkable peculiarity; namely, the *perforation* of the shell by a large number of canals, which generally pass nearly perpendicularly from one surface to the other (as is shown vertical sections, Fig. 393), and terminate internally by open orifices (Fig. 392, A), whilst externally they are covered by the periostracum (B). Their diameter is greatest towards the external surface, where they sometimes expand suddenly, so as to become trumpet-shaped; and it is usually narrowed rather suddenly, when, as sometimes happens, a new internal layer is formed as a lining to the preceding (Fig. 393, A, d d). Hence the diameter of these canals, as shown in different transverse sections of one and the same shell, will vary according to the part of its thickness which the section happens to traverse.—The shells of different species of perforated *Brachiopods*,

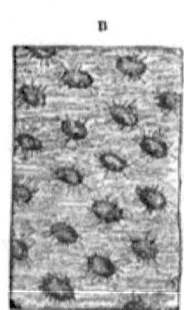

FIG. 392.

A, Internal surface (a), and oblique section (b), of Shell of *Terebratula* (Waldheimia) *australis*; B, external surface of the same.

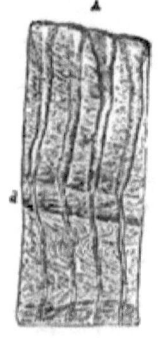

FIG. 393.

Vertical Sections of Shell of *Terebratula* (Waldheimia) *australis*; showing at A the canals opening by large trumpet-shaped orifices on the outer surface, and contracting at d, d, into narrow tubes; and showing at B a bifurcation of the canals.

however, present very striking diversities in the size and closeness of their canals, as shown by sections taken in corresponding parts; three examples of this kind are given for the sake of comparison in Figs. 394-396. These canals are occupied in the living state by tubular prolongations of the mantle, whose interior is filled with a fluid containing minute cells and granules, which, from its corresponding in appearance with the fluid contained in the great sinuses of the mantle, may perhaps be considered to be the animal's blood. Of their special function in the economy of the animal, it is difficult to form any probable idea; but it is interesting to remark (in connection with the hypothseis of a relationship between Brachiopods and Polyzoa) that they seem to have their parallel in extensions of the peri-visceral cavity of many species of *Flustra, Eschara, Lepralia*, etc., into passages excavated in the walls of the cells of the polyzoary.

571. In the Family *Rhynchonellidæ*, which is represented by only two recent species (the *Rh. psittacea* and *Rh. nigricans*, both formerly ranking as Terebratulæ), but which contains a very large proportion of fossil Brachiopods, these canals are almost entirely absent; so that the uniformity of their presence in the Terebratulidæ, and their general absence in the Rhynchonellidæ, supplies a character of great value in the discrimination of the fossil shells belonging to these two groups respectively. Great caution is necessary, however, in applying this test; *mere surface-markings cannot be relied-on;* and no statement on this point is worthy of reliance, which is not based on a Microscopic examination of thin *sections* of the shell.—In the Families *Spiriferidæ* and *Strophomenidæ*, on the other hand, some species possess the perforations, whilst others are desti-

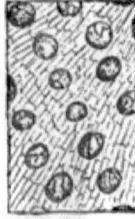  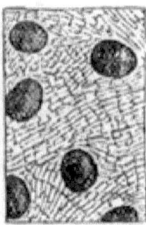

FIG. 394. FIG. 395. FIG. 396.

FIG. 394. Horizontal section of Shell of *Terebratula bullata* (fossil, Oolite).
FIG. 395. Ditto . . . . of *Megerlia lima* (fossil, Chalk).
FIG. 396. Ditto . . . . of *Spiriferina rostrata* (Triassic).

tute of them; so that their presence or absence *there* serves only to mark-out subordinate groups. This, however, is what holds-good in regard to characters of almost every description, in other departments of Natural History; a character which is of fundamental importance from its close relation to the general plan of organization in one group, being, from its want of constancy, of far less account in another.[1]

572. There is not by any means the same amount of diversity in the structure of the Shell in the class of *Gasteropods;* a certain typical plan of construction being common to by far the greater number of them. The small proportion of animal matter contained in most of these shells, is a very marked feature in their character; and it serves to render other

---

[1] For a particular account of the Author's researches on this group, see his Memoir on the subject, forming part of the introduction of Mr. Davidson's "Monograph of the British Fossil Brachiopoda," published by the Palæontographical Society.—A very remarkable example of the importance of the presence or absence of the perforations, in distinguishing shells whose internal structure shows them to be generically different, whilst from their external conformation they would be supposed to be not only *generically* but *specifically identical*, will be found in the "Annals of Natural History," Ser. 3, Vol. xx. (1867), p. 68.

features indistinct, since the residuum left after the removal of the calcareous matter is usually so imperfect, as to give no clue whatever to the explanation of the appearances shown by sections. Nevertheless, the structure of these shells is by no means homogeneous, but always exhibits indications, more or less clear, of a definite arrangement. The 'porcellanous' shells are composed of three layers, all presenting the same kind of structure, but each differing from the others in the mode in which this is disposed. For each layer is made-up of an assemblange of thin laminæ placed side-by-side, which separate one from another, apparently in the planes of rhomboidal cleavage, when the shell is fractured; and as was first pointed out by Mr. Bowerbank, each of these laminæ consists of a series of elongated spicules (considered by him as prismatic cells filled with carbonate of lime) lying side-by-side in close apposition; and, these series are disposed alternately in contrary directions, so as to intersect each other nearly at right angles, though still lying in parallel planes. The direction of the planes is different, however, in the three layers of the shell, bearing the same relation to each other as have those three sides of a cube which meet each other at the same angle; and by this arrangement, which is better seen in the fractured edge of the *Cypræa* or any similar shell, than in thin sections, the strength of the shell is greatly augmented.—A similar arrangement, obviously answering the same purpose, has been shown by Mr. Tomes to exist in the enamel of the teeth of Rodentia.

573. The principal departures from this plan of structure are seen in *Patella, Chiton, Haliotis, Turbo* and its allies, and in the 'naked Gasteropods, many of which last, both terrestrial and marine, have some rudiment of a shell. Thus in the common Slug, *Limax rufus*, a thin oval plate of calcareous texture is found imbedded in the shield-like fold of the mantle covering the fore-part of its back; and if this be examined in an early stage of its growth, it is found to consist of an aggregation of minute calcareous nodules, generally somewhat hexagonal in form, and sometimes quite transparent, whilst in other instances it presents an appearance closely resembling that delineated in Fig. 390.—In the epidermis of the mantle of some species of *Doris*, on the other hand, we find long calcareous spicules, generally lying in parallel directions, but not in contact with each other, giving firmness to the whole of its dorsal portion; and these are sometimes covered with small tubercles, like the spicules of Gorgonia (Fig. 363). They may be separated from the soft tissue in which they are imbedded, by means of caustic potash; and when treated with dilute acid, whereby the calcareous matter is dissolved-away, an organic basis is left, retaining in some degree the form of the original spicule. This basis cannot be said to be a true cell; but it seems to be rather a cell in the earliest stage of its formation, being an isolated particle of sarcode without wall or cavity; and the close correspondence between the appearance presented by thin sections of various Univalve shells, and the forms of the spicules of *Doris*, seems to justify the conclusion that even the most compact shells of this group are constructed out of the like elements, in a state of closer aggregation and more definite arrangement, with the occasional occurrence of a layer of more spheroidal bodies of the same kind, like those forming the rudimentary shell of *Limax*.

574. The structure of Shells generally is best examined by making *sections* in different planes as nearly parallel as may be possible to the surfaces of the shell, and other sections at right angles to these: the

former may be designated as *horizontal*, the latter as *vertical*. Nothing need here be added to the full directions for making such Sections, which have already been given (§§ 192–194). Many of them are beautiful and interesting objects for the Polariscope.—Much valuable information may also be derived from the examination of the surfaces presented by *fracture*. The membranous residua left after the decalcification of the shell by dilute acid, may be mounted in weak spirit or in Goadby's solution.

575. The animals composing the class of *Cephalopoda* (cuttle-fish and nautilus tribe) are for the most part unpossessed of shells; and the structure of the few that we meet-with in the genera *Nautilus*, *Argonauta* ('paper-nautilus'), and *Spirula*, does not present any peculiarities that need here detain us. The rudimentary shell or *sepiostaire* of the common Cuttle-fish, however, which is frequently spoken-of as the 'cuttle-fish bone,' exhibits a very beautiful and remarkable structure, such as causes sections of it to be very interesting Microscopic objects. The outer shelly portion of this body consists of horny layers, alternating with calcified layers, in which last may be seen a hexagonal arrangement somewhat corresponding with that in Fig. 390. The soft friable substance that occupies the hollow of this boat-shaped shell, is formed of a number of delicate calcareous plates, running across it from one side to the other in parallel directions, but separated by intervals several times wider than the thickness of the plates; and these intervals are in great part filled-up by what appear to be fibres or slender pillars, passing from one plate or floor to another. A more careful examination shows, however, that instead of a large number of detached pillars, there exists a comparatively small number of very thin sinuous laminæ, which pass from one surface to the other, winding and doubling upon themselves, so that each lamina occupies a considerable space. Their precise arrangement is best seen by examining the parallel plates, after the sinuous laminæ have been detached from them; the lines of junction being distinctly indicated upon these. By this arrangement each layer is most effectually supported by those with which it is connected above and below; and the sinuosity of the thin intervening laminæ, answering exactly the same purpose as the 'corrugation' given to iron plates for the sake of diminishing their flexibility, adds greatly to the strength of this curious texture; which is at the same time lightened by the large amount of open space between the parallel plates, that intervenes among the sinuosities of the laminæ. The best method for examining this structure, is to make sections of it with a sharp knife in various directions, taking care that the sections are no thicker than is requisite for holding-together; and these may be mounted on a Black Ground as opaque objects, or in Canada balsam as transparent objects, under which last aspect they furnish very beautiful objects for the Polariscope.

576. *Palate of Gasteropod Mollusks.*—The organ which is sometimes referred to under this designation, and sometimes as the 'tongue,' is one of a very singular nature; and cannot be likened to either the tongue or the palate of higher animals. For it is a tube that passes backward and downwards beneath the mouth, closed at its hinder end, whilst in front it opens obliquely upon the floor of the mouth, being (as it were) slit-up and spread-out so as to form a nearly flat surface. On the interior of the tube, as well as on the flat expansion of it, we find numerous transverse rows of minute teeth, which are set upon flattened plates; each principal tooth sometimes having a basal plate of its own, whilst in

other instances one plate carries several teeth.—Of the former arrangement we have an example in the palate of many terrestrial Gasteropods, such as the snail (*Helix*) and Slug (*Limax*), in which the number of plates in each row is very considerable (Figs. 397, 398), amounting to 180 in the large garden Slug (*Limax maximus*); whilst the latter prevails in many marine Gasteropods, such as the common Whelk (*Buccinum undatum*), the palate of which has only three plates in each row, one bearing the small central teeth, and the two others the large lateral teeth (Fig.

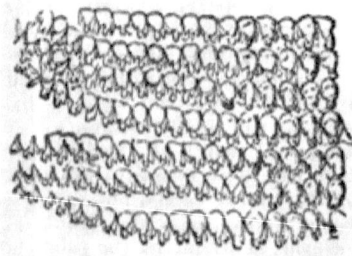

Fig. 397.

Portion of the left half of the Palate of *Helix hortensis*; the rows of teeth near the edge separated from each other to show their form.

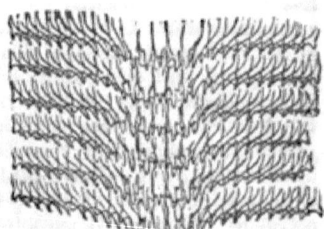

Fig. 398.

Palate of *Zonites cellarius*.

401). The length of the palatal tube, and the number of rows of teeth, vary greatly in different species. Generally speaking, the tube of the terrestrial Gasteropods is short, and is contained entirely within the nearly globular head; but the rows of teeth being closely set together are usually very numerous, there being frequently more than 100, and in

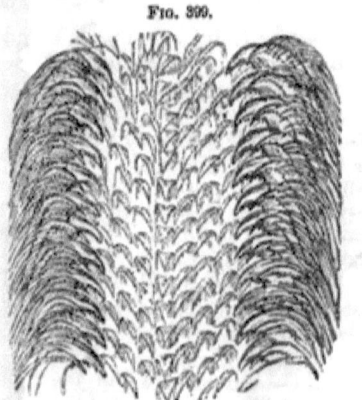

Fig. 399.

Palate of *Trochus zizyphinus*.

Fig. 400.

Palate of *Doris tuberculata*.

some species as many as 160 or 170; so that the total numer of teeth may mount-up, as in *Helix pomatia*, to 21,000, and in *Limax maximus*, to 26,800. The transverse rows are usually more or less curved, as shown in Fig. 398, whilst the longitudinal rows are quite straight: and the curvature takes its departure on each side from a central longitudinal row, the teeth of which are symmetrical, whilst those of the lateral portions of each transverse row present a modification of that symmetry, the prominences on the *inner* side of each tooth being suppressed, whilst those on the outer side are increased; this modification being observed to augment in degree, as we pass from the central line towards the edges.

577. The palatal tube of the marine Gasteropods is generally longer, and its teeth larger; and in many instances it extends far beyond the head, which may, indeed, contain but a small part of it. Thus in the common Limpet (*Patella*), we find the principal part of the tube to lie folded-up, but perfectly free, in the abdominal cavity, between the intestine and the muscular foot; and in some species its length is twice or even three times as great as that of the entire animal. In a large proportion of cases, these palates exhibit a very marked separation between the central and the lateral portions (Figs. 399, 401); the teeth of the central band being frequently small and smooth at their edges, whilst those of the lateral are large and serrated. The palate of *Trochus zizyphinus*, represented in Fig. 399, is one of the most beautiful examples of this form; not only the large teeth of the lateral bands, but the delicate leaf-like teeth of the central portion, having their edges minutely serrated. A yet more complex type, however, is found in the palate of *Haliotis*; in which there is a central band of teeth having nearly straight edges instead of points: then, on each side, a lateral band consisting of large teeth shaped like those of the Shark; and beyond this, again, another lateral band on either side, composed of several rows of smaller teeth.—Very curious differences also present themselves among the different species of the same genus. Thus in *Doris pilosa*, the central band is almost entirely wanting, and each lateral band is formed of a single row of very large hooked teeth, set obliquely like those of the lateral band in Fig. 399; whilst in *Doris tuberculata*, the central band is the part most developed, and contains a number of rows of conical teeth, standing almost perpendicularly, like those of a harrow (Fig. 400).

578. Many other varieties might be described, did space permit; but we must be content with adding, that the form and arrangement of the teeth of these 'palates' afford characters of great value in classification, as was first pointed out by Prof. Lovèn (of Stockholm) in 1847, and has been since very strongly urged by Dr. J. E. Gray, who considers that the structure of these organs is one of the best guides to the natural affinities of the species, genera, and families of this group, since any important alteration in the form or position of the teeth must be accompanied by some corresponding peculiarity in the habits and food of the animal.'[1] Hence a systematic examination and delineation of the structure and arrangement of these organs, by the aid of the Microscope and Camera Lucida, would be of the greatest service to this department of Natural History. The short thick tube of *Limax* and other terrestrial Gasteropods, appears adapted for the trituration of the food previously to its passing into the œsophagus; for in these animals we find the roof of the mouth furnished with a large strong horny plate, against which the flat end of the tongue can work. On the other hand, the flattened portion of the palate of *Buccinum* (whelk) and its allies is used by these animals as a file, with which they bore holes through the shells of the

[1] "Annals of Natural History," Ser. 2, Vol. x. (1852), p. 413.

Mollusks that serve as their prey; this they are enabled to effect by everting that part of the probosis-shaped mouth whose floor is formed by the flattened part of the tube, which is thus brought to the exterior, and by giving a kind of sawing-motion to the organ by means of the alternate action of two pairs of muscles,—a protractor, and a retractor,—which put-forth and draw-back a pair of cartilages whereon the tongue is supported, and also elevate and depress its teeth. Of the use of the long blind tubular part of the palate in these Gasteropods, however, scarcely any probable guess can be made; unless it be a sort of 'cavity of reserve,' from which a new toothed surface may be continually supplied as the old one is worn-away, somewhat as the front teeth of the Rodents are constantly being regenerated from the surface of the pulps which occupy their hollow conical bases, as fast as they are rubbed-down at their edges.

579. The preparation of these Palates for the Microscope can, of course, be only accomplished by carefully dissecting them from their attachments within the head; and it will be also necessary to remove the membrane that forms the sheath of the tube, when this is thick enough to interfere with its transparence. The tube itself should be slit up with a pair of fine scissors through its entire length; and should be so opened out, that its expanded surface may be a continuation of that which forms the floor of the mouth. The mode of mounting it will depend upon the manner in which it is to be viewed. For the ordinary purposes of Microscopic examination, no method is so good as mounting in fluid; either weak Spirit or Goadby's solution answering very well. But many of these palates, especially those of the marine Gasteropods, become most beautiful objects for the Polariscope when they are mounted in Canada balsam; the form and arrangement of the teeth being very strongly brought-out by it (Fig. 401), and a gorgeous play of colors being exhibited when a selenite plate is placed behind the object, and the analyzing prism is made to rotate.[1]

580. *Development of Mollusks.*—Leaving to the scientific Embryologist the large field of study that lies open to him in this direction,[2] the ordinary Microscopist will find much to interest him in the observation of certain special phenomena of which a general account will be here given. Attached to the gills of fresh-water Mussels (*Unio* and *Anodon*) there are often found minute bodies, which, when first observed, were described as parasites, under the name of *Glochidia*, but are now known to be their own progeny in an early phase of development. When a Fish is near, they are expelled from between the valves of their parent, and attach themselves in a peculiar manner to its fins and gills (Fig. 402, A). In this stage of the existence of the young *Anodon*, its valves are provided with curious barbed or serrated hooks (D, *b*), and are continually snapping together (so as to remind the observer of the *avicularia* of Polyzoa, § 554), until they have inserted their hooks into the skin of the Fish, which seems so to retain the barbs as to prevent the reopening of the valves. In this stage of its existence no internal organ is definitely formed, except the strong 'adductor muscle' (c, *a*) which draws the valves together, and the long, slender, byssus-filament (B, *a*, D) which makes its appearance while the embryo is still within the egg membrane,

---

[1] For additional details on the organization of the Palate and Teeth of the Gasteropod Mollusks, see Mr. W. Thomson, in "Cyclop. of Anat. and Physiol." Vol. iv., pp. 1142, 1143; and in "Ann. of Nat. Hist.," Ser. 2 Vol. vii., p. 86.

[2] See Balfour's "Comparative Embryology," Chap. ix.

lying coiled-up between the lateral lobes. The hollow of each valve is filled with a soft granular-looking mass, in which are to be distinguished what are perhaps the rudiments of the branchiæ and of oral tentacles; but their nature can only be certainly determined by further observation, which is rendered difficult by the opacity of the valves. By keeping a supply of Fish, however, with these embryoes attached, the entire history of the development of the fresh-water Mussel may be worked out.[1]

581. In certain members of the Class *Gasteropods*, the history of embryonic development presents numerous phenomena of great interest. The eggs (save among the terrestrial species) are usually deposited in aggregate masses, each inclosed in a common protective envelope or *nidamentum*. The nature of this envelope, however, varies greatly: thus, in the common *Limnæus stagnalis* or 'water-snail' of our ponds

FIG. 401.   FIG. 402.

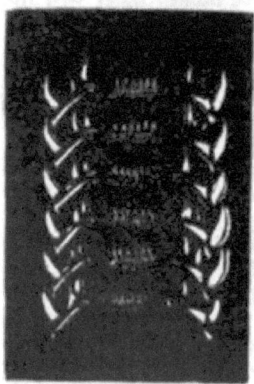
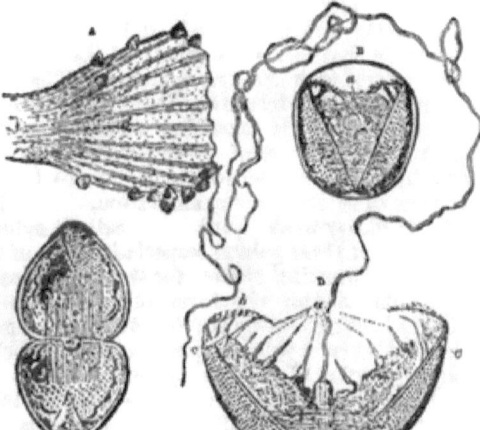

Palate of *Buccinum undatum* as seen under Polarized Light.

Parasitic Larva (*Glochidium*) of *Anodon*:—A, glochidia attached to the tail of a Stickleback; B, side view of glochidium still inclosed in the egg-membrane, showing the hooks of its valves and the byssus-filament *a*; C, glochidium with its valves widely opened, showing the adductor-muscle *a*; D, side view of glochidium; with the valves opened to show the origin of the byssus-filament and the three pairs of tentacular (?) organs, the barbed hooks *b*, and the muscular or membranous folds *c*, *c*, connected with them.

and ditches, it is nothing else than a mass of soft jelly about the size of a sixpence, in which from 50 to 60 eggs are imbedded, and which is attached to the leaves or stems of aquatic plants; in the *Buccinum undatum*, or common Whelk, it is a membranous case, connected with a considerable number of similar cases by short stalks, so as to form large globular masses which may often be picked-up on our shores especially between April and June; in the *Purpura lapillus*, or 'rock-whelk,' it is a little flask-shaped capsule, having a firm horny wall, which is attached by a short stem to the surface of rocks between the tide-marks, great numbers being often found standing erect side by side; whilst in the

---

[1] See the Rev. W. Houghton 'On the Parasitic Nature of the Fry of the *Anodonta cygnea*,' in "Quart. Journ. of Microsc. Sci.," N.S., Vol. ii. (1861), p. 162; and Balfour, *op. cit.*, pp. 220–223.

Nudibranchiate order generally (consisting of the *Doris, Eolis,* and other 'sea-slugs') it forms a long tube with a membranous wall, in which immense numbers of eggs (even half a million or more) are packed closely together in the midst of a jelly-like substance, this tube being disposed in coils of various forms, which are usually attached to the Sea-weeds or Zoophytes.—The course of development, in the first and last of these instances, may be readily observed from the very earliest period down to that of the emersion of the embryo; owing to the extreme transparence

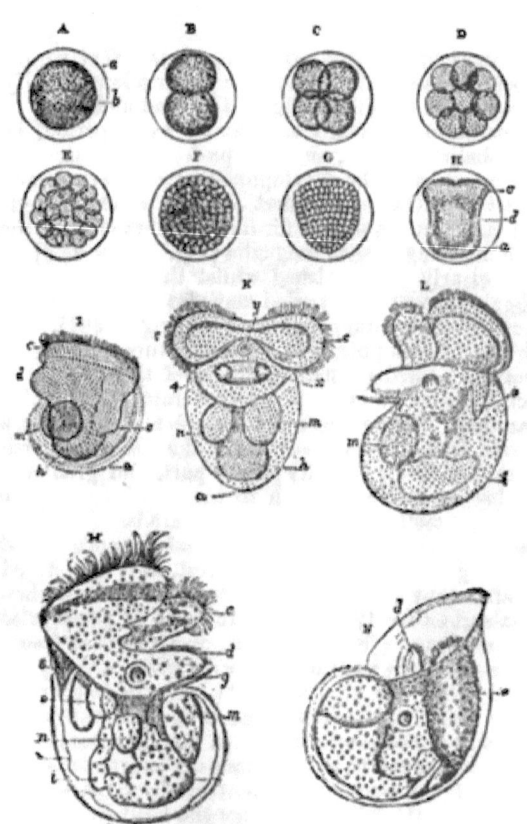

Fig. 403.

Embryonic development of *Doris bilamellata*:—A, Ovum, consisting of enveloping membrane *a* and yolk *b*; B, C, D, E, F, successive stages of segmentation of yolk; G, first marking-out of the shape of the embryo; H, embryo on the 8th day; I, the same on the 9th day; K, the same on the 12th day, seen on the left side at L; M, still more advanced embryo, seen at N as retracted within its shell:—*a*, superficial layer of yolk-segments coalescing to give origin to the shell; *c, c*, ciliated lobes; *d*, foot; *g*, hard plate or operculum attached to it; *h*, stomach; *i*, intestine; *m, n*, masses (glandular ?) at the sides of the œsophagus; *o*, heart (?); *s*, retractor muscle (?); *t*, situation of funnel; *v*, membrane enveloping the body; *z*, auditory vesicles; *y*, mouth.

of the nidamentum and of the egg-membranes themselves. The first change which will be noticed by the ordinary observer, is the 'segmentation' of the yolk-mass, which divides itself (after the manner of a cell undergoing binary subdivision) into two parts, each of these two into two others, and so on until a *morula* or mulberry-like mass of minute yolk-segments is produced (Fig. 403, A-F), which is converted by 'invagination' into a 'gastrula' (§ 391), whose form is shown at G. This 'gastrula' soon begins to exhibit a very curious alternating rotation within the egg, two or three turns being made in one direction, and the same number in a reverse direction: this movement is due to the cilia fringing a sort of fold of the ectoderm termed the *velum*, which afterwards usually gives origin to a pair of large ciliated lobes (H-L, *c*) resembling those of Rotifers. The velum is so little developed in *Limnæus*, however, that its existence has been commonly overlooked until recognized by Prof. Ray Lankester,[1] who also has been able to distinguish its fringe of minute cilia. This, however, has only a transitory existence; and the later rotation of the embryo, which presents a very curious spectacle when a number of ova are viewed at once under a low magnifying power, is due to the action of the cilia fringing the head and foot.

582. A separation is usually seen at an early period, between the anterior or 'cephalic' portion, and the posterior or 'visceral' portion, of the embryonic mass; and the development of the former advances with the greater activity. One of the first changes which is seen in it, consists in its extension into a sort of fin-like membrane on either side, the edges of which are fringed with long cilia (Fig. 403, H-L, *c*), whose movements may be clearly distinguished whilst the embryo is still shut-up within the egg; at a very early period may also be discerned the 'auditory vesicles' (K, *x*) or rudimentary organs of hearing (§ 587), which scarcely attain any higher development in these creatures during the whole of life; and from the immediate neighborhood of these is put-forth a projection, which is afterwards to be evolved into the 'foot' or muscular disk of the animal. While these organs are making their appearance, the shell is being formed on the surface of the posterior portion, appearing first as a thin covering over its hinder part, and gradually extending itself until it becomes large enough to inclose the embryo completely, when this contracts itself. The ciliated lobes are best seen in the embryoes of *Nudibranchs;* and the fact of the universal presence of a shell in the embryoes of that group is of peculiar interest, as it is destined to be cast-off very soon after they enter upon active life. These embryoes may be seen to move-about as freely as the narrowness of their prison permits, for some time previous to their emersion; and when set free by the rupture of the egg-cases, they swim forth with great activity by the action of their ciliated lobes,—these, like the 'wheels' of Rotifera, serving also to bring food to the mouth, which is at that time unprovided with the reducing apparatus subsequently found in it. The same is true of the embryo of *Lymnæus*, save that its swimming movements are less active, in consequence of the non-development of the ciliated lobes; and the currents produced by the cilia that fringe the head and the orifice of the respiratory sac, seem to have reference chiefly to the provision of supplies of food, and of aërated water for respiration. The disappearance of the cilia has been observed by Mr. Hogg to be coincident with the development of the teeth to a degree sufficient to enable the young water-snail to crop its vegetable food; and he has further ascertained that if the growing animal be kept in fresh water alone for some time, without vege-

---

[1] See his valuable 'Observations on the Development of *Limnæus stagnalis*, and on the other stages of other Mollusca,' in "Quart. Journ. Microsc. Science," Oct. 1874. See also Lereboullet, 'Recherches sur le Développement du Limnée,' in "Ann. des Sci. Nat. Zool.," 4ième Sér., Tom. xviii., p. 47.

table matter of any kind, the gastric teeth are very imperfectly developed, and the cilia are still retained.[1]

583. A very curious modification of the ordinary plan of development is presented in the *Purpura lapillus;* and it is probable that something of the same kind exists also in *Buccinum,* as well as in other Gasteropods of the same extensive Order (*Pectinibranchiata*).—Each of the capsules already described (§ 581) contains from 500 to 600 egg-like bodies (Fig. 404, A), imbedded in a viscid gelatinous substance; but only from 12 to 30 embroyes usually attain complete development; and it is obvious from the large comparative size which these attain (Fig. 405, B), that each of them must include an amount of substance equal to that of a great number of the bodies originally found within the capsule. The explanation of this fact (long since noticed by Dr. J. E. Gray, in regard to Buccinum) seems to be as follows:—Of those 500 or 600 egg-like bodies, only a small part are fertile *ova,* the remainder being unfertilized eggs, the yolk-

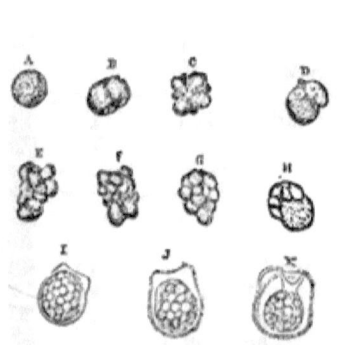

Fig. 404.

Early stages of Embryonic Development of *Purpura lapillus:*—A, egg-like spherule; B, C, E, F, G, successive stages of segmentation of yolk-spherules; D, H, I, J, K, successive stages of development of early embryoes.

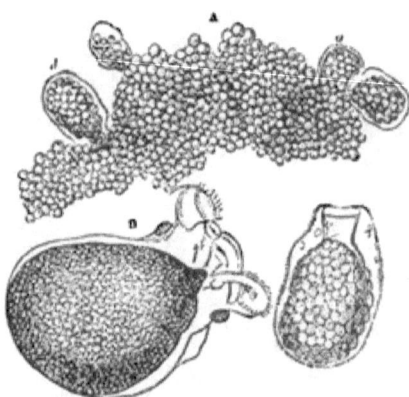

Fig. 405.

Later stages of embryonic Development of *Purpura lapillus:*—A, conglomerate mass of vitelline segments, to which were attached the embryoes, a, b, c, d, e:—B, full-size embryo, in more advanced stage of development.

material of which serves for the nutrition of the embryoes in the later stages of their intra-capsular life. The distinction between them manifests itself at a very early period, even in the first segmentation; for while the latter divide into two equal hemispheres (Fig. 404, B), the fertilized ova divide into a larger and a smaller segment (D); in the cleft between these are seen the minute 'directive vesicles,' which appear to be always double or even triple, although from being seen 'end on,' only one may be visible; and near these is generally to be seen a clear space in each segment. The difference is still more strongly marked in the subsequent divisions; for whilst the cleavage of the infertile eggs goes-on irregularly, so as to divide each into from 14 to 20 segments, having no definiteness of arrangement (C, E, F, G), that of the fertile ova takes place in such a manner as to mark-out the distinction already alluded-to between the

---

[1] See "Transact. of Microsc. Soc.," 2d Ser., Vol. ii. (1854), p. 93.

'cephalic' and the 'visceral' portions of the mass (H); and the evolution of the former into distinct organs very speedily commences. In the first instance, a narrow transparent border is seen around the whole embryonic mass, which is broader at the cephalic portion (I); next, this border is fringed with short cilia, and the cephalic extension into two lobes begins to show itself; and then between the lobes a large mouth is formed, opening through a short, wide œsophagus, the interior of which is ciliated, into the visceral cavity, occupied as yet only by the yolk-particles originally belonging to the ovum (K).

584. Whilst these developmental changes are taking place in the embryo, the whole aggregate of segments formed by the yolk-cleavage of the infertile eggs coalesces into one mass, as shown at A, Fig. 405; and the embryoes are often, in the first instance, so completely buried within this, as only to be discoverable by tearing its portions asunder: but some of them may commonly be found upon its exterior; and those contained in one capsule very commonly exhibit the different stages of development represented in Fig. 404, H-K. After a short time, however, it becomes apparent that the most advanced embryoes are beginning to *swallow* the yolk-segments of the conglomerate mass; and capsules will not unfrequently be met-with, in which embryoes of various sizes, as a, b, c, d, e (Fig. 405, A), are projecting from its surface, their difference of size not being accompanied by advance in development, but merely depending upon the amount of this 'supplemental' yolk which the embryoes have respectively gulped-down. For during the time in which they are engaged in appropriating this additional supply of nutriment, although they increase in *size*, yet they scarcely exhibit any other change; so that the large embryo, Fig. 405, e, is not apparently more advanced as regards the formation of its organs, than the small embryo, Fig. 404, K. So soon as this operation has been completed, however, and the embryo has attained its full bulk, the evolution of its organs takes-place very rapidly; the ciliated lobes are much more highly developed, being extended in a long sinuous margin, so as almost to remind the observer of the 'wheels' of Rotifera (§ 445), and being furnished with very long cilia (Fig. 405, B); the auditory vesicles, the tentacula, the eyes, and the foot, successively make their appearance; a curious rhythmically-contractile vesicle is seen, just beneath the edge of the shell in the region of the neck, which may, perhaps, serve as a temporary heart; a little later, the real heart may be seen pulsating beneath the dorsal part of the shell; and the mass of yolk-segments of which the body is made-up, gradually shapes itself into the various organs of digestion, respiration, etc., during the evolution of which (and while they are as yet far from complete) the capsule thins-away at its summit, and the embryoes make their escape from it.'

585. It happens not unfrequently that one of the embryoes which a capsule contains does not acquire its 'supplemental' yolk in the manner now described, and can only proceed in its development as far as its original yolk will afford it material; and thus, at the time when the other embryoes have attained their full size and maturity, a strange-looking

---

[1] The Author thinks it worth while to mention the method which he has found most convenient for examining the contents of the egg-capsules of *Purpura;* as he believes that it may be advantageously adopted in many other cases. This consists in cutting off the two ends of the capsule (taking care not to cut far into its cavity), and in then forcing a jet of water through it, by inserting the end of a fine-pointed syringe (§ 127) into one of the orifices thus made, so as to drive the contents of the capsule before it through the other. These should be received into a shallow cell, and first examined under the Simple Microscope.

creature, consisting of two large ciliated lobes with scarcely the rudiment of a body, may be seen in active motion among them. This may happen, indeed, not only to one but to several embryoes within the same capsule, especially if their number should be considerable; for it sometimes appears as if there were not food enough for all, so that whilst some attain their full dimensions and complete development, others remain of unusually small size, without being deficient in any of their organs, and others again are more or less completely abortive,—the supply of supplemental yolk which they have obtained having been too small for the development of their viscera, although it may have afforded what was needed for that of the ciliated lobes, eyes, tentacles, auditory vesicles, and even the foot,—or, on the other hand, no additional supply whatever having been acquired by them, so that their development has been arrested at a still earlier stage.—These phenomena are of so remarkable a character, that they furnish an abundant source of interest to any Microscopist who may happen to be spending the months of August and September in a locality in which the *Purpura* abounds; since, by opening a sufficient number of capsules, no difficulty need be experienced in arriving at all the facts which have been noticed in this brief summary.[1] It is much to be desired that such Microscopists as possess the requisite opportunity, would apply themselves to the study of the corresponding history in other Pectinibranchiate Gasteropods, with a view of determining how far the plan now described prevails through the Order. And now that these Mollusks have been brought not only to live, but to breed, in artificial *aquaria*, it may be anticipated that a great addition to our knowledge of this part of their life-history will ere long be made.

586. *Ciliary Motions on Gills.*—There is no object that is better suited to exhibit the general phenomena of Ciliary motion (§ 435), than a portion of the gill of some bivalve Mollusk. The *Oyster* will answer the purpose sufficiently well; but the cilia are much larger on the gills of the *Mussel*,[2] as they are also on those of the *Anodon* or common 'freshwater mussel' of our ponds and streams. Nothing more is necessary than to detach a small portion of one of the riband-like bands, which will be seen running parallel with the edge of each of the valves when the shell is opened; and to place this, with a little of the liquor contained within the shell, upon a slip of glass,—taking care to spread it out sufficiently with needles to separate the *bars* of which it is composed, since it is on the edges of these, and round their knobbed extremities, that the ciliary movement presents itself,—and then covering it with a thin-glass disk. Or it will be convenient to place the object in the Aquatic-box (§ 122), which will enable the observer to subject it to any degree of pressure that

---

[1] Fuller details on this subject will be found in the Author's account of his researches, in "Transactions of the Microscopical Society," 2d Ser., Vol. iii. (1855), p. 17. His account of the process was called in question by MM. Koren and Danielssen, who had previously given an entirely different version of it, but was fully confirmed by the observations of Dr. Dyster; see "Ann. of Nat. Hist." 2d Ser., Vol. xx. (1857), p. 16. The independent observations of M. Claparède on the development of *Neritina fluviatilis* (Müller's "Archiv," 1857, p. 109, and abstract in "Ann. of Nat. Hist.," 2d Ser., Vol. xx. (1857), p. 196, showed the mode of development in that species to be the same in all essential particulars as that of Purpura. The subject has again been recently studied with great minuteness by Selenka, "Niederländisches Archiv für Zoologie," Bd. i., July, 1862

[2] This Shell-fish may be obtained, not merely at the sea-side, but likewise at the shops of the fishmongers who supply the humbler classes, even in midland towns.

he may find convenient. A magnifying power of about 120 diameters is amply sufficient to afford a general view of this spectacle; but a much greater amplification is needed to bring into view the peculiar mode in which the stroke of each cilium is made. Few spectacles are more striking to the unprepared mind, than the exhibition of such wonderful activity as will then become apparent, in a body which to all ordinary observation is so inert. This activity serves a double purpose; for it not only drives a continual current of water over the surface of the gills themselves, so as to effect the aëration of the blood, but also directs a portion of this current (as in the *Tunicata*, § 555) to the mouth, so as to supply the digestive apparatus with the aliment afforded by the *Diatomaceæ*, *Infusoria*, etc., which it carries-in with it.

587. *Organs of Sense of Mollusks.*—Some of the minuter and more rudimentary forms of the special organs of sight, hearing, and touch, which the Molluscous series presents, are very interesting objects of Microscopic examination. Thus, just within the margin of each valve of *Pecten*, we see (when we observe the animal in its living state, under water) a row of minute circular points of great brilliancy, each surrounded by a dark ring; these are the eyes, with which this creature is provided, and by which its peculiarly-active movements are directed. Each of them, when their structure is carefully examined, is found to be protected by a sclerotic coat with a transparent cornea in front; and to possess a colored iris (having a pupil) that is continuous with a layer of pigment lining the sclerotic, a crystalline lens and vitreous body, and a retinal expansion proceeding from an optic nerve which passes to each eye from the trunk that runs along the margin of the mantle.[1]—Eyes of still higher organization are borne upon the head of most Gasteropod Mollusks, generally at the base of one of the pairs of tentacles, but sometimes, as in the *Snail* and *slug*, at the points of these organs. In the latter case, the tentacles are furnished with a very peculiar provision for the protection of the eyes; for when the extremity of either of them is touched, it is drawn-back into the basal part of the organ, much as the finger of a glove may be pushed-back into the palm. The retraction of the tentacle is accomplished by a strong muscular band, which arises within the head, and proceeds to the extremity of the tentacles; whilst its protrusion is effected by the agency of the circular bands with which the tubular wall of the tentacle is itself furnished, the inverted portion being (as it were) squeezed-out by the contraction of the lower part in which it has been drawn back. The structure of the eyes, and the curious provision just described, may easily be examined by snipping-off one of the eye-bearing tentacles with a pair of scissors.—None but the Cephalopod Mollusks have distinct organs of hearing; but rudiments of such organs may be found in most Gasteropods (Fig. 403, K, x), attached to some part of the nervous collar that surrounds the œsophagus; and even in many Bivalves, in connection with the nervous ganglion imbedded in the base of the foot. These 'auditory vesicles,' as they are termed, are minute sacculi, each of which contains a fluid, wherein are suspended a number of minute calcareous particles (named *otoliths* or ear-stones), which are kept in a state of continual movement by the action of cilia lining the vesicles. This "wonderful spectacle," as it was truly designated by its discoverer Siebold, may be brought into view without any dissection, by submitting the head of any small

---

[1] See Mr. S. J. Hickson on 'The Eye of Pecten,' in "Quart. Journ. Microsc. Sci.," Vol. xx., N.S. (1880), p. 443.

and not very thick-skinned Gasteropod, or the young of the larger forms, to gentle compression under the Microscope, and transmitting a strong light through it. The very early appearance of the auditory vesicles in the embryo Gasteropod has been already alluded-to (§ 582).—Those who have the opportunity of examining young specimens of the common *Pecten*, will find it extremely interesting to watch the action of the very delicate tentacles which they have the power of putting-forth from the margin of their mantle, the animal being confined in a shallow cell, or in the zoophyte-trough; and if the observer should be fortunate enough to obtain a specimen so young that the valves are quite transparent, he will find the spectacle presented by the ciliary movement of the gills, as well as the active play of the foot (of which the adult can make no such use), to be worthy of more than a cursory glance.

588. *Chromatophores of Cephalopods.*—Almost any species of Cuttle-fish (*Sepia*) or Squid (*Loligo*) will afford the opportunity of examining the very curious provision which their skin contains for changing its hue. This consists in the presence of numerous large 'pigment-cells,' containing coloring-matter of various tints; the prevailing color, however, being that of the fluid of the ink-bag. These pigment-cells may present very different forms, being sometimes nearly globular, whilst at other times they are flattened and extended into radiating prolongations; and, by the peculiar contractility with which they are endowed, they can pass from one to the other of these conditions, so as to spread their colored contents over a comparatively-large surface, or to limit them within a comparatively small area. Very commonly there are different layers of these pigment-cells, their contents having different hues in each layer and thus a great variety of coloration may be given, by the alteration in the form of the cells of which one or another layer is made-up. It is curious that the changes in the hue of the skin appear to be influenced, as in the case of the Chameleon, by the color of the surface with which it may be in proximity. The alternate contractions and extensions of these pigment-cells or *chromatophores* may be easily observed in a piece of skin detached from the living animal and viewed as a transparent object; since they will continue for some time, if the skin be placed in sea-water. And they may also be well seen in the embryo cuttle-fish, which will sometimes be found in a state of sufficient advancement in the grape-like eggs of these animals attached to Sea-weeds, Zoophytes, etc.—The eggs of the small cuttle-fish termed the *Sepiola*, which is very common on our southern coasts, are imbedded, like those of the Doris, in gelatinous masses, which are attached to Sea-weeds, Zoophytes, etc.; and their embryoes, when near maturity, are extremely beautiful and interesting objects, being sufficiently transparent to allow the action of the heart to be distinguished, as well as to show most advantageously the changes incessantly occurring in the form and hue of the 'chromatophores.'

CHAPTER XVII.

ANNULOSA, OR WORMS.

589. UNDER the general designation of 'Annulose' animals, or Worms, may be grouped-together all that lower portion of the great *Articulated* Sub-kingdom, in which the division of the body into longitudinally-arranged segments is not distinctly marked-out, and there is an absence of those 'articulated' or jointed limbs that constitute so distinct a feature of Insects and their allies. This group includes the classes of *Entozoa* or Intestinal Worms, *Rotifera* or wheel-animalcules, *Turbellaria*, and *Annelida;* each of which furnishes many objects for Microscopic examination, that are of the highest scientific interest. As our business, however, is less with the professed Physiologist, than with the general inquirer into the minute wonders and beauties of Nature, we shall pass over these classes (the Rotifera having been already treated-of in detail, Chap. XI.) with only a notice of such points as are likely to be specially deserving the attention of observers of the latter order.

590. ENTOZOA.—This class consists almost entirely of animals of a very peculiar plan of organization, which are parasitic within the bodies of other animals, and which obtain their nutriment by the absorption of the juices of these,—thus bearing a striking analogy to the parasitic Fungi (§§ 312–316). The most remarkable feature in their structure consists in the entire absence or the extremely low development of their nutritive system, and the extraordinary development of their reproductive apparatus. Thus, in the common *Tænia* ('tape-worm'), which may be taken as the type of the Cestoid group, there is neither mouth nor stomach, the so-called 'head' being merely an organ for attachment, whilst the segments of the 'body' contain repetitions of a complex generative apparatus, the male and female sexual organs being so united in each as to enable it to fertilize and bring to maturity its own very numerous eggs; and the chief connection between these segments is established by two pairs of longitudinal canals, which, though regarded by some as representing a digestive apparatus, and by others as a circulating system, appear really to represent the 'water-vascular system,' whose simplest condition has been noticed in the wheel-animalcule (§ 449).—Few among the recent results of Microscopic inquiry have been more curious, than the elucidation of the real nature of the bodies formerly denominated *Cystic* Entozoa, which had been previously ranked as a distinct group. These are not found, like the preceding, in the cavity of the alimentary canal of the animals they infest; but always occur in the substance of solid organs, such as the glands, muscles, etc. They present themselves to the eye as bags or vesicles of various sizes, sometimes occurring singly, some-

times in groups; but upon careful examination each vesicle is found to bear upon some part a 'head' furnished with hooklets and suckers; and this may be either single, as in *Cysticercus* (the entozoon whose presence gives to pork what is known as the 'measly' disorder), or multiple, as in *Cœnurus*, which is developed in the brain, chiefly of sheep, giving rise to the disorder known as 'the staggers.' Now in none of these Cystic forms has any generative apparatus ever been discovered, and hence they are obviously to be considered as imperfect animals. The close resemblance between the 'heads' of certain *Cysticerci* and that of certain *Tæniæ* first suggested that the two might be different states of the same animal; and experiments made by those who have devoted themselves to the working-out of this curious subject have led to the assured conclusion, that the Cystic Entozoa are nothing else than Cestoid Worms, whose development has been modified by the peculiarity of their position,—the large bag being formed by a sort of dropsical accumulation of fluid when the young are evolved in the midst of solid tissues, whilst the very same bodies, conveyed into the alimentary canal of some carnivorous animal which has fed upon the flesh infested with them, begin to bud-forth the generative segments, the long succession of which, united end-to-end, gives to the entire series a Worm-like aspect.

591. The higher forms of Entozoa, belonging to the *Nematoid* or thread-like Order,—of which the common *Ascaris* may be taken as a type, one species of it (the *A. lumbricoides*, or 'round worm') being a common parasite in the small intestine of man, while another (the *A. vermicularis*, or thread-worm') is found rather in the lower bowel,—approach more closely to the ordinary type of conformation of Worms; having a distinct alimentary canal, which commences with a mouth at the anterior extremity of the body, and which terminates by an anal orifice near the other extremity; and also possessing a regular arrangement of circular and longitudinal muscular fibres, by which the body can be shortened, elongated, or bent in any direction. The smaller species of *Ascaris*, by some or other of which almost every Vertebrated animal is infested, are so transparent that every part of their internal organization may be made-out, especially with the assistance of the Compressor (§ 125) without any dissection; and the study of the structure and actions of their Generative apparatus has yielded many very interesting results, especially in regard to the first formation of the ova, the mode of their fertilization, and the history of their subsequent development.—Some of the Worms belonging to this group are not parasitic in the bodies of other animals, but live in the midst of dead or decomposing Vegetable matter. The *Gordius* or 'hair worm,' which is peculiar in not having any perceptible anal orifice, seems to be properly a parasite in the intestines of water-insects; but it is frequently found in large knot-like masses (whence its name) in the water or mud of the pools inhabited by such insects, and may apparently be developed in these situations. The *Anguillulæ* are little eel-like worms of which one species, *A. fluviatilis*, is very often found in fresh water amongst *Desmidieæ, Confervæ*, etc., also in wet moss and moist earth, and sometimes also in the alimentary canals of snails, frogs, fishes, insects, and larger worms; whilst another species, *A. tritici*, is met-with in the ears of Wheat affected with the blight termed the 'cockle;' another, the *A. glutinis*, is found in sour paste; and another, the *A. aceti*, was often found in stale vinegar, until the more complete removal of mucilage and the addition of sulphuric acid, in the course of the manufacture, rendered this liquid a less favorable 'habitat' for these little creatures. A

writhing mass of any of these species of 'eels,' is one of the most curious spectacles which the Microscopist can exhibit to the unscientific observer; and the capability which they all possess (in common with Rotifers and Tardigrades, § 452), of revival after desiccation, at however remote an interval, enables him to command the spectacle at any time. A grain of wheat within which these worms (often erroneously called *Vibriones*) are being developed, gradually assumes the appearance of a black peppercorn; and if it be divided in two, the interior will be found almost complete filled with a dense white cottony mass, occupying the place of the flour, and leaving merely a small place for a little glutinous matter. The cottony substance seems to the eye to consist of bundles of fine fibres closely packed-together; but on taking-out a small portion, and putting it under the Microscpe with a little water under a thin glass-cover, it will be found after a short time (if not immediately) to be a wriggling mass of life, the apparent fibres being really *Anguillulæ*, or 'eels' of the Microscopist. If the seeds be soaked in water for a couple of hours before they are laid open, the eels will be found in a state of activity from the first; their movements, however, are by no means so energetic as those of the *A. glutinis* or 'paste-eel.' This last frequently makes its appearance spontaneously in the midst of paste that is turning sour; but the best means of securing a supply for any occasion, consists in allowing a portion of any mass of paste in which they may present themselves to dry up, and then, laying this by so long as it may not be wanted, to introduce it into a mass of fresh paste, which if it be kept warm and moist, will be found after a few days to swarm with these curious little creatures.

592. Besides the foregoing Orders of Entozoa, the *Trematode* group must be named; of which the *Distoma hepaticum* or 'fluke,' found in the livers of Sheep affected with the 'rot,' is a typical example. Into the details of the structure of this animal, which has the general form of a sole, there is no occasion for us here to enter; it is remarkable, however, for the branching form of its digestive cavity, which extends throughout almost the entire body, very much as in Planariæ (Fig. 406); and also for the curious phenomena of its development, several distinct forms being passed through between one sexual generation and another. These have been especially studied in the Distoma, which infests the *Lymnæus;* the ova of which are not developed into the likeness of their parents, but into minute worm-like bodies, which seem to be little else than masses of cells inclosed in a contractile integument, no formed organs being found in them; these cells, in their turn, are developed into independent zooids, which escape from their containing cyst in the condition of free ciliated Animalcules; in this condition they remain for some time, and then imbed themselves in the mucus that covers the tail of the Mollusk, in which they undergo a gradual development into true Distomata; and having thus acquired their perfect form, they penetrate the soft integument, and take-up their habitation in the interior of the body. Thus a considerable number of Distomata may be produced from a single ovum, by a process of cell-multiplication in an early stage of its development. In some instances the free ciliated larva possesses distinct eyes; although these organ are wanting in the fully developed Distoma, the peculiar 'habitat' of which would render them useless.

593. TURBELLARIA.—This group of animals, which is distinguished by the presence of cilia over the entire surface of the body, seems intermediate in some respects between the 'trematode' Entozoa and the Leech-tribe among Annelida. It deserves special notice here, chiefly on account

ANNULOSA, OR WORMS.

of the frequency with which the worms of the *Planarian* tribe present themselves among collections both of marine and of fresh-water animals (particular species inhabiting either locality), and on account of the curious organization which many of these possess. Most of the members of this tribe have elongated flattened bodies, and move by a sort of gliding or crawling action over the surfaces of aquatic Plants and Animals. Some of the smaller kind are sufficiently transparent to allow of their internal structure being seen by transmitted light, especially when they are slightly compressed; and the accompanying figure (Fig. 406) displays the general conformation of their principal organs, as thus shown. The body has the flattened sole-like shape of the Trematode Entozoa; its mouth, which which is situated at a considerable distance from the anterior extremity of the body, is surrounded by a circular sucker that is applied to the living surface from which the animal draws its nutriment; and the buccal cavity (*b*) opens into a short œsophagus (*c*), which leads at once to the cavity of the stomach. In the true *Planariæ* the mouth is furnished with a sort of long funnel-shaped proboscis; and this, even when detached from the body, continues to swallow anything presented to it. The cavity of the stomach does not give origin to any intestinal tube, nor is it provided with any second orifice; but a large number of ramifying canals are prolonged from it, which carry its contents into every part of the body. This seems to render unnecessary any system of vessels for the circulation of nutritive fluid; and the two principal trunks, with connecting and ramifying branches, which may be observed in them, are probably to be regarded in the light of a water-vascular system, the function of which is essentially respiratory. Both sets of sexual organs are combined in the same individuals; though the congress of two, each impregnating the ova of the other, seems to be generally necessary. The ovaria, as in the Entozoa, extend through a large part of the body, their ramifications proceeding from the two oviducts (*k, k*), which have a dilatation (*l*) at their point of junction.—There is still much obscurity about the history of the embryonic development of these animals; as the accounts given of it by different observers by no means harmonize with each other.[1]— The Planariæ, however, do not multiply by eggs alone; for they occasionally undergo spontaneous fission in a transverse direction, each segment becoming a perfect animal; and an artificial division into two or even

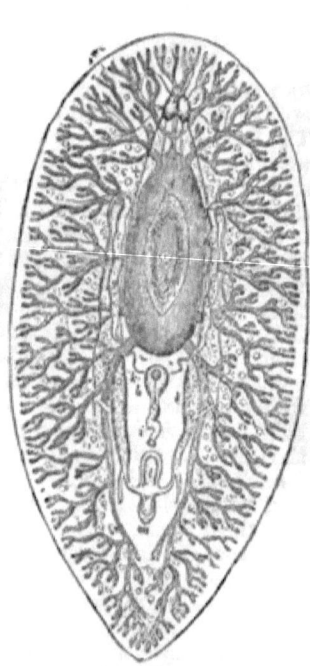

Fig. 406.

Structure of *Polycelis levigatus* (a Planarian worm).—*a*, Mouth surrounded by its circular sucker; *b*, buccal cavity; *c*, œsophageal orifice; *d*, stomach; *e*, ramifications of gastric canals; *f*, cephalic ganglia and their nervous filaments; *g, g*, testes; *h*, vesicula seminalis; *i*, male genital canal; *k, k*, oviducts; *l*, dilatation at their point of junction; *m*, female genital orifice.

---

[1] See Balfour's "Comparative Embryology," Vol. i., pp. 159–162.

more parts may be practised with a like result. In fact, the power of the Planariæ to reproduce portions which have been removed, seems but little inferior to that of the Hydra (§ 515); a circumstance which is peculiarly remarkable, when the much higher character of their organization is borne in mind. They possess a distinct pair of nervous ganglia (f, f), from which branches proceed to various parts of the body; and in the neighborhood of these are usually to be observed a number (varying from 2 to 40) of *ocelli* or rudimentary eyes, each having its refracting body or crystalline lens, its pigment-layer, its nerve bulb, and its cornea-like bulging of the skin. The integument of many of these animals is furnished with 'thread-cells' or 'filiferous capsules,' very much resembling those of Zoophytes (§ 528).

594. ANNELIDS.—This Class includes all the higher kinds of Worm-like animals, the greater part of which are marine, though there are several species which inhabit fresh water, and some which live on land. The body in this class is usually very long, and nearly always presents a well-marked segmental division, the segments being for the most part similar and equal to each other, except at the two extremities; but in the lower forms, such as the Leech and its allies, the segmental division is very indistinctly seen, on account of the general softness of the integument. A large proportion of the marine Annelids have special respiratory appendages, into which the fluids of the body are sent for aëration; and these are situated upon the head (Fig. 407), in those species which (like the *Serpula, Terebella, Sabellaria*, etc.) have their bodies inclosed by tubes, either formed of a shelly substance produced from their own surface, or built up by the agglutination of grains of sand, fragments of shell, etc.; whilst they are distributed along the two sides of the body in such as swim freely through the water, or crawl over the surfaces of rocks, as is the case with the *Nereidæ*, or simply bury themselves in the sand, as the *Arenicola* or 'lob-worm.' In these respiratory appendages the circulation of the fluids may be distinctly seen by Microscopic examination; and these fluids are of two kinds,—first, a colorless fluid, containing numerous cell-like corpuscles, which can be seen in the smaller and more transparent species to occupy the space that intervenes between the outer surface of the alimentary canal and the inner wall of the body, and to pass from this into canals which often ramify extensively in the respiratory organs, but are never furnished with a returning series of passages, —and second, a fluid which is usually red, contains few floating particles, and is inclosed in a system of proper vessels that communicates with a central propelling organ, and not only carries away the fluid away from this, but also brings it back again. In *Terebella* we find a distinct provision for the aëration of both fluids; for the first is transmitted to the tendril-like tentacles which surround the mouth (Fig. 407, *b, b*), whilst the second circulates through the beautiful aborescent gill-tufts (*k, k*), situated just behind the head. The former are covered with cilia, the action of which continually renews the stratum of water in contact with them, whilst the latter are destitute of these organs; and this seems to be the general fact as to the several appendages to which these two fluids are respectively sent for aëration, the nature of their distribution varying greatly in the different members of the class. The red fluid is commonly considered as blood, and the tubes through which it circulates as blood-vessels; but the Author has elsewhere given his reasons[1] for coinciding in the opinion of Prof. Huxley, that the colorless

---

[1] See his "Principles of Comparative Physiology," 4th Edit., §§ 218, 219, 292.

corpusculated fluid which moves in the peri-visceral cavity of the body and in its extensions, is that which really represents the blood of other Articulated animals; and that the system of vessels carrying the red fluid is to be likened on the one hand to the 'water-vascular system' of the inferior Worms, and on the other to the tracheal apparatus of Insects (§ 634).—In the observation of the beautiful spectacle presented by the respiratory circulation of the various kinds of Annelids which swarm on most of our shores, and in the examination of what is going on in the interior of their bodies (where this is rendered possible by their transparence), the Microscopist will find a most fertile source of interesting occupation; and he may easily, with care and patience, make many valuable additions to our present stock of knowledge on these points. There are many of these marine Annelids, in which the appendages of various kinds put forth from the sides of their bodies furnish very beautiful microscopic objects; as do also the different forms of teeth, jaws, etc., with which the mouth is commonly armed in the free or non-tubicolar species, these being eminently carnivorous.

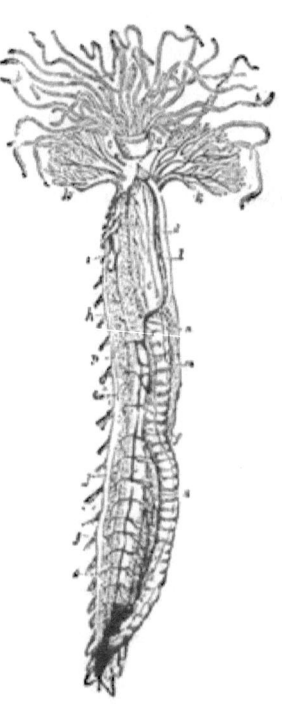

Fig. 407.

595. The early history of the Development of Annelids, too, is extremely curious; for they come forth from the egg in a condition very little more advanced than the ciliated gemmules of Polypes, consisting of a globular mass of untransformed cells, certain parts of whose surface are covered with cilia; in a few hours, however, this embryonic mass elongates, and the indications of a segmental division become apparent, the head being (as it were) marked off in front, whilst behind this is a large segment thickly covered with cilia, then a narrower and non-ciliated segment, and lastly the caudal or tail-segment, which is furnished with cilia. A little later, a new segment is seen to be interposed in front of the caudal; and the dark internal granular mass shapes itself into the outline of an alimentary canal.[1] The number of segments progressively increases by the interposition of new ones between the caudal

Circulating Apparatus of *Terebella conchilega*:—*a*, labial ring; *b, b*, tentacles; *c*, first segment of the trunk; *d*, skin of the back; *e*, pharynx; *f*, intestine; *g*, longitudinal muscles of the inferior surface of the body; *h*, glandular organ (liver ?); *i*, organs of generation; *j*, feet; *k, k*, branchiæ; *l*, dorsal vessel acting as a respiratory heart; *m*, dorso-intestinal vessel; *n*, venous sinus surrounding œsophagus; *n'*, inferior intestinal vessel; *o*, ventral trunk; *p*, lateral vascular branches.

and its preceding segments; the various internal organs become more and more distinct, eye-spots make their appearance, little bristly appendages are

---

[1] A most curious transformation once occurred within the Author's experience in the larva of an Annelid, which was furnished with a broad collar or disk fringed with very long cilia, and showed merely an appearance of segmentation in its hinder part; for in the course of a few minutes, during which it was not under observation, this larva assumed the ordinary form of a marine Worm three

put forth from the segments, and the animal gradually assumes the likeness of its parent; a few days being passed by the tubicolar kinds, however, in the actively moving condition, before they settle down to the formation of a tube.'

596. To carry out any systematic observations on the embryonic development of Annelids, the eggs should be searched for in the situations which these animals haunt; but in places where Annelids abound, free-swimming larvæ are often to be obtained at the same time and in the same manner as small Medusæ (§ 522); and there is probably no part of our coasts, off which some very curious forms may not be met with. The following may be specially mentioned as departing widely from the ordinary type, and as in themselves extremely beautiful objects.—The *Actinotrocha* (Fig. 408) bears a strong resemblance in many particulars to the 'bipinnarian' larva of a Star-fish (§ 543), having an elongated body, with a series of ciliated tentacles (*d*) symmetrically arranged; these tentacles, however, proceed from a sort of disk which somewhat resembles the 'lophophore' of certain Polyzoa (§ 549). The mouth (*e*) is concealed by a broad but pointed hood or 'epistome' (*a*), which sometimes close down upon the tentacular disk, but is sometimes raised and extended forwards. The nearly cylindrical body terminates abruptly at the other extremity, where the anal orifice of the intestine (*b*) is surrounded by a circlet of very large cilia. This animal swims with great activity, sometimes by the tentacular cilia, sometimes by the anal circlet, sometimes by both combined; and besides its movement of progression, it frequently doubles itself together, so as to bring the anal extremity and the epistome almost in contact. It is so transparent that the whole of its alimentary canal may be as distinctly seen as that of Laguncula (§ 549); and, as in that Polyzoon, the alimentary masses often to be seen within the stomach (*c*) are kept in a continual whirling movement by the agency of cilia with which its walls are clothed. This very interesting creature was for a long time a puzzle to Zoologists; since, although there could be little doubt of its being a larval form, there was no clue to the nature of the adult produced from it, until this was discovered by Krohn in 1858 to be a Gephyrean Worm.² An even more extraordinary departure from the ordinary type is presented by the larva which has received the name *Pilidium* (Fig. 409); its shape being that of a helmet, the plume of which is replaced by a single long bristle-like appendage that is in continual motion, its point moving round and round in a circle. This curious organism, first noticed by Müller, has been since ascertained to be the larva of the well-known *Nemertes*, a Turbellarian

---

or four times its previous length, and the ciliated disk entirely disappeared. An accident unfortunately prevented the more minute examination of this Worm, which the Author would have otherwise made; but he may state that he is certain that there was no fallacy as to the fact above stated; this larva having been placed by itself in a cell, on purpose that it might be carefully studied, and having been only laid aside for a short time whilst other selections were being made from the same gathering of the Tow-net.

¹ For further information on this subject, see Balfour's "Comparative Embryology," Chap. xii., and the Memoirs there cited.

² 'Ueber *Pilidium* und *Actinotrocha*' in "Müller's Archiv," 1858, p. 293.—For more recent observations upon this interesting creature, see Balfour's "Comparative Embryology," Vol. i., pp. 299-302, and a paper on 'The Origin and Significance of the Metamorphosis of *Actinotrocha*,' by Mr. E. B. Wilson (of Baltimore), in "Quart. Journ. Microsc. Sci." April, 1881.

worm of enormous length, which is commonly found entwining itself among the roots of Algæ.[1]

597. Among the animals captured by the Tow-net, the marine Zoologist will be not unlikely to meet with an Annelid which, although by no means Microscopic in its dimensions, is an admirable subject for Microscopic observation, owing to the extreme transparence of its entire body, which is such as to render it difficult to be distinguished when swimming in a glass jar, except by a very favorable light. This is the *Tomopteris*, so named from the division of the lateral portions of its body into a succession of wing-like segments (Plate XXIII., B), each of them carrying at its extremity a pair of pinnules, by the movements of which it is rapidly

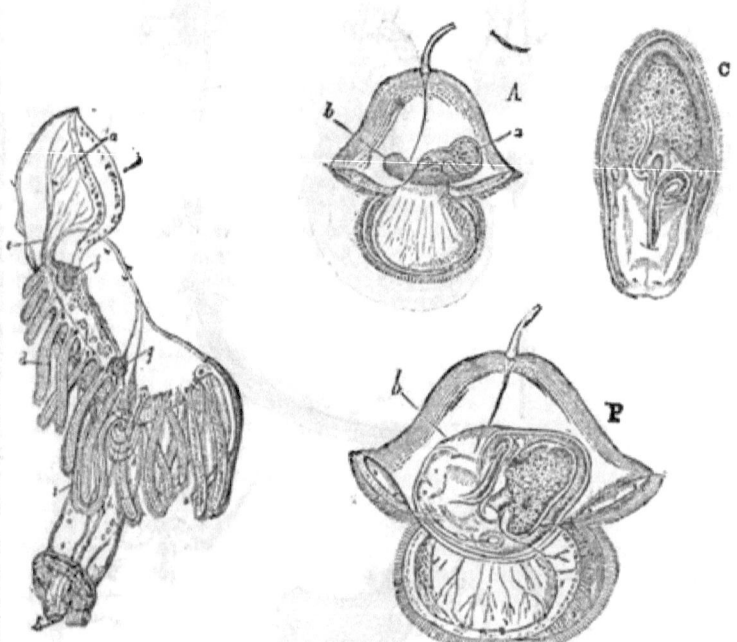

FIG. 408.   FIG. 409.

Actinotrocha branchiata:—*a*, Epistome or hood; *b*, anus; *c*, stomach; *d*, ciliated tentacles; *e*, mouth.

Pilidium gyrans:—A, young, showing at *a* the alimentary canal, and at *b* the rudiment of the Nemertid; B, more advanced stage of the same; c, newly-freed Nemertid.

propelled through the water. The full-grown animal, which measures nearly an inch in length, has first a curious pair of 'frontal horns' projecting laterally from the head, so as to give the animal the appearance of a 'hammer-headed' Shark; behind these there is a pair of very long antennæ, in each of which we distinguish a rigid bristle-like stem or *seta*, inclosed in a soft sheath, and moved at its base by a set of muscles contained within the lateral protuberances at the head. Behind these are

---

[1] See especially Leuckart and Pagenstecher's 'Untersuchungen über niedere Seethiere,' in Müller's "Archiv," 1853, p. 569, and Balfour, *op. cit.*, p. 165. The Author has frequently met with *Pilidium* in Lamlash Bay.

PLATE XXIII.

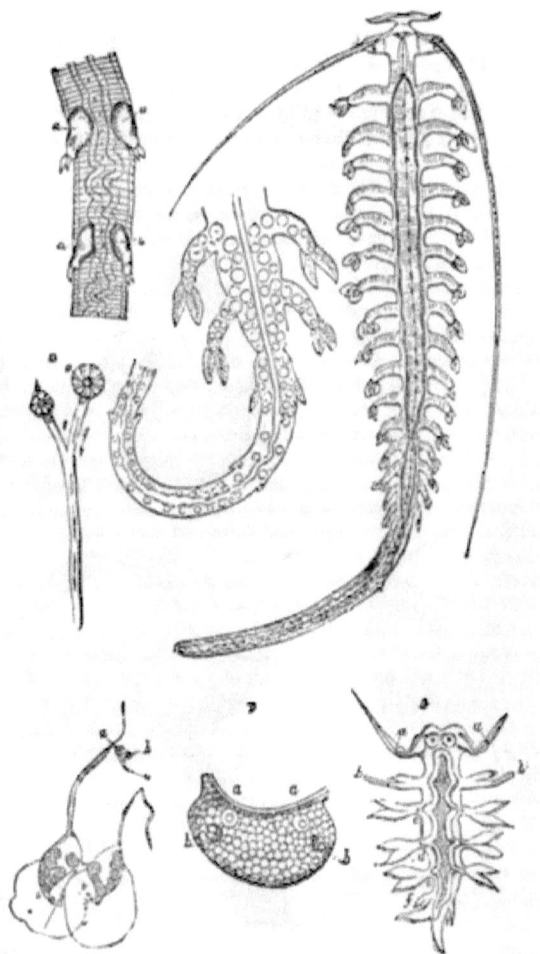

STRUCTURE AND DEVELOPMENT OF TOMOPTERIS ONISCIFORMIS (Original).

A. Portion of caudal prolongations, containing the spermatic sacs, *a, a*.
B. Adult Male specimen.
C. Hinder part of adult Female specimen, more enlarged, showing ova lying freely in the perivisceral cavity and its caudal prolongation.
D. Ciliated canal, commencing externally in the larger and smaller rosette-like disks, *a, b*.
E. One of the pinnulated segments, showing the position of the ciliated canal, *c*, and its rosette-like disks, *a, b*; showing also the incipient development of the ova, *d*, at the extremity of the segment.
F. Cephalic Ganglion, with its pair of auditory (?) vesicles, *a, a*, and its two ocelli, *b, b*.
G. Very young *Tomopteris*, showing at *a, a* the larval antennæ; *b, b*, the incipient long antennæ of the adult; *c, d, e, f*, four pairs of succeeding pinnulated segment, followed by the bifid tail.

about sixteen pairs of the ordinary pinnulated segments, of which the hinder ones are much smaller than those in front, gradually lessening in size until they become almost rudimentary; and where these cease, the body is continued onwards into a tail-like prolongation, the length of which varies greatly according as it is contracted or extended. This prolongation, however, bears four or five pairs of very minute appendages, and the intestine is continued to its very extremity; so that it is really to be regarded as a continuation of the body. In the head we find, between the origins of the antennæ, a ganglionic mass, the component cells of which may be clearly distinguished under a sufficient magnifying power, as shown at F; seated upon this are two pigment-spots (b, b), each bearing a double pellucid lens-like body, which are obviously rudimentary eyes: whilst imbedded in its anterior portion are two peculiar nucleated vesicles, a, a, which are probably the rudiments of some other sensory organs. On the under side of the head is situated the mouth, which, like that of many other Annelids, is furnished with a sort of proboscis that can be either projected or drawn-in; a short œsophagus leads to an elongated stomach, which, when distended with fluid, occupies the whole cavity of the central portion of the body, as shown in fig. B, but which is sometimes so empty and contracted as to be like a mere cord, as shown in fig. C. In the caudal appendage, however, it is always narrowed into an intestinal canal; this, when the appendage is in extended state as at C, is nearly straight; but when the appendage is contracted, as seen at B, it is thrown into convolutions. The perivisceral cavity is occupied by fluid in which some minute corpuscles may be distinguished; and these are kept in motion by cilia which clothe some parts of the outer surface of the alimentary canal and line some part of the wall of the body. No other more special apparatus, either for the circulation or for the aëration of the nutrient fluid, exists in this curious Worm; unless we are to regard as subservient to the respiratory function the ciliated canal which may be observed in each of the lateral appendages except the five anterior pairs. This canal commences by two orifices at the base of the segment, as shown at fig. E, b, and on a larger scale at fig. D; each of these orifices (D, a, b) is surrounded by a sort of rosette; and the rosette of the larger one (a) is furnished with radiating ciliated ridges. The two branches incline towards each other, and unite into a single canal, that runs along for some distance in the wall of the body, and then terminates in the perivisceral cavity; and the direction of the motion of the cilia which line it, is from without inwards.

598. The Reproduction and Developmental history of this Annelid present many points of great interest. The sexes appear to be distinct, ova being found in some individuals, and spermatozoa in others. The development of the ova commences in certain 'germ-cells' situated within the extremities of the pinnulated segments, where they project inwards from the wall of the body; these, when set free, float in the fluid of the perivisceral cavity, and multiply themselves by self-division; and it is only after their number has thus been considerably augmented, that they begin to increase in size and to assume the characteristic appearance of ova. In this stage they usually fill the perivisceral cavity not only of the body, but of its caudal extension, as shown at C; and they escape from it through transverse fissures which form in the outer wall of the body, at the third and fourth segments. The male reproductive organs, on the other hand, are limited to the caudal prolongation, where the sperm-cells are developed within the pinnulated appendages, as the germ-cells of the

female are within the appendages of the body. Instead of being set free, however, into the perivisceral cavity, they are retained within a saccular envelope forming a testis (A, *a*, *a*) which fills up the whole cavity of each appendage; and within this the spermatozoa may be observed, when mature, in active movement. They make their escape externally by a passage that seems to communicate with the smaller of the two just-mentioned rosettes; but they also appear to escape into the perivisceral cavity by an aperture that forms itself when the spermatozoa are mature. Whether the ova are fertilized while yet within the body of the female, by the entrance of spermatozoa through the ciliated canals, or after they have made their escape from it, has not yet been ascertained.—Of the earliest stages of embryonic development nothing whatever is yet known; but it has been ascertained that the animal passes through a larval form, which differs from the adult not merely in the number of the segments of the body (which successively augment by additions at the posterior extremity), but also in that of the antennæ. At G is represented the earliest larva hitherto met-with, enlarged as much as ten times in proportion to the adult at B; and here we see that the head is destitute of the frontal horns, but carries a pair of setigerous antennæ, *a*, *a*, behind which there are five pairs of bifid appendages, *b*, *c*, *d*, *e*, *f*, in the first of which, *b*, one of the pinnules is furnished with a *seta*. In more advanced larvæ having eight or ten segments, this is developed into a second pair of antennæ resembling the first; and the animal in this stage has been described as a distinct species, *T. quadricornis*. At a more advanced age, however, the second pair attains the enormous development shown at B; and the first or larval antennæ disappear, the setigerous portions separating at a sort of joint (G, *a*, *a*), whilst the basal projections are absorbed into the general wall of the body.—This beautiful creature has been met-with on so many parts of our coast, that it cannot be considered at all uncommon; and the Microscopist can scarcely have a more pleasing object for study.[1] Its elegant form, its crystal clearness, and its sprightly, graceful movements render it attractive even to the unscientific observer; whilst it is of special interest to the Physiologist, as one of the simplest examples yet known of the Annelid type.

599. To one phenomenon of the greatest interest, presented by various small Marine Annelids, the attention of the Microscopist should be specially directed; this is their *luminosity*, which is not a steady glow like that of the Glow-worm or Fire-fly, but a series of vivid scintillations (strongly resembling those produced by an electric discharge through a tube spotted with tin-foil), that pass along a considerable number of segments, lasting for an instant only, but capable of being repeatedly excited by any irritation applied to the body of the animal. These scintillations may be discerned under the Microscope, even in separate segments, when they are subjected to the irritation of a needle-point or to a gentle pressure; and it has been ascertained by the careful observations of M. de Quatrefages, that they are given out by the muscular fibres in the act of contraction.[2]

600. Among the fresh-water Annelids, those most interesting to the Microscopist are the worms of the *Nais* tribe, which are common in our

---

[1] See the Memoirs of the Author and M. Claparède in Vol. xxii. of the "Linnæan Transactions," and the authorities there referred to; also a recent Memoir by Dr. F. Vejdovsky in "Zeitschrift f. wiss. Zool.," Bd. xxxi., 1880.
[2] See his Memoirs on the Annelida of La Manche, in "Ann. des Sci. Nat.," Ser. 2, Zool., Tom. xix., and Ser. 3, Zool., Tom. xiv.

rivers and ponds, living chiefly amidst the mud at the bottom, and especially among the roots of aquatic plants. Being blood-red in color, they give to the surface of the mud, when they protrude themselves from it in large numbers and keep the protruded portion of their bodies in constant undulation, a very peculiar appearance; but if disturbed, they withdraw themselves suddenly and completely. These Worms, from the extreme transparency of their bodies, present peculiar facilities for Microscopic examination, and especially for the study of the internal circulation of the red liquid commonly considered as blood. There are here no external respiratory organs; and the thinness of the general integument appears to supply all needful facility for the aëration of the fluids. One large vascular trunk (dorsal) may be seen lying above the intestinal canal, and another (ventral) beneath it; and each of these enters a contractile dilatation, or heart-like organ, situated just behind the head. The fluid moves forwards in the dorsal trunk as far as the heart, which it enters and dilates; and when this contracts, it propels the fluid partly to the head, and partly to the ventral heart, which is distended by it. The ventral heart, contracting in its turn, sends the blood backwards along the ventral trunk to the tail, whence it passes towards the head as before. In this circulation, the stream branches-off from each of the principal trunks into numerous vessels proceeding to different parts of the body, which then return into the other trunk; and there is a peculiar set of vascular coils, hanging down in the perivisceral cavity that contains the corpusculated liquid representing the true blood, which seems specially destined to convey to it the aërating influence received by the red fluid in its circuit, thus acting (so to speak) like internal gills.—The *Naiad*-worms have been observed to undergo spontaneous division during the summer months; a new head and its organs being formed for the posterior segment behind the line of constriction, before its separation from the anterior. It has been generally believed that each segment continues to live as a complete worm; but it is asserted by Dr. T. Williams that from the time when the division occurs, neither half takes-in any more food, and that the two segments only retain vitality enough to enable them to be (as it were) the 'nurses' of the eggs which both include.—In the *Leech* tribe, the dental apparatus with which the mouth is furnished, is one of the most curious among their points of minute structure; and the common 'medicinal' Leech affords one of the most interesting examples of it. What is commonly termed the 'bite' of the leech, is really a saw-cut, or rather a combination of three saw-cuts, radiating from a common centre. If the mouth of a leech be examined with a hand-magnifier, or even with the naked eye, it will be seen to be a triangular aperture in the midst of a sucking disk; and on turning back the lips of that aperture, three little white ridges are brought into view. Each of these is the convex edge of a horny semicircle, which is bordered by a row of eighty or ninety minute hard and sharp teeth; whilst the straight border of the semicircle is imbedded in the muscular substance of the disk, by the action of which it is made to move backwards and forwards in a saw-like manner, so that the teeth are enabled to cut into the skin to which the suctorial disk has affixed itself.[1]

---

[1] Among the more recent sources of information as to the Anatomy and Physiology of the *Annelids*, the following may be specially mentioned:—The "Histoire Naturelle des Annelés Marins et d'Eau douce" of M. de Quatrefages, forming part of the "Suites à Buffon;" the successive admirable Monographs of the late Prof. Ed. Claparède, "Recherches Anatomiques sur les Annélides, Turbellariés,

Opalines, et Grégarines, observés dans les Hébrides" (Geneva, 1861); "Recherches Anatomiques sur les Oligochètes" (Geneva, 1862); "Beobachtungen über Anatomie und Entwickelungsgeschichte Wirbelloser Thiere an der Küste von Normandie" (Leipzig, 1863); and "Les Annélides Chétopodes du Golfe de Naples" (Geneva, 1868–70); the Monograph of Dr. Ehler's, "Die Borstenwürmer (Annelida Chætopoda)," 1864–8; and lastly, Dr. Macintosh's "Monograph of the British Annelids," now in course of publication by the Ray Society.

CHAPTER XVIII.

CRUSTACEA.

601. PASSING from the lower division of the Articulated series to that of *Arthropods*, in which the body is furnished with distinctly articulated or jointed limbs, we come first to the Class of *Crustacea*, which includes (when used in its most comprehensive sense) all those animals belonging to this group, which are fitted for aquatic respiration. It thus comprehends a very extensive range of forms; for although we are accustomed to think of the Crab, Lobster, Cray-fish, and other well-known species of the order *Decapoda* (ten-footed) as its typical examples, yet all these belong to the highest of its many orders; and among the lower are many of a far simpler structure, and not a few which would not be recognized as belonging to the class at all, were it not for the information derived from the study of their development as to their real nature, which is far more apparent in their early than it is in their adult condition. Many of the inferior kinds of Crustacea are so minute and transparent, that their whole structure may be made-out by the aid of the Microscope without any preparation; this is the case, indeed, with nearly the whole group of *Entomostraca* (§ 603), and with the larval forms even of the *Crab* and its allies (§ 614); and we shall give our first attention to these, afterwards noticing such points in the structure of the larger kinds as are likely to be of general interest.

602. A curious example of the reduction of an elevated type to a very simple form is presented by the group of *Pycnogonida*, some of the members of which may be found by attentive search in almost every locality where sea-weeds abound; it being their habit to crawl (or rather to sprawl) over the surfaces of these, and probably to imbibe as food the gelatinous substance with which they are invested.[1] The general form of their bodies (Fig. 410) usually reminds us of that of some of the long-legged Crabs; the abdomen being almost or altogether deficient, whilst the head is very small, and fused (as it were) into the thorax; so that the last-named region, with the members attached to it, constitutes nearly the whole bulk of the animal. The head is extended in front into a probosis-like projection, at the extremity of which is the narrow orifice of the mouth; which seems to be furnished with vibratile cilia, that serve to draw into it the semi-fluid aliment. Instead of being furnished (as in the higher Crustaceans) with two pairs of antennæ and numerous pairs of 'feet-jaws,' it has but a single pair of either; it also bears four minute *ocelli*, or rudimentary eyes, set at a little distance from each other on a

---

[1] It is remarkable that very large forms of this group, sometimes extending to more than twelve inches across, have been brought up from great depths of the sea.

sort of tubercle. From the thorax proceed four pairs of legs, each composed of several joints, and terminated by a hooked claw; and by these members the animal drags itself slowly along, instead of walking actively upon them like a crab. The mouth leads to a very narrow œsophagus (*a*), which passes back to the central stomach (*b*) situated in the midst of the thorax, from the hinder end of which a narrow intestine (*c*) passes-off, to terminate at the posterior extremity of the body. From the central stomach five pairs of cæcal prolongations radiate; one pair (*d*) entering the feet-jaws, the other four (*e, e*) penetrating the legs, and passing along them as far as the last joint but one; and those extensions are covered with a layer of brownish-yellow granules, which are probably to be regarded as a diffused and rudimentary condition of the liver. The stomach and its cæcal prolongations are continually executing peristaltic movements of a very curious kind; for they contract and dilate with an irregular alternation, so that a flux and reflux of their contents is constantly taking place between the central portion and its radiating extensions, and between one of these extensions and another. The perivisceral space between the widely-extended stomach and the walls of the body and limbs is occupied by a transparent liquid, in which are seen floating a number of minute transparent corpuscles of irregular size; and this fluid, which represents the blood, is kept in continual motion, not only by the general movements of the animal, but also by the actions of the digestive apparatus; since, whenever the cæcum of any one the legs undergoes dilatation, a part of the circumambient liquid will be pressed-out from the cavity of that limb, either into the thorax, or into some other limb whose stomach is contracting. The fluid must obtain its aëration through the general surface of the body, as there are no special organs of respiration. The nervous system consists of a single ganglion in the head (formed by the coalescence of a pair), and of another in the thorax (formed by the coalescence of four pairs), with which the cephalic ganglion is connected in the usual mode, namely, by two nervous cords which diverge from each other to embrace the œsophagus.— In the study of the very curious phenomena exhibited by the digestive apparatus, as well as of the various points of internal conformation which have been described, the Achromatic Condenser will be found use-

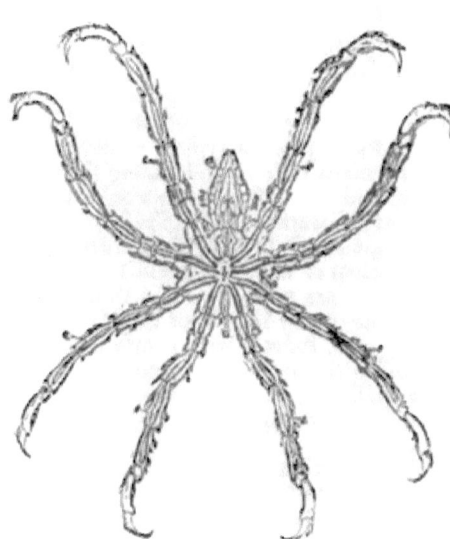

Fig. 410.

Ammothea pycnogonoides:—*a,* narrow œsophagus; *b,* stomach; *c,* intestine; *d,* digestive cæca of the feet-jaws; *e, e,* digestive cæca of the legs.

ful, even with the 1 inch, 2-3d inch, or ½ inch Objective; for the imperfect transparence of the bodies of these animals renders it of importance to drive a large quantity of light through them, and to give to this light such a quality as shall sharply define the internal organs.[1]

603. ENTOMOSTRACA.—This group of Crustaceans, nearly all the existing members of which are of such minute size as to be only just visible to the naked eye, is distinguished by the inclosure of the entire body within a horny or shelly casing; which sometimes closely resemble a bivalve shell in form and in the mode of junction of its parts, whilst in other instances it is formed of only a single piece, like the hard envelope of certain Rotifera (§ 453, III.). The segments into which the body is divided, are frequently very numerous, and are for the most part similar to each other; but there is a marked difference in regard to the appendages which they bear, and to the mode in which these minister to the locomotion of the animals. For in the *Lophyropoda*, or 'bristly-footed' tribe, the number of legs is small, not exceeding five pairs, and their function is limited to locomotion, the respiratory organs being attached to the parts in the neighborhood of the mouth; whilst in the *Branchiopoda*, or 'gill-footed' tribe, the same members (known as 'fin-feet') serve both for locomotion and for respiration, and the number of these is commonly large, being in *Apus* not less than sixty pairs. The character of their movements differ accordingly; for whilst all the members of the first-named tribe dart through the water in a succession of jerks, so as to have acquired the common name of 'water-fleas,' those among the latter which possess a great number of 'fin-feet,' swim with an easy gliding movement, sometimes on their back alone (as in the case with *Branchipus*), and sometimes with equal facility on the back, belly, or sides (as is done by *Artemia salina*, the 'brine shrimp').—Some of the most common forms of both tribes will now be briefly noticed.

604. The tribe of *Lophyropoda* is divided into two Orders; of which the first, *Ostracoda*, is distinguished by the complete inclosure of the body in a bivalve shell, by the small number of legs, and by the absence of an external ovary. One of the best known examples is the little *Cypris*, which is a common inhabitant of pools and streams: this may be recognized by its possession of two pairs of antennæ, the first having numerous joints with a pencil-like tuft of filaments, and projecting forwards from the front of the head, whilst the second has more the shape of legs, and is directed downwards; and by the limitation of its legs to two pairs, of which the posterior does not make its appearance outside the shell, being bent upwards to give support to the ovaries. The valves are generally opened widely enough to allow the greater part of both pairs of antennæ and of the front pair of legs to pass-out between them; but when the animals are alarmed, they draw these members within the shell, and close the valves firmly. They are very lively creatures, being almost constantly seen in motion, either swimming by the united action of their foot-like antennæ and legs, or walking upon plants and other solid bodies floating in the water.—Nearly allied to the preceding is the *Cythere*, whose body is furnished with three pairs of legs, all projecting out of the shell, and whose superior antennæ are destitute of the filamentous brush; this genus

---

[1] Certain points of resemblance borne by *Pycnogonida* to Spiders, makes the careful study of their development a matter of special interest and importance; as there is some reason to regard them rather as *Arachnida* adapted to a marine *habitat*, than as Crustacea.—See Balfour's "Comparative Embryology," pp. 448, 449, and the authorities there referred to.

is almost entirely marine, and some species of it may almost invariably be met-with in little pools among the rocks between the tide-marks, creeping about (but not swimming) amongst Confervæ and Corallines.—There is abundant evidence of the former existence of Crustacea of this group, of larger size than any now existing, to an enormous extent; for in certain fresh-water strata, both of the Secondary and Tertiary series, we find layers, sometimes of great extent and thickness, which are almost entirely composed of the fossilized shells of *Cyprides;* whilst in certain parts of the Chalk, which was a marine deposit, the remains of bivalve shells resembling those of *Cythere* present themselves in such abundance as to form a considerable part of its substance.

605. In the order *Copepoda,* there is a jointed shell forming a kind of buckler or carapace that almost entirely incloses the head and thorax, an opening being left beneath, through which the members project; and there are five pairs of legs, mostly adapted for swimming, the fifth pair, however, being rudimentary in the genus *Cyclops,* the commonest example of the group. This genus receives its name from possessing only a single eye, or rather a single cluster of ocelli; which character, however, it has in common with the two genera already named, as well as with Daphnia (§ 606), and with many other Entomostraca. It contains numerous species, some of which belong to fresh-water, whilst others are marine. The Fresh-water species often abound in the muddiest and most stagnant pools, as well as in the clearest springs; the ordinary water with which London is supplied frequently contains large numbers of them. Of the marine species, some are to be found in the localities in which the Cythere is most abundant, whilst others inhabit the open ocean, and must be collected by the Tow-net. The body of the Cyclops is soft and gelatinous, and it is composed of two distinct parts, a thorax (Fig. 411, *a*) and an abdomen (*b*), of which the latter, being comparatively slender, is commonly considered as a tail, though traversed by the intestine which terminates near its extremity. The head, which coalesces with the thorax, bears one very large pair of antennæ (*c*), possessing numerous articulations and furnished with bristly appendages, and another small pair (*d*); it is also furnished with a pair of mandibles or true jaws, and with two pairs of 'feet-jaws,' of which the hinder pair is the longer and more abundantly supplied with bristles. The legs (*e*) are all beset with plumose tufts, as is also the tail (*f, f*) which is borne at the extremity of the abdomen. On either side of the abdomen of the female, there is often to be seen an egg-capsule or external ovarium (B);

Fig. 411.

A, Female of *Cyclops quadricornis:*—*a*, body; *b*, tail; *c*, antenna; *d*, antennule; *e*, feet; *f*, plumose setæ of tail;—B, tail, with external egg-sacs:—C, D, E, F, G, successive stages of development of young.

within which the ova, after being fertilized, undergo the earlier stages of their development.—The Cyclops is a very active creature, and strikes the water in swimming, not merely with its legs and tail, but also with its antennæ. The rapidly-repeated movements of its feet-jaws serve to create a whirlpool in the surrounding water, by which minute animals of various kinds, and even its own young, are brought to its mouth to be devoured.

606. The tribe of *Branchiopoda* also is divided into two Orders, of which the *Cladocera* present the nearest approach to the preceding, having a bivalve carapace, no more than from four to six pairs of legs, two pairs of antennæ, of which one is large and branched and adapted for swimming, and a single eye. The commonest form of this is the *Daphnia pulex*, sometimes called the 'arborescent water-flea,' from the branching form of its antennæ. It is very abundant in many ponds and ditches, coming to the surface in the mornings and evenings and in cloudy weather, but seeking the depths of the water during the heat of the day. It swims by taking short springs; and feeds on minute particles of vegetable substances, not, however, rejecting animal matter when offered. Some of the peculiar phenomena of its reproduction will be presently described (§ 609).

607. The other order, *Phyllopoda*, includes those Branchiopoda whose body is divided into a great number of segments, nearly all of which are furnished with leaf-like members, or 'fin-feet.' The two Families which this order includes, however, differ considerably in their conformation; for in that of which the genera *Apus* and *Nebalia* are representatives, the body is inclosed in a shell, either shield-like or bivalve, and the feet are generally very numerous; whilst in that in which contains *Branchipus* and *Artemia*, the body is entirely unprotected, and the number of pairs of feet does not exceed eleven. The *Apus cancriformis*, which is an animal of comparatively large size, its entire length being about $2\frac{1}{2}$ inches, is an inhabitant of stagnant waters; but although occasionally very abundant in particular pools or ditches, it is not to be met-with nearly so commonly as the Entomostraca already noticed. It is recognized by its large oval carapace, which covers the head and body like a shield; by the nearly cylindrical form of its body, which is composed of thirty articulations; and by the multiplication of its legs, which amount to about sixty pairs. The number of joints in these and in the other appendages is so great, that in a single individual they may be safely estimated at not less than two millions. These organs, however, are for the most part small; and the instruments chiefly used by the animal for locomotion are the first pair of feet, which are very much elongated (bearing such a resemblance to the principal antennæ of other Entomostraca, as to be commonly ranked in the same light), and are distinguished as *rami* or oars. With these they can swim freely in any position; but when the rami are at rest and the animal floats idly on the water, its fin-feet may be seen in incessant motion, causing a sort of whirlpool in the water, and bringing to the mouth the minute animals (chiefly the smaller Entomostraca inhabiting the same localities) that serve for its food.—The *Branchipus stagnalis* has a slender, cylindriform, and very transparent body of nearly an inch in length, furnished with eleven pairs of fin-feet, but is destitute of any protecting envelope; its head is furnished with a pair of very curious prehensile organs (which are really modified antennæ), whence it has received the name of *Cheirocephalus;* but these are not used by it for the seizure of prey, the food of this animal being vegetable, and their function is to clasp the female in the act of copulation. The

Branchipus or Cheirocephalus is certainly the most beautiful and elegant of all the Entomostraca, being rendered extremely attractive to the view by "the uninterrupted undulatory wavy motion of its graceful branchial feet, slightly tinged as they are with a light reddish hue, the brilliant mixture of transparent bluish-green and bright red of its prehensile antennæ, and its bright red tail with the plumose setæ springing from it;" unfortunately, however, it is a comparatively rare animal in this country.—The *Artemia salina* or 'brine shrimp' is an animal of very similar organization, and almost equally beautiful in its appearance and movements, but of smaller size, its body being about half an inch in length. Its 'habitat' is very peculiar; for it is only found in the salt-pans or brine-pits in which sea-water is undergoing concentration as at Lymington); and in these situations it is sometimes so abundant as to communicate a red tinge to the liquid.

608. Some of the most interesting points in the history of the *Entromostraca* lie in the peculiar mode in which their generative function is performed, and their tenacity of life when desiccated, in which last respect they correspond with many Rotifers (§ 452). By this provision they escape being completely exterminated, as they might otherwise soon be, by the drying-up of the pools, ditches, and other small collections of water which constitute their usual 'habitats.' It does not appear, however, that the adult Animals can bear a *complete* desiccation, although they will preserve their vitality in mud that holds the smallest quantity of moisture; but their eggs are more tenacious of life, and there is ample evidence that these will become fertile on being moistened, after having remained for a long time in the condition of fine dust. Most Entomostraca, too, are killed by severe cold, and thus the whole race of adults perishes every winter; but their eggs seem unaffected by the lowest temperature, and thus continue the species, which would be otherwise exterminated.—Again, we frequently meet in this group with that *agamic* reproduction, which we have seen to prevail so extensively among the lower radiata and Mollusca. In many species there is a double mode of multiplication, the sexual and the non-sexual. The former takes-place at certain seasons only; the males (which are often so different in conformation from the females, that they would not be supposed to belong to the same species, if they were not seen in actual congress) disappearing entirely at other times. The latter, on the other hand, continues at all periods of the year, so long as warmth and food are supplied; and is repeated many times (as in the Hydra) so as to give origin to as many successive 'broods.' Further, a single act of impregnation serves to fertilize not merely the ova which are then mature or nearly so, but all those subsequently produced by the same female, which are deposited at considerable intervals. In these two modes, the multiplication of these little creatures is carried-on with great rapidity, the young animal speedily coming to maturity and beginning to propagate; so that according to the computation of Jurine, founded upon data ascertained by actual observation, a single fertilized female of the common *Cyclops quadricornis* may be the progenitor in one year of 4,442,189,120 young.

609. The eggs of some Entomostraca are deposited freely in the water, or are carefully attached in clusters to aquatic Plants; but they are more frequently carried for some time by the parent in special receptacles developed from the posterior part of the body; and in many cases they are retained there until the young are ready to come-forth, so that these animals may be said to be ovo-viviparous. In *Daphnia*, the eggs

are received into a large cavity between the back of the animal and its shell, and there the young undergo almost their whole development, so as to come-forth in a form nearly resembling that of their parent. Soon after their birth, a moult or exuviation of the shell takes-place; and the egg-coverings are cast-off with it. In a very short time afterwards, another brood of eggs is seen in the cavity, and the same process is repeated, the shell being again exuviated after the young have been brought to maturity. At certain times, however, the *Daphnia* may be seen with a dark opaque substance within the back of the shell, which has been called the *ephippium*, from its resemblance to a saddle. This, when carefully examined, is found to be of dense texture, and to be composed of a mass of hexagonal cells; and it contains two oval bodies, each consisting of an ovum covered with a horny casing, enveloped in a capsule which opens like a bivalve shell. From the observations of Sir J. Lubbock,[1] it appears that the ephippium is really only an altered portion of the carapace; its outer valve being a part of the outer layer of the epidermis, and its inner valve the corresponding part of the inner layer. The development of the ephippial eggs takes-place at the posterior part of the ovaries, and is accompanied by the formation of a greenish-brown mass of granules; and form this situation the eggs pass into the receptacle formed by the new carapace, where they become included between the two layers of the ephippium. This is cast-off, in process of time, with the rest of the skin, from which, however, it soon becomes detached; and it continues to envelop the eggs, generally floating on the surface of the water until they are hatched with the returning warmth of spring. This curious provision obviously affords protection to the eggs which are to endure the severity of winter cold; and an approach to it may be seen in the remarkable firmness of the envelopes of the ' winter eggs' of some Rotifera (§ 451). There seems a strong probability, from the observations of Sir J. Lubbock, that the 'ephippial' eggs are true sexual products, since males are to be found at the time when the ephippia are developed; whilst it is certain that the ordinary eggs can be produced non-sexually, and that the young which spring from them can multiply the race in like manner. The young produced from the ephippial eggs seem to have the same power of continuing the race by non-sexual reproduction, as the young developed under ordinary circumstances.

610. In most Entomostraca, the young at the time of their emersion from the egg differ considerably from the parent, especially in having only the thoracic portion of the body as yet evolved, and in possessing but a small number of locomotive appendages (see Fig. 411, c–g); the visual organs, too, are frequently wanting at first. The process of development, however, takes place with great rapidity; the animal at each successive moult (which process is very commonly repeated at intervals of a day or two) presenting some new parts, and becoming more and more like its parent, which it very early resembles in its power of multiplication, the female laying eggs before she has attained her own full size. Even when the Entomostraca have attained their full growth, they continue to exuviate their shell at short intervals during the whole of life; and this repeated moulting seem to prevent the animal from being injured, or its movements obstructed, by the over-growth of parasitic Animalcules and Confervæ; weak and sickly individuals being frequently

---

[1] An account of the two methods of Reproduction in *Daphnia*, and of the structure of the Ephippium,' in " Philosophical Transactions," 1857, p. 79.

seen to be so covered with such parasites, that their motion and life are soon arrested, apparently because they have not strength to cast-off and renew their envelopes. The process of development appears to depend in some degree upon the influence of light, being retarded when the animals are secluded from it; but its rate is still more influenced by heat; and this appears also to be the chief agent that regulates the time which elapses between the moultings of the adult, these, in *Daphnia*, taking-place at intervals of two days in warm summer weather, whilst several days intervene between them when the weather is colder. The cast shell carries with it the sheaths not only of the limbs and plumes, but of the most delicate hairs and setæ which are attached to them. If the animal have previously sustained the loss of a limb, it is generally renewed at the next moult, as in higher Crustacea.[1]

611. Closely connected with the Entomostracous group is the tribe of *suctorial* Crustacea; which for the most part live as parasites upon the exterior of other animals (especially Fish), whose juices they imbibe by means of the peculiar proboscis-like organ which takes in them the place of the jaws of other Crustaceans; whilst other appendages, representing the feet-jaws, are furnished with hooks, by which these parasites attach themselves to the animals from whose juices they derive their nutriment. Many of the suctorial Crustacea bear a strong resemblance, even in their adult condition, to certain Entomostraca; but more commonly it is between the earlier forms of the two groups that the resemblance is the closest, most of the *suctoria* undergoing such extraordinary changes in their progress towards the adult condition, that, if their complete forms were alone attended-to, they might be excluded from the class altogether, as has (in fact) been done by many Zoologists.—Among those Suctorial Crustacea which present the nearest approach to the ordinary Entomostracous type, may be specially mentioned the *Argulus foliaceus*, which attaches itself to the surfaces of the bodies of fresh-water Fish, and is commonly known under the name of the 'fish louse.' This animal has its body covered with a large firm oval shield, which does not extend, however, over the posterior part of the abdomen. The mouth is armed with a pair of styliform mandibles; and on each side of the proboscis there is a large short cylindrical appendage, terminated by a curious sort of sucking-disk, with another pair of longer jointed members, terminated by prehensile hooks. These two pairs of appendages, which are probably to be considered as representing the feet-jaws, are followed by four pairs of legs, which, like those of the Branchiopods, are chiefly adapted for swimming; and the tail, also, is a kind of swimmeret. This little animal can leave the fish upon which it feeds, and then swims freely in the water, usually in a straight line, but frequently and suddenly changing its direction, and sometimes turning over and over several times in succession. The stomach is remarkable for the large cæcal prolongations which it sends out on either side, immediately beneath the shell; for these subdivide and ramify in such a manner, that they are distributed almost as minutely as the cæcal prolongations of the stomach of the *Planaria* (Fig. 406). The proper alimentary canal, however, is continued backwards from the central cavity of the stomach, as an intestinal tube, which terminates in an anal orifice at the extremity of the abdomen.—A far more marked departure from the typical form of the class

---

[1] For a systematic and detailed account of this group, see Dr. Baird's "Natural History of the British Entomostraca," published by the Ray Society.

is shown in the *Lernæa*, which is found attached to the gills of Fishes. This creature has a long suctorial proboscis; a short thorax, to which is attached a single pair of legs, which meet at their extremities, where they bear a sucker which helps to give attachment to the parasite; a large abdomen; and a pair of pendent egg-sacs. In its adult condition it buries its anterior portion in the soft tissue of the animal it infests, and appears to have little or no power of changing its place. But the young, when they come forth from the egg, are as active as the young of *Cyclops* (Fig. 411, C, D), which they much resemble; and only attain the adult form after a series of metamorphoses, in which they cast-off their locomotive members and eyes. It is curious that the original form is retained with comparatively slight change by the males, which increase but little in size, and are so unlike the females that no one would suppose the two to belong to the same family, much less to the same species, but for the Microscopic study of their development.[1]

612. From the parasitic Suctorial Crustacea, the transition is not really so abrupt as it might at first sight appear to the group of *Cirrhipeda*, consisting of the *Barnacles* and their allies: for these, like many of the Suctoria, are fixed to one spot during the adult portion of their lives, but come into the world in a condition that bears a strong resemblance to the early state of many of the true Crustacea. The departure from the ordinary Crustacean type in the adults, is, in fact, so great that it is not surprising that Zoologists in general should have ranked them in a distinct Class; their superficial resemblance to the Mollusca, indeed, having caused most systematists to place them in that series, until due weight was given to those structural features which mark their 'articulated' character. We must limit ourselves, in our notice of this group, to that very remarkable part of their history, the Microscopic study of which has contributed most essentially to the elucidation of their real nature. The observations of Mr. J. V. Thompson,[2] with the extensions and rectifications which they have subsequently received from others (especially Mr. Spence Bate[3] and Mr. Darwin[4]) show that there is no essential difference between the early forms of the *sessile* (Balanidæ or 'acorn-shells') and of the *pedunculated* Cirrhipeds (Lepadidæ or 'barnacles'); for both are active little animals (Fig. 412, A), possessing three pairs of legs, and a pair of compound eyes, and having the body covered with an expanded carapace, like that of many Entomostracous Crustaceans, so as in no essential particular to differ from the larva of *Cyclops* (Fig. 411, C). After going through a series of Metamorphoses, one stage of which is represented in Fig. 412, B, C, these larvæ come to present a form, D, which reminds us strongly of that of *Daphnia;* the body being inclosed in a shell composed of two valves, which are united along the back, whilst they are free along their lower margin, where they separate for the protrusion of a large and strong anterior pair of prehensile limbs provided with an adhesive sucker and hooks, and of six pairs of posterior legs adapted for swimming. This Bivalve shell, with the

---

[1] As the group of Suctorial Crustacea is rather interesting to the professed Naturalist than to the amateur Microscopist, even an outline view of it would be unsuitable to the present work; and the Author would refer such of his readers as may desire to study it, to the excellent Treatise by Dr. Baird already referred to.

[2] "Zoological Researches," No. iv., 1830, and Philos. Transact., 1835, p. 355.

[3] 'On the Development of the Cirripedia,' in "Ann. of Nat. Hist.," Ser. 2, Vol. viii. (1851), p. 324.

[4] "Monograph of the Sub-Class *Cirripedia*," published by the Ray Society.

members of both kinds, is subsequently thrown-off; the animal then attaches itself by its *head*, a portion of which, in the Barnacle, becomes excessively elongated into the 'peduncle' of attachment, whilst in Balanus it expands into a broad disk of adhesion; the first thoracic segment sends backwards a prolongation which arches over the rest of the body so as completely to inclose it, and of which the exerior layer is consolidated into the 'multivalve' shell; whilst from the other thoracic segments are evolved the six pairs of *cirrhi*, from whose peculiar character the name of the group is derived. These are long, slender, many-jointed, tendril-like appendages, fringed with delicate filaments covered with cilia, whose action serves both to bring food to the mouth, and to maintain aërating currents in the water. The Balani are peculiarly interesting objects in the Aquarium, on account of the pumping action of their beautiful feathery appendages, which may be watched through a Tank-

FIG. 412.

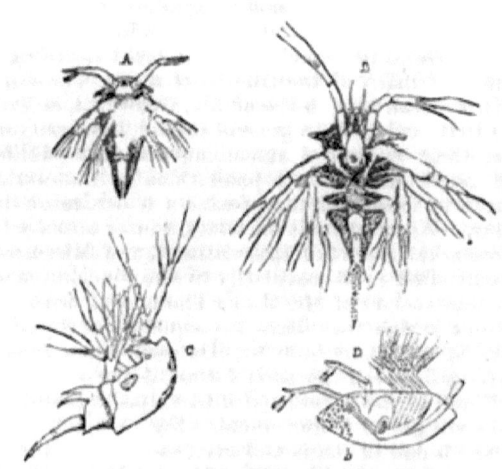

Development of *Balanus balanoides*;—A, earliest form; B, larva after second moult; C, side view of the same; D, stage immediately preceding the loss of activity; *a*, stomach (?); *b*, nucleus of future attachment (?).

Microscope; and their cast skins, often collected by the Tow-net, are well worth mounting.

613. MALACOSTRACA.—The chief points of interest to the Microscopist in the more highly organized forms of Crustacea, are furnished by the structure of the *shell*, and by the phenomena of *metamorphosis*, both which may be best studied in the commonest kinds.—The Shell of the *Decapods* in its most complete form consists of three strata—namely, 1, a horny structureless layer covering the exterior; 2, an areolated stratum; and 3, a laminated tubular substance. The innermost and even the middle layers, however, may be altogether wanting; thus, in the *Phyllosoma* or 'glass-crabs,' the envelope is formed by the transparent horny layer alone; and in many of the small crabs belonging to the genus *Portuna*, the whole substance of the carapace beneath the horny investment presents the areolated structure. It is in the large thick-shelled Crabs that we find the three layers most differentiated.

Thus, in the common *Cancer pagurus*, we may easily separate the structureless horny covering after a short maceration in dilute acid; the areolated layer, in which the pigmentary matter of the colored parts of the shell is chiefly contained, may be easily brought into view by grinding away from the *inner* side as flat a piece as can be selected, having first cemented the outer surface to the glass slide, and by examining this with a magnifying power of 250 diameters, driving a strong light through it with the Achromatic Condenser; whilst the tubular structure of the thick inner layer may be readily demonstrated, by means of sections parallel and perpendicular to its surface. This structure, which resembles that of *dentine* (§ 655), save that the tubuli do not branch, but remain of the same size through their whole course, may be particularly well seen in the black extremity of the claw, which (apparently from some peculiarity in the molecular arrangement of its mineral particles) is much denser than the rest of the shell; the former having almost the semi-transparence of ivory, whilst the latter has a chalky opacity. In a transverse section of the claw, the tubuli may be seen to radiate from the central cavity towards the surface, so as very strongly to resemble their arrangement in a tooth; and the resemblance is still further increased by the presence, at tolerably regular intervals, of minute sinuosities corresponding with the laminations of the shell, which seem, like the 'secondary curvatures' of the dentinal tubuli, to indicate successive stages in the calcification of the animal basis. In thin sections of the areolated layer it may be seen that the apparent walls of the areolæ are merely translucent spaces from which the tubuli are absent, their orifices being abundant in the intervening spaces.[1] The tubular layer rises up through the pigmentary layer of the Crab's shell in little papillary elevations, which seem to be concretionary nodules; and it is from the deficiency of the pigmentary layer at these parts, that the colored portion of the shell derives its minutely-speckled appearance.—Many departures from this type are presented by the different species of Decapods; thus, in the *Prawns*, there are large stellate pigment-spots (resembling those of Frogs, Fig. 465, *c*), the colors of which are often in remarkable conformity with those of the bottom of the rock-pools frequented by these creatures; whilst in the *Shrimps* there is seldom any distinct trace of the areolated layer, and the calcareous portion of the skeleton is disposed in the form of concentric rings, which seem to be the result of the concretionary aggregation of the calcifying deposit (§ 713).

614. It is a very curious circumstance, that a strongly-marked difference exists between Crustaceans that are otherwise very closely allied, in regard to the degree of change to which their young are subject in their progress towards the adult condition. For whilst the common *Crab*, *Lobster*, *Spiny Lobster*, *Prawn*, and *Shrimp* undergo a regular metamorphosis, the young of the *Cray-fish* and some *Land-crabs* come forth from the egg in a form which corresponds in all essential particulars with that of their parents. Generally speaking, a strong resemblance exists among the young of all the species of Decapods which undergo a

---

[1] The Author is now quite satisfied of the correctness of the interpretation put by Prof. Huxley (see his Article, 'Tegumentary Organs,' in the "Cyclop. of Anat. and Phys.," Vol. v., p. 487) and by Prof. W. C. Williamson ('On some Histological Features in the Shells of Crustacea,' in "Quart. Journ. of Microsc. Science," Vol. viii., 1860, p. 38), upon the appearances which he formerly described ("Reports of British Association" for 1847, p. 128) as indicating a cellular structure in this layer.

metamorphosis, whether they are afterwards to belong to the *macrourous* (long-tailed) or to the *brachyourous* (short-tailed) division of the group; and the forms of these larvæ are so peculiar, and so entirely different from any of those into which they are ultimately to be developed, that they were considered as belonging to a distinct genus, *Zoea*, until their real nature was first ascertained by Mr. J. V. Thompson. Thus, in the earliest state of *Carcinus mænas* (small edible Crab), we see the head and thorax, which form the principal bulk of the body, included within a large carapace or shield (Fig. 413, A) furnished with a long projecting spine, beneath which the fin-feet are put forth: whilst the abdominal segments, narrowed and prolonged, carry at the end a flattened tail-fin, by the strokes of which upon the water, the propulsion of the animal is chiefly effected. Its condition is hence comparable, in almost all essential particulars, to that of *Cyclops* (§ 605). In the case of the Lobster, Prawn, and other 'macrourous' species, the metamorphosis chiefly consists in the separation of the locomotive and respiratory organs; true legs being developed from the thoracic segments for the former, and true gills (concealed within a special chamber formed by an extension of the carapace beneath the body) for the latter; while the abdominal segments

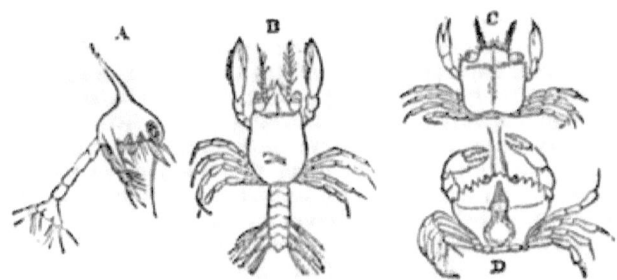

Fig. 413.

Metamorphosis of *Carcinus Mænas*:—A, first or *Zoea* stage; B, second or *Megalopa* stage; C, third stage, in which it begins to assume the adult form; D, perfect form.

increase in size, and become furnished with appendages (false feet) of their own. In the Crabs, or 'brachyourous' species, on the other hand, the alteration is much greater; for besides the change first noticed in the thoracic members and respiratory organs, the thoracic region becomes much more developed at the expense of the abdominal, as seen at B, in which stage the larva is remarkable for the large size of its eyes, and hence received the name of *Megalopa* when it was supposed to be a distinct type. In the next stage, C, we find the abdominal portion reduced to an almost rudimentary condition, and bent under the body; the thoracic limbs are more completely adapted for walking, save the first pair, which are developed into *chelæ* or pincers; and the little creature entirely loses the active swimming habits which it originally possessed, and takes on the mode of life peculiar to the adult.[1]

615. In collecting minute Crustacea, the Ring-net should be used for the fresh-water species, and the Tow-net for the marine. In localities favorable for the latter, the same 'gathering' will often contain multi-

[1] On the Metamorphosis of *Crustacea* and *Cirripedia*, see especially the recent "Untersuchungen über Crustaceen" of Prof. Claus; Vienna, 1876.

tudes of various species of Entomostraca, accompanied, perhaps, by the larvæ of higher Crustacea, Echinoderm larvæ, Annelid larvæ, and the smaller Medusæ. The water containing these should be put into a large glass jar, freely exposed to the light; and, after a little practice, the eye will become so far habituated to the general appearance and modes of movement of these different forms of animal life, as to be able to distinguish them one from the other. In selecting any specimen for Microscopic examination, the Dipping-tube (§ 126) will be found invaluable. The collector will frequently find *Megalopa* larvæ, recognizable by the brightness of their two black eye-spots, on the surface of floating leaves of *Zostera*.—The study of the Metamorphosis will be best prosecuted, however, by obtaining the fertilized eggs which are carried about by the females, and watching the history of their products.—For preserving specimens, whether of Entomostraca, or of larvæ of the higher Crustacea, the Author would recommend Glycerine-jelly as the best medium.

CHAPTER XIX.

INSECTS AND ARACHNIDA.

616. THERE is no Class in the whole Animal Kingdom which affords to the Microscopist such a wonderful variety of interesting objects, and such facilities for obtaining an almost endless succession of novelties, as that of Insects. For, in the first place, the number of different kinds that may be brought together (at the proper time) with extremely little trouble, far surpasses that which any other group of animals can supply to the most painstaking collector; then again, each specimen will afford, to him who knows how to employ his materials, a considerable number of Microscopic objects of very different kinds; and thirdly, although some of these objects require much care and dexterity in their preparation, a large proportion may be got out, examined, and mounted, with very little skill or trouble. Take, for example, the common House-fly:—its *eyes* may be easily mounted, one as a transparent, the other as an opaque object (§ 626); its *antennæ*, although not such beautiful objects as those of many other Diptera, are still well worth examination (§ 628); its *tongue* or 'proboscis' (§ 629) is a peculiarly interesting object, though requiring some care in its preparation; its *spiracles*, which may be easily cut off from the sides of its body, have a very curious structure (§ 635); its alimentary canal affords a very good example of the minute distribution of the *tracheæ* (§ 634); its *wing*, examined in a living specimen newly come forth from the pupa state, exhibits the circulation of the blood in the 'nervules' (§ 633), and when dead shows a most beautiful play of iridescent colors, and a remarkable areolation of surface, when examined by light reflected from its surface at a particular angle (§ 638); its *foot* has a very peculiar conformation, which is doubtless connected with its singular power of walking over smooth surfaces in direct opposition to the force of gravity, and on the action of which additional light has lately been thrown (§ 640); while the structure and physiology of its *sexual* apparatus, with the history of its development and metamorphoses, would of itself suffice to occupy the whole time of an observer who should desire thoroughly to work it out, not only for months but for years.[1] Hence, in treating of this department in such a work as the present, the Author labors under the *embarras des richesses;* for to enter into such a description of the parts of the structure of Insects most interesting to the Microscopist, as should be at all comparable in fulness with the accounts which it has been thought desirable to give of other Classes, would swell out the volume to an inconvenient bulk; and no course seems open, but to limit the treatment of the subject to a notice of the

---

[1] See Mr. Lowne's valuable Treatise on "The Anatomy and Physiology of the Blow-fly," 1870.

kinds of objects which are likely to prove most generally interesting, with a few illustrations that may serve to make the descriptions more clear, and with an enumeration of some of the sources whence a variety of specimens of each class may be most readily obtained. And this limitation is the less to be regretted, since there already exist in our language numerous elementary treatises on Entomology, wherein the general structure of Insects is fully explained, and the conformation of their minute parts as seen with the Microscope is adequately illustrated.

617. A considerable number of the smaller Insects—especially those belonging to the Orders *Coleoptera* (Beetles), *Neuroptera* (Dragon-fly, May-fly, etc.), *Hymenoptera* (Bee, Wasp, etc.), and *Diptera* (two-winged Flies)—may be mounted entire as opaque objects for low magnifying powers; care being taken to spread out their legs, wings, etc., so as adequately to display them, which may be accomplished, even after they have dried in other positions, by softening them by stepping them in hot water, or, where this is objectionable, by exposing them to steam. Full directions on this point, applicable to small and large Insects alike, will be found in all Text-books of Entomology. There are some, however, whose translucence allows them to be viewed as transparent objects; and these are either to be mounted in Canada balsam or in Deane's medium, Glycerine-jelly, or Farrant's gum, according to the degree in which the horny opacity of their integument requires the assistance of the balsam to facilitate the transmission of light through it, or the softness and delicacy of their textures render an aqueous medium more desirable. Thus, an ordinary *Flea* or *Bug* will best be mounted in balsam; but the various parasites of the *Louse* kind, with some or other of which almost every kind of animal is affected, should be set-up in some of the 'media.' Some of the aquatic larvæ of the Diptera and Neuroptera, which are so transparent that their whole internal organization can be made-out without dissection, are very beautiful and interesting objects when examined in the living state, especially because they allow the Circulation of the blood and the action of the dorsal vessel to be discerned (§ 632). Among these, there is none preferable to the larva of the *Ephemera marginata* (Day-fly), which is distinguished by the possession of a number of beautiful appendages on its body and tail, and is, moreover, an extremely common inhabitant of our ponds and streams. This insect passes two or even three years in its larval state, and during this time it repeatedly throws-off its skin; the cast skin, when perfect, is an object of extreme beauty, since, as it formed a complete sheath to the various appendages of the body and tail, it continues to exhibit their outlines with the utmost delicacy; and by keeping these larvæ in an Aquarium, and by mounting the entire series of their cast skins, a record is preserved of the successive changes they undergo. Much care is necessary, however, to extend them upon slides, in consequence of their extreme fragility; and the best plan is to place the slip of glass under the skin whilst it is floating on water, and to lift the object out upon the slide.—Thin *sections* of Insects, Caterpillars, etc., which bring the internal parts into view in their normal relations, may be cut with the Microtome (§ 184), by first soaking the body (as suggested by Dr. Halifax) in thick gum-mucilage, which passes into its substance, and gives support to its tissues, and then inclosing it in a casing of melted paraffin, made to fit the cavity of the Section-instrument.

618. *Structure of the Integument.*—In treating of those separate parts of the organization of Insects which furnish the most interesting objects

of Microscopic study, we may most appropriately commence with their Integument and its appendages (scales, hairs, etc). The body and members are closely invested by a hardened skin, which acts as their skeleton, and affords points of attachment to the muscles by which their several parts are moved; being soft and flexible, however, at the joints. This skin is usually more or less horny in its texture, and is consolidated by the animal substance termed *Chitine*, as well as, in some cases, by a small quantity of mineral matter. It is in the *Coleoptera* that it attains its greatest development; the 'dermo-skeleton' of many Beetles being so firm as not only to confer upon them an extraordinary power of passive resistance, but also to enable them to put forth enormous force by the action of the powerful muscles which are attached to it. It may be stated as a general rule, that the outer layer of this dermo-skeleton is always cellular, taking the place of an epidermis; and that the cells are straight-sided and closely-fitted together, so as to be polygonal (usually hexagonal) in form. Of this we have a very good example in the *superficial* layers (Fig. 427, B) of the thin horny lamellæ or blades which constitute the terminal portion of the antenna of the *Cockchafer* (Fig. 426); this layer being easily distinguished from the intermediate portion (A) of the lamina by careful focussing. In many Beetles, the hexagonal areolation of the surface is distinguishable when the light is reflected from it at a particular angle, even when not discernible in transparent sections. The integument of the common *Red Ant* exhibits the hexagonal cellular arrangement very distinctly throughout; and the broad flat expansion of the leg of the *Crabro* ('sand-wasp') affords another beautiful example of a distinctly-cellular structure in the outer layer of the integument. The inner layer, however, which constitutes the principal part of the thickness of the horny casing of the Beetle-tribe, seldom exhibits any distinct organization; though it may be usually separated into several lamellæ, which are sometimes traversed by tubes that pass into them from the inner surface, and extend towards the outer without reaching it.

619. *Tegumentary Appendages.*—The surface of Insects is often beset, and is sometimes completely covered, with *appendages*, having either the form of broad flat Scales, or that of Hairs more or less approaching the cylindrical shape, or some form intermediate between the two.—The *scaly* investment is most complete among the *Lepidoptera* (Butterfly and Moth tribe); the distinguishing character of the insects of this order being derived from the presence of a regular layer of scales upon each side of their large membranous wings. It is to the peculiar coloration of the scales that the various hues and figures are due, by which these wings are so commonly distinguished; all the scales of one patch (for example) being green, those of another red, and so on: for the subjacent membrane remains perfectly transparent and colorless, when the scales have been brushed off from its surface. Each scale seems to be composed of two or more membranous lamellæ, often with an intervening deposit of pigment, on which, especially in *Lepidoptera*, their color depends. Certain scales, however, especially in the Beetle-tribe, have a metallic lustre, and exhibit brilliant colors that vary with the mode in which the light glances from them; and this 'iridescence,' which is specially noteworthy in the scales of the *Curculio imperialis* ('diamond-beetle'), seems to be a purely optical effect, depending either (like the prismatic hues of a soap-bubble) on the extreme thinness of the membranous lamellæ, or (like those of 'mother-of-pearl,' § 565) on a lineation of surface produced by their corrugation. Each scale is

furnished at one end with a sort of handle or 'pedicle' (Figs. 414, 415), by which it is fitted into a minute socket attached to the surface of the insect; and on the wings of *Lepidoptera* these sockets are so arranged that the scales lie in very regular rows, each row overlapping a portion of the next, so as to give to their surface, when sufficiently magnified, very much the appearance of being *tiled* like the roof of a house. Such an arrangement is said to be 'imbricated.' The forms of these scales are often very curious, and frequently differ a good deal on the several parts of the wings and of the body of the same individual; being usually more expanded on the former, and narrower and more hair-like on the latter. A peculiar type of scale, which has been distinguished by the designation *plumule*, is met with among the *Pieridæ*, one of the principal families of the Diurnal Lepidoptera. The 'plumules' are not flat, but cylindrical or bellows-shaped, and are hollow; they are attached to the wing by a bulb, at the end of a thin elastic peduncle that differs in length in different species, and proceeds from the broader, not from the narrower end of the scale; whilst the free extremity usually tapers off, and ends in a kind of brush, though sometimes it is broad and has its edge fringed with minute filaments. These 'plumules,' which are peculiar to the males, are found on the upper surface of the wings, partly between and partly under the ordinary scales. They seem to be represented among the *Lycænidæ* by the 'battledore' scales to be presently described (§ 621).[1]

620. The peculiar markings exhibited by many of these Scales, very early attracted the attention of Opticians engaged in the application of Achromatism to the Microscope (§ 15); for, as the clearness and strength with which they could be shown, were found to depend on the degree to which the angular aperture of an Objective could be opened without sacrifice of perfect correction for spherical and chromatic aberration, such scales proved very serviceable as 'tests.' The Author can well remember the time when those of *Morpho menelaus* (Fig. 414), the ordinary and 'battledore' scales of the *Polyommatus argus* (Figs. 415, 416), and the scales of the *Lepisma saccharina* (Fig. 417), which are now only used for testing Objectives of *low* or *medium* power (§ 159, I., II.), were the recognized tests for objectives of *high* power; while the exhibition of alternating light and dark bands on a *Podura*-scale was regarded as a first-rate performance. The resolution of these bands into the 'notes of admiration' (Plate II., fig. 2) now clearly shown by every good 'Student's' 1-4th, marked the next step in advance; and though the introduction of the Diatom-tests greatly promoted the enlargement of angular aperture, yet the Author has the authority of the ablest constructors of high-power Objectives in this county for stating, that they still regard the Podura-scale as the best test for *definition*, and consequently for that *combination* of qualities which is most required in Objectives to be used for Biological investigations of the greatest difficulty (§ 158, VI.).[2] As the real structure of this scale, of which the 'notes of admiration,' or the 'exclamation-markings' constitute the optical expression, has been a matter of much controversy, the question requires special consideration;

---

[1] See Mr. Watson's Memoirs 'On the Scales of Battledore Butterflies,' in "Monthly Microscopical Journal," Vol. ii., pp. 73, 314.

[2] The Author is assured that it is by no means an uncommon experience, on first putting together an Objective of wide aperture, to find it capable of resolving a difficult Diatom, whilst, when tested on a *Podura-scale*, it utterly fails, on account of its imperfect 'definition.'

and in discussing it, regard should be had to what we are taught by the study of the larger and more strongly marked forms of Insect-scales, as to *what scales are.*—That they are in reality flattened *cells*, analogous to the Epidermic cells of higher animals (§ 671), can scarcely be doubted by any Physiologist. Their ordinary flattening is simply the result of their drying up; and the exception presented by the 'plumules' and 'battledore' scales (Fig. 416), which have the two surfaces separated by a considerable cavity, helps to prove the rule. It is perfectly clear in some of these, that the membranous wall of the cell is strengthened by longitudinal ribs, which diverge from the peduncle; as is particularly well seen in the plumules of two West African butterflies, *Pieris Agathina* and *Pieris Phloris*, in which the plumules are as much as 1·300th of an inch in length (large enough to be studied under the Binocular Microscope), and are of cylindrical form, save that they are drawn in as

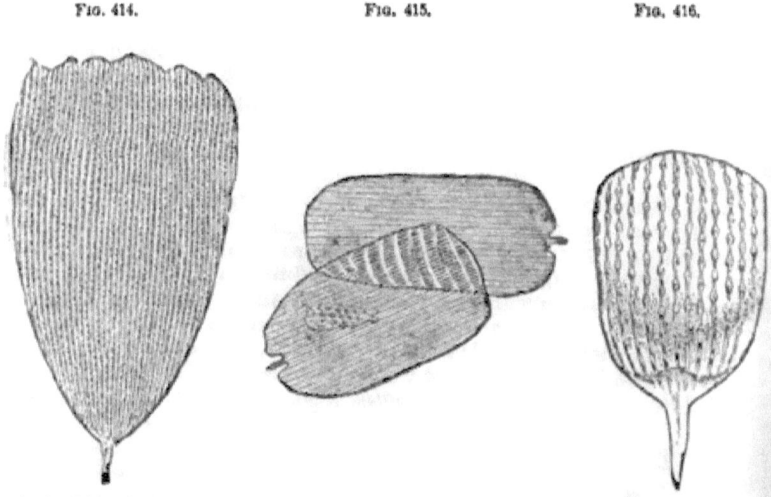

Fig. 414.   Fig. 415.   Fig. 416.

Scale of *Morpho Menelaus*.   Scales of *Polyommatus argus* (Azure-blue);—*a*, battledore-scale; *b*, Inerference striæ.   Battledore Scale of *Polyommatus argus* (Azure-blue).

if by a cord at about one-half or one-third of their length, the ribs curving inwards to this constriction.¹ In ordinary scales we find similar ribs, sometimes running parallel to each other, or nearly so (Figs. 414, 415), and occasionally connected by distinct cross-bars (Fig. 418), but sometimes diverging from the 'quill;' and where, as in *Lepisma* (Fig. 417), the ribs are parallel on one surface and divergent on the other, a very curious set of appearances is presented by their optical intersection, which throws considerable light on the meaning of the *Podura*-markings.

621. The easier test-scales are furnished by the order *Lepidoptera* (Butterflies and Moths); and among the most beautiful of these, both for color and for regularity of marking, are those of the *Morpho Menelaus* (Fig. 414). These are of a rich blue tint, and exhibit strong longi-

---

¹ See Watson, *loc. cit.*, p. 75.

tudinal striæ, which seem due to ribbed elevations of one of the superficial layers. There is also an appearance of transverse striation, which cannot be seen at all with an inferior objective, but becomes very decided with a good objective of medium focus; and this is found, when submitted to the test of a high power and good illumination, to depend upon the presence of transverse thickenings or corrugations (Fig. 414), probably on the internal surface of one of the membranes.—The large scales of the *Polyommatus argus* ('azure-blue' butterfly) resemble those of the Menelaus in form and structure, but are more delicately marked (Fig. 415). Their ribs are more nearly parallel than those of the *Menelaus* scale, and do not show the same transverse striation. When one of these scales lies partly over another, the effect of the optical intersection of the two sets of ribs at an oblique angle is to produce a set of interrupted striations (*b*), very much resembling those of the *Podura*-scale. The same Butterfly furnishes smaller scales, which are commonly termed the 'battledore' scales, from their resemblance in form to that object (Fig. 415, *a*). These scales, which occur in the males of several genera of the family *Lycænidæ*, and present a considerable variety of shape,' are marked by narrow longitudinal ribbings, which at intervals seem to expand into rounded or oval elevations that give to the scales a dotted appearance (Fig. 416); at the lower part of the scale, however, these dots are wanting. Dr. Anthony describes and figures them as elevated bodies, somewhat resembling dumb-bells or shirt-studs, ranged along the ribs, and standing out from the general surface.[2] Other good observers, however, whilst recognizing the stud-like bodies described by Dr. Anthony, regard them as not projecting from the external surface of the scale, but as interposed between its two lamellæ;[3] and this view seems to the author to be more conformable than Dr. Anthony's to general probability.

622. The more difficult 'test-scales' are furnished by little wingless insects ranked together by Latreille in the order *Thysanura*, but now separated by Sir John Lubbock,[4] on account of important differences in internal structure, into the two groups *Collembola* and true *Thysanura*. Of the former of these, the *Lepismidæ* constitute the typical family; and the scale of the common *Lepisma saccharina*, or 'sugar-louse,'[5] very early attracted the attention of Microscopists on account of its beautiful shell-like sculpture. When viewed under a low magnifying power, it presents a beautiful 'watered silk' appearance, which, with higher amplification, is found to depend (as Mr. R. Beck first pointed out)[6] upon the intersection of two sets of striæ, representing the different structural arrangements of its two superficial membranes. One of its surfaces (since ascertained by Mr. Joseph Beck[7] to be the *under* or attached surface of the

---
[1] See Watson, *loc. cit.*
[2] 'The Markings on the Battledore Scales of some of the *Lepidoptera*,' in "Monthly Microsc. Journal," Vol. vii. (1872), pp. 1, 250.
[3] See "Proceedings of the Microscopical Society," *op. cit.*, p. 278.
[4] See his "Monograph of the *Collembola* and *Thysanura*," published by the Ray Society, 1872.
[5] This insect may be found in most old houses, frequenting damp warm cupboards, and especially such as contain sweets; it may be readily caught in a small pill-box, which should have a few pin-holes in the lid; and if a drop of chloroform be put over the holes, the inmate will soon become insensible, and may be then turned out upon a piece of clean paper, and some of its scales transferred to a slip of glass by simply pressing this gently on its body.
[6] "The Achromatic Microscope," p. 50.
[7] See his Appendix to Sir John Lubbock's "Monograph."

scale) is raised, either by corrugation or thickening, into a series of strongly-marked longitudinal ribs, which run nearly *parallel* from one end of the scale to the other, and are particularly distinct at its margins and at its free extremity; whilst the other surface (the free or *outer*, according to Mr. J. Beck) presents a set of less definite corrugations, *radiating* from the pedicle, where they are strongest, towards the sides and free extremity of the scale, and therefore crossing the parallel ribs at angles more or less acute (Fig. 417). It was further pointed out by Mr. R. Beck, that the intersection of these two sets of corrugations at different angles produces most curious effects upon the appearances which optically represent them. For where the diverging ribs cross the longitudinal ribs very obliquely, as they do near the free extremity of the

FIG. 417.

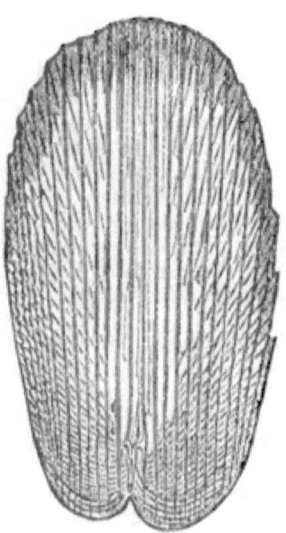

Scale of *Lepisma saccharina*.

FIG. 418.

[Scale of *Machilis polypoda*.

scale, the longitudinal ribs seem broken up into a series of 'exclamation-markings,' like those of the Podura; but where the crossing is transverse or nearly so, as at the sides of the scale, an appearance is presented as of successions of large bright beads. The conclusion drawn by the Messrs. Beck, that these interrupted appearances are "produced by two sets of uninterrupted lines on different surfaces," has been confirmed by the careful investigations of Mr. Morehouse.[1] The minute beaded structure observed by Dr. Royston-Pigott[2] alike in the ribs and in the intervening spaces, may now be pretty certainly regarded as an optical effect of diffraction (§ 156). In the scale of a type nearly allied to Lepisma, the *Machilis polypoda*, the very distinct ribbing (Fig. 418) is produced by the

---

[1] "Monthly Microsc. Journal," Vol. xi. (1874), p. 13, and Vol. xviii. (1877), p. 31.
[2] "Monthly Microsc. Journ.," Vol. ix. (1873), p. 63.

corrugation of the under membranous lamina alone; the upper or exposed lamina being smooth, with the exception of slight undulations near the pedicle; and the cross-markings being due to structure between the superposed membranes, probably a deposit on the interior surface of one or both of them.[1]

623. Although the *Poduridæ* and *Lepismidæ* now rank as distinct Families, yet they approximate sufficiently in general organization, as well as in habits, to justify the expectation that their scales would be framed upon the same plan. The *Poduridæ* are found amidst the sawdust of wine-cellars, in garden tool-houses, or near decaying wood; and derive their popular name of 'spring-tails' from the possession by many of them of a curious caudal appendage, by which they can leap like fleas. This is particularly well developed in the species now designated *Lepidocyrtus curvicollis*, which furnishes what are ordinarily known as 'Podura'-

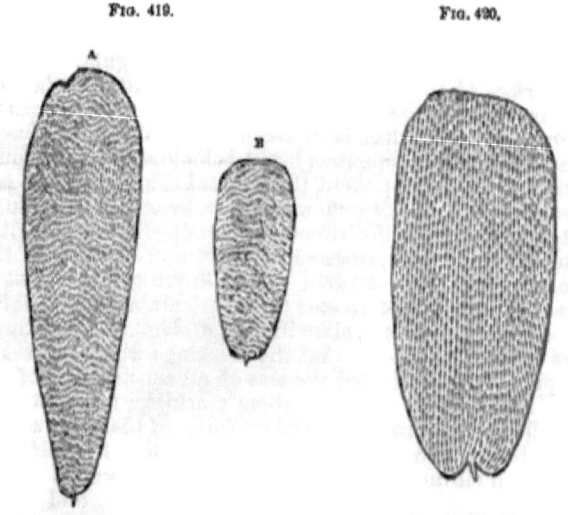

FIG. 419.   FIG. 420.

Test-scales of *Lepidocyrtus curvicollis*:— A, large, strongly-marked scale; B, small scale, more faintly marked.

Ordinary scale of *Lepidocyrtus curvicollis*.

scales. "When full-grown and unrubbed," says Sir John Lubbock, "this species is very beautiful, and reflects the most gorgeous metallic tints." Its scales are of different sizes and of different degrees of strength of marking (Fig. 419, A, B), and are therefore by no means of uniform value as tests. The general appearance of their surface, under a power not sufficient to resolve their markings, is that of watered silk,—light and dark bands passing across it with wavy irregularity; but a well-corrected Objective of very moderate angular aperture now suffices to resolve every dark band into a row of distinct 'exclamation marks' (Plate II., fig. 2). If, however, they are illuminated by oblique light from above (the scales being placed under the objective without any cover, so as to avoid the loss of light by reflection from its surface), the appearances presented are those shown in fig. 4 when the markings are at right angles

---

[1] See Mr. Joseph Beck, in Sir J. Lubbock's "Monograph," p. 253.

to the direction of the light, and in fig. 5 when they lie in the same direction as the light with their narrow ends pointing to it. When this last direction is reversed, the light from the points is so slight, that the scales appear to have lost their markings altogether. If moisture should insinuate itself between the scale and the covering-glass, the markings disappear entirely, as shown in fig. 3; and this is true also of the scale of *Lepisma*. A certain longitudinal continuity may be traced between the 'exclamation-marks' in the ordinary test-scale; but this is much more apparent in other scales from the same species (Fig. 420), as well as in the scales of various allied types, which were carefully studied by the late Mr. R. Beck.[1] In certain other types, indeed, the scales have very distinct longitudinal parallel ribs, sometimes with regularly disposed crossbars; these ribs, being confined to one surface only (that which is in contact with the body), are not subject to any such interference with their optical continuity as has been shown to occur in *Lepisma;* but more or less distinct indications of radiating corrugations often present themselves. The appearance of the interrupted 'exclamation-marks' Mr. J. Beck (*op. cit.*, p. 254) considers to be due "to irregular corrugations of the outer surface of the under membrane, to slight undulations on the outer surface of the upper membrane, and to structure between the superposed membranes." It has been recently stated by Mr. Joseph Beck, that the scales of a Lepidopterous insect belonging to the genus *Mormo*, which under a low power present the watered-silk appearance seen in the Podura-scale, under a 1-5th show the 'exclamation' markings, whilst under a 1-10th they exhibit distinct ribs from pedicle to apex; thus showing in one scale how the appearances run from one scale into the other.[2] On the other hand, we are assured by Dr. Royston-Pigott, not only that what a lens most perfectly corrected for spherical aberration ought to show, is a minute beaded structure, alike in the 'exclamation-markings' and in the spaces between them; but that the markings whose perfect definition had been previously considered the aim of all constructors of high-power Objectives, are altogether illusory, these markings representing nothing else than the manner in which the *rouleaux* of beads lie with reference to one another.[3] The Author has fully satisfied himself by his own study, under an oil-immersion 1-25th of Messrs. Powell and Lealand, of a *Podura*-scale illuminated by the 'immersion paraboloid' (which gives a view of it entirely different than any that can be obtained either by transmitted or reflected light), that the 'exclamation-markings' are—as maintained by the Messrs. Beck—the optical expression of a corrugated or ribbed arrangement of the lower membrane of the scale, slightly modified by the internal structure of the upper membrane, and probably also (as confirmed by Mr. Wenham) by a structure interposed between the two membranes.[4] And this conclusion is borne out, in opposition to the doctrine of Dr. Royston Pigott, by two unrivalled Photographs taken of the *Podura*-scale by Col. Dr. Woodward. One of these, taken with a magnifying power of 3200 diameters, central monochromatic light, immersion 1-16th, and amplifier, shows the 'exclamation-marks' better

---

[1] "Trans. of Microsc. Soc.," N.S., Vol. x. (1862), p. 83. See also Mr. Joseph Beck, in the Appendix to Sir John Lubbock's "Monograph," and in "Monthly Microsc. Journ.," Vol. iv., p. 253.
[2] "Journ. Roy. Microsc. Soc.," Vol. ii. (1879), p. 810.
[3] See his paper 'On High Power Definition,' in "Monthly Microscopical Journal," Vol. ii. (1869), p. 295, and several subsequent papers.
[4] "Monthly Journ. of Microsc. Sci.," Vol. xi. (1874), p. 75.

than any photographic representation previously obtained; and it is clear that Dr. Woodward regards this as the *truest* view. "Immediately afterwards," he says, "with the same optical combination and magnifying power, without any change in the cover-correction, by simply rendering the illuminating pencil oblique, and slightly withdrawing the objective from its first focal position, I obtained a negative which displays the 'bead-like' or varicose appearance of the ribbing more satisfactorily than I had previously been able to do."[1] The beaded appearance shown in this photograph, a copy of a portion of which is given in Fig. 421, corresponds so entirely with that which Dr. Woodward afterwards found to be producible in the scale of the Gnat by a like alteration in the illumination (§ 156), that the Author feels fully justified in adhering to his original opinion that it does not represent real structure, but is an optical effect of diffraction.[2]

624. The *Hairs* of many Insects, and still more of their larvæ, are very interesting objects for the microscope, on account of their branched or tufted conformation; this being particularly remarkable in those

Fig. 421.

Portion of a *Podura-scale*, from a Photograph by Col. Dr. Woodward.

with which the common hairy Caterpillars are so abundantly beset. Some of these afford very good tests for the perfect correction of Objectives. Thus the hair of the *Bee* is pretty sure to exhibit strong prismatic colors, if the Chromatic aberration should not have been exactly neutralized; and that of the larva of a *Dermestes* (commonly but erroneously termed the 'bacon-beetle') was once thought a very good test of defining power, and is still useful for this purpose. It has a cylindrical shaft (Fig. 422, B) with closely-set whorls of spiny protuberances, four or five

---

[1] "Monthly Microscopical Journal," Vol. v., p. 246.
[2] The successive Volumes of the "Monthly Microscopical Journal," from the 2d (in which Dr. Royston-Piggott's views were first promulgated) to the present date, teem with Papers on this subject from Mr. Jos. Beck, Mr. McIntire, Dr. Maddox, Dr. Royston-Piggott, Mr. Wenham, and Col. Dr. Woodward; which, with a Paper by Mr. Slack in "The Student," Vol. v., p. 49, and a Paper by Mr. Morehouse, giving the results of his examination of the scales of *Lepisma* and *Podura* as opaque objects, under very high immersion objectives, with Beck's Vertical Illuminator ("Monthly Microsc. Journ.," Vol. xviii., 1877, p. 31), should be consulted by such as wish to follow out the inquiry.

in each whorl; the highest of these whorls is composed of mere knobby spines: and the hair is surmounted by a curious circle of six or seven large filaments, attached by their pointed ends to its shaft, whilst at their free extremities they dilate into knobs. An approach to this structure is seen in the hairs of certain *Myriapods* (centipedes, gally-worms, etc.), of which an example is shown in Fig. 422, A; and some minute forms of this class are most beautiful objects under the Binocular Microscope, on account of the remarkable structure and regular arrangement of their hairs.

625. In examining the Integument of Insects, and its appendages, parts of the surface may be viewed either by reflected or transmitted light, according to their degree of transparence and the nature of their covering. The Beetle and Butterfly tribes furnish the greater number of the specimens suitable to be viewed as *opaque* objects: and nothing is easier than to mount portions of the *elytra* of the former (which are usually the most showy parts of their bodies), or of the wings of the latter, in the manner described in § 175. The tribe of *Curculionidæ*, in which the surface of the body is beset with scales having the most varied and lustrous hues, is distinguished among Coleoptera for the brilliancy of the objects it affords; the most remarkable in this respect being the well-known *Curculio imperialis*, or 'diamond-beetle' of South America, parts of whose elytra, when properly illuminated and looked-at with a low power, show like clusters of jewels flashing against a dark velvet ground. In many of the British Curculionidæ, which are smaller and less brilliant, the scales lie at the bottom of little depressions of the surface; and if the elytra of the 'diamond beetle' be carefully examined, it will be found that each of the clusters of scales which are arranged upon it in rows, seems to rise out of a deep pit which sinks-in by its side. The transition from Scales to Hairs is extremely well seen by comparing the different parts of the surface of the diamond-beetle with each other. The beauty and brilliancy of many objects of this kind are increased by mounting them in cells in Canada balsam, even though they are to be viewed with reflected light; other objects, however, are rendered less attractive by this treatment; and in order to ascertain whether it is likely to improve or to deteriorate the specimen, it is a good plan first to test some other portion of the body having scales of the same kind, by touching it with turpentine, and then to mount the part selected as an object, either in balsam or dry, according as the turpentine increases or diminishes the brilliancy of the scales on the spot to which it was applied. Portions of the wings of Lepidoptera are best mounted as opaque objects, without any other preparation than gumming them flat down to the disk of the wooden slide (§ 175); care being taken to avoid disturbing the arrangement of the scales, and to keep the objects, when mounted, as secluded as possible from dust. In selecting such portions, it is well to choose those which have the brightest and the most contrasted colors, exotic butterflies being in this respect usually preferable to British; and before attaching them to slides, care should be taken to ascertain in what position, with

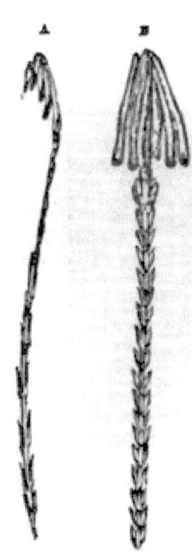

Fig. 422.

A, Hair of *Myriapod*.
B, Hair of *Dermestes*.

the arrangement of light ordinarily used, they are seen to the best advantage, and to fix them there accordingly.—Whenever portions of the integument of Insects are to be viewed as *transparent* objects, for the display of their intimate structure, they should be mounted in Canada balsam, after soaking for some time in turpentine; since this substance has a peculiar effect in increasing their translucence. Not only the horny casings of perfect Insects of various orders, but also those of their pupæ, are worthy of this kind of study; and objects of great beauty (such as the chrysalis case of the Emperor-moth), as well as of scientific interest, are sure to reward such as may prosecute it with any assiduity. Further information may often be gained by softening such parts in potash, and viewing them in fluid.—The *scales* of the wings of Lepidoptera, etc., are best transferred to the slide, by simply pressing a portion of the wing either upon the slip of glass or upon the cover; if none should adhere, the glass may first be gently breathed-on. Some of them are best seen when examined 'dry,' whilst others are more clear when mounted in fluid; and for the determination of their exact structure, it is well to have

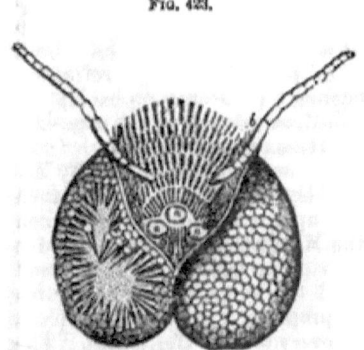

Fig. 423.

Head and Compound Eyes of the *Bee*, showing the ocellites *in situ* on one side (A), and displaced on the other (B); *a, a, a,* stemmata, *b, b,* antennæ.

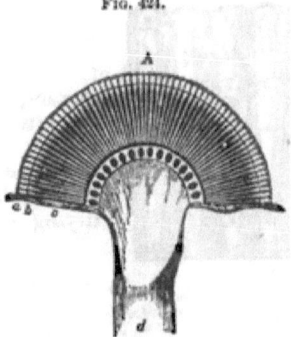

Fig. 424.

Section of the Composite Eye of *Melolontha vulgaris* (Cockchafer):—*a,* facets of the cornea; *b,* transparent pyramids surrounded with pigment; *c,* fibres of the optic nerve; *d,* trunk of the optic nerve.

recourse to both these methods. Hairs, on the other hand, are best mounted in Balsam.

626. *Parts of the Head.*—The *eyes* of Insects, situated upon the upper and outer part of the head, are usually very conspicuous organs, and are frequently so large as to touch each other in front (Fig. 423). We find in their structure a remarkable example of that multiplication of similar parts, which seems to be the predominating 'idea' in the conformation of Articulated animals; for each of the large protuberant bodies which we designate as *an eye,* is really a 'compound' eye, made up of many hundred or even many thousand minute conical *ocellites* (B). Approaches to this structure are seen in Annelida and Entomastraca; but the number of 'ocellites' thus grouped-together is usually small. In the higher Crustacea, however, the 'ocellites' are very numerous; and their compound eyes are constructed upon the same general plan as those of Insects, though their shape and position are often very peculiar (Fig. 491). The individual ocellites are at once recognized, when the 'compound

eyes' are examined under even a low magnifying power, by the 'facetted' appearance of the surface (Fig. 423, A), which is marked-out by very regular divisions either into hexagons or into squares: each facet is the 'corneule' of a separate ocellite, and has a convexity of its own; hence by counting the facets, we can ascertain the number of ocellites in each 'compound eye.' In the two eyes of the common *Fly*, there are as many as 4,000; in those of the *Cabbage Butter-fly* there are about 17,000; in the *Dragon-fly*, 24,000; and in the *Mardella Beetle*, 25,000. Behind each 'corneule' is a layer of dark pigment, which takes the place and serves the purpose of the 'iris' in the eyes of vertebrate animals; and this is perforated by a central aperture or 'pupil,' through which the rays of light that have traversed the corneule gain access to the interior of the eye. The further structure of these bodies is best examined by vertical sections (Fig. 424); and these show that the shape of each ocellite (*b*) is conical, or rather pyramidal, the corneule forming its base (*a*), whilst its apex abuts upon the extremity of a fibre (*c*) proceeding from the termination of the optic nerve (*d*). The details of the structure of each ocellite are shown in Fig. 425; in which it is shown that each corneule is a double-convex lens, made up by the junction of two plano-convex lenses, *a a* and *a' a'*, which have been found by Dr. Hicks to possess different refractive powers; by this arrangement (it seems probable) the aberrations are diminished, as they are by the combination of 'humors' in the Human eye. That each 'corneule' acts as a distinct lens, may be shown by detaching the entire assemblage by maceration, and then drying it (flattened out) upon a slip of glass; for when this is placed under the Microscope, if the point of a knife, scissors, or any similar object, be interposed between the mirror and the stage, the image of this point will be seen, by a proper adjustment of the focus of the microscope, in every one of the lenses. The focus of each 'corneule' has been ascertained by experiment to be equivalent to the length of the pyramid behind it; so that the image which it produces will fall upon the extremity of the filament of the optic nerve which passes to the latter. The pyramids (*b*, *b*) consist of a transparent substance, which may be considered as representing the 'vitreous humor;' and they are separated from each other by a layer of dark pigment *d' d'*, which closes-in at *d d* between their bases and the corneules, leaving a set of pupillary apertures *c, c*, for the entrance of the rays which pass to them from the 'corneules.' After traversing these pyramids, the rays reach the bulbous extremities *e, e* of the fibres of the optic nerve, which are surrounded, like the pyramid, by pigmentary substance. Thus the rays which have passed through the several 'corneules' are prevented from mixing with each other; and no rays, save those which pass in the axes of the pyramids, can reach the fibres of the optic nerve. Hence, it is evident, that, as no two ocellites on the same side (Fig. 424) have exactly the same axis, no two can receive their rays from the same point of an object; and thus, as each compound eye is immovably fixed upon the head, the combined action of the entire aggregate will probably only afford but a single image, resembling that which *we* obtain by means of our

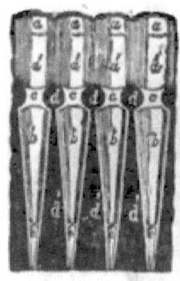

Fig. 425.

Minute structure of the Eye of the *Bee*:—*a a*, anterior lenses of corneule; *a' a'*, its posterior lenses; *c c*, pupillary apertures, separated by intervening pigment *d d*; *b b*, pyramids separated by pigment *d' d'*, and abutting on *e e*, bulbous extremities of nerve-fibres.

single eyes.—Although the foregoing may be considered as the typical structure of the Eyes of Insects, yet there are various departures from it (most of them slight) in the different members of the Class. Thus in some cases the posterior surface of each 'corneule' is concave; and a space is left between it and the iris-like diaphragm, which seems to be occupied by a watery fluid or 'aqueous humor;' in other instances again, this space is occupied by a double-convex body, which seems to represent the 'crystalline-lens;' and this body is sometimes found behind the iris, the number of ocellites being reduced, and each one being larger, so that the cluster presents more resemblance to that of Spiders, etc.—Besides their 'compound' eyes, Insects usually possess a small number of 'simple' eyes (termed *ocelli* or *stemmata*) seated upon the top of the head (Fig. 423, *a, a, a*). Each of these consists of a single very convex corneule; to the back of which proceeds a bundle of rods that are in connection with fibrils of the optic nerve. Such ocelli are the only visual organs of the Larvæ of insects that undergo complete metamorphosis; the 'compound' eyes being only developed towards the end of the Pupa-stage.[1]

627. Various modes of preparing and mounting the Eyes of Insects may be adopted, according to the manner wherein they are to be viewed. For the observation of their external facetted surface by reflected light, it is better to lay down the entire head, so as to present a front-face or a side-face, according to the position of the eyes; the former giving a view of *both* eyes, when they approach each other so as nearly or quite to meet (as in Fig. 423); whilst the latter will best display *one*, when the eyes are more situated at the sides of the head. For the minuter examination of the 'corneules,' however, these must be separated from the hemispheroidal mass whose exterior they form, by prolonged maceration; and the pigment must be carefully washed away, by means of a fine camel-hair brush, from the inner or posterior surface. In flattening them out upon the glass-slide, one of two things must necessarily happen; either the margin must tear when the central portion is pressed-down to a level; or, the margin remaining entire, the central portion must be thrown into plaits, so that its corneules overlap one another. As the latter condition interferes with the examination of the structure much more than the former does, it should be avoided by making a number of slits in the margin of the convex membrane before it is flattened-out. Vertical sections, adapted to demonstrate the structure of the ocelli and their relations to the optic nerve, can be only made when the insect is fresh, or or has been preserved in strong spirit. Mr. Lowne (*loc. cit.*) recommends that the head should be hardened in a 2 per cent solution of chromic acid, and then imbedded in cacoa-butter; the sections must be cut *very* thin, and should be mounted in Canada balsam. The following are some of the Insects whose eyes are best adapted for Microscopic preparations:— *Coleoptera*, Cicindela, Dytiscus, Melolontha (Cockchafer), Lucanus (Stag-beetle) ;—*Orthoptera*, Acheta (House and Field Crickets), Locusta;— *Hemiptera*, Notonecta (Boat-fly);—*Neuroptera*, Libellula (Dragon-fly), Agrion;—*Hymenoptera*, Vespidæ (Wasps) and Apidæ (Bees) of all kinds; —*Lepidoptera*, Vanessa (various species of Butterflies), Sphinx ligustri (Privet Hawk-moth), Bombyx (Silk-worm moth, and its allies);—*Diptera*, Tabanus (Gad-fly), Asilus, Eristalis (Drone-fly), Tipula (Crane-fly), Musca (House-fly), and many others.

---

[1] For minute details as to the structure of the Eyes of Insects, see the admirable Memoir by Mr. Lowne, in "Phil. Trans.," 1878, p. 577.

628. The *Antennæ*, which are the two jointed appendages arising from the upper part of the head of Insects (Fig. 423, *b b*), present a most wonderful variety of conformation in the several tribes of Insects; often differing considerably in the several species of one genus, and even in the two sexes of the same species. Hence the characters which they afford are extremely useful in classification; especially since their structure must almost necessarily be in some way related to the habits and general economy of the creatures to which they belong, although our imperfect acquaintance with their function may prevent us from clearly discerning this relation. Thus among the Coleoptera we find one large family, including the Glow-worm, Fire-fly, Skip-jack, etc., distinguished by the toothed or serrated form of the antennæ, and hence called *Serricornes;* in another, of which the Burying-beetle is the type, the antennæ are terminated by a club-shaped enlargement, so that these beetles are termed

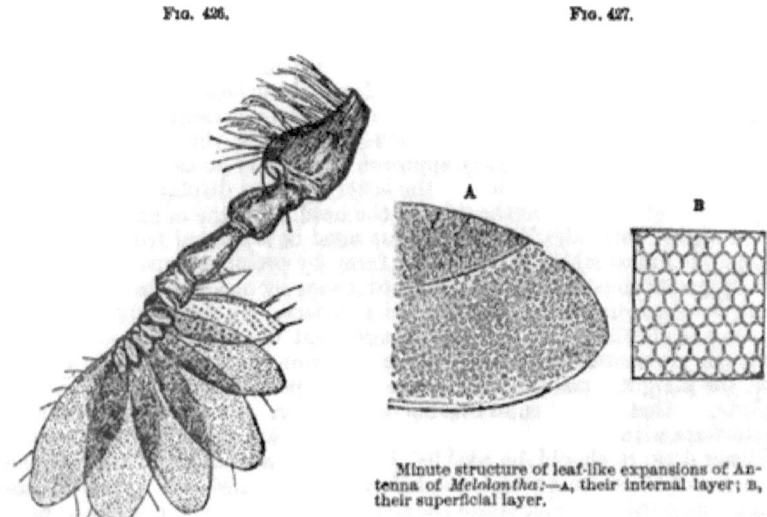

FIG. 426.

Antenna of *Melolontha* (Cockchafer).

FIG. 427.

Minute structure of leaf-like expansions of Antenna of *Melolontha:*—A, their internal layer; B, their superficial layer.

Clavicornes; in another, again, of which the *Hydrophilus* or large Water-beetle is an example, the antennæ are never longer and are commonly shorter than one of the pairs of palpi, whence the name of *Palpicornes* is given to this group; in the very large family that includes the *Lucani* or Stag-beetles with the *Scarabæi*, of which the Cockchafer is the commonest example, the antennæ terminate in a set of leaf-like appendages, which are sometimes arranged like a fan or the leaves of an open book (Fig. 426), are sometimes parallel to each other like the teeth of a comb, and sometimes fold one over the other, thence giving the name *Lamellicornes;* whilst another large family is distinguished by the appellation *Longicornes*, from the great length of the antennæ, which are at least as long as the body, and often longer. Among the Lepidoptera, again, the conformation of the antennæ, frequently enables us at once to distinguish the group to which any specimen belongs. As every treatise on Entomology contains figures and descriptions of the principal types of conformation of these organs, there is no occasion here to dwell upon them longer than to specify such as are most interesting to the Microscopist:—*Coleoptera*, Brachinus, Calathus, Harpalus, Dytiscus, Staphylinus, Philonthus, Elater, Lampyris, Silpha, Hydrophilus, Aphodius, Melolontha, Cetonia, Curculio;—*Orthoptera*, Forficula (Earwig), Blatta (Cockroach) ;—*Lepidoptera*, Sphinges (Hawk-moth), and Nocturna (Moths) of various kinds, the large 'plumed' antennæ of the latter being peculiarly beautiful objects under a low magnifying power;—*Diptera*, Culicidæ (Gnats of various kinds), Tipulidæ (Crane-flies and Midges), Tabanus, Eristalis, and Muscidæ (Flies of various kinds). All the larger antennæ, when not mounted 'dry' as opaque objects, should be put up in Balsam, after being soaked for some time in turpentine; but the small feathery antennæ of Gnats and Midges are so liable to distortion when thus mounted, that it is better to set them up in fluid, the head with its pair of antennæ being thus preserved together when not too large.— A curious set of organs has been recently discovered in the antennæ of many Insects, which have been supposed to constitute collectively an apparatus for Hearing. Each consists of a cavity hollowed out in the horny integument, sometimes nearly spherical, sometimes flask-shaped, and sometimes prolonged into numerous extensions formed by the folding of its lining membrane; the mouth of the cavity seems to be normally closed-in by a continuation of this membrane, though its presence cannot always be satisfactorily determined; whilst to its deepest part a nerve-fibre may be traced. The expanded lamellæ of the antennæ of *Melolontha* present a great display of these cavities, which are indicated in Fig. 427, A, by the small circles that beset almost their entire area; their form, which is very peculiar, can here be only made out by vertical sections; but in many of the smaller antennæ, such as those of the Bee, the cavities can be seen sideways without any other trouble than that of bleaching the specimen to render it more transparent.[1]

629. The next point in the organization of Insects to which the attention of the Microscopist may be directed, is the structure of the *mouth*. Here, again, we find almost infinite varieties in the details of conformation; but these may be for the most part reduced to a small number of types or plans, which are characteristic of the different orders of Insects. It is among the *Coleoptera*, or Beetles, that we find the several parts of which the mouth is composed, in their most distinct form; for although some of these parts are much more highly developed in other Insects, other parts may be so much altered or so little developed as to be scarcely recognizable. The Coleoptera present the typical conformation of the *mandibulate* mouth, which is adapted for the prehension and division of solid substances; and this consists of the following parts:—1, a pair of jaws, termed *mandibles*, frequently furnished with

---

[1] See the Memoir of Dr. Hicks 'On a new Structure in the Antennæ of Insects,' in "Trans. of Linn. Soc.," Vol. xxii., p. 147; and his 'Further Remarks,' at p. 383 of the same volume. See also the Memoir of M. Lespès, 'Sur l'Appareil Auditif des Insectes,' in "Ann. des Sci. Nat.," Sér. 4, Zool., Tom. ix., p. 258; and that of M. Claparède, 'Sur les prétendus Organes Auditifs des coléoptères lamellicornes et autres Insectes,' in "Ann. des. Sci. Nat.," Sér. 4, Zool., Tom. x., p. 236. Dr. Hicks lays great stress on the 'bleaching process,' as essential to success in this investigation; and he gives the following directions for performing it:—Take of Chlorate of Potass a drachm, and of Water a drachm and a half; mix these in a small wide bottle containing about an ounce; wait five minutes, and then add about a drachm and a half of strong Hydrochloric Acid. Chlorine is thus slowly developed; and the mixture will retain its bleaching power for some time.

powerful teeth, opening *laterally* on either side of the mouth, and serving as the chief instruments of manducation; 2, a second pair of jaws, termed *maxillæ*, smaller and weaker than the preceding, beneath which they are placed, and serving to hold the food, and to convey it to the back of the mouth; 3, an upper lip, or *labrum;* 4, a lower lip or *labium;* 5, one or two pairs of small jointed appendages termed *palpi*, attached to the maxillæ, and hence called *maxillary palpi;* 6, a pair of *labial palpi.* The labium is often composed of several distinct parts; its basal portion being distinguished as the *mentum* or chin, and its anterior portion being sometimes considerably prolonged forwards, so as to form an organ which is properly designated the *ligula*, but which is more commonly known as the 'tongue,' though not really entitled to that designation, the real *tongue* being a soft and projecting organ which forms the floor of the

Fig. 428.

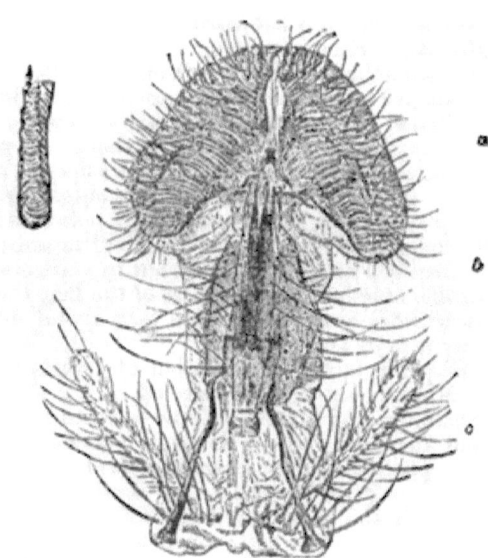

Tongue of common *Fly:—a*, lobes of ligula; *b*, portion inclosing the lancets, formed by the metamorphosis of the maxillæ; *c*, maxillary palpi:—A, portion of one of the pseudotracheæ enlarged.

mouth, and which is only found as a distinct part in a comparatively small number of Insects, as the Cricket.—This *ligula* is extremely developed in the *Fly* kind, in which it forms the chief part of what is commonly called the 'proboscis' (Fig. 428);[1] and it also forms the 'tongue' of the *Bee* and its allies (Fig. 429). The ligula of the common Fly presents a curious modification of the ordinary tracheal structure

---

[1] The representation given in the figure is taken from one of the ordinary preparations of the Fly's proboscis, which is made by slitting it open, flattening it out, and mounting it in Balsam. For representations of the true relative positions of the different parts of this wonderful organ, and for minute descriptions of them, the reader is referred to Mr. Suffolk's Memoir 'On the Proboscis of the Blow-fly,' in "Monthly Microsc. Journ.," Vol. i., p. 331; and to Mr. Lowne's Treatise on "The Anatomy and Physiology of the Blow-fly," p. 41.

(§ 634), the purpose of which is not apparent; for instead of its tracheæ being kept pervious, after the usual fashion, by the winding of a continuous spiral fibre through their interior, the fibre is broken into rings, and **these rings do** not surround the whole tube, but are terminated by a set **of arches that pass from one to** another (Fig. 428, A).[1]— In the *Diptera* or two-winged Flies generally, the labrum, maxillæ, mandibles, and the internal tongue (where it exists) are converted into delicate lancet-shaped organs termed *setæ*, which, when closed-together, are received into a hollow on the upper side **of** the labium (Fig. 428, *b*), but which are capable of being used to make punctures in the skin of Animals or the epidermis **of** Plants, whence the juices may be drawn forth by the proboscis. Frequently, however, two or more of these organs may be wanting, so that their number is reduced from six, to four, three, or two.—In the *Hymenoptera* (Bee and Wasp tribe), the labrum and the mandibles (Fig. 429, *b*) much resemble those of Mandibulate Insects, and are used for corresponding purposes; the maxillæ (*c*) are greatly elongated, **and** form, when closed, a tubular sheath for the *Ligula* or 'tongue,' through which the honey is drawn up; the labial palpi (*d*) also are greatly developed, and fold together, like the maxillæ, so as to form an inner sheath for the 'tongue;' while the 'ligula' itself (*e*) is a long tapering muscular organ, marked by an immense number of short annular divisions, and densely covered over its own length with long hairs (B). It is not tubular, as some have stated, but is solid; when actively employed in taking food, it is extended to a great distance beyond the other parts of the mouth; but when at rest it is closely packed-up and concealed between the maxillæ. "The manner," says Mr. Newport, "in which the honey is obtained when the organ is plunged into it at the bottom of **a flower, is** by 'lapping,' or a constant succession of short and quick extensions **and contractions of** the organ, which occasion the fluid to accumulate upon it **and to ascend** along its upper surface, until it reaches the orifice of the tube **formed by** the approximation of the maxillæ above, **and of** the labial palpi and this part of the ligula below."

FIG. 429.

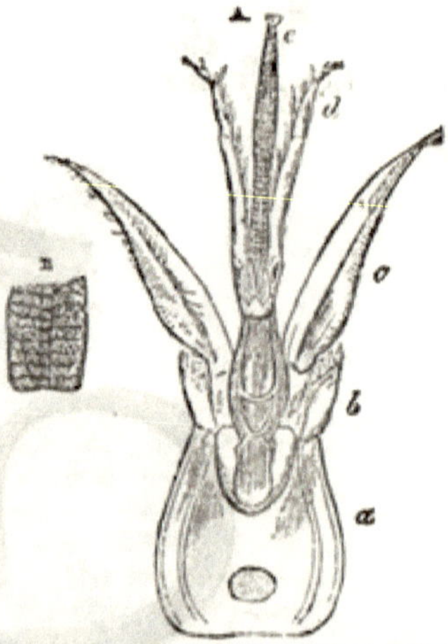

A, Parts of the Mouth of *Apis mellifica* (Honey-bee):—*a*, mentum; *b*, mandibles; *c*, maxillæ; *d*, labial palpi; *e*, ligula, or prolonged labium, commonly termed the tongue:—B, portion of the surface of the ligula, more highly magnified.

630. By the plan of conformation just described, we are led to that

---

[1] According to Dr. Anthony ("Monthly Micros. Journ.," Vol. xi., p. 242), these 'pseudo-tracheæ' are suctorial organs, which can take-in liquid alike at their extremities and through the whole length of the fissure caused by the interruption of the rings; the edges of this fissure being formed by the alternating series of 'ear-like appendages,' connected with the terminal 'arches,' the closing-together of which converts the pseudo-trachea into a complete tube. Dr. A. considers each of these ear-like appendages to be a minute sucker, "either for the adhesion of the fleshy tongue, or for the imbibition of fluids, or perhaps for both purposes." —The point is well worthy of further investigation.

which prevails among the *Lepidoptera* or Butterfly tribe, and which, being pre-eminently adapted for suction, is termed the *haustellate* mouth. In these Insects, the labrum and mandibles are reduced to three minute triangular plates; whilst the maxillæ are immensely elongated, and are united together along the median line to form the *haustellium* or true 'proboscis,' which contains a tube formed by the junction of the two grooves that are channelled out along their mutually applied surfaces, and which serves to pump-up the juices of deep cup-shaped flowers, into which the size of their wings prevents these insects from entering. The length of this haustellium varies greatly: thus in such Lepidoptera as take no food in their perfect state, it is a very insignificant organ; in some of the white Hawk-moths, which hover over blossoms without alighting, it is nearly two inches length; and in most Butterflies and Moths it is about as long as the body itself. This 'haustellium,' which, when not in use, is coiled-up in a spiral beneath the mouth, is an extremely beautiful Microscopic object, owing to the peculiar banded arrangement it exhibits (Fig. 430), which is probably due to the disposition of its muscles. In many instances, the two halves may be seen to be locked together

Fig. 430.

Haustellium (proboscis) of *Vanessa*.

by a set of hooked teeth, which are inserted into little depressions between the teeth of the opposite side. Each half, moreover, may be ascertained to contain a trachea or air-tube (§ 634); and it is probable, from the observations of Mr. Newport, that the sucking-up of the juices of a flower through the proboscis (which is accomplished with great rapidity) is effected by the agency of the respiratory apparatus. The proboscis of many Butterflies is furnished, for some distance from its extremity, with a double row of small projecting barrel-shaped bodies (shown in Fig. 430), which are surmised by Mr. Newport (whose opinion is confirmed by the kindred inquiries of Dr. Hicks, § 628) to be organs of taste.—Numerous other modifications of the structure of the mouth, existing in the different tribes of Insects, are well worthy of the careful study of the Microscopist; but as detailed descriptions of most of these will be found in every Systematic Treatise on Entomology, the foregoing general account of the principal types must suffice.

631. *Parts of the Body.*—The conformation of the several divisions of the *alimentary canal* presents such a multitude of diversities, not only

in different tribes of Insects, but in different states of the same individual, that it would be utterly vain to attempt here to give even a general idea of it; more especially as it is a subject of far less interest to the ordinary Microscopist, than to the professed Anatomist. Hence we shall only stop to mention that the 'muscular gizzard' in which the œsophagus very commonly terminates, is often lined by several rows of strong horny teeth for the reduction of the food, which furnish very beautiful objects, especially for the Binocular. These are particularly developed among the Grasshoppers, Crickets, and Locusts, the nature of whose food causes them to require powerful instruments of its reduction.

632. The *Circulation of Blood* may be distinctly **watched in many** of the more transparent larvæ, and may sometimes **be observed in the** perfect insect. It is kept up, not by an ordinary heart, but by a 'dorsal vessel' (so named from the position it always occupies along the middle of the back), which really consists of a succession of muscular hearts or contractile cavities, one for each segment, opening one into another from behind forwards, so as to form a continuous trunk divided by valvular partitions. In many larvæ, however, these partitions are very indistinct; and the walls of the 'dorsal vessel' are so thin and transparent, that it can with difficulty be made-out, a limitation of the light by **the diaphragm being often necessary.** The blood which moves through this trunk, and which is distributed by it to the body, is a transparent and nearly-colorless fluid, carrying with it a number of 'oat shaped' corpuscles, by the motion of which its flow can be followed. The current enters the 'dorsal vessel' at its posterior extremity, and is propelled forwards by the contractions of the successive chambers, being prevented from moving in the opposite direction by the valves between the chambers, which only open forwards. Arrived at the anterior extremity of the 'dorsal vessel,' the blood is distributed in three principal channels; a central one, namely, passing to the head, and a lateral one to either side; descending so as to approach the lower surface of the body. It is from the two lateral currents that the secondary streams diverge, which pass into the legs and wings, and then return back to the main stream; and it is from these also, that, in the larva of the *Ephemera marginata* (Day-fly), the extreme transparence of which renders it one of the best of all subjects for the observation of Insect Circulation, the smaller currents diverge into the gill-like appendages with which the body is furnished (§ 636). The blood-currents seem rather to pass through channels excavated among the tissues, than through vessels with distinct walls; but it is not improbable that in the perfect Insect the case may be different. In many aquatic larvæ, especially those of the *Culicidæ* (Gnat tribe), the body is almost entirely occupied by the visceral cavity; and the blood may be seen to move backwards in the space that surrounds the alimentary canal, which here serves the purpose of the channels usually excavated through the solid tissues, and which freely communicates at each end with the 'dorsal vessel.' This condition strongly resembles that found in many Annelida.[1]

633. The circulation may be easily seen in the wings of many Insects in their *pupa* state, especially in those of the Neuroptera (such as Dragonflies, and Day-flies), which pass this part of their lives under water in a

---

[1] See the Memoirs on *Corethra plumicornis*, by Prof. Rymer Jones, in "Transact. of Microsc. Soc.," N.S., Vol. xv. (1867), p. 99; by Prof. E. Ray Lankester, in the "Popular Science Review" for October, 1865; and by Dr. A. Weissmann, in "Siebold and Kölliker's Zeitschrift," Bd. xvi., p. 45.

condition of activity; the pupa of *Agrion puella*, one of the smaller dragon-flies, being a particularly favorable subject for such observations. Each of the 'nervures' of the wings contains a 'trachea' or air-tube (§ 634), which branches-off from the tracheal system of the body; and it is in a space around the trachea that the blood may be seen to move, when the hard framework of the nervure itself is not too opaque. The same may be seen, however, in the wings of pupæ of Bees, Butterflies, etc., which remain shut-up motionless in their cases; for this condition of apparent torpor is one of great activity of their nutritive system,—those organs, especially, which are peculiar to the perfect Insect, being then in a state of rapid growth, and having a vigorous circulation of blood through them. In certain insects of nearly every order, a movement of fluid may be seen in the wings for some little time after their last metamorphosis; but this movement soon ceases, and the wings dry-up. The common *Fly* is as good a subject for this observation as can be easily found; it must be caught within a few hours or days of its first appearance; and the circulation may be most conveniently brought into view by inclosing it (without water) in the aquatic box, and pressing-down the cover sufficiently to keep the body at rest without doing it any injury.

634. The *Respiratory apparatus* of Insects affords a very interesting series of Microscopic objects; for, with great uniformity in its general plan there is almost infinite variety in its details. The aëration of the blood in this class is provided-for, not by the transmission of the fluid to any special organ representing the *lung* of a Vertebrated animal (§ 692) or the *gill* of a Mollusk (§ 586), but by the introduction of air into every part of the body, through a system of minutely-distributed *tracheæ* or air-tubes, which penetrate even the smallest and most delicate organs. Thus, as we have seen, they pass into the *haustellium* or 'proboscis' of the Butterfly (§ 630), and they are minutely distributed in the elongated *labium* or 'tongue' of the Fly (Fig. 428). Their general distribution is shown in Fig. 431; where we see two long trunks (*f*) passing from one end of the body to the other, and connected with each other by a transverse canal in every segment; these trunks communicate on the one hand, by short wide passages, with the 'stigmata,' 'spiracles,' or 'breathing pores' (*g*), through which the air enters and is discharged; whilst they give off branches to the different segments, which divide again and again into ramifications of extreme minuteness. They usually communicate also with a pair of air-sacs (*h*) which is situated in the thorax; but the size of these (which are only found in the perfect Insect, no trace of them existing in the larvæ) varies greatly in different tribes, being usually greatest in those insects which (like the Bee) can sustain the longest and most powerful flight, and least in such as habitually live upon the ground or upon the surface of the water. The structure of the air-tubes reminds us of that of the 'spiral vessels' of Plants, which seem destined (in part at least) to perform a similar office (§ 362); for within the membrane that forms their outer wall, an elastic fibre winds round and round, so as to form a spiral closely resembling in its position and functions the spiral wire-spring of flexible gas-pipes; within this again, however, there is another membranous wall to the air-tubes, so that the spire winds between their innner and outer coats.—When a portion of one of the great trunks with some of the principal branches of the tracheal system has been dissected-out, and so pressed in mounting that the sides of the tubes are flattened against each other (as has happened in the specimen represented in Fig. 432), the spire forms two layers which are brought into close

apposition; and a very beautiful appearance, resembling that of watered silk, is produced by the crossing of the two sets of fibres, of which one overlies the other. That this appearance, however, is altogether an optical illusion, may be easily demonstrated by carefully following the course of any one of the fibres, which will be found to be perfectly regular.

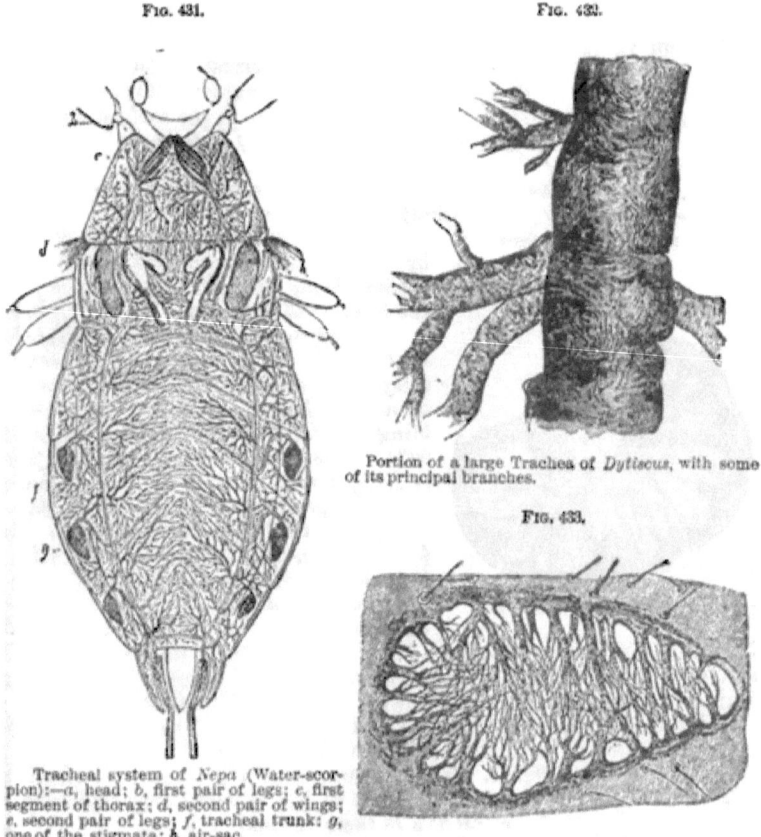

Fig. 431.

Tracheal system of *Nepa* (Water-scorpion):—*a*, head; *b*, first pair of legs; *c*, first segment of thorax; *d*, second pair of wings; *e*, second pair of legs; *f*, tracheal trunk; *g*, one of the stigmata; *h*, air-sac.

Fig. 432.

Portion of a large Trachea of *Dytiscus*, with some of its principal branches.

Fig. 433.

Spiracle of Common Fly.

635. The 'stigmata' or 'spiracles' through which the air enters the tracheal system, are generally visible on the exterior of the body of the insect (especially on the abdominal segments) as a series of pores along each margin of the under surface. In most larvæ, nearly every segment is provided with a pair; but in the perfect insect several of them remain closed, especially in the thoracic region, so that their number is often considerably reduced. The structure of the spiracles varies greatly in regard to complexity in different insects; and even where the general plan is the same, the details of conformation are peculiar, so that perhaps in scarcely any two species are they alike. Generally speaking they are furnished with some kind of sieve at their entrance, by which particles of

dust, soot, etc., which would otherwise enter the air-passages, are filtered out; and this sieve may be formed by the interlacement of the branches of minute arborescent growths from the border of the spiracle, as in the common *Fly* (Fig. 433), or in the *Dytiscus*; or it may be a membrane perforated with minute holes, and supported upon a framework of bars that is prolonged in like manner from the thickened margin of the aperture (Fig. 434), as in the larva of the *Melolontha* (Cockchafer). Not unfrequently, the centre of the aperture is occupied by an impervious disk, from which radii proceed to its margin, as is well seen in the spiracle of *Tipula* (Crane-fly).—In those aquatic Larvæ which breathe air, we often find one of the spiracles of the last segment of the abdomen prolonged into a tube, the mouth of which remains at the surface while the body is immersed; the larvæ of the *Gnat* tribe may frequently be observed in this position.

636. There are many aquatic Larvæ, however, which have an entirely-different provision for respiration; being furnished with external leaf-like or brush-like appendages into which the tracheæ are prolonged, so that, by absorbing air from the water that bathes them, they may convey this into the interior of the body. We cannot have a better example of this than is afforded by the larva of the common *Ephemera* (Day-fly), the body of which is furnished with a set of branchial appendages resembling the 'fin-feet' of Branchiopods (§ 603), whilst the three-pronged tail also is fringed with clusters of delicate hairs which appear to minister to the same function. In the larva of the *Libellula* (Dragon-fly), the extension of the surface for aquatic respiration takes place within the termination of the intestine; the lining membrane of which is folded into an immense number of plaits, each containing a minutely ramified system of tracheæ; the water, slowly drawn-in through the anus for bathing this surface, is ejected with such violence that the body is impelled in the opposite direction; and the air taken-up by its tracheæ is carried, through the system of the air-tubes of which they form-part, into the remotest organs. This apparatus is a peculiarly interesting object for the Microscope, on account of the extraordinary copiousness of the distribution of the tracheæ in the intestinal folds.

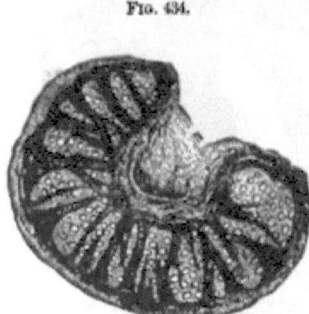

Fig. 434.

Spiracle of Larva of *Cockchafer*.

637. The main trunks of the tracheal system, with their principal ramifications, may generally be got-out with little difficulty, by laying-open the body of an Insect or Larva under water in a Dissecting-trough (§ 180), and removing the whole visceral mass, taking care to leave as many as possible of the branches which will be seen proceeding to this from the two great longitudinal tracheæ, to whose position these branches will serve as a guide. Mr. Quekett recommends the following as the most simple method of obtaining a perfect system of tracheal tubes from a larva:—a small opening having been made in its body, this is to be placed in strong acetic acid, which will soften or decompose all the viscera; and the tracheæ may then be well-washed with the syringe, and removed from the body with the greatest facility, by cutting away the connections of the main tubes with the spiracles by means of fine pointed

scissors. In order to mount them, they should be floated upon the slide, on which they should then be laid-out in the position best adapted for displaying them. If they are to be mounted in Canada balsam, they should be allowed to dry upon the slide, and should then be treated in the usual way; but their natural appearance is best preserved by mounting them in fluid (weak spirit or Goadby's solution), using a shallow cell to prevent pressure. The finer ramifications of the tracheal system may generally be seen particularly well in the membranous wall of the stomach or intestine; and this, having been laid-out and dried upon the glass, may be mounted in balsam so as to keep the tracheæ full of air (whereby they are much better displayed), if care be taken to use balsam that has been previously thickened, to drop this on the object without liquefying it more than is absolutely necessary, and to heat the slide and the cover (the heat may be advantageously applied directly to the cover, after it has been put-on, by turning-over the slide so that its upper face shall look downward) only to such a degree as to allow the balsam to spread and the cover to be pressed-down.—The spiracles are easily dissected-out by means of a pointed knife or a pair of fine scissors; they should be mounted in glycerine-jelly when their texture is soft, and in balsam when the integument is hard and horny.

638. *Wings*.—These organs are essentially composed of an extension of the external membranous layer of the integument, over a framework formed by prolongations of the inner horny layer, within which prolongations tracheæ are nearly always to be found, whilst they also include channels through which blood circulates during the growth of the wing and for a short time after is completion (§ 633). This is the simple structure presented to us in the Wings of *Neuroptera* (Dragon-flies, etc.), *Hymenoptera* (Bees and Wasps), *Diptera* (two-winged-Flies, and also of many *Homoptera* (Cicadæ and Aphides); and the principal interest of these wings as Microscopic objects lies in the distribution of their 'veins' or 'nervures' (for by both names are the ramifications of their skeleton known), and in certain points of accessory structure. The venation of the wings is most beautiful in the smaller Neuroptera; since it is the distinguishing feature of this order that the veins, after subdividing, reunite again, so as to form a close network; whilst in the Hymenoptera and Diptera such reunions are rare, especially towards the margin of the wings, and the areolæ are much larger. Although the membrane of which these wings are composed appears perfectly homogeneous when viewed by transmitted light, even with a high magnifying power, yet, when viewed by light reflected obliquely from their surfaces, an appearance of cellular areolation is often discernible; this is well seen in the common *Fly*, in which each of these areolæ has a hair in its centre. In order to make this observation, as well as to bring-out the very beautiful iridescent hues which the wings of many minute Insects (as the Aphides) exhibit when thus viewed, it is convenient to hold the wing in the Stage-forceps for the sake of giving it every variety of inclination; and when that position has been found which best displays its most interesting features, it should be set up as nearly as possible in the same. For this purpose it should be mounted on an opaque slide; but instead of being laid down upon its surface, the wing should be raised a little above it, its 'stalk' being held in the proper position by a little cone of soft wax, in the apex of which it may be imbedded.—The wings of most Hymenoptera are remarkable for the peculiar apparatus by which those of the same side are connected together, so as to constitute in flight but one large wing; this consists of

a row of curved hooklets on the anterior margin of the posterior wing, which lay hold of the thickened and doubled-down posterior edge of the anterior wing. These hooklets are sufficiently apparent in the wings of the common *Bee*, when examined with even a low magnifying power; but they are seen better in the *Wasp*, and better still in the *Hornet*.—The peculiar scaly covering of the wings of the Lepidoptera has already been noticed (§ 619); but it may here be added that the entire wings of many of the smaller and commoner insects of this order, such as the *Tineidæ* or 'clothes-moths,' form very beautiful opaque objects for low powers; the most beautiful of all being the divided wings of the *Fissipennes* or 'plumed moths,' especially those of the genus *Pterophorus*.

639. There are many Insects, however, in which the Wings are more or less consolidated by the interposition of a layer of horny substance between the two layers of membrane. This plan of structure is most fully carried-out in the *Coleoptera* (Beetles), whose anterior wings are metamorphosed into *elytra* or 'wing-cases;' and it is upon these that the brilliant hues by which the integument of many of these insects is distinguished are most strikingly displayed. In the anterior wings of the *Forficulidæ* or Earwig-tribe (which form the connecting link between this order and the Orthoptera), the cellular structure may often be readily distinguished when they are viewed by transmitted light, especially after having been mounted in Canada balsam. The anterior wings of the *Orthoptera* (Grasshoppers, Crickets, etc.), although not by any means so solidified as those of Coleoptera, contain a good deal of horny matter; they are usually rendered sufficiently transparent, however, by Canada balsam, to be viewed with transmitted light; and many of them are so colored as to be very showy objects (as are also the posterior fan-like wings) for the Electric or Gas-microscope, although their large size, and the absence of any minute structure, prevent them from affording much interest to the ordinary Microscopist.—We must not omit mention, however, the curious Sound-producing apparatus which is possessed by most **insects of** this order, and especially by the common *House-cricket*. This **consists** of the 'tympanum' or drum, which is a space on each of the upper wings, scarcely crossed by veins, but bounded externally by a large dark vein provided with three or four longitudinal ridges; and of the 'file' or 'bow,' which is a transverse horny ridge in front of the tympanum, furnished with numerous teeth: and it is believed that the sound is produced by the rubbing of the two bows across each other, while its intensity is increased by the sound-board action of the tympanum.—The wings of the *Fulgoridæ* (Lantern-flies) have much the same texture with those of Orthoptera, and possess about the same value as Microscopic objects; differing considerably from the purely membranous wings of the Cicadæ and Aphides, which are associated with them in the order *Homoptera*. In the order *Hemiptera*, to which belong various kinds of land and water Insects that have a suctorial mouth resembling that of the common *bug*, the wings of the anterior pair are usually of parchmenty consistence, though membranous near their tips, and are often so richly colored as to become very beautiful objects, when mounted in Balsam and viewed by transmitted light; this is the case especially with the terrestrial vegetable-feeding kinds, such as the *Pentatoma* and its allies, some of the tropical forms of which rival the most brilliant of the Beetles. The British species are by no means so interesting; and the aquatic kinds, which, next to the bed-bugs, are the most common, always have a dull brown or almost black hue: even among these last, however,—of which

the *Notonecta* (water-boatman) and the *Nepa* (water-scorpion) are well-known examples,—the wings are beautifully variegated by differences in the depth of that hue. The *halteres* of the Diptera, which are the representatives of the posterior wings, have been shown by Dr. J. B. Hicks to present a very curious structure, which is found also in the elytra of Coleoptera and in many other situations; consisting in a multitude of vesicular projections of the superficial membrane, to each of which there proceeds a nervous filament, that comes to it through an aperture in the tegumentary wall on which it is seated. Various considerations are stated by Dr. Hicks, which lead him to the belief that this apparatus, when developed in the neighborhood of the spiracles or breathing-pores, essentially ministers to the sense of *smell*, whilst, when developed upon the palpi and other organs in the neighborhood of the mouth, it ministers to the sense of *taste*.[1]

640. *Feet.*—Although the feet of Insects are formed pretty much on one general plan, yet that plan is subject to considerable modifications, in accordance with the habits of life of different species. The entire limb usually consists of five divisions, namely the *coxa* or hip, the *trochanter*, the *femur* or thigh, the *tibia* or shank, and the *tarsus* or foot; and this last part is made up of several successive joints. The typical number of these joints seems to be *five*; but that number is subject to reduction; and the vast order Coleoptera is subdivided into primary groups, according as the tarsus consists of *five, four,* or *three* segments. The last joint of the tarsus is usually furnished with a pair of strong hooks or claws (Figs. 435, 436); and these are often serrated (that is, furnished with saw-like teeth), especially near the base. The under-surface of the other joints is frequently beset with tufts of hairs, which are arranged in various modes, sometimes forming a complete 'sole;' this is

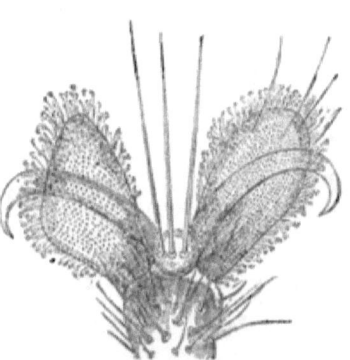

Foot of Fly.

especially the case in the family *Curculionidæ;* so that a pair of the feet of the 'diamond-bettle,' mounted so that one shows the upper surface made resplendent by its jewel-like scales, and the other the hairy cushion beneath, is a very interesting object. In many Insects, especially of the *Fly* kind, the foot is furnished with a pair of membranous expansions termed *pulvilli* (Fig. 435); and these are beset with numerous hairs, each of which has a minute disk at its extremity. This structure is evidently connected with the power which these Insects possess of walking over smooth surfaces in opposition to the force of gravity; yet there is still considerable uncertainty as to the precise mode in which it ministers to

---

[1] See his Memoir 'On a new Organ in Insects,' in "Journal of Linnæan Society," Vol. i. (1856), p. 136; his 'Further Remarks on the Organs found on the bases of the Halteres and Wings of Insects,' in "Trans. of the Linn. Society," Vol. xxii., p. 141; and his Memoir 'On certain Sensory Organs in Insects, hitherto undescribed,' in "Trans of Linn. Soc.," Vol. xxiii., p. 189.

this faculty. Some believe that the disks act as suckers, the Insect being held-up by the pressure of the air against their upper surface, when a vacuum is formed beneath; whilst others maintain that the adhesion is the result of the secretion of a viscid liquid from the under side of the foot. The careful observations of Mr. Hepworth have led him to a conclusion which seems in harmony with all the facts of the case—namely, that each hair is a tube conveying a liquid from a glandular sacculus situated in the tarsus; and that when the disk is applied to a surface, the pouring-forth of this liquid serves to make its adhesion perfect. That this adhesion is not produced by atmospheric pressure alone, is proved by the fact that the feet of Flies continue to hold-on to the interior of an exhausted receiver; whilst, on the other hand, that the feet pour-forth a secreted fluid, is evidenced by the marks left by their attachment on a clean surface of glass. Although, when all the hairs have the strain put upon them equally, the adhesion of their disks suffices to support the insect, yet each row may be detached separately by the gradual raising of the tarsus and pulvilli, as when we remove a piece of adhesive plaster by lifting it from the edge or corner. Flies are often found adherent to window-panes in the autumn, their reduced strength not being sufficient to enable them to detach their tarsi.'—A similar apparatus on a far larger scale, presents itself on the foot of the *Dytiscus* (Fig. 436, A). The first joints of the tarsus of this insect are widely expanded, so as to form a nearly-circular plate: and this is provided with a very remarkable apparatus of suckers, of which one disk (*a*) is extremely large, and is furnished with strong radiating fibres, a second (*b*) is a smaller one formed on the same plan (a third, of the like kind, being often present), whilst the greater number are comparatively small tubular club-shaped bodies, each having a very delicate membranous sucker at its extremity, as shown on a larger scale at B. These all have essentially the same structure; the large suckers being furnished, like the hairs of the Fly's foot, with secreting sacculi, which pour forth fluid through the tubular footstalks that carry the disks, whose adhesion is thus secured; whilst the small suckers form the connecting link between the larger suckers and the hairs of many beetles, especially *Curculionidæ*.² The leg and foot of the Dytiscus, if

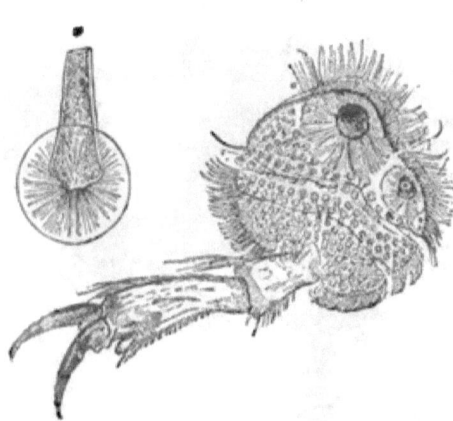

Fig. 436.

A, Foot of *Dytiscus*, showing its apparatus of suckers; *a*, *b*, large suckers; *c*, ordinary suckers:—B, one of the ordinary suckers more highly magnified.

---

[1] See Mr. Hepworth's communications to the "Quart. Journ. of Microsc. Science," Vol. ii. (1854), p. 158, and Vol. iii. (1855), p. 312. See also Mr. Tuffen West's Memoir 'On the Foot of the Fly,' in "Transact. of Linnæan Society," Vol. xxii., p. 393, and Mr. Lowne's "Anatomy of the Blow-fly," p. 19.

[2] See Mr. Lowne 'On the so-called Suckers of Dytiscus and the Pulvilli of Insects,' in "Monthly Microscopical Journal," Vol. v., p. 267.

mounted without compression, furnish a peculiarly beautiful object for the Binocular Microscope.—The Feet of Caterpillars differ considerably from those of perfect Insects. Those of the first three segments, which are afterwards to be replaced by true legs, are furnished with strong horny claws; but each of those of the other segments, which are termed 'pro-legs,' is composed of a circular series of comparatively slender curved hooklets, by which the Caterpillar is enabled to cling to the minute roughnesses of the surface of the leaves, etc., on which it feeds. This structure is well seen in the pro-legs of the common Silkworm.

641. *Stings and Ovipositors.*—The insects of the order *Hymenoptera* are all distinguished by the prolongation of the last segment of the abdomen into a peculiar organ, which in one division of the order is a 'sting,' and in the other is an 'ovipositor' or instrument for the deposition of the eggs, which is usually also provided with the means of boring a hole for their reception. The former group consists of the Bees, Wasps, Ants, etc.; the latter of the Saw-flies, Gall-flies, Ichneumon-flies, etc. These two sets of instruments are not so unlike in structure, as they are in function.—The 'sting' is usually formed of a pair of darts, beset with barbed teeth at their points, and furnished at their roots with powerful muscles, whereby they can be caused to project from their sheath, which is a horny case formed by the prolongation of the integument of the last segment, slit into two halves, which separate to allow the protrusion of the of the sting; whilst the peculiar 'venom' of the sting is due to the ejection, by the same muscular action, of a poisonous liquid, from a bag situated near the root of the sting, which passes down a canal excavated between the darts, so as to be inserted into the puncture which they make. The stings of the common Bee, Wasp, and Hornet, may all be made to display this structure without much difficulty in the dissection. —The 'ovipositor' of such insects as deposit their eggs in holes ready-made, or in soft animal or vegetable substances (as is the case with the *Ichneumonidæ*), is simply a long tube, which is inclosed, like the sting, in a cleft sheath. In the Gall-flies (*Cynipidæ*), the extremity of the ovipositor has a toothed edge, so as to act as a kind of saw whereby harder substances may be penetrated; and thus an aperture is made in the leaf, stalk, or bud of the plant or tree infested by the particular species, in which the egg is deposited, together with a drop of fluid that has a peculiarly irritating effect upon the vegetable tissues, occasioning the production of the 'galls,' which are new growths that serve not only to protect the larvæ, but also to afford them nutriment. The oak is infested by by several species of these Insects, which deposit their eggs in different parts of its fabric; and some of the small 'galls' which are often found upon the surface of oak-leaves, are extremely beautiful objects for the lower powers of the Microscope. In the *Tenthredinidæ*, or 'saw-flies,' and in their allies the *Siricidæ*, the ovipositor is furnished with a still more powerful apparatus for penetration, by means of which some of these Insects can bore into hard timber. This consists of a pair of 'saws' which are not unlike the 'stings' of Bees, etc., but are broader, and toothed for a greater length, and are made to slide along a firm piece that supports each blade, like the 'back' of a carpenter's 'tenon-saw;' they are worked alternately (one being protruded while the other is drawn back) with great rapidity; but, when not in use, they lie in a fissure beneath a sort of arch formed by the terminal segment of the body. When a slit has been made by the working of the saws, they are withdrawn into this sheath; the ovipositor is then protruded from the end of the abdo-

men (the body of the insect being curved downwards); and, being guided into the slit by a pair of small hairy feelers, there deposits an egg.[1]— Many other insects, especially of the order *Diptera*, have very prolonged ovipositors, by means of which they can insert their eggs into the integuments of animals, or into other situations in which the larvæ will obtain appropriate nutriment. A remarkable example of this is furnished by the Gad-fly (*Tabanus*), whose ovipositor is composed of several joints, capable of being drawn together or extended like those of a telescope, and is terminated by boring instruments; and the egg being conveyed by its means, not only *into* but *through* the integument of the Ox, so as to be imbedded in the tissue beneath, a peculiar kind of inflammation is set-up there, which (as in the analogous case of the gall-fly) forms a nidus appropriate both to the protection and to the nutrition of the larva. Other insects which deposit their eggs in the ground, such as the *Locusts*, have their ovipositors so shaped as to answer for digging holes for their reception.—The preparations which serve to display the foregoing parts, are best seen when mounted in Balsam; save in the case of the muscles and poison-apparatus of the sting, which are better preserved in fluid or in glycerine-jelly.

642. The Sexual organs of Insects furnish numerous objects of extreme interest to the Anatomist and Physiologist; but as an account of them would be unsuitable to the present work, a reference to a copious source of information respecting one of their most curious features, and to a list of the species that afford good illustrations, must here suffice.[2]— The *eggs* of many Insects are objects of great beauty, on account of the regularity of their form, and the symmetry of the markings on their surface (Fig. 437). The most interesting belong for the most part to the order *Lepidoptera;* and there are few among these that are not worth examination, some of the commonest (such as those of the Cabbage butterfly, which are found covering large patches of the leaves of that plant) being as remarkable as any. Those of the Puss moth (*Cirura vinula*), the Privet hawk-moth (*Sphinx ligustri*), the small Tortoiseshell butterfly (*Vanessa urticæ*), the Meadow-brown butterfly (*Hipparchia janira*), the Brimstone-moth (*Rumia cratægata*), and the Silkworm (*Bombyx mori*), may be particularly specified: and from other orders, those of the Cockroach (*Blatta orientalis*), Field Cricket (*Acheta campestris*), Water-scorpion (*Nepa ranatra*), Bug (*Cimex lectularius*), Cow-dung-fly (*Scatophaga stercoraria*), and Blow-fly (*Musca vomitoria*). In order to preserve these eggs, they should be mounted in fluid in a cell; since they will otherwise dry up and may lose their shape.—They are very good objects for the 'conversion of relief' effected by Nachet's Stereo-pseudoscopic Binocular (§ 38).

643. The remarkable mode of Reproduction that exists among the *Aphides* must not pass unnoticed here, from its curious connection with

---

[1] The above is the account of the process given by Mr. J. W. Gooch; who has informed the Author that he has repeatedly verified the statement formerly made by him ("Science Gossip," Feb. 1, 1873), that the eggs are deposited, not as originally stated by Reaumur, by means of a tube formed by the coaptation of the saws, but through a separate ovipositor, protruded when the saws have been withdrawn.

[2] See the Memoirs of M. Lacaze-Dutheirs, 'Sur l'armure genitale des Insectes,' in "Ann. des Sci. Nat.," Sér. 3, Zool., Tomes xii., xiv., xvii., xviii., xix.; and M. Ch. Robin's "Mémoire sur les Objets qui peuvent être conservés en Préparations Microscopiques" (Paris, 1856), which is peculiarly full in the enumeration of the objects of interest afforded by the Class of Insects.

the non-sexual reproduction of *Entomostraca* (§ 609) and *Rotifera* (§ 451) as also of *Hydra* (§ 515) and *Zoophytes* generally; all of which fall specially, most of them exclusively, under the observation of the Microscopist. The Aphides which may be seen in the spring and early summer, and which are commonly but not always wingless, are all of one sex, and give birth to a brood of similar Aphides, which come into the world alive, and before long go through a like process of multiplication. As many as from seven to ten successive broods may thus be produced in the course of a single season; so that from a single Aphis, it has been calculated that no fewer than ten thousand million millions, may be evolved within that period. In the latter part of the year, however, some of these viviparous Aphides attain their full development into males and females; and these perform the true Generative process, whose products are eggs, which, when hatched in the succeeding spring, give origin to a new viviparous brood that repeat the curious life-history of their predecessors. It appears from the observations of Prof. Huxley,[1] that the broods of viviparous Aphides originate in *ova* which are not to be distinguished from those deposited by the perfect winged female. Nevertheless, this non-sexual or *agamic* reproduction must be considered analogous rather to the 'gemmation' of other Animals and Plants, than to their sexual

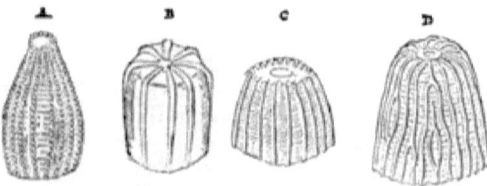

Fig. 437.

Eggs of Insects, magnified;—A, *Pontia napi;* B, *Vanessa urticæ;* C, *Hipparchia tithous;* D, *Argynnis Lathonia*.

'generation;' for it is favored, like the gemmation of *Hydra*, by warmth and copious sustenance, so that by appropriate treatment the viviparous reproduction may be caused to continue (as it would seem) indefinitely, without any recurrence to the sexual process. Further, it seems now certain that this mode of reproduction is not at all peculiar to the Aphides, but that many other Insects ordinarily multiply by 'agamic' propagation, the production of males and the performance of the true generative act being only occasional phenomena; and the researches of Prof. Siebold have led him to conclude that even in the ordinary economy of the Hive-bee the same double mode of reproduction occurs. The queen, who is the only perfect female in the hive, after impregnation by one of the drones (or males), deposits eggs in the 'royal' cells, which are in due time developed into young queens; others in the drone-cells, which become drones; and others in the ordinary cells, which become workers or neuters. It has long been known that these last are really undeveloped females, which, under certain conditions, might become queens; and it has been observed by bee-keepers that worker-bees, in common with

---

[1] 'On the Agamic Reproduction and Morphology of Aphis,' in "Transact. of Linn. Soc.," Vol. xxii., p. 193.

virgin or unimpregnated queens, occasionally lay eggs, from which eggs none but drones are ever produced. From careful Microscopic examination of the drone eggs laid even by impregnated queens, Siebold drew the conclusion that they have not received the fertilizing influence of the male fluid, which is communicated to the queen-eggs and worker eggs alone; so that the products of sexual generation are always female, the male being developed from these by a process which is essentially one of gemmation.[1]

644. The Embryonic Development of Insects is a study of peculiar interest, from the fact that it may be considered as divided (at least in such as undergo a 'complete metamorphosis') into two stages that are separated by the whole active life of the larva; that, namely, by which the Larva is produced within the egg, and that by which the Imago of perfect insect is produced within the body of the Pupa. Various circumstances combine, however, to render the study a very difficult one; so that it is not one to be taken up by the inexperienced Microscopist. The following summary of the process in the common Blow-fly, however, will probably be acceptable.—A *gastrula* with two membranous lamellæ (§ 391) having been evolved in the first instance, the outer lamella very rapidly shapes itself into the form of the larva, and shows a well-marked segmental division. The alimentary canal, in like-manner, shapes itself from the inner lamella; at first being straight and very capacious, including the whole yolk; but gradually becoming narrow and tortous, as additional layers of cells are developed between the two primitive lamellæ, from which the other internal organs are evolved. When the larva comes forth from the egg, it still contains the remains of the yolk; it soon begins, however, to feed voraciously; and in no long period it grows to many thousand times it original weight, without making any essential progress in development, but simply accumulating material for future use. An adequate store of nutriment (analogous to the 'supplemental yolk' of *Purpura*, § 584) having thus been laid up within the body of the larva, it resumes (so to speak) its embryonic development, its passage into the pupa state, from which the imago is to come forth, involving a degeneration of all the larval tissues: whilst the tissues and organs of the imago "are re-developed from cells which originate from the disintegrated parts of the larva, under conditions similar to those appertaining to the formation of the embryonic tissues from the yolk." The development of the segments of the head and body in insects generally proceeds from the corresponding larval segments; but according to Dr. Weissman, there is a marked exception in the case of the *Diptera* and other insects whose larvæ are unfurnished with legs —their head and thorax being newly formed from 'imaginal disks' which adhere to the nerves and tracheæ of the anterior extremity of the larva;[2] and, strange as this assertion may seem, it has been confirmed by the subsequent investigations of Mr. Lowne.

645. ARACHNIDA.—The general remarks which have been made in regard to Insects, are equally applicable to this Class; which includes, along with the *Spiders* and *Scorpions*, the tribe of *Acarida*, consisting of *Mites* and *Ticks*. Many of these are parasitic, and are popularly

---

[1] See Prof. Siebold's Memoir "On true Parthenogenesis in Moths and Bees," translated by W. S. Dallas: London, 1857.
[2] See his 'Entwickelung der Dipteren,' in "Kölliker and Siebold's Zeitschrift," Bande xiv.-xvi.; and Mr. Lowne's "Anatomy of the Blow-fly," pp. 6-9, 113–121.

associated with the wingless parasitic Insects, to which they bear a strong general resemblance, save in having *eight* legs instead of *six*. The true 'mites' (*Acarinæ*) generally have the legs adapted for walking, and some of them are of active habits. The common *cheese-mite*, as seen by the naked-eye, is familiar to every one; yet few who have not seen it under a Microscope have any idea of its real conformation and movements; and a cluster of them, cut out of the cheese they infest, and placed under a magnifying power sufficiently low to enable a large number to be seen at once, is one of the most amusing objects that can be shown to the young. There are many other species, which closely resemble the Cheese-mite in structure and habits, but which feed upon different substances; and some of these are extremely destructive. To this group belongs a small species, the *Sarcoptes scabiei*, whose presence appears to be the occasion of one of the most disgusting diseases of the skin—the itch—and which is hence commonly termed the 'itch-insect.' It is not found in the pustule itself, but in a burrow which passes-off from one side of it, and which is marked by a red line on the surface; and if this burrow be carefully examined, the creature will very commonly, but not always, be met-with. It is scarcely visible to the naked eye; but when examined under the microscope, it is found to have an oval body, a mouth of conical form, and eight feet, of which the four anterior are terminated by small suckers, whilst the four posterior end in very prolonged bristles. The male is only about half the size of the female. The *Ricinia* or 'ticks' are usually destitute of eyes, but have the mouth provided with lancets, that enable them to penetrate more readily the skins of animals whose blood they suck. They are usually of a flattened, round, or oval form; but they often acquire a very large size by suction, and become distended like a blown-bladder. Different species are parasitic upon different animals; and they bury their suckers (which are often furnished with minute recurved hooks) so firmly in the skins of these, that they can hardly be detached without pulling away the skin with them. It is probably the young of a species of this group, which is commonly known as the 'harvest-bug,' and which is usually designated as the *Acarus autumnalis;* this is very common in the autumn upon grass or other herbage, and insinuates itself into the skin at the roots of the hair, producing a painful irritation; like other Acarida, it possesses only six legs for some time after its emersion from the egg (the other pair being only acquired after the first moult), so that its resemblance to parasitic Insects becomes still stronger.—It is probable that to this group also belongs the *Demodex folliculorum*, a creature which is very commonly found parasitic in the sebaceous follicles of the Human skin, especially in those of the nose. In order to obtain it, pressure should be made upon any one of these that appears enlarged and whitish with a terminal black spot; the matter forced-out will consist principally of the accumulated sebaceous secretion, having the parasites with their eggs and young mingled with it. These are to be separated by the addition of oil, which will probably soften the sebaceous matter sufficiently to set free the animals, which may be then removed with a pointed brush; but if this mode should not be effectual, the fatty matter may be dissolved-away by digestion in a mixture of alcohol and ether. The pustules in the skin of a Dog affected with the 'mange' were found by Mr. Topping to contain a Demodex, which seems only to differ from that of the human sebaceous follicles in its somewhat smaller size; and M. Gruby is said to have given to a dog a disease resembling the mange, if not identical with it, by inoculating it with the Human parasite.—The *Acarida* are best preserved, as Microscopic objects, by mounting in one or other of the 'media' described in § 206.

646. The number of objects of general interest furnished to the Microscopist by the *Spider* tribe, is by no means considerable. Their Eyes exhibit a condition intermediate between that of Insects and Crustaceans, and that of Vertebrata; for they are simple, like the 'stemmata' of the former (§ 626), usually number from six to eight, are sometimes clustered together in one mass, though sometimes disposed separately; while they present a decided approach in internal structure to the type characteristic of the visual organs of the latter.—The structure of the Mouth is always mandibulate, and is less complicated than that of the 'mandibulate' insects.—The Respiratory apparatus, which, where developed at all among the Acarida, is tracheary like that of Insects, is here constructed upon a very different plan; for the 'stigmata' which are usually four in number on each side, open into a like number of respiratory sacculi, each of which contains a series of leaf-like folds of its lining membrane, upon which the blood is distributed so as to afford a large surface to the air.—In the structure of the limbs, the principal point

Fig. 438.

Foot, with comb-like claws, of the common *Spider* (Epeira).

worthy of notice is the peculiar appendage with which they usually terminate; for the strong claws, with a pair of which the last joint of the Foot is furnished, have their edges cut into comb-like teeth (Fig. 438), which seem to be used by the animal as cleansing-instruments.

647. One of the most curious parts of the organization of the Spiders, is the 'spinning-apparatus' by means of which they fabricate their elaborately constructed webs. This consists of the 'spinnerets,' and of the glandular organs in which the fluid that hardens into the thread is elaborated. The usual number of the spinnerets, which are situated at the posterior extremity of the body, is six; they are little teat-like prominences, beset with hairy appendages; and it is through a certain set of these appendages, which are tubular and terminate in fine-drawn points, that the glutinous secretion is forced-out in a multitude of streams of extreme minuteness. These streams harden into fibrils immediately on coming into contact with the air; and the fibrils proceeding from all the apertures of each spinneret coalesce into a single thread. It is doubtful, however, whether all the spinnerets are in action at once, or whether those of different pairs may not have dissimilar functions; for whilst the radiating threads of a spider's web are simple (Fig. 439, A) those which

lie across these, forming its concentric circles, or rather polygons, are studded at intervals with viscid globules (B), which appear to give to these threads their peculiarly adhesive character; and it does not seem by any means unlikely that each kind of thread should be produced by its own pair of spinnerets. It was observed by Mr. R. Beck, that these viscid threads are of uniform thickness when first spun; but that undulations soon appear in them, and that the viscid matter then accumulates in globules at regular intervals.—The total number of spinning-tubes varies greatly, according to the species of the Spider, and the sex and age of

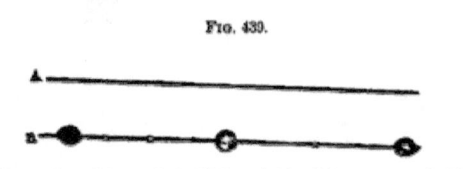

Fig. 439.

Ordinary thread (A), and viscid thread (B), of the common *Spider*.

the individual; being more than 1000 in some cases, and less than 100 in others. The size and complexity of the secreting glandulæ vary in like manner:—Thus in the Spiders which are most remarkable for the large dimensions and regular construction of their webs, they occupy a large portion of the abdominal cavity, and are composed of slender branching tubes whose length is increased by numerous convolutions; whilst in those which have only occasional use for their threads, the secreting organs are either short and simple follicles, or undivided tubes of moderate length.

CHAPTER XX.

VERTEBRATED ANIMALS.

648. WE are now arrived at the highest division of the Animal Kingdom, in which the bodily fabric attains its greatest development, not only as to completeness, but also as to size; and it is in most striking contrast with the Class we have been last considering. Since not only the entire bodies of Vertebrated animals, but, generally speaking, the smallest of their integral parts, are far too large to be viewed as Microscopic objects, we can study their structure only by a separate examination of their component elements; and it seems, therefore, to be a most appropriate course to give under this head a sketch of the microscopic characters of those *Primary Tissues* of which their fabric is made-up, and which, although they may be traced with more or less distinctness in the lower tribes of Animals, attain their most complete development in this group.[1]—For some time after Schwann first made public the remarkable results of his researches, it was very generally believed that all the Animal tissues are formed, like those of Plants, by a metamorphosis of *cells;* an exception being taken, however, by some Physiologists in regard to the 'simple fibrous' tissues (§ 668). There can be no longer any doubt, however, that this doctrine must be greatly modified;[2] so that, whilst the *Vegetable* Physiologist may rightly treat the most highly organized Plant as a mere aggregation of *cells*, analogous in all essential particulars to those which singly constitute the 'unicellular' *Protophytes* (§ 227), the *Animal* Physiologist does wrong in seeking a cellular origin for all the component parts of the Animal fabric; and may best interpret the phenomena of tissue-formation in the most complicated organisms, by the study of the behavior of that apparently-homogenous 'protoplasm' of which the simplest *Protozoa* are made up, and by tracing the progressive 'differentiation' which presents itself as we pass from this through the ascending series of Animal forms.[3]

---

[1] This sketch is intended, not for the Professional student, but only for the amateur Microscopist, who wishes to gain some general idea of the elementary structure of his own body and of that of Vertebrate animals generally. Those who wish to go more deeply into the inquiry are referred to the following as the most recent and elaborate Treatises that have appeared in this country:—The translation of Striker's "Manual of Histology," published by the New Sydenham Society; the "Handbook for the Physiological Laboratory," by Drs. Burdon-Sanderson, Michael Foster, Brunton, and Klein; the translation of the 4th edition of Prof. Frey's "Histology and Histo-chemistry of Man;" the 'General Anatomy' of the 8th edition of "Quain's Anatomy" (1874); and the "Atlas of Histology," by Prof. Klein and Mr. Noble Smith (1880–1).

[2] The important 'Review of the Cell-Theory,' by Prof. Huxley, in the "Brit. and For. Med.-Chir. Review," Vol. xii. (Oct, 1853), p. 285, may be considered the starting-point of many later inquiries.

[3] The study of Comparative Histology, prosecuted on this basis, promises to be

649. Although there would at first sight appear but little in common between the simple bodies of those humble *Monerozoa* which constitute the lowest types of the Animal series (§ 392), and the complex fabric of Man or other Vertebrates, yet it appears from recent researches, that in the latter, as in the former, the process of 'formation' is essentially carried-on by the instrumentality of *protoplasmic substance*, universally diffused through it in such a manner as to bear a close resemblance to the pseudopodial network of the Rhizopod (Fig. 283); whilst the *tissues* produced by its agency lie, as it were, on the outside of this, bearing the same kind of relation to it as the Foraminiferal shell (Fig. 314) does to the sarcodic substance which fills its cavities and extends itself over its surface. For, as was first pointed out by Dr. Beale,[1] the smallest living 'elementary' part' of every organized fabric is composed of organic matter in two states: the protoplasmic (which he termed *germinal matter*), possessing the power of selecting pabulum from the blood, and of transforming this either into the material of its own extension, or into some product which it elaborates; whilst the other, which may be termed *formed material*, may present every gradation of character from a mere inorganic deposit to a highly organized structure, but is in every case altogether incapable of self-increase. A very definite line of demarcation can be generally drawn between these two substances, by the careful use of the staining-process (§ 200); but there are many instances in which there is the same gradation between the one and the other, as we have have formerly noticed between the 'endosarc' and the 'ectosarc' of the *Amœba* (§ 403).—Thus it is on the protoplasmic component that the existence of every form of Animal organization essentially depends; since it serves as the instrument by which the nutrient material furnished by the blood is converted into the several forms of tissue. Like the sarcodic substance of the Rhizopods, it seems capable of indefinite extension; and it may divide and subdivide into independent portions, each of which may act as the instrument of formation of an 'elementary part.' Two principal forms of such elementary parts present themselves in the fabric of the higher Animals—namely, *cells* and *fibres;* and it will be desirable to give a brief notice of these, before proceeding to describe those more complex tissues which are the products of a higher elaboration.

650. The *cells* of which many Animal tissues are essentially composed, consist, when fully and completely formed, of the same parts as the typi-

---

exceedingly fertile in results of this most interesting character. Thus Dr N. Kleinenberg, in his admirable "Anatomische entwickelungsgeschichtliche Untersuchung" (1872), on *Hydra*, gives strong reason for regarding a particular set of cells in the body of that animal as combining the functions of Nerve and Muscle. And the Author has been led by his study of *Comatula* to recognize the most elementary type of Nerve-trunk in a simple protoplasmic cord, not yet separated into distinct fibres with insulating sheaths.

[1] Prof. Beale's views are most systematically expounded in his lectures 'On the Structure of the simple Tissues of the Human Body," 1861; in his "How to Work with the Microscope," 5th edition, 1880; and in the Introductory portion of his new edition of "Todd and Bowman's Physiological Anatomy," 1867. The principal results of the inquiries of German Histologists on this point are well stated in a Paper by Dr. Duffin on 'Protoplasm, and the part it plays in the actions of Living Beings,' in " Quart. Journ. of Microsc. Science," Vol. iii., N.S., (1863), p. 251.—The Author feels it necessary, however, to express his dissent from Prof. Beale's views in one important particular—viz., his denial of 'vital' endowments to the 'formed material' of any of the tissues; since it seems to him illogical to designate contractile muscular fibre (for example) as 'dead,' merely because it has not the power of self-reparation.

cal cell of the Plant (§ 223);—viz., a definite 'cell-wall,' inclosing cell-contents (of which the nature may be very diverse), and also including a 'nucleus,' which is the seat of its formative activity. It is of such cells, retaining more or less of their characteristic spheroidal shape, that every mass of *fat*, whether large or small, is chiefly made up (Fig. 468). And the internal cavities of the body are lined by a layer of *epithelium-cells* (Fig. 466), which, although of flattened form, present the like combination of components. But there is a large number of cases in which the cell shows itself in a form of much less complete development; the 'elementary part' being a corpuscle of protoplasm, of which the exterior has undergone a slight consolidation, like that which constitutes the 'primordial utricle' of the Vegetable cell (§ 223) or the 'ectosarc' of the *Amœba* (§ 403), but in which there is no proper distinction between 'cell-wall' and 'cell-contents.' This condition, which is characteristically exhibited by the nearly globular *colorless corpuscles* of the Blood (§ 666), appears to be common to all cells in the incipient stage of their formation: and the progress of their development consists in the gradual *differentiation* of their parts, the 'cell-wall' becoming distinctly separated from the 'cell-contents,' and these from the 'nucleus;' and the original protoplasm being very commonly replaced more or less completely by some special product (such as fat in the cells of adipose tissue, or hæmoglobin in the red corpuscles of the blood), in which cases the nucleus often disappears altogether.—In the earlier stages of cell-development, multiplication takes place with great activity by a duplicative sub-division that corresponds in all essential particulars with that of the Plant-cell (§ 226); as is well seen in Cartilage, a section of which will often exhibit in one view the successive stages of the process[1] (compare Fig. 470 with Fig. 139). Whether 'free' cell-multiplication ever takes place in the higher Animals, is at present uncertain.

651. A large part of the fabric of the higher Animals, however, is made up of *fibrous* tissues, which serve to bind together the other components, and which, when consolidated by calcareous deposit, constitute the substance of the skeleton. In these, the relation of the 'germinal matter' and the 'formed material' presents itself under an aspect which seems at first sight very different from that just described. A careful examination, however, of those 'connective-tissue-corpuscles' (Fig. 461) that have long been distinguished in the midst of the fibres of which these tissues are made up, shows that they are the equivalents of the corpuscles of 'germinal matter,' which in the previous instance came to constitute cell-nuclei; and that the fibres hold the same relation to them, that the 'walls' and 'contents' of cells do to their germinal corpuscles. The transition from the one type to the other is well seen in Fibro-cartilage, in which the so-called 'intercellular substance' is often as fibrous as tendon. The difference between the two types, in fact, seems essentially to consist in this—that, whilst the segments of 'germinal matter' which form the cell-nuclei in cartilage (Fig. 470) and in other cellular tissues, are completely isolated from each other, each being completely sur-

---

[1] Great attention has lately been given by many able observers, to the changes which take place in the *nucleus* before and during its cleavage. A full account of these is contained in the recently-published *third* Edition of Prof. Strassburger's "Zeilbildung und Zelltheilung" (1880). See also Dr. Klein's 'Observations on the Structure of Cells and Nuclei,' in "Quart. Journ. Microsc. Science," N.S., Vol. xviii. (1878), p. 315, and Vol. xix. (1879), pp. 125, 404; and Chap. xliv. of his "Atlas of Histology."

rounded by the product of its own elaborating action, those which form the 'connective-tissue-corpuscles' are connected together by radiating prolongations (Fig. 461) that pass between the fibres, so as to form a continuous network closely resembling that formed by the pseudopodia of the Rhizopod (Fig. 283).—Of this we have a most beautiful example in Bone; for whilst its solid substance may be considered as connective tissue solidified by calcareous deposit, the 'lacunæ' and 'canaliculi' which are excavated in it (Fig. 441) give lodgment to a set of radiating corpuscles closely resembling those just described; and these are centres of 'germinal matter,' which appear to have an active share in the formation and subsequent nutrition of the osseous texture. In Dentine (or tooth-substance) we seem to have another form of the same thing; the walls of its 'tubuli' and the 'intertubular substance' (§ 655) being the 'formed material' that is produced from thread-like prolongations of 'germinal matter' issuing from its pulp, and continuing during the life of the tooth to occupy its tubes; just as in the *Foraminifera* we have seen a minutely-tubular structure to be formed around the individual threads of sarcode which proceded from the body of the contained animal (Figs. 314, 335). It may now be stated, indeed, with considerable confidence, that the bodies of even the highest Animals are everywhere penetrated by that sarcodic substance of which those of the lowest and simplest are entirely composed; and that this substance, which forms a continuous network through almost every portion of the fabric, is the main instrument of the Formation, Nutrition, and Reparation of the more specialized or differentiated Tissues.—As it is the purpose of this work not to instruct the professional student in Histology (or the Science of the Tissues), but to supply scientific information of general interest to the ordinary Microscopist, no attempt will here be made to do more than describe the most important of those distinctive characters which the principal tissues present when subjected to Microscopic examination; and as it is of no essential consequence what order is adopted, we may conveniently begin with the structure of the *skeleton*,[1] which gives support and protection to the softer parts of the fabric.

652. *Bone.*—The Microscopic characters of osseous tissue may sometimes may be seen in a very thin natural plate of bone, such as in that forming the scapula (shoulder-blade) of a Mouse; but they are displayed more perfectly by artificial sections, the details of the arrangement being dependent upon the nature of the specimen selected, and the direction in which the section is made. Thus when the shaft of a 'long' bone of a Bird or Mammal is cut-across in the middle of its length, we find it to consist of a hollow cylinder of dense bone, surrounding a cavity which is occupied by an oily marrow; but if the section be made nearer its extremity, we find the outside wall gradually becoming thinner, whilst the interior, instead of forming one large cavity, is divided into a vast number of small chambers, partially divided by a sort of 'lattice-work' of osseous fibres, but communicating with each other and with the cavity of the shaft, and filled, like it, with marrow. In the bones of Reptiles and Fishes, on the other hand, this 'cancellated' structure usually extends throughout the shaft, which is not so completely differentiated into solid bone and medullary cavity as it is in the higher Vertebrata.

---

[1] This term is used in its most general sense, as including not only the proper *vertebral* or internal skeleton, but also the hard parts protecting the exterior of the body, which form the *dermal* skeleton.

In the most developed kind of 'flat' bones, again, such as those of the head, we find the two surfaces to be composed of dense plates of bone, with a 'cancellated' structure between them; whilst in the less perfect type presented to us in the lower Vertebrata, the whole thickness is usually more or less 'cancellated,' that is, divided-up into minute medullary cavities.—When we examine, under a low magnifying power, a *longitudinal* section of a long bone, or a section of a flat bone *parallel* to its surface, we find it traversed by numerous canals, termed *Haversian* after their discoverer Havers, which are in connection with the central cavity, and are filled, like it, with marrow: in the shafts of 'long' bones these canals usually run in the direction of their length, but are connected here and there by cross-branches; whilst in the flat-bones they form an irregular network.—On applying a higher magnifying power to a thin *transverse* section of a long bone, we observe that each of the canals whose orifices present themselves in the field of view (Fig. 440), is the centre of a rod of bony tissue (1), usually more or less circular in its

FIG. 440.   FIG. 441.

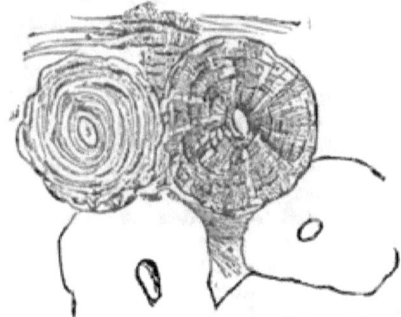
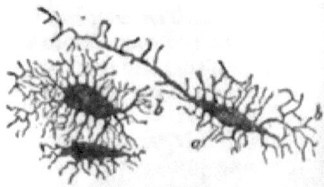

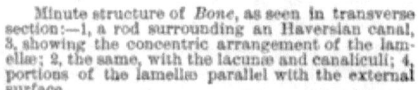

Minute structure of *Bone*, as seen in transverse section:—1, a rod surrounding an Haversian canal, showing the concentric arrangement of the lamellæ; 2, the same, with the lacunæ and canaliculi; 4, portions of the lamellæ parallel with the external surface.

Lacunæ of Osseous substance:—*a*, central cavity; *b*, its ramifications

form, which is arranged around it in concentric rings, resembling those of an Exogenus stem (Fig. 254). These rings are marked out and divided by circles of little dark spots, which, when closely examined (2), are seen to be minute flattened cavities excavated in the solid substance of the bone, from the two flat sides of which pass forth a number of extremely minute tubules, one set extending inwards, or in the direction of the centre of the system of rings, and the other outwards, or in the direction of its circumference; and by the inosculation of the tubules (or *canaliculi*) of the different rings with each other, a continuous communication is established between the central Haversian canal and the outermost part of the bony rod that surrounds it, which doubtless ministers to the nutrition of the texture. Blood-vessels are traceable into the Haversian canals, but the 'canaliculi' are far too minute to carry blood-corpuscles; they are occupied, however, in the living bone, by threads of sarcodic substance, which bring the segments of 'germinal matter' contained in the lacunæ into communication with the walls of the blood-vessels.

653. The minute cavities or *lacunæ* (sometimes, but erroneously termed 'bone-corpuscles,' as if they were solid bodies), from which the canaliculi proceed (Fig. 441), are highly characteristic of the true osseous structure; being never deficient in the minutest parts of the bones of the higher Vertebrata, although those of Fishes are occasionally destitute of them. The dark appearance which they present in sections of a dried bone is not due to opacity, but is simply an optical effect, dependent (like the blackness of air-bubbles in liquids) upon the dispersion of the rays by the highly refracting substance that surrounds them (§ 153). The size and form of the lacunæ differ considerably in the several Classes of Vertebrata, and even in some instances in the Orders; so as to allow of the determination of the tribe to which a bone belonged, by the Microscopic examination of even a minute fragment of it (§ 705). The following are the average dimensions of the lacunæ, in characteristic examples drawn from the four principal Classes expressed in fractions of an inch:—

	Long Diameter.		Short Diameter.
Man	1-1440 to 1-2400	....	1-4000 to 1-8000
Ostrich	1-1333 to 1-2250	....	1-5425 to 1-9650
Turtle	1-375 to 1-1150	....	1-4500 to 1-5840
Conger-eel	1-550 to 1-1135	....	1-4500 to 1-8000

Fig. 442.

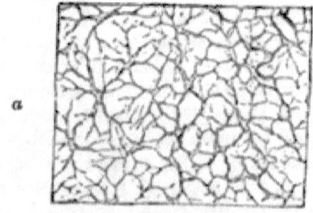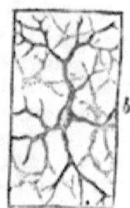

Section of the Bony Scale of *Lepidosteus*:—*a*, showing the regular distribution of the lacunæ and of the connecting canaliculi; *b*, small portion more highly magnified.

The lacunæ of *Birds* are thus distinguished from those of *Mammals* by their somewhat greater length and smaller breadth; but they differ still more in the remarkable tortuosity of their canaliculi, which wind backwards and forwards in a very irregular manner. There is an extraordinary increase in length in the lacunæ of *Reptiles*, without a corresponding increase in breadth; and this also seen in some *Fishes*, though in general the lacunæ of the latter are remarkable for their angularity of form and the fewness of their radiations,—as shown in Fig. 442, which represents the lacunæ and canaliculi in the bony scale of the *Lepidosteus* ('bony pike' of the North American lakes and rivers), with which the bones of its internal skeleton perfectly agree in structure. The dimensions of the lacunæ in any bone do not bear any relation to the size of the animal to which it belonged; thus there is little or no perceptible difference between their size in the enormous extinct Iguanodon, and in the smallest Lizard now inhabiting the earth. But they bear a close relation to the size of the Blood-corpuscles in the several Classes; and this relation is particularly obvious in the 'perennibranchiate' Batrachia, the extraordinary size of whose blood-corpuscles will be presently noticed (§ 665):—

	Long Diameter.		Short Diameter.
Proteus	1-570 to 1-980	....	1-885 to 1-1200
Siren	1-290 to 1-480	....	1-540 to 1-975
Menopoma	1-450 to 1-700	....	1-1300 to 1-2100
Lepidosiren	1-375 to 1-494	....	1-980 to 1-2200
Pterodactyle	1-445 to 1-1185	....	1-4000 to 1-5225 [1]

654. In preparing Sections of Bone, it is important to avoid the penetration of the Canada balsam into the interior of the lacunæ and canaliculi; since, when these are filled by it, they become almost invisible. Hence it is preferable not to employ this cement at all, except it may be, in the first instance; but to rub-down the section beneath the finger, guarding its surface with a slice of cork or a slip of gutta-percha (§ 196); and to give it such a polish that it may be seen to advantage even when mounted dry. As the polishing, however, occupies much time, the benefit which is derived from covering the surfaces of the specimen with Canada balsam may be obtained, without the injury resulting from the penetration of the balsam into its interior, by adopting the following method:—a quantity of balsam proportioned to the size of the specimen is to be spread upon a glass slip, and to be rendered stiffer by boiling, until it becomes nearly solid when cold; the same is to be done to the thin-glass cover; next, the specimen being placed on the balsamed surface of the slide, and being overlaid by the balsamed cover, such a degree of warmth is to be applied as will suffice to liquefy the balsam without causing it to flow freely; and the glass-cover is then to be quickly pressed-down, and the slide to be rapidly cooled, so as to give as little time as possible for the penetration of the liquefied balsam into the lacunar system.—The same method may be employed in making sections of Teeth. The study of the organic basis of Bone (commonly, but erroneously termed cartilage) should be pursued by macerating a fresh bone in dilute Nitro-hydrochloric acid, then macerating it for some time in pure water, and then tearing thin shreds from the residual substance, which will be found to consist of an imperfectly-fibrillated material, allied in its essential constitution to the 'white fibrous' tissue (§ 668).

655. *Teeth.*—The intimate structure of the Teeth in the several Classes and Orders of Vertebrata, presents differences which are no less remarkable than those of their external form, arrangement, and succession. It will obviously be impossible here to do more than sketch some of the most important of these varieties.—The principal part of the substance of all teeth is made-up of a solid tissue that has been appropriately termed *dentine*. In the Shark tribe, as in many other Fishes, the general structure of this dentine is extremely analogous to that of bone; the tooth being traversed by numerous canals, which are continuous with the Haversian canals of the subjacent bone, and receive blood-vessels from them (Fig. 443); and each of these canals being surrounded by a system of tubuli (Fig. 444), which radiate into the surrounding solid substance. These tubuli, however, do not enter lacunæ, nor is there any concentric annular arrangement around the medullary canals; but each system of tubuli is continued onwards through its own

---

[1] See Prof. J. Quekett's Memoir on this subject, in the Transac. of the Microsc. Soc.," Ser. 1, Vol. ii.; and his more ample illustration of it in the "Illustrated Catalogue of the Histological Collection in the Museum of the Royal College of Surgeons," Vol. ii.

[2] Some useful hints on the mode of making these preparations will be found in the "Quart. Journ. of Microsc. Science," Vol. vii. (1859), p. 258.

division of the tooth, the individual tubes sometimes giving-off lateral branches, whilst in other instances their trunks bifurcate. This arrangement is peculiarly well displayed, when sections of teeth constructed upon this type are viewed as opaque objects (Fig. 445).—In the teeth of the higher Vertebrata, however, we usually find the centre excavated into a single cavity (Fig. 446), and the remainder destitute of vascular canals; but there are intermediate cases (as in the teeth of the great fossil Sloths) in which the inner portion of the dentine is traversed by prolongations of this cavity, conveying blood-vessels, which do not pass into the exterior layers. The tubuli of the 'non-vascular' dentine, which exists by itself in the teeth of nearly all Mammalia, and which in the Elephant is known as 'ivory,' all radiate from the central cavity, and pass towards the surface of the tooth in a nearly parallel course. Their diameter at their largest part averages 1-10,000th of an inch; their smallest branches are immeasurably fine. The tubuli in their course present greater and lesser undulations; the former are few in number;

FIG. 443.   FIG. 444.

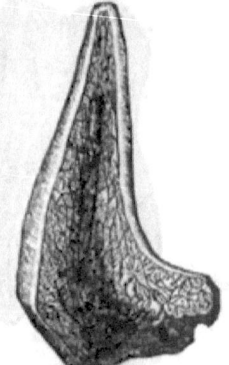
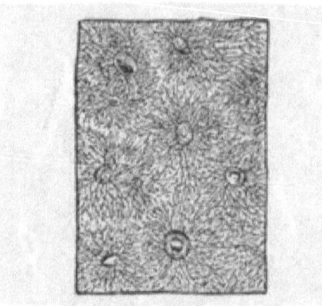

Fig. 443. Perpendicular section of Tooth of *Lamna*, moderately enlarged, showing network of medullary canals.

Fig. 443. Transverse section of portion of Tooth of *Pristis*, more highly magnified, showing orifices of medullary canals, with systems of radiating and inosculating tubuli.

but the latter are numerous, and as they occur at the same part of the course of several contiguous tubes they give rise to the appearance of lines concentric with the centre of radiation. These 'secondary curvatures' probably indicate, in dentine, as in the Crab's shell (§ 613), successive stages of calcification.—The tubuli are occupied, during the life of the tooth, by delicate threads of protoplasmic substance, extending into them from the central pulp.

656. In the Teeth of Man and most other Mammals, and in those of many Reptiles and some Fishes, we find two other substances, one of them harder, and the other softer, than dentine; the former is termed *enamel;* and the latter *cementum* or *crusta petrosa*.—The *enamel* is composed of long prisms, closely resembling those of the 'prismatic' Shell-substance formerly described (§ 563), but on a far more minute scale; the diameter of the prisms not being more in Man than 1-5600th of an inch. The length of the prisms corresponds with the thickness of the layer of enamel; and the two surfaces of this layer present the ends of

the prisms, the form of which usually approaches the hexagonal. The course of the enamel-prisms is more or less wavy; and they are marked by numerous transverse striæ, resembling those of the prismatic shell-substance, and probably originating in the same cause,—the coalescence of a series of shorter prisms to form the lengthened prism. In Man and in Carnivorous animals the enamel covers the crown of the tooth only, with a simple cap or superficial layer of tolerably uniform thickness (Fig. 446, *a*), which follows the surface of the dentine in all its inequalities; and its component prisms are directed at right angles to that surface, their inner extremities resting in slight but regular depressions on the exterior of the dentine. In the teeth of many Herbivorous animals, however, the enamel forms (with the cementum) a series of vertical plates, which dip down into the substance of the dentine, and present their edges alternately with it, at the grinding surface of the tooth; and there is in such teeth no continuous layer of enamel over the crown. This arrangement provides, by the unequal *wear* of these three substances (of which the enamel is the hardest, and the cementum the softest), for the constant

FIG. 445.   FIG. 446.

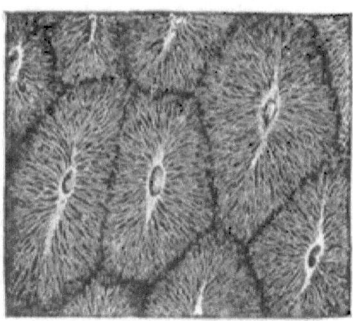 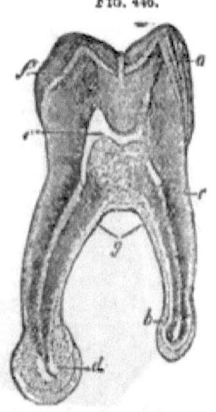

Transverse section of Tooth of *Myliobates* (Eagle Ray) viewed as an opaque object.

Vertical section of *Human Molar Tooth*: *a*, enamel; *b*, cementum or crusta petrosa; *c*, dentine or ivory; *d*, osseous excrescence, arising from hypertrophy of cementum; *e*, pulp-cavity; *f*, osseous lacunæ at outer part of dentine.

maintenance of a rough surface, adapted to triturate the tough vegetable substances on which these animals feed. The enamel is the least constant of the dental tissues. It is more frequently absent than present in the teeth of Fishes; it is entirely wanting in the teeth of Serpents; and it forms no part of those of the Edentata[1] (sloths, etc.) and Cetacea (whales) among Mammals.—The *cementum*, or *crusta petrosa*, has the characters of true bone; possessing its distinctive stellate lacunæ and radiating canaliculi. Where it exists in small amount, we do not find it traversed by medullary canals; but, like dentine, it is occasionally furnished with them, and thus resembles bone in every particular. These medullary canals enter its substance from the exterior of the tooth,

[1] It has been shown by Mr. Charles Tomes, however, that the 'enamel organ' is originally present within the tooth-capsule of the *Armadillo*, though it undergoes an early degeneration;—a fact of no little interest in connection with the general doctrine of ' Descent with modification.'

and consequently pass towards those which radiate from the central cavity in the direction of the surface of the dentine, where this possesses a similar vascularity,—as was remarkably the case in the teeth of the great extinct *Megatherium*. In the Human tooth, however, the cementum has no such vascularity; but forms a thin layer (Fig. 446, *b*), which envelops the root of the tooth, commencing near the termination of the cap of enamel. In the teeth of many herbivorous Mammals, it dips down with the enamel to form the vertical plates of the interior of the tooth; and in the teeth of the Edentata, as well as of many Reptiles and Fishes, it forms a thick continuous envelope over the whole surface, until worn-away at the crown.

657. *Dermal skeleton.*—The skin of Fishes, of most Reptiles, and of a few Mammals, is strengthened by plates of a horny, cartilaginous, bony, or even enamel-like texture; which are sometimes fitted-together at their edges, so as to form a continuous box-like envelope; whilst more commonly they are so arranged as partially to overlie one another, like the tiles on a roof; and it is in this latter case that they are usually known as *scales*. Although we are accustomed to associate in our minds the 'scales' of Fishes with those of Reptiles, yet they are essentially-different structures; the former being developed in the *substance* of the true skin (with a layer of which, in addition to the epidermis, they are always covered), and bearing a resemblance to cartilage and bone in their texture and composition; whilst the latter are formed upon the surface of the true skin, and are to be considered as analogous to nails, hoofs, etc., and other 'epidermic appendages.' In nearly all the existing Fishes the scales are flexible, being but little consolidated by calcareous deposit; and in some species they are so thin and transparent, that, as they do not project obliquely from the surface of the skin, they can only be detected by raising the superficial layer of the skin, and searching beneath it, or by tearing off the entire thickness of the skin, and looking for them near its under surface. This is the case, for example, with the common *Eel*, and with the *viviparous Blenny;* of either of which fish the skin is a very interesting object when dried and mounted in Canada balsam, the scales being seen imbedded in its substance, whilst its outer surface is studded with pigment-cells. Generally speaking, however, the posterior extremity of each scale projects obliquely from the general surface, carrying before it the thin membrane that incloses it, which is studded with pigment-cells; and a portion of the skin of almost any Fish, but especially of such as have scales of the *ctenoid* kind (that is, furnished at their posterior extremities with comb-like teeth, Fig. 448), when dried with its scales *in situ*, is a very beautiful opaque object for the low powers of the Microscope (Fig. 447), especially with the Binocular arrangement. Care must be taken, however, that the light is made to glance upon it in the most advantageous manner; since the brilliance with which it is reflected from the comb-like projections entirely depends upon the angle at which it falls upon them. The only appearance of structure exhibited by the thin flat scale of the Eel, when examined microscopically, is the presence of a layer of isolated spheroidal transparent bodies, imbedded in a plate of like transparence; these, from the researches of Prof. W. C. Williamson[1] upon other scales, appear not to be cells (as they might readily be

---

[1] See his elaborate Memoirs 'On the Microscopic Structure of the Scales and Dermal Teeth of some Ganoid and Placoid Fish,' in "Philos. Transact.," 1849; and 'Investigations into the Structure and Development of the Scales and Bones of Fishes,' in "Philos. Transact.," 1851.

supposed to be), but concretions of Carbonate of Lime. When the scale of the Eel is examined by Polarized light, its surface exhibits a beautiful St. Andrew's cross; and if a plate of Selenite be placed behind it, and the analyzing prism be made to revolve, a remarkable play of colors is presented.

658. In studying the structure of the more highly developed scales, we may take as an illustration that of the *Carp;* in which two very distinct layers can be made-out by a vertical section, with a third but incomplete layer interposed between them. The outer layer is composed of several concentric laminæ of a structureless transparent substance, like that of cartilage; the outermost of these laminæ is the smallest, and the size of the plates increases progressively from without inwards, so that their margins appear on the surface as a series of concentric lines; and their surfaces are thrown into ridges and furrows, which commonly have a radiating direction. The inner layer is composed of numerous laminæ

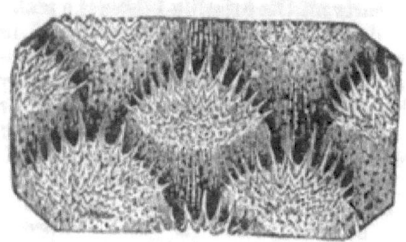

Portion of Skin of *Sole*, viewed as an opaque object.

Scale of *Sole*, viewed as a transparent object.

of a fibrous structure, the fibres of each lamina being inclined at various angles to those of the lamina above and below it. Between these two layers is interposed a stratum of calcareous concretions, resembling those of the scale of the Eel: these are sometimes globular or spheroidal, but more commonly 'lenticular,' that is, having the form of a double-convex lens. The scales which resemble those of the Carp in having a form more or less circular, and in being destitute of comb-like prolongations, are called *cycloid;* and such are the characters of those of the Salmon, Herring, Roach, etc. The structure of the *ctenoid* scales (Fig. 448), which we find in the Sole, Perch, Pike, etc., does not differ essentially from that of the cycloid, save as to the projection of the comb-like teeth from the posterior margin; and it does not appear that the strongly-marked divisions which Prof. Agassiz has attempted to establish between the 'cycloid' and the 'ctenoid' Orders of Fishes, on the basis of this difference, is in harmony with their general organization. Scales of every kind may become consolidated to a considerable extent by the calcification of their soft substance; but still they never present any approach to the true Bony structure, such as is shown in the two Orders to be next adverted-to.

659. In the *ganoid* Scales, on the other hand, the whole substance of the scale is composed of a substance which is essentially bony in its nature: its intimate structure being always comparable to that of one or other of the varieties which present themselves in the bones of the Vertebrate skeleton; and being very frequently identical with that of the bones of the same fish, as is the case with the *Lepidosteus* (Fig. 442), one of the few existing representatives of this Order, which, in former ages of the Earth's history, comprehended a large number of important families. Their name (from γάνος, splendor) is bestowed on account of the smoothness, hardness, and high polish of the outer surface of the scales; which is due to the presence of a peculiar layer that has been likened (though erroneously) to the enamel of teeth, and is now distinguished as *ganoin*. The scales of this order are for the most part angular in their form; and are arranged in regular rows, the posterior edges of each slightly overlapping the anterior ones of the next, so as to form a very complete defensive armor to the body.—The scales of the *placoid* type, which characterizes the existing Sharks and Rays, with their fossil allies, are irregular in their shape, and very commonly do not come into mutual contact, but are separately imbedded in the skin, projecting from its surface under various forms. In the Rays each scale usually consists of a flattened plate of a rounded shape, with a hard spine projecting from its centre; in the Sharks (to which tribe belongs the 'dog-fish' of our own coast) the scales have more of the shape of teeth. This resemblance is not confined to external form; for their intimate structure strongly resembles that of dentine, their dense substance being traversed by tubuli, which extend from their centre to their circumference in minute ramifications, without any trace of osseous lacunæ. These tooth-like scales are often so small as to be invisible to the naked eye; but they are well seen by drying a piece of the skin to which they are attached, and mounting it in Canada balsam; and they are most brilliantly shown by the assistance of polarized light.—A like structure is found to exist in the 'spiny rays' of the dorsal fin, which, also, are parts of the dermal skeleton; and these rays usually have a central cavity filled with medulla, from which the tubuli radiate towards the circumference. This structure is very well seen in thin sections of the fossil 'spiny rays,' which, with the teeth and scales, are often the sole relics of the vast multitudes of Sharks that must have swarmed in the ancient seas, their cartilaginous internal skeletons having entirely decayed away.—In making sections of bony Scales, Spiny rays, etc., the method must be followed which has been already detailed under the head of Bone (§ 654).

660. The *scales* of Reptiles, the *feathers* of Birds, and the *hairs, hoofs, nails, claws*, and *horns* (when not bony) of Mammals, are all *epidermic* appendages; that is, they are produced upon the surface, not within the substance, of the true Skin, and are allied in structure to the Epidermis (§ 671); being essentially composed of aggregations of cells filled with horny matter, and frequently much altered in form. This structure may generally be made-out in horns, nails, etc., with little difficulty; by treating thin sections of them with a dilute solution of soda; which after a short time causes the cells that had been flattened into scales, to resume their globular form. The most interesting modifications of this structure are presented to us in Hairs and in Feathers; which forms of clothing are very similar to each other in their essential nature, and are developed in the

same manner—namely, by an increased production of epidermic cells at the bottom of a flask-shaped follicle, which is formed in the substance of the true skin, and which is supplied with abundance of blood by a special distribution of vessels to its walls. When a hair is pulled-out 'by its root,' its base exhibits a bulbous enlargement, of which the exterior is tolerably firm, whilst its interior is occupied by a softer substance, which is known as the 'pulp;' and it is to the continual augmentation of this pulp in the deeper part of the follicle, and to its conversion into the peculiar substance of the hair when it has been pushed upwards to its narrow neck, that the growth of the hair is due.—The same is true of feathers, the stems of which are but hairs on a larger scale; for the 'quill' is the part contained within the follicle answering to the 'bulb' of the hair; and whilst the outer part of this is converted into the peculiarly-solid horny substance forming the 'barrel' of the quill, its anterior is occupied, during the whole period of the growth of the feather, with the soft pulp,

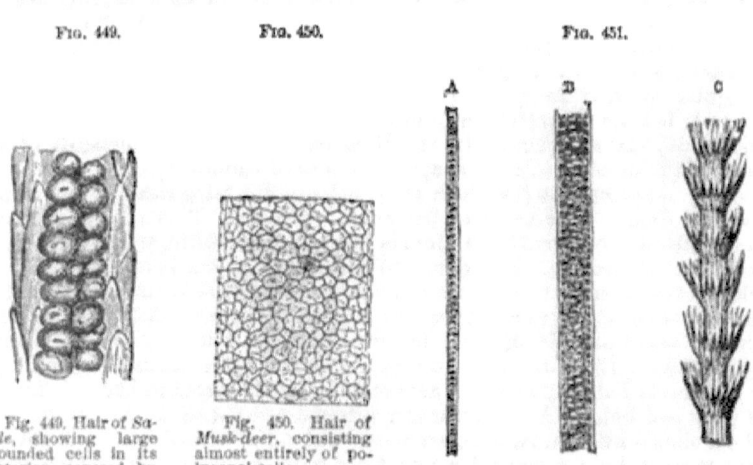

Fig. 449. Fig. 450. Fig. 451.

Fig. 449. Hair of *Sable*, showing large rounded cells in its interior, covered by imbricated scales or flattened cells.

Fig. 450. Hair of *Musk-deer*, consisting almost entirely of polygonal cells.

A, Small Hair of *Squirrel*:—B, Large Hair of *Squirrel*:—C, Hair of *Indian Bat*.

only the shrivelled remains of which, however, are found within it after the quill has ceased to grow.

661. Although the *hairs* of different Mammals differ greatly in the appearances they present, we may generally distinguish in them two elementary parts—namely, a *cortical* or investing substance, of a dense horny texture, and a *medullary* or pith-like substance, usually of a much softer texture, occupying the interior. The former can sometimes be distinctly made out to consist of flattened scales arranged in an imbricated manner, as in some of the hairs of the *Sable* (Fig. 449); whilst, in the same hairs, the medullary substance is composed of large spheroidal cells. In the *Musk-deer*, on the other hand, the cortical substance is nearly undistinguishable; and almost the entire hair seems made up of thin-walled polygonal cells (Fig. 450). The hair of the *Reindeer*, though much larger, has a very similar structure; and its cells, except near the root, are occupied with hair alone, so as to seem black by transmitted light, except when penetrated by the fluid in which they are mounted. In the hair of the *Mouse, Squirrel*, and other small Rodents (Fig. 451, A, B), the corti-

cal substance forms a tube, which we see crossed at intervals by partitions that are sometimes complete, sometimes only partial; these are the walls of the single or double line of cells, of which the medullary substance is made-up. The hairs of the *Bat* tribe are commonly distinguished by the projections on their surface, which are formed by extensions of the component scales of the cortical substance: these are particularly well seen in the hairs of one of the Indian species, which has a set of whorls of long narrow leaflets (so to speak) arranged at regular intervals on its stem (c).

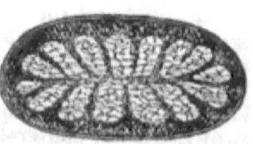

Transverse section of Hair of *Pecari*.

In the hair of the *Pecari* (Fig. 452), the cortical envelope sends inwards a set of radial prolongations, the interspaces of which are occupied by the polygonal cells of the medullary substance; and this, on a larger scale, is the structure of the 'quills' of the *Porcupine;* the radiating partitions of which, when seen through the more transparent parts of the cortical sheath, give to the surface of the latter a fluted appearance. The hair of the *Ornithorhynchus* is a very curious object; for whilst the lower part of it resembles the fine hair of the Mouse or Squirrel, this thins away and then dilates again into a very thick fibre, having a central portion composed of polygonal cells, inclosed in a flattened sheath of a brown fibrous substance.

662. The structure of the *human* Hair is in certain respects peculiar.

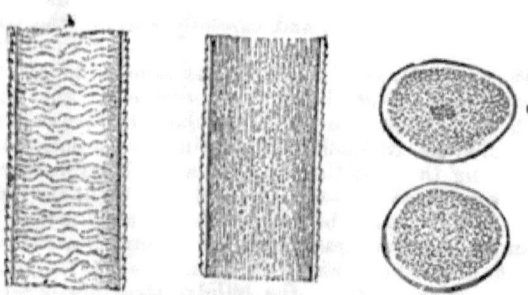

Structure of *Human Hair:*—A, external surface of the shaft, showing the transverse striæ and jagged boundary caused by the imbrications of the cuticular layer; B, longitudinal section of the shaft, showing the fibrous character of the cortical substance, and the arrangement of the pigmentary matter; c, transverse section, showing the distinction between the cuticular envelope, the cylinder of cortical substance, and the medullary centre; D, another transverse section, showing deficiency of the central cellular substance.

When its outer surface is examined, it is seen to be traversed by irregular lines (Fig. 453, A), which are most strongly marked in fœtal hairs; and these are the indications of the imbricated arrangement of the flattened cells or scales which form the cuticular layer. This layer, as is shown by transverse sections (C, D), is a very thin and transparent cylinder; and it incloses the peculiar fibrous substance that constitutes the principal part of the shaft of the hair. The constituent fibres of this substance, which are marked-out by the delicate striæ that may be traced in longitudinal sections of the hair (B), may be separated from each other by crushing the hair, especially after it has been macerated for some time in sulphuric acid; and each of them, when completely isolated from its fellows, is found to be a long spindle-shaped cell. In the axis of this fibrous cylinder there is very commonly a band which is formed of spheroidal cells; but this 'medullary' substance is usually deficient in the fine hair scattered over the general surface of the body, and is not always present in those of the head. The hue of the Hair is due partly to the presence of pigmentary granules, either collected into patches, or diffused through its substance; but partly also to the existence of a multitude of minute air-spaces, which cause it to appear dark by transmitted and white by reflected light. The cells of the medullary axis in particular, are very commonly found to contain air, giving it the black appearance shown at c. The difference between the blackness of pigment and that of air-spaces may be readily determined by attending to the characters of the latter as already laid-down (§§ 153, 154); and by watching the effects of the penetration of Oil of Turpentine or other liquids, which do not alter the appearance of pigment-spots, but obliterate all the markings produced by air-spaces, these returning again as the hair dries.—In mounting Hairs as Microscopic preparations, they should in the first instance be cleansed of all their fatty matter by maceration in ether; and they may then be put up either in weak Spirit or in Canada balsam, as may be thought preferable, the former menstruum being well adapted to display the characters of the finer and more transparent hairs, while the latter allows the light to penetrate more readily through the coarser and more opaque. Transverse sections of Hairs are best made by gluing or gumming several together, and then putting them into the Microtome; those of Human hair may be easily obtained, however, by shaving a second time, very closely, a part of the surface over which the razor has already passed more lightly, and by picking-out from the lather, and carefully washing the sections thus taken-off.

663. The stems of *feathers* exhibit the same kind of structure as Hairs; their cortical portion being the horny sheath that envelops the shaft, and their medullary portion being the pith-like substance which that sheath includes. In small feathers, this may usually be made very plain by mounting them in Canada balsam; in large feathers, however, the texture is sometimes so altered by the drying up of the pith (the cells of which are always found to be occupied by air alone), that the cellular structure cannot be demonstrated save by boiling thin slices in a dilute solution of potass, and not always even then. In small feathers, especially such as have a downy character, the cellular structure is very distinctly seen in the lateral *barbs*, which are sometimes found to be composed of single files of pear-shaped cells, laid end-to-end; but in larger feathers it is usually necessary to increase the transparence of the barbs, especially when these are thick and but little pervious to light, either by soaking them in turpentine, mounting them in Canada balsam, or boiling them in a weak solution of potass. In feathers which are destined to strike the air with great force in the act of flight, we find each barb fringed on either side with slender flattened filaments or 'barbules;' the barbules of one side of each barb are furnished with curved hooks, whilst those of the other side have thick turned-up edges; and as the two sets of barbules that spring from two adjacent barbs cross one another at an angle, and as each hooked barbule of one locks into the thickened edge of several barbules of the other, the barbs are connected very firmly, in a mode very similar to that in which the anterior and posterior wings of certain Hy-

menopterous Insects are locked together (§ 638).—Feathers or portions of feathers of Birds distinguished by the splendor of their plumage are very good objects for low magnifying powers, when illuminated on an opaque ground; but care must be taken that the light falls upon them at the angle necessary to produce their most brilliant reflection into the axis of the Microscope; since feathers which exhibit the most splendid metallic lustre to an observer at one point, may seem very dull to the eye of another in a different position. The small feathers of Humming-birds, portions of the feathers of the Peacock, and others of a like kind, are well worthy of examination; and the scientific Microscopist, who is but little attracted by mere gorgeousness, may well apply himself to the discovery of the peculiar structure which imparts to these objects their most remarkable character.

664. Sections of *horns, hoofs, claws*, and other like modifications of Epidermic structure,—which can be easily made by the Microtome (§ 184), the substance to be cut having been softened, if necessary, by soaking in warm water,—do not in general afford any very interesting features when viewed in the ordinary mode; but there are no objects on which Polarized light produces more remarkable effects, or which display a more beautiful variety of colors when a plate of Selenite is placed behind them and the analyzing prism is made to rotate. A curious modification of the ordinary structure of Horn is presented in the appendage borne by the *Rhinoceros* upon its snout, which in many points resembles a bundle of hairs, its substance being arranged in minute cylinders around a number of separate centres, which have probably been formed by independent papillæ (Fig. 454). When transverse sections of these cylinders are viewed by polarized light, each of them is seen to be marked by a cross, somewhat resembling that of Starch-grains; and the

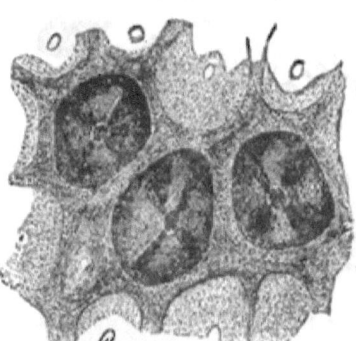

Fig. 454.

Transverse section of Horn of *Rhinoceros*, viewed by Polarized Light.

light and shadow of this cross are replaced by contrasted colors when the Selenite plate is interposed.—The substance commonly but erroneously termed *whalebone*, which is formed from the surface of the membrane that lines the mouth of the Whale, and has no relation to its true bony skeleton, is almost identical in structure with Rhinoceros horn, and is similarly affected by polarized light. The central portion of each of its component threads, like the medullary substance of Hairs, contains cells that have been so little altered as to be easily recognized; and the outer or cortical portion also may be shown to have a like structure, by macerating it in a solution of potass, and then in water.—Sections of any of the Horny tissues are best mounted in Canada balsam.

665. *Blood.*—Carrying our Microscopic survey, now, to the elementary parts of which those softer tissues are made up, that are subservient to the active life of the body rather than to its merely-mechanical requirements, we shall in the first place notice the isolated floating cells contained in the Blood, which are known as Blood-corpuscles. These

are of two kinds; the 'red,' and the 'white' or 'colorless.'—The *red* present, in every instance, the form of a flattened disk, which is circular in Man and most Mammalia (Fig. 456), but is oval in Birds, Reptiles (Fig. 455), and Fishes, as also in a few Mammals (all belonging to the *Camel* tribe). In the one form, as in the other, these corpuscles seem to be flattened cells, the walls of which, however, are not distinctly differentiated from the ground-substance they contain; as appears from the changes of form which they spontaneously undergo when kept by means of a 'warm stage'[1] at a temperature of about 100°, and from the effects of pressure in breaking them up. The red corpuscles in the blood of Oviparous Vertebrata are distinguished by the presence of a central spot or *nucleus;* this is most distinctly brought into view by treating the blood-disks with acetic acid, which causes the nucleus to shrink and become more opaque, whilst rendering the remaining portion extremely transparent (Fig. 455, *d*). By examining unaltered red corpuscles of the Frog or Newt under a sufficiently high magnifying power, the nucleus is seen to be traversed by a network of filaments, which extends from it

FIG. 455. FIG. 456.

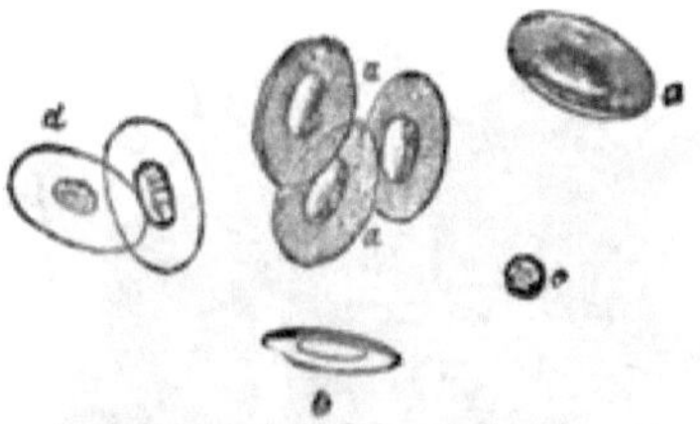

Red Corpuscles of *Frog's* Blood:—*a a*, their flattened face; *b*, particle turned nearly edgeways; *c*, colorless corpuscle; *d*, red corpuscles altered by diluted acetic acid.

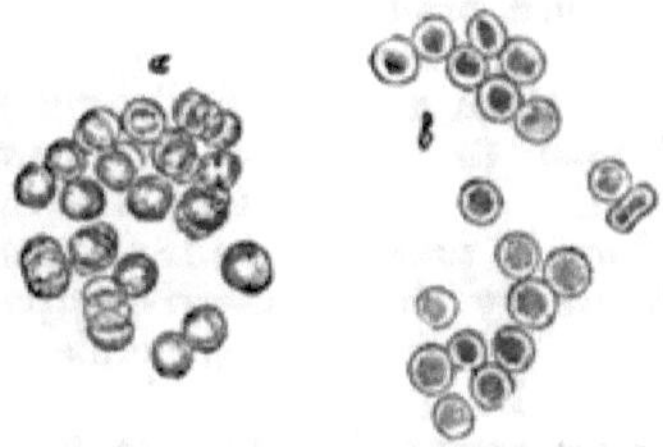

Red Corpuscles of *Human* Blood; represented at *a*, as they are seen when rather within the focus of the Microscope, and at *b* as they appear when precisely in the focus.

throughout the ground-substance of the corpuscle, constituting an intra-cellular reticulation.—The red corpuscles of the blood of Mammals, however, possess no distinguishable nucleus; the dark spot which is seen in their centre (Fig. 456, *b*) being merely an effect of refraction, consequent upon the double-concave form of the disk. When these corpuscles are treated with water, so that their form becomes first flat, and then double-convex, the dark spot disappears; whilst, on the other hand, it is made

---

[1] A very simple mode of applying continued warmth to an object under observation, is to lay the slide on a thin plate of brass or tin, about 3 inches longer than the breadth of the stage, and about 2 inches broad; which must be perforated with a hole about 1-4th inch in diameter, at the distance of half the breadth of the stage from one end of it. When this plate is laid on the stage, and its hole is brought into the optic axis, so as to allow the light reflected upwards from the mirror to pass to the slide laid upon it, the plate will project about 3 inches on one side of the stage,—preferably the right. By placing a small lamp beneath this projection and keeping the finger of the left hand on the part of the plate close to the object (so as to feel the degree of warmth imparted to it), the heat given by the lamp may be regulated by varying its position.—For more exact and continuous regulation of the temperature, recourse may be had to the 'warm stage' devised by Prof. Schäfer and made by Mr. Casella, which is traversed by a current of warm water. See "Quart. Journ. of Microsc. Sci.," N.S., Vol. xiv. (1874), p. 394.

more evident when the concavity is increased by the partial shrinkage of the corpuscles, which may be brought about by treating them with fluids of greater density than their own substance. When floating in a sufficiently thick stratum of blood drawn from the body, and placed under a cover-glass, the red corpuscles show a marked tendency to approach one another, adhering by their discoidal surfaces so as to present the aspect of a pile of coins; or, if the stratum be too thin to admit of this, partially overlapping, or simply adhering by their edges, which then become polygonal instead of circular. The size of the red corpuscles is not altogether uniform in the same blood; thus it varies in that of Man from about the 1-4000th to the 1-2800th of an inch. But we generally find that there is an average size, which is pretty constantly maintained among the different individuals of the same species; that of Man may be stated at about 1-3200th of an inch. The following Table[1] exhibits the average dimensions of some of the most interesting examples of the red corpuscles in the four classes of Vertebrated Animals, expressed in fractions of an inch. Where two measurements are given, they are the long and the short diameters of the same corpuscles. (See also Fig. 457).

MAMMALS.

Man	1-3200	Camel 1-3254, 1-5921
Dog	1-3542	Llama 1-3361, 1-6294
Whale	1-3099	Java Musk-Deer............1-12325
Elephant	1-2745	Caucasian Goat............ 1-7045
Mouse	1-3814	Two-toed Sloth............ 1-2865

BIRDS.

Golden Eagle	1-1812, 1-3832	Ostrich............1-1649, 1-3000
Owl	1-1830, 1-3400	Cassowary............1-1455, 1-2800
Crow	1-1961, 1-4000	Heron 1-1913, 1-3491
Blue-Tit	1-2313, 1-4128	Fowl............1-2102, 1-3466
Parrot	1-1898, 1-4000	Gull............1-2097, 1-4000

REPTILES AND BATRACHIA.

Turtle	1-1231, 1-1882	Frog............1-1108, 1-1821
Crocodile	1-1231, 1-2286	Water-Newt............1-8014, 1-1246
Green Lizard	1-1555, 1-2743	Siren 1-420, 1-760
Slow-worm	1-1178, 1-2666	Proteus............ 1-400, 1-727
Viper	1-1274, 1-1800	Amphiuma............ 1-345, 1-561

FISHES.

Perch	1-2099, 1-2824	Pike............1-2000, 1-3555
Carp	1-2142, 1-3429	Eel............1-1745, 1-2842
Gold-Fish	1-1777, 1-2824	Gymnotus............1-1745, 1-2599

Thus it appears that the *smallest* red corpuscles known are those of the *Musk-deer;* whilst the *largest* are those of that curious group of Batrachia (Frog-tribe) which retain the gills through the whole of life; and one of the oval blood-disks of the *Proteus*, being more than 30 times as long and 17 times as broad as those of the Musk-deer, would cover no fewer than 510 of them.—Those of the *Amphiuma* are still larger.[2]—According to the estimate of Vierordt, a cubic inch of Human blood con-

---

[1] These measurements are chiefly selected from those given by Mr. Gulliver, in his edition of Hewson's Works, p. 236 *et seq.*

[2] A very interesting account of the 'Structure of the Red Corpuscles of the *Amphiuma tridactylum*' has been given by Dr. H. D. Schmidt, of New Orleans, in the "Journ. of the Royal Microsc. Society," Vol. i. (1879), pp. 57, 97.

tains upwards of *eighty millions* of red corpuscles, and nearly a *quarter of a million* of the colorless.

666. The *white* or 'colorless' corpuscles are more readily distinguished in the blood of Reptiles than in that of Man; being in the former case of much smaller size, as well as having a circular outline (Fig. 455, *c*); whilst in the latter their size and contour are so nearly the same, that, as the red corpuscles themselves, when seen in a single layer, have but a very pale hue, the deficiency of color does not sensibly mark their difference of nature. The proportion of *white* to *red* corpuscles being scarcely ever greater (in a healthy Man) than 1 to 250, and often as low as from one-half to one-quarter of that ratio, there are seldom many of them to be seen in the field at once; and these may be recognized rather by their isolation than their color, especially if the glass cover be moved a little on the slide, so as to cause the red corpuscles to become aggregated into rows and irregular masses.—It is remarkable that, notwithstanding the great variations in the sizes of the red corpuscles in different species of Vertebrated animals, the size of the white is extremely constant throughout, their diameter being seldom much greater or less than 1-3000th of an inch in the warm-blooded classes, and 1-2500th in Reptiles. Their ordinary form is globular; but their aspect is subject to considerable variations, which seem to depend in great part upon their phase of development. Thus, in their early state, in which they seem to be identical with the corpuscles found floating in *chyle* and *lymph*, they seem to be nearly homogeneous particles of protoplasmic substance; but in their more advanced condition, according to Dr. Klein, their substance consists of a reticulation of very fine contractile protoplasmic fibres, termed the 'intra-cellular network;' in the meshes of which a hyaline interstitial material is included; and which is continuous with a similar network that can be discerned in the substance of the single or double nucleus, when this comes into view after the withdrawal of these corpuscles from the body. In their living state, however, whilst circulating in the vessels, the white corpuscles, although clearly distinguishable in the slow-moving stratum in contact with their walls (the red corpuscles rushing rapidly through the centre of the tube), do not usually show a distinct nucleus. This may be readily brought into view by treating the corpuscles with water, which causes them to swell up, become granular, and at last disintegrate, with the emission of

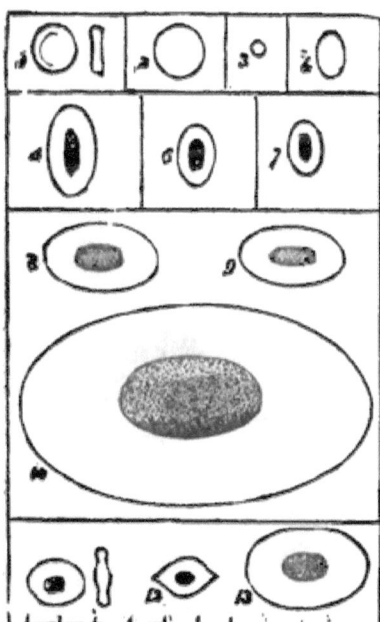

Fig. 457.

Comparative sizes of Red Blood-Corpuscles:—1. Man; 2. Elephant; 3. Musk-Deer; 4. Dromedary; 5. Ostrich; 6. Pigeon; 7. Humming Bird; 8. Crocodile; 9. Python; 10. Proteus, 11. Perch; 2. Pike; 13. Shark.

granules which may have been previously seen in active molecular movement within the corpuscle.—When the white corpuscles in a drop of freshly drawn blood are carefully watched for a short time, they may be observed to undergo changes of form, and even to move from place to place, after the manner of *Amœbæ* (§ 403). When thus moving, they engulf particles which lie in their course—such as granules of vermilion that have been injected into the blood-vessels of the living animal,—and afterwards eject these, in the like fashion. Such movements will continue for some time in the colorless corpuscles of cold-blooded animals, but still longer if they are kept in a temperature of about 75°. The movement will speedily come to an end, however, in the white corpuscles of Man or other warm-blooded animals, unless the slide is kept on a warm stage at the temperature of about 100° F. A remarkable example of an extreme change of form in a white corpuscle of Human blood, is represented in Fig. 458. Similar changes have been observed also in the corpuscles floating in the circulating fluid of the higher Invertebrata, as the Crab, which resemble the 'white' corpuscles of Vertebrated blood, rather than its 'red' corpuscles,— these last, in fact, being altogether peculiar to the circulating fluid of Vertebrated animals.

667. In examining the Blood microscopically, it is, of course, important to obtain as thin a stratum of it as possible, so that the corpuscles may not overlie one another. This is best accomplished by selecting a piece of thin glass of perfect flatness, and then, having received a small drop of Blood upon a glass slide, to lay the thin-glass cover *not upon* this, but with its edge just touching the edge of the drop; for the blood will then be drawn-in by capillary attraction, so as to spread in a uniformly-thin layer between the two glasses. Such thin films may be preserved in the liquid state by applying a cover-glass and cementing it with gold size before evaporation has taken place; but it is preferable first to expose the drop to the vapor of Osmic acid, and then to apply a drop of a weak solution of Acetate of Potass; after which a cover-glass may be put on, and secured with gold-size in the usual way. It is far simpler, however, to allow such films to dry without any cover, and then merely to cover them for protection; and in this condition the general characters of the corpuscles can be very well made-out, notwithstanding that they have in some degree shrivelled by the desiccation they have undergone. And this method is particularly serviceable, as affording a fair means of comparison, when the assistance of the Microscopist is sought in determining, for Medico-legal purposes, the source of suspicious blood-stains; the average dimensions of the dried blood-corpuscles of the several domestic animals being sufficiently different from each other, and from those of Man, to allow the nature of any specimen to be pronounced-upon with a high degree of probability.

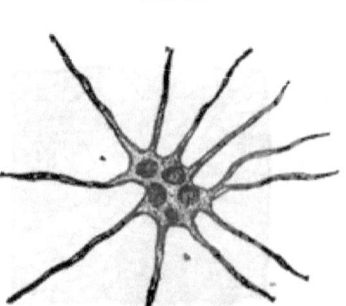

Fig. 458.

Altered White Corpuscles of Blood, an hour after having been drawn from the finger.

668. *Simple Fibrous Tissues.*—A very beautiful example of a tissue of this kind is furnished by the membrane of the common Fowl's egg;

which (as may be seen by examining an egg whose shell remains soft for want of consolidation by calcareous particles) consists of two principal layers, one serving as the basis of the shell itself, and the other forming that lining to it which is know as the *membrana putaminis*. The latter may be separated by careful tearing with needles and forceps, after prolonged maceration in water, into several matted lamellæ resembling that represented in Fig. 459; and similar lamellæ may be readily obtained from the shell itself, by dissolving away its lime by dilute acid.[1]—The simply-fibrous structures of the body generally, however, belong to one of two very definite kinds of tissue, the 'white' and the 'yellow,' whose appearance, composition, and properties are very different. The *white* fibrous tissue, though sometimes apparently composed of distinct fibres, more commonly presents the aspect of bands, usually of a flattened form, and attaining the breadth of 1-500th of an inch, which are marked by numerous longitudinal streaks, but can seldom be torn-up into minute fibres of determinate size. The fibres and bands are occasionally somewhat **wavy** in their direction; and they have a peculiar tendency to fall into

Fig. 459. Fig. 460.

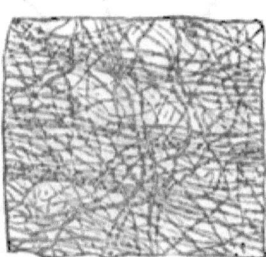
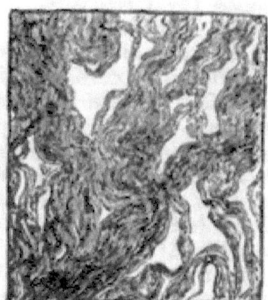

Fibrous membrane from Egg-shell.   *White Fibrous Tissue* from Ligament.

undulations, when it is attempted to tear them apart from each other (Fig. 460). This tissue is easily distinguished from the other by the effect of Acetic acid, which swells it up and renders it transparent, at the same time bringing into view certain oval nuclear particles of 'germinal matter,' which are known as 'connective-tissue corpuscles' (§ 651). These are relatively much larger, and their connections more distinct, in the earlier stages of the formation of this tissue (Fig. 461). It is perfectly inelastic; and we find it in such parts as tendons, ordinary ligaments, fibrous capsules, etc., whose function it is to resist tension without yielding to it. It constitutes, also, the organic **basis or matrix of** bone; for although the substance which is left when a bone has been macerated sufficiently long in dilute acid for all its Mineral components to be removed, is commonly designated as cartilage, this is shown by careful Microscopic analysis not to be a correct description of it; since it does not show any of the characteristic structure of cartilage, but is capable of being torn into lamellæ, in which, if sufficiently thin, the ordinary structure of a fibrous membrane can be distinguished.—The *yellow*

---

[1] For an account of the curious form in which the Carbonate of Lime is disposed in the Egg shell, see § 710.

fibrous tissue exists in the form of long, single, elastic, branching filaments with a dark decided border; which are disposed to curl when not put on the stretch (Fig. 462), and frequently anastomose, so as to form a network. They are for the most part between 1-5000th and 1-10,000th of an inch in diameter; but they are often met with both larger and smaller. This tissue does not undergo any change, when treated with Acetic acid. It exists alone (that is without any mixture of the white) in parts which require a peculiar elasticity, such as the middle coat of arteries, the 'vocal cords,' 'ligamentum nuchæ' of Quadrupeds, the elastic ligament which holds together the valves of a Bivalve shell, and that by which the claws of the Feline tribe are retracted when not in use; and it enters largely into the composition of *areolar* or connective tissue.

669. The tissue formerly known to Anatomists as 'cellular,' but now more properly designated *connective* or *areolar* tissue, consists of a net-

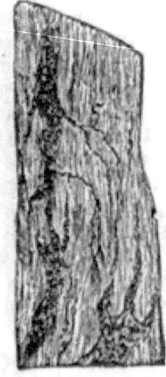

Fig. 461.

Portion of young *Tendon*, showing the corpuscles of Germinal Matter, with their stellate prolongations, interposed among its fibres.

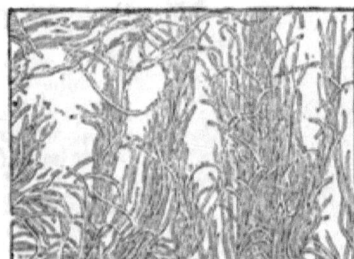

Fig. 462.

Yellow Fibrous Tissue from Ligamentum Nuchæ of Calf.

work of minute fibres and bands, which are interwoven in every direction, so as to leave innumerable *areolæ* or little spaces that communicate freely with one another. Of these fibres, some are of the 'yellow' or elastic kind, but the majority are composed of the 'white' fibrous tissue; and, as in that form of elementary structure, they frequently present the condition of broad flattened bands or membranous shreds in which no distinct fibrous arrangement is visible. The proportion of the two forms varies, according to the amount of elasticity, or of simple resisting power, which the endowments of the part may require. We find this tissue in a very large proportion of the bodies of higher Animals; thus it binds together the ultimate muscular fibres into minute fasciculi, unites these fasciculi into larger ones, these again into still larger ones which are obvious to the eye, and these into the entire muscle; whilst it also forms the membranous divisions between distinct muscles. In like manner it unites the elements of nerves, glands, etc., binds together the fat-cells into minute masses (Fig. 468), these into large ones, and so on; and in this

18

way penetrates and forms part of all the softer organs of the body. But whilst the fibrous structures of which the 'formed tissue' is composed have a purely mechanical function, there is good reason to regard the 'connective-tissue-corpuscles' which are everywhere dispersed among them, as having a most important function in the first production and subsequent maintenance of the more definitely organized portions of the fabric (§ 650). In these corpuscles, distinct *movements*, analogous to those of the sarcodic extensions of Rhizopods, have been recognized in transparent parts, such as the cornea of the eye and the tail of the young Tadpole, by observations made on these parts whilst living.—For the display of the characters of the fibrous tissues, small and thin shreds may be cut with the curved scissors (§ 183) from any part that affords them; and these must be torn asunder with needles under the simple Microscope, until the fibres are separated to a degree sufficient to enable them to be examined to advantage under a higher magnifying power. The difference between the 'white' and the 'yellow' components of connective tissue is at once made apparent by the effect of Acetic acid; whilst the 'connective-tissue-corpuscles' are best distinguished by the staining-process (§ 200), especially in the early stage of the formation of these tissues (Fig. 461).

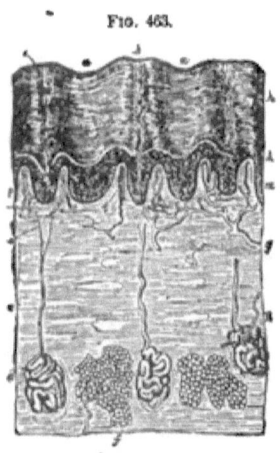

FIG. 463.

Vertical Section of Skin of Finger:—A, *epidermis*, the surface of which shows depressions *a, a*, between the eminences *b, b*, on which open the perspiratory ducts *s*, at *m* is seen the deeper layer of the epidermis, or stratum Malpighii:—B, *cutis vera*, in which are imbedded the perspiratory glands *d*, with their ducts *e*, and aggregations of fat-cells *f*; *g*, arterial twig supplying the vascular papillæ *p*; *t*, one of the tactile papillæ with its nerves.

670. *Skin; Mucous and Serous Membranes.*—The Skin which forms the external envelope of the body, is divisible into two principal layers; the *cutis vera* or 'true skin,' which usually makes up by far the larger part of its thickness, and the 'cuticle,' 'scarf-skin,' or *epidermis*, which covers it. At the mouth, nostrils, and the other orifices of the *open* cavities and canals of the body, the skin passes into the membrane that lines these, which is distinguished as the *mucous* membrane, from the peculiar glairy secretion of mucus by which its surface is protected. But those great *closed* cavities of the body, which surround the heart, lungs, intestines, etc., are lined by membranes of different kind; which, as they secrete only a thin serous fluid from their surfaces, are known as *serous* membranes. Both Mucous and Serous membranes consist, like the skin, of a proper membranous basis, and of a thin cuticular layer, which, as it differs in many points from the epidermis, is distinguished as the Epithelium (§ 673).—The substance of the 'true skin' and of the 'mucous' and 'serous' membranes is principally composed of the fibrous tissues last described; but the skin and the mucous membranes are very copiously supplied with Blood-vessels and with Glandulæ of various kinds; and in the skin we also find abundance of Nerves and Lymphatic vessels, as well as, in some parts, of Hair-follicles. The general appearance ordinarily presented by a thin vertical section of the skin of a part furnished with numerous sensory *papillæ* (§ 682), is shown in Fig. 463: where we see in deeper layers of the *cutis vera* little clumps of fat-cells, *f*, and the perspiratory glandulæ,

d, d, whose ducts, *e, e,* pass upwards: whilst on its surface we distinguish the *vascular* papillæ, *p,* supplied with loops of blood-vessels from the trunk, *g,* and a *tactile* papilla, *t,* with its nerve twig. The spaces between the papillæ are filled-up by the soft ' Malpighian layer,' *m,* of the epidermis, A, in which its coloring matter is chiefly contained, whilst this is covered by the horny layer, *h,* which is traversed by the spirally twisted continuations of the perspiratory ducts, opening at *s* upon the surface, which presents alternating depressions, *a,* and elevations, *b.*—The distribution of the blood-vessels in the Skin and Mucous membranes, which is one of the most interesting features in their structure, and which is intimately connected with their several functions, will come under our notice hereafter (Figs. 479, 482, 483). In Serous membranes, on the other hand, whose function is simply protective, the supply of Blood-vessels is more scanty.

671. *Epidermic and Epithelial Cell-layers.*—The Epidermis or 'cuticle' covers the whole exterior of the body, as a thin semi-transparent pellicle, which is shown by Microscopic examination to consist of a series of layers of cells, that are continually wearing-off at the external surface, and being renewed at the surface of the true skin; so that the newest and deepest layers gradually become the oldest and most superficial, and are at last thrown-off by slow desquamation. In their progress from the internal to the external surface of the epidermis, the cells undergo a series of well-marked changes. When we examine the innermost layer, we find it soft and granular; consisting of germinal corpuscles in various stages of development into cells, held together by a tenacious semi-fluid substance. This was formerly considered as a distinct tissue, and was supposed to be the peculiar seat of the color of the skin; it received the designation of Malpighian layer or *rete mucosum.* Passing outwards, we find the cells more completely formed; at first nearly spherical in shape, but becoming polygonal where they are flattened one against another. As we proceed further towards the surface, we perceive that the cells are gradually more and more flattened until they become mere horny scales, their cavity being obliterated; their origin is indicated, however, by the nucleus in the centre of each. This change in form is accompanied by a change in the chemical composition of the tissue, which seems to be due to the metamorphosis of the contents of the cells into a horny substance identical with that of which hair, horn, nails, hoofs, etc., are composed.—Mingled with the epidermic cells, we find others which secrete coloring matter instead of horn; these, which are termed 'pigment-cells,' are especially to be noticed in the epidermis of the Negro and other dark races, and are most distinguishable in the Malpighian layer, their color appearing to fade as they pass towards the surface.—The most remarkable development of pigment cells in the higher animals, however, is on the inner surface of the choroid coat of the Eye, where they have a very regular arrangement, and form several layers, known as the *pigmentum nigrum.* When examined separately, these cells are found to have a polygonal form (Fig. 464, *a*), and to have a distinct nucleus (*b*) in their interior. The black color is given by the accumulation, within each cell, of a number of flat rounded or oval granules, of extreme minuteness, which exhibit an active movement when set free from the cell, and even whilst inclosed within it. The pigment-cells are not always, however, of this simply rounded or polygonal form; they sometimes present remarkable stellate prolongations, under which form they are well seen in the skin of the Frog (Fig. 478,

c, c). The gradual formation of these prolongations may be traced in the pigment-cells of the Tadpole during its metamorphosis (Fig. 465). Similar varieties of form are to be met-with in the pigmentary cells of Fishes and small Crustacea, which also present a great variety of hues; and these seem to take the color of the bottom over which the animal may live, so as to serve for its concealment.

672. The structure of the Epidermis may be examined in a variety of ways. If it be removed by maceration from the true Skin, the cellular nature of its under surface is at once recognized, when it is subjected to a magnifying power of 200 or 300 diameters, by light transmitted through it, with this surface uppermost; and if the epidermis be that of a Negro or any other dark-skinned race, the pigment-cells will be very distinctly seen. This under-surface of the epidermis is not flat, but

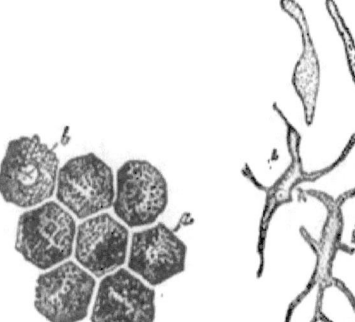

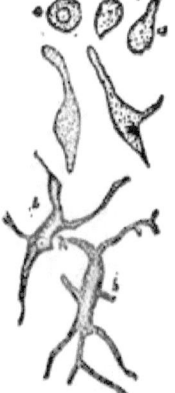

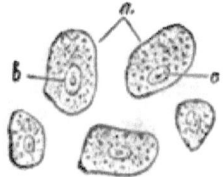

Detached Epithelium-cells: *a*, with nuclei *b*, and nucleoli *c*, from Mucous Membrane of mouth.

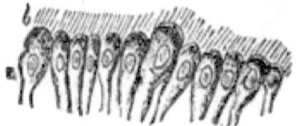

Ciliated Epithelium; *a*, nucleated cells resting on their smaller extremities; *b*, cilia.

Cells from *Pigmentum Nigrum:*—*a*, pigmentary granules concealing the nucleus; *b*, the nucleus.

Pigment-cells from tail of *Tadpole:*—*a*, *a*, simple forms of recent origin; *b*, *b*, more complex forms subsequently assumed.

is excavated into pits and channels for the reception of the papillary elevations of the true Skin; an arrangement which is shown on a large scale in the thick cuticular covering of the Dog's foot, the subjacent papillæ being large enough to be distinctly seen (when injected) with the naked eye. The cellular nature of the newly-forming layers is best seen by examining a little of the soft film that is found upon the surface of the true Skin, after the more consistent layers of the cuticle have been raised by a blister. The alteration which the cells of the external layers have undergone, tends to obscure their character; but if any fragment of epidermis be macerated for a little time in a weak solution of Soda of Potass, its dry scales become softened, and are filled-out by imbibition into rounded or polygonal cells. The same mode of treatment enables us to make out the cellular structure in warts and corns, which are epidermic growths from the surface of papillæ enlarged by hypertrophy.

673. The *Epithelium* may be designated as a delicate cuticle, cover-

ing all the free *internal* surfaces of the body, and thus lining all its cavities, canals, etc. Save in the mouth and other parts in which it approximates to the ordinary cuticle both in locality and in nature, its cells (Fig. 466) usually form but a single layer; and are so deficient in tenacity of mutual adhesion, that they cannot be detached in the form of a continuous membrane. Their shape varies greatly. Sometimes they are broad, flat, and scale-like, and their edges approximate closely to each other, so as to form what is termed a 'pavement' or 'tessellated' epithelium: such cells are observable on the web of a Frog's foot, or on the tail of a Tadpole; for, though covering an external surface, the soft moist cuticle of these parts has all the characters of an epithelium. In other cases the cells have more of the form of cylinders, standing erect side-by-side; one extremity of each cylinder forming part of the free surface, whilst the other rests upon the membrane to which it serves as a covering. If the cylinders be closely pressed together, their form is changed into prisms; and such epithelium is often known as 'prismatic.' On the other hand, if the surface on which it rests be convex, the bases or lower ends of the cylinders become smaller than their free extremities; and thus each has the form of a truncated cone rather than of a cylinder, and such epithelium (of which that covering the *villi* of the intestine, Fig. 479, is a peculiarly-good example) is termed 'conical.' But between these primary forms of epithelial cells, there are several intermediate gradations; and one often passes almost insensibly into the other.—Any of these forms of epithelium may be furnished with *cilia;* but these appendages are more commonly found attached to the elongated, than to the flattened forms of epithelium cells (Fig. 467). Ciliated epithelium is found upon the lining membrane of the air-passages in all air-breathing Vertebrata: and it also presents itself in many other situations, in which a propulsive power is needed to prevent an accumulation of mucous or other secretions. Owing to the very slight attachment that usually exists between the epithelium and the membranous surface whereon it lies, there is usually no difficulty whatever in examining it; nothing more being necessary than to scrape the surface of the membrane with a knife, and to add a little water to what has been thus removed. The ciliary action will generally be found to persist for some hours or even days after death, if the animal has been previously in full vigor;[1] and the cells that bear the cilia, when detached from each other, will swim freely about in water. If the thin fluid that is copiously discharged from the nose in the first stage of an ordinary 'cold in the head,' be subjected to microscopic examination, it will commonly be found to contain a great number of ciliated epithelium-cells, which have been thrown-off from the lining membrane of the nasal passages.

674. *Fat.*—One of the best examples which the bodies of higher animals afford, of a tissue composed of an aggregation of cells, is presented by Fat; the cells of which are distinguished by their power of drawing into themselves oleaginous matter from the blood. Fat-cells are sometimes dispersed in the interspaces of areolar tissue; whilst in other cases they are aggregated in distinct masses, constituting the proper Adipose substance. The individual fat-cells always present a

---

[1] Thus it has been observed in the lining of the windpipe of a decapitated criminal, as much as seven days after death; and in that of the river Tortoise it has been seen fifteen days after death, even though putrefaction had already far advanced.

nearly spherical or spheroidal form; sometimes, however, when they are closely pressed together, they become somewhat polyhedral, from the flattening of their walls against each other (Fig. 468). Their intervals are traversed by a minute network of blood-vessels (Fig. 480), from which they derive their secretion; and it is probably by the constant moistening of their walls with a *watery* fluid, that their contents are retained without the least transudation, although these are quite fluid at the temperature of the living body. Fat-cells, when filled with their characteristic contents, have the peculiar appearance which has been already described as appertaining to oil-globules (§ 154), being very bright in their centre, and very dark towards their margin, in consequence of their high refractive power; but if, as often happens in preparations that have been long mounted, the oily contents should have escaped, they then look like any other cells of the same form. Although the fatty matter which fills these cells (consisting of a solution of Stearine or Margarine in Oleine) is liquid at the ordinary temperature

Fig. 468.

Fig. 469.

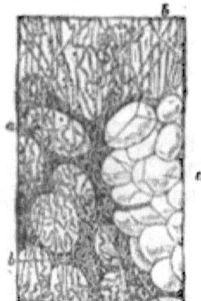
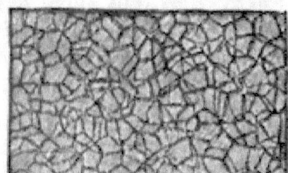

Areolar and *Adipose* tissue; *a, a,* fat-cells; *b, b,* fibres of areolar tissue.

Cellular Cartilage of Mouse's ear.

of the body of a warm-blooded animal, yet its harder portion sometimes crystallizes on cooling; the crystals shooting from a centre, so as to form a star-shaped cluster.—In examining the structure of Adipose tissue, it is desirable, where practicable, to have recourse to some specimen in which the fat-cells lie in single layers, and in which they can be observed without disturbing or laying them open; such a condition is found, for example, in the mesentery of the Mouse; and it is also occasionally met with in the fat-deposits which present themselves at intervals in the connective tissues of the muscles, joints, etc. Small collections of fat-cells exist in the deeper layers of the true skin, and are brought into view by vertical sections of it (Fig. 463, *f*). And the structure of large masses of fat may be examined by thin sections, these being placed under water in thin cells, so as to take-off the pressure of the glass-cover from their surface, which would cause the escape of the oil-particles. No method of mounting (so far as the Author is aware) is successful in causing these cells permanently to retain their contents.

675. *Cartilage.*—In the ordinary forms of Cartilage, also, we have an example of a tissue essentially composed of cells; but these are commonly

separated from each other by an 'intercellular substance,' which is so closely adherent to the outer walls of the cells as not to be separable from them. The thickness of this substance differs greatly in different kinds of cartilage, and even in different stages of the growth of any one. Thus in the cartilage of the external ear of a bat or mouse (Fig. 469), the cells are packed as closely together as are those of an ordinary Vegetable parenchyma (Fig. 236, A); and this seems to be the early condition of most cartilages that are afterwards to present a different aspect. In the ordinary cartilages, however, that cover the extremities of the bones, so as to form smooth surfaces for the working of the joints, the amount of intercellular substance is usually considerable; and the cartilage-cells are commonly found imbedded there in clusters of two, three, or four (Fig. 470), which are evidently formed by a process of 'binary subdivision.' The substance of these *cellular* cartilages is entirely destitute of blood-vessels; being nourished solely by imbibition from the blood brought to the membrane covering their surface. Hence they may be compared, in regard

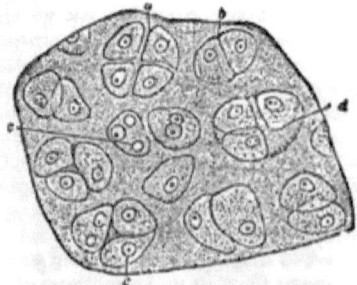

Fig. 470.

Section of the branchial *Cartilage* of Tadpole:—*a*, group of four cells, separating from each other; *b*, pair of cells in apposition; *c, c,* nuclei of cartilage-cells; *d,* cavity containing three cells (the fourth probably behind).

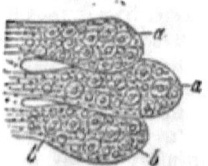

Fig. 471.

Ultimate Follicles of Mammary Gland, with their secreting cells *a, a,* containing nuclei *b, b*.

to their grade of organization, with the larger Algæ; which consist, like them, of aggregations of cells held together by intercellular substance, without vessels of any kind, and are nourished by imbibition through their whole surface.—There are many cases, however, in which the structureless intercellular substance is replaced by bundles of fibres, sometimes elastic, but more commonly non-elastic; such combinations, which are termed *fibro*-cartilages, are interposed in certain joints, wherein tension as well as pressure has to be resisted, as, for example, between the vertebræ of the spinal column and the bones of the pelvis.—In examining the structure of Cartilage, nothing more is necessary than to make very thin sections with a sharp razor or scalpel, or, if the specimen be large and dense (as the cartilage of the ribs), with the Microtome. These sections may be mounted in weak spirit, Goadby's solution, or glycerine-jelly; but in whatever way they are mounted, they undergo a gradual change by lapse of time, which renders them less fit to display the characteristic features of their structure.

676. *Structure of the Glands.*—The various Secretions of the body (as

the saliva, bile, urine, etc.), are formed by the instrumentality of organs termed Glands; which are, for the most part, constructed on one fundamental type, whatever be the nature of their product. The simplest idea of a gland is that which we gain from an examination of the 'follicles' or little bags imbedded in the wall of the stomach; some of which secrete mucus for the protection of its surface, and others gastric juice. These little bags are filled with cells of a spheroidal form, which may be considered as constituting their epithelial lining; these cells, in the progress of their development, draw into themselves from the blood the constituents of the particular product they are to secrete; and they then seem to deliver it up, either by the bursting or by the melting-away of their walls, so that this product may be poured-forth from the mouth of the bag into the cavity in which it is wanted. The Liver itself, in the lowest animals wherein it is found, presents this condition. Some of the cells that form the lining of the stomach in the Hydra and Actinia, seem to be distinguished from the rest by their power of secreting bile, which gives them a brownish-yellow tinge; in many Polyzoa, Compound Tunicata, and Annelida, these biliary cells can be seen to occupy follicles in the walls of the stomach; in Insecta these follicles are few in number, but are immensely elongated so as to form biliary tubes, which lie loosely within the abdominal cavity, frequently making many convolutions within it, and discharge their contents into the commencement of the intestinal canal; whilst in the higher Mollusca, and in Crustacea, the follicles are vastly multiplied in number, and are connected with the ramifications of gland-ducts, like grapes upon the stalks of their bunch, so as to form a distinct mass which now becomes known as the Liver. The examination of the biliary tubes of the Insect, or of the biliary follicles of the Crab, which may be accomplished with the utmost facility, is well adapted to give an idea of the essential nature of glandular structure. Among Vertebrated animals the Salivary glands, the Pancreas (sweet-bread), and the Mammary glands, are well adapted to display the follicular structure (Fig. 471); nothing more being necessary than to make sections of these organs, thin enough to be viewed as transparent objects. The Liver of Vertebrata, however, presents certain peculiarities of structure, which are not yet fully understood; for although it is essentially composed, like other glands, of secreting cells, yet it has not been determined beyond doubt whether these cells are contained within any kind of membranous investment. The Kidneys of Vertebrated animals are made-up of elongated tubes, which are straight, and are lined with a pavement-epithelium in the inner or 'medullary' portion of the kidney, whilst they are convoluted and filled with a spheroidal epithelium in the outer or 'cortical.' Certain flask-shaped dilatations of these tubes include curious little knots of blood-vessels, which are known as the 'Malpighian bodies' of the kidney; these are well displayed in injected preparations.—For such a full and complete investigation of the structure of these organs as the Anatomist and Phosiologist require, various methods must be put in practice which this is not the place to detail. It is perfectly easy to demonstrate the cellular nature of the substance of the Liver, by simply scraping a portion of its cut surface; since a number of its cells will be then detached. The general arrangement of the cells in the lobules may be displayed by means of sections thin enough to be transparent; whilst the arrangement of the blood-vessels can only be shown by means of Injections (§ 687). Fragments of the tubules of the Kidney, sometimes having the Malpighian capsules in connection with them, may also be detached by scraping its cut surface; but the true relations of these parts can only be shown by thin transparent sections, and by injections of the blood-vessels and tubuli. The simple follicles contained in the walls of the Stomach are brought into view by vertical sections; but they may be still better examined by leaving small portions of the lining membrane for a few days in dilute nitric acid (one part to four of water), whereby the fibrous tissue will be so softened, that the clusters of glandular epithelium lining the follicles (which are but very little altered) will be readily separated.

677. *Muscular Tissue.*—Although we are accustomed to speak of this tissue as consisting of 'fibres,' yet the ultimate structure of the 'muscular fibre' is very different from that of the 'simple fibrous tissues' already described. When we examine an ordinary muscle (or piece of 'flesh') with the naked eye, we observe that it is made-up of a number of *fasciculi* or bundles of fibres (Fig. 472), which are arranged side-by-side with great regularity, in the direction in which the muscle is to act, and are united by connective tissue. These fasciculi may be separated into smaller parts, which appear like simple fibres; but when these are

FIG. 472.   FIG 473.

Fasciculus of striated Muscular Fibre, showing at *a* the transverse striæ, and at *b* its junction with the tendon.

Striated *Muscular Fibre*, separating into fibrillæ.

examined by the Microscope, they are found to be themselves fasciculi, composed of minuter fibres bound together by delicate filaments of connective tissue. By carefully separating these, we may obtain the ultimate muscular fibre. This fibre exists under two forms, the *striated* and the *non-striated*. The former is chiefly distinguished by the transversely-striated appearance which it presents (Fig. 473), and which is due to an alternation of light and dark spaces along its whole extent; the breadth and distance of these striæ vary, however, in different fibres, and even in different parts of the same fibre, according to their state of contraction or relaxation. Longitudinal striæ are also frequently visible, which are due to a partial separation between the component fibrillæ into which the fibre may be broken up.—When a fibre of this kind is more closely examined, it is seen to be inclosed within a delicate tubular sheath, which is quite distinct on the one hand from the connective tissue that binds the fibres into fasciculi, and equally distinct from the internal substance of the fibre. This membranous tube, which is termed the *sarcolemma*, is not perforated by capillary vessels, which therefore lie *outside* the ultimate elements of the muscular substance; whether it is penetrated by the ultimate fibres of **nerves,** is a point not yet certainly ascertained.—The diameter of the fibres varies greatly in different kinds of Vertebrated animals. Its average is greater in Reptiles and Fishes than in Birds and Mammals, and its extremes also are wider; thus its dimensions vary in the Frog from 1-100th to 1-1000th of an inch, and in the Skate from 1-65th to 1-300th; whilst in the Human subject the average is about 1-400th of an inch, and the extremes about 1-200th and 1-600th.

678. The substance of the fibre, when broken up by 'teazing' with needles, is found to consist of very minute fibrillæ, which, when examined under a magnifying power of from 250 to 400 diameters, are seen to present a slightly-beaded form, and to show the same alternation of light and dark spaces as when the fibrillæ are united into fibres or into small bundles (Fig. 473). The dark and light spaces are usually of nearly equal length: each light space is divided by a transverse line, called 'Dobie's line;' while each dark space is crossed by a lighter band, known as 'Hensen's stripe.' It has been generally supposed that these markings indicate differences in the *composition* of the fibre; but Mr. J. B. Haycroft has recently revived an idea which originated with Mr. Bowman, that they are the optical expressions of its *shape*. The borders of the striated fibre (he truly states) present wavy margins, indicative of a transverse ridging and furrowing; the whole fibre (or a single fibril) thus consisting of a succession of convex bead-like projections with intermediate concave depressions. When the *axis* of the fibre is in true focus, Dobie's line, D, crosses the deepest part of the concavity, while Hensen's stripe, H, crosses the most projecting part of the convexity; and it can be shown, both theoretically and experimentally, that this alternation of lights and shades will be produced by the passage of light through a similarly-shaped homogenous rod of any transparent substance. If, on the other hand, the *surface* of the fibre be brought into focus, the convex ribbings appear light and the intervening depressions dark,—which is the aspect originally represented by Bowman. The appearances are the same in the extended and contracted states of the fibre; with the exception that the alternation of light and dark striæ is closer in the contracted state, while the breadth (representing the thickness) of the fibre is correspondingly increased.[1]

679. In the examination of Muscular tissue, a small portion may be cut-out with the curved scissors; this should be torn up into its component fibres; **and these,** if possible, should be separated into their fibrillæ, by dissection with a pair of needles under the Simple Microscope. The general characters of the *striated* fibre are admirably shown in the large fibres of the Frog; and by selecting a portion in which these fibres spread themselves out to unite with a broad tendinous expansion, they may often be found so well displayed in a single layer, as not only to exhibit all their characters without any dissection, but also to show their mode of connection with the 'simple fibrous' tissue of which that expansion is formed. As the ordinary characters of the fibre are but little altered by boiling, recourse may be had to this process for their more ready separation, especially in the case of the tongue. Dr. Beale recommends Glycerine for the preparation, and Glycerine-media for the preservation, of objects of this class; and states that the alternation of light and dark spaces in the fibrillæ is rendered more distinct by such treatment. The fibrillæ are often more readily separable when the muscle has been

---

[1] "Quart. Journ. Microsc. Science," N.S., Vol. xxi., p. 307.

macerated in a weak solution of Chromic acid.—The shape of the fibres can only be properly seen in cross sections; and these are best made by the Freezing Microtome (§ 191).—Striated fibres, separable with great facility into their component fibrillæ, are readily obtainable from the limbs of Crustacea and of Insects; and their presence is also readily distinguishable in the bodies of Worms, even of very low organization; so that it may be regarded as characteristic of the Articulated series generally. On the other hand, the Molluscous classes are for the most part distinguished by the non-striation of their fibre; there are, however, two remarkable exceptions, strongly striated fibre having been found in the *Terebratula* and other *Brachiopods* (where, however, it is limited to the *anterior* adductor muscles of the shell), and also in many *Polyzoa*. Its presence seems related to energy and rapidity of movement; the non-striated presenting itself where the movements are slower and feebler in their character.

FIG. 474.

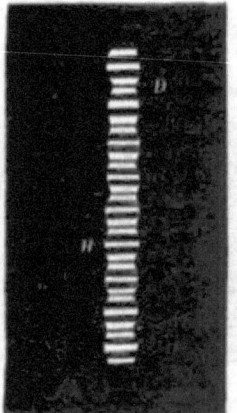

Diagram of Striated Fibrilla.

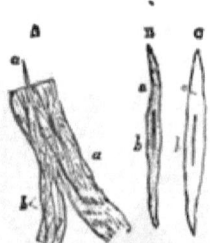

Structure of *non-striated* Muscular Fibre:— A, portion of tissue showing fusiform cells *a, a*, with elongated nuclei *b, b*;—B, a single cell isolated and more highly magnified; C, a similar cell treated with acetic acid.

680. The 'smooth' or *non-striated* form of Muscular fibre, which is especially found in the walls of the stomach, intestines, bladder, and other similar parts, is composed of flattened bands whose diameter is usually between 1-2000th and 1-3000th of an inch; and these bands are collected into fasciculi, which do not lie parallel with each other, but cross and interlace. By macerating a portion of such muscular substance, however, in dilute nitric acid (about one part of ordinary acid to three parts of water) for two or three days, it is found that the bands just mentioned may be easily separated into elongated fusiform cells, not unlike 'woody fibre' in shape (Fig. 474, *a, a*); each distinguished, for the most part, by the presence of a long staff-shaped nucleus, *b*, brought into view by the action of acetic acid, *c*. These cells, in which the distinction between cell-wall and cell-contents can by no means be clearly seen, are composed of a soft yellow substance often containing small pale granules, and sometimes yellow globules of fatty matter. In the coats of the Blood-vessels are found cells having the same general characters, but shorter and wider in form; and although some of these approach very

closely in their general appearance to epithelium-cells, yet they seem to have quite a different nature, being distinguished by their elongated nuclei, as well as by their contractile endowments.

681. *Nerve-substance.*—Wherever a distinct Nervous System can be made out, it is found to consist of two very different forms of tissue— namely, the *cellular*, which are the essential components of the ganglionic centres, and the *fibrous*, of which the connecting trunks consist. The typical form of the nerve-cells or 'ganglion-globules' may be regarded as globular; but they often present an extension into one or more long processes, which give them a 'caudate' or 'stellate' aspect. These processes have been traced into continuity, in some instances, with the axis-cylinders of nerve-tubes (Fig. 475); whilst in other cases they seem to inosculate with those of other vesicles. The cells, which do not seem to possess a definite cell-wall, are for the most part composed of a finely-granular substance, which extends into their prolongations; and in the midst of this is usually to be seen a large well-defined nucleus. They also

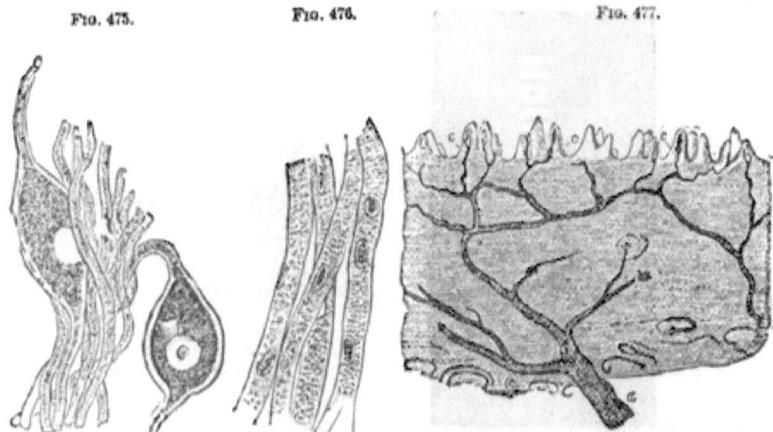

Fig. 475. Ganglion-cells and Nerve-fibres from a ganglion of Lamprey.

Fig. 476. Gelatinous Nerve-fibres, from Olfactory Nerve.

Fig. 477. Vertical Section of Skin of Finger, showing the branches of the cutaneous nerves, *a, b,* inosculating to form a plexus, of which the ultimate fibres pass into the cutaneous papillæ, *c, c.*

generally contain pigment-granules, which give them a reddish or yellowish-brown color, and thus impart to collections of ganglionic cells in the warm-blooded Vertebrata that peculiar hue, which causes it be known as the *cineritious* or *gray* matter, but which is commonly absent among the lower animals.—Each of the tubular nerve-fibres, on the other hand, of which the trunks are made up, consists, in its fully developed form, of a delicate membranous sheath, within which is a hollow cylinder of a material known as the 'white substance of Schwann,' whose outer and inner boundaries are marked-out by two distinct lines, giving to each margin of the nerve-tube what is described as a 'double contour.' The contents of the membranous envelope are very soft, yielding to slight pressure: and they are so quickly altered by the contact of water or of any liquids which are foreign to their nature, that their characters can only be properly judged-of when they are quite fresh. The centre or axis of the tube is then found to be occupied by a transparent substance which is known as the 'axis-cylinder:' and there is reason to believe that this last, which is a protoplasmic substance, is the *essential* component of the nerve-fibre, while the function of the hollow cylinder that surrounds it, which is composed of a combination of fat and albuminous matter, is simply protective. The diameter of the nerve-tubes differs in different nerves; being sometimes as great as 1-1500th of an inch, and as small in other instances as 1-12,000th.—In many of the lower Invertebrata, such as *Medusæ* (§ 523) and *Comatulæ* (§ 546), we seem fully justified by physiological evidence in regarding as Nerves certain protoplasmic fibres which do not possess the characteristic structure of 'nerve-tubes;' and fibres destitute of the 'double contour' are found also in certain parts of the body of even the highest Vertebrates. These fibres, which are known as 'gelatinous,' are considerably smaller than the preceding, and do not exhibit any differentiation of parts (Fig. 476). They are flattened, soft, and homogenous in their appearance, and contain numerous nuclear particles which are brought into view by acetic acid. They can sometimes be seen to be continuous with the axis-cylinders of the ordinary fibres, and also with the radiating prolongations of the ganglion-cells; so that their nervous character, which has been questioned by some anatomists, seems established beyond doubt.

682. The ultimate distribution of the Nerve-fibres is a subject on which there has been great divergence of opinion, and which can only be successfully investigated by observers of great experience. The Author believes that it may be stated as a general fact, that in both the motor and the sensory nerve-tubes, as they approach their terminations in the muscles and in the skin respectively, the protoplasmic axis-cylinder is continued beyond its envelopes; often then breaking-up into very minute fibrillæ, which inosculate with each other so as to form a network closely resembling that formed by the pseudopodial threads of *Rhizopods* (Fig. 283.) Recent observers have described the fibrillæ of motor nerves as terminating in 'motorial end-plates' seated upon or in the muscular fibres; and these seem analogous to the little 'islets' of sarcodic substance, into which those threads often dilate.—Where the Skin is specially endowed with tactile sensibility, we find a special *papillary* apparatus, which in the skin may be readily made out in thin vertical sections treated with solution of soda (Fig. 477). It was formerly supposed that all the cutaneous papillæ are furnished with nerve-fibres, and minister to sensation: but it is now known that a large proportion (at any rate) of those that are furnished with loops of blood-vessels (Figs. 463, *p*, 483), being destitute of nerve-fibres, must have for their special office the production of Epidermis; whilst those which, possessing nerve-fibres, have sensory functions, are usually destitute of blood-vessels. The greater part of the interior of each sensory papilla (Fig. 477, *c, c*) of the skin is occupied by a peculiar 'axile body,' which seems to be merely a bundle of ordinary connective tissue, whereon the nerve-fibre appears to terminate. The nerve-fibres are more readily seen, however, in the 'fungiform' papillæ of the Tongue, to each of which several of them proceed; these bodies, which are very transparent, may be well seen by snipping-off minute portions of the tongue of the Frog; or by snipping-off the papillæ themselves from the surface of the living Human tongue, which can be readily done by a dexterous use of the curved scissors, with no more pain than the prick of a pin would give. The transparence of these papillæ also is increased by treating them with a weak solution of soda.—Nerve-fibres have also been found to terminate on sensory surfaces in minute 'end-bulbs' of spheroidal shape and about 1-600th of an inch in diameter; each of them being

composed of a simple outer capsule of connective tissue, filled with clear soft matter, in the midst of which the nerve-fibre, after losing its dark border, ends in a knob. The 'Pacinian corpuscles,' which are best seen in the mesentery of the Cat, and are from 1-15th to 1-10 of an inch long, seem to be more developed forms of these 'end-bulbs.'

683. For the sake of obtaining a general acquaintance with the Microscopic characters of these principal forms of Nerve-substance, it is best to have recourse to minute nerves and ganglia. The small nerves which are found between the skin and the muscles of the back of the Frog, and which become apparent when the former is being stripped-off, are extremely suitable for this purpose; but they are best seen in the *Hyla* or 'tree-frog,' which is recommended by Dr. Beale as being much superior to the common Frog for the general purposes of minute histological investigation. If it be wished to examine the natural appearance of the nerve-fibres, no other fluid should be used than a little blood-serum; but if they be treated with strong acetic acid, a contraction of their tubes takes place, by which the axis-cylinders are forced-out from their cut extremities, so as to be made more apparent than they can be in any other way. On the other hand, by immersion of the tissue in a dilute solution of Chromic acid (about one part of the solid crystals to two hundred of water), the nerve-fibres are rendered firmer and more distinct. Again, the axis-cylinders are brought into distinct view by the staining-process (§ 202 *a*), being dyed much more quickly than their envelopes; and they may thus be readily made-out by reflected light, in transverse sections of nerves that have been thus treated. The *gelatinous* fibres are found in the greatest abundance in the Sympathetic nerves; and their characters may be best studied in the smaller branches of that system.—So, for the examination of the ganglionic cells, and of their relation to the nerve-tubes, it is better to take some minute ganglion as a whole (such as one of the sympathetic ganglia of the Frog, Mouse, or other small animal), than to dissect the larger ganglionic masses, whose structure can only be successfully studied by such as are proficient in this kind of investigation. The nerves of the orbit of the eyes of Fishes, with the ophthalmic ganglion and its branches, which may be very readily got-at in the Skate, and of which the components may be separated without much difficulty, form one of the most convenient objects for the demonstration of the principal forms of nerve-tissue, and especially for the connection of nerve-fibres and ganglion-cells.—For minute inquiries, however, into the ultimate distribution of the nerve-fibres in Muscles, and Sense-organs, certain special methods must be followed, and very high magnifying powers must be employed. Those who desire to follow out this inquiry should acquaint themselves with the methods which have been found most successful in the hands of the able Histologists whose works have been already referred to.

684. *Circulation of the Blood.*—One of the most interesting spectacles that the Microscopist can enjoy, is that which is furnished by the Circulation of the Blood in the *capillary* blood-vessels which distribute the fluid through the tissues it nourishes. This, of course, can only be observed in such parts of Animal bodies as are sufficiently thin and transparent to allow of the transmission of light through them, without any disturbance of their ordinary structure; and the number of these is very limited. The web of the Frog's foot is perhaps the most suitable for ordinary purposes, more especially since this animal is to be easily obtained in almost every locality; and the following is the simple

arrangement preferred by the Author:—A piece of thin Cork is to be obtained, about 9 inches long and 3 inches wide (such pieces are prepared by Cork-cutters, as soles), and a hole about 3-8th of an inch in diameter is to be cut at about the middle of its length, in such a position that, when the cork is secured upon the stage, this aperture may correspond with the axis of the Microscope. The body of the Frog is then to be folded in a piece of wet calico, one leg being left free, in such a manner as to confine its movements, but not to press to tightly upon its body; and being then laid down near one end of the cork-plate, the free leg is to be extended, so that the foot can be laid over the central aperture. The spreading-out of the foot over the aperture is to be accomplished, either by passing pins through the edge of the web into the cork beneath, or by tying the ends of the toes with threads to pins stuck into the cork at a small distance from the aperture; the former method is by far the least troublesome, and it may be doubted whether it is really the source of more suffering to the animal than the latter, the confinement being obviously that which is most felt. A few turns of tape, carried *loosely* around the calico bag, the projecting leg, and the cork, serve to prevent any sudden start; and when all is secure, the cork-plate is to be laid down upon the stage of the Microscope, where a few more turns of the tape will serve to keep it in place. The web being moistened with water (a precaution which should be repeated as often as the membrane exhibits the least appearance of dryness), and an adequate light being reflected through the web from the mirror, this wonderful spectacle is brought into view on the adjustment of the focus (a power of from 75 to 100 diameters being the most suitable for ordinary purposes), provided that no obstacle to the movement of the blood be produced by undue pressure upon the body or leg of the animal. It will not unfrequently be found, however, that the current of blood is nearly or altogether stagnant for a time; this seems occasionally due to the animal's alarm at its new position, which weakens or suspends the action of its heart, the movement recommencing again after the lapse of a few minutes, although no change has been made in any of the external conditions. But if the movement should not renew itself, the tape which passes over the body should be slackened; and if this does not produce the desired effect, the calico envelope also must be loosened. When everything has once been properly adjusted, the animal will often lie for hours without moving, or will only give an occasional twitch; and even this may be avoided by previously subjecting it to the influence of chloroform, which may be renewed from time to time whilst it is under observation.—The movement of the Blood will be distinctly seen by that of its corpuscles (Fig. 478), which course after one another through the network of Capillaries that intervenes between the smallest arteries and the smallest veins; in those tubes which pass most directly from the veins to the arteries, the current is always in the same direction; but in those which pass across between these, it may not unfrequently be seen that the direction of the movement changes from time to time. The larger vessels with which the capillaries are seen to be connected, are almost always *veins*, as may be known from the direction of the flow of blood in them from the branches (b, b) towards their trunk (a); the *arteries*, whose ultimate subdivisions discharge themselves into the capillary network, are for the most part restricted to the immediate borders of the toes. When a power of 200 or 250 diameters is employed, the visible area is of course greatly reduced; but the individual vessels and their contents are much more plainly seen: and it may then be observed that whilst the 'red' corpuscles (§ 655) flow at a very rapid rate along the centre of each tube, the 'white' corpuscles (§ 666), which are occasionally discernible, move slowly in the clear stream near its margin.

685. The Circulation may also be displayed in the *tongue* of the Frog, by laying the animal (previously chloroformed) on its back, with its head close to the hole in the cork-plate, and, after securing the body in this position, drawing-out the tongue with the forceps, and fixing it on the other side of the hole with pins. So, again, the circulation may be examined in the *lungs*—where it affords a spectacle of singular beauty, —or in the *mesentery*, of the living Frog, by laying open its body, and, drawing forth either organ; the animal having previously been made insensible by chloroform. The *tadpole* of the Frog, when sufficiently young, furnishes a good display of the capillary circulation in its tail; and the difficulty of keeping it quiet during the observation may be overcome by gradually mixing some warm water with that in which it is

Fig. 478.

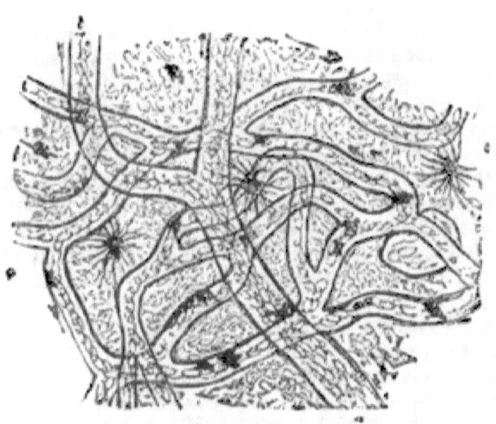

Capillary Circulation in a portion of the web of a *Frog's* foot:—*a*, trunk of vein; *b, b*, its branches; *c, c*, pigment-cells.

swimming, until it becomes motionless; this usually happens when it has been raised to a temperature of between 100° and 110°; and notwithstanding that the muscles of the body are thrown into a state of spasmodic rigidity by this treatment, the heart continues to pulsate, and the circulation is maintained.[1] The *larva of the Water-newt*, when it can be obtained, furnishes a most beautiful display of the circulation, both in its external gills and in its delicate feet. It may be inclosed in a large Aquatic-box or in a shallow cell, gentle pressure being made upon its body, so as to confine its movements without stopping the heart's action. —The circulation may also be seen in the tails of small Fish, such as the *minnow* or the *stickleback*, by confining these animals in tubes, or in shallow cells, or in a large Aquatic-box;[2] but although the extreme

---

[1] A special form of Live-box for the observation of living Tadpoles, etc., contrived by F. E. Schultze, of Rostock, is described and figured in the "Quart. Journ. of Microsc. Science," N.S., Vol. vii. (1867), p. 261.

[2] A convenient Trough for this purpose is described in the "Quart. Journ. of Microsc. Science," Vol. vii. (1859), p. 113.

transparence of these parts adapts them well for this purpose in one respect, yet the comparative scantiness of their blood-vessels prevents them from being as suitable as the Frog's web in another not less important particular.—One of the most beautiful of all displays of the circulation, however, **is that** which may be seen upon the *yolk-bag* of young Fish (such as **the** Salmon or Trout) soon after they have been hatched; and as it is their habit to remain almost entirely motionless at this stage of their existence, the observation can be made with the greatest facility by means of the Zoophyte-trough, provided that the subject of **it can** be obtained. Now that the artificial breeding of these Fish is largely practised for **the** sake of stocking rivers and fish-ponds, there can **seldom** be much difficulty in procuring specimens at the proper period. The store **of** yolk which the yolk-bag supplies for the nutrition of the embryo, not being exhausted in the Fish (as it is in the bird) previously to the hatching of the egg, this bag hangs-down from the belly of the little creature on its emersion; and continues to do so until its contents have been absorbed into the body, which does not take place for some little time afterwards. **And** the blood is distributed over it in copious streams, partly that it may draw into itself fresh nutritive material, and partly that it **may be** subjected to the aërating influence of the surrounding water.

686. The Tadpole serves, moreover, for the display, under **proper** management, **not** only of the capillary but of the *general* Circulation; and if this be studied under the Binocular Microscope, **the** observer not only enjoys the gratification of witnessing a most **wonderful** spectacle, but may also obtain a more accurate notion of the **relations of the** different parts of the circulating system than is otherwise **possible.**[1] The Tadpole, as every naturalist is aware, is **essentially a Fish** in the early period of its existence, breathing by gills **alone, and having its circulating** apparatus arranged accordingly: but as its **limbs are developed and its tail** becomes relatively shortened, its lungs **are gradually evolved in preparation** for its terrestrial life, and the **course of the blood is considerably** changed. In the Tadpole as it **comes forth from the** egg, the gills are *external*, forming **a** pair of fringes hanging **at the** sides **of the head** (Plate XXIV., fig. 1); and **at the** bases of these, concealed **by operculs or** gill-flaps resembling those of Fishes, are seen the rudiments **of the** *internal* gills, which soon begin **to be** developed in the stead **of the preceding.** The *external* gills reach their highest development on the **fourth or fifth** day after emersion; and they then wither so rapidly (whilst being **at the** same time drawn-in by the growth of the animal), that by the end **of the** first week only a remnant of the right gill can be seen under the **edge of** the operculum (fig. 2, *c*), though the left gill (*b*) is somewhat **later in its** disappearance. Concurrently with this change, the *internal* gills are undergoing rapid development; and **the** beautiful arrangement of their vascular tufts, which originate from **the roots** of the arteries of the external gills, as seen at *g*, fig. 5, is shown **in fig. 4.** It is requisite that the Tadpole subjected to observation should **not** be so far advanced as to have lost its early transparence of skin; and **it is** further essential to the trac-

---

[1] See Mr. Whitney's account of 'The Circulation in the Tapdole,' in "Transact. of Microsc. Soc.," N. S., Vol. x. (1862), p. 1, and his subsequent paper 'On the Changes which accompany the Metamorphosis of the Tadpole' in the same Transactions, Vol. xv., p. 43.—In the first of these Memoirs Mr. W. described the internal gills as lungs, an error which he corrected in the second.

PLATE XXIV.

CIRCULATION IN THE TADPOLE (after Whitney).

Fig. 1. Anterior portion of young Tadpole, **showing** the external gills, with the incipient tufts of the internal gills, and the pair of minute tubes **between** the heart and the spirally-coiled intestine, which are the rudiments of the future lungs.

2. More advanced Tadpole, in which the external gills have almost disappeared:—*a*, remnant of external gills on the left side; *b*, operculum; *c*, remnant of external gill on the right side, turned in.

3. Advanced Tadpole, showing the course of the general Circulation:—*a*, heart; *b*, branchial arteries; *c*, pericardium; *d*, internal gill; *e*, **first** or cephalic trunk; *f*, branch to lip; *g*, branches to head; *h*, second or branchial trunk; *i*, **third trunk**, uniting with its fellow to form the abdominal aorta, which is continued as **the** caudal artery *k*, **to** the extremity of the tail; *l*, caudal vein; *m*, kidney; *n*, **vena** cava; *o*, liver; *p*, vena portæ; *q*, sinus venosus, receiving the jugular vein, *r*, and the abdominal veins, *t*, *u*, as also the branchial vein, *v*.

4. The branchial Circulation on a larger scale:—A, B, C, three primary branches of the branchial artery; *a*, cartilaginous arches; *b*, additional framework; *c*, *e*, twigs of branchial artery; *d*, *f*, rootlets of branchial vein.

5 Origin of the vessels **of the internal gills,** *g*, **from the** roots of those of the external.

6. The heart, systemic **arteries, pulmonary arteries and** veins, and lungs, in the adult Frog; the **heart** being turned **up in the right-hand figure, to show the** junction of the Pulmonary veins and **their** entrance into **the left auricle.**

ing-out the course of the abdominal vessels, that the creature should have been kept without food for some days, so that the intestine may empty itself. This starving process reduces the quantity of red corpuscles, and thus renders the blood paler; but this, although it makes the smaller branches less obvious, brings the circulation in the larger trunks into more distinct view. " Placing the Tadpole on his back," says Mr. Whitney, " we look, as through a pane of glass, into the chamber of the chest. Before us is the beating heart, a bulbous-looking cavity, formed of the most delicate transparent tissues, through which are seen the globules of the blood, perpetually, but alternately, entering by one orifice and leaving it by another. The heart (Plate XXIV., fig. 3, *a*) appears to be slung, as it were, between two arms or branches, extending right and left. From these trunks (*b*) the main arteries arise. The heart is inclosed within an envelope or pericardium (*c*), which is, perhaps, the most delicate, and is, certainly, the most elegant beauty in the creature's organism. Its extreme fineness makes it often elude the eye under the single Microscope, but under the Binocular its form is distinctly revealed. Then it is seen as a canopy or tent, inclosing the heart, but of such extreme tenuity that its *folds* are really the means by which its existence is recognized. Passing along the course of the great vessels to the right and left of the heart, the eye is arrested by a large oval body (*d*) of a more complicated structure and dazzling appearance. This is the internal gill, which, in the Tadpole, is a cavity formed of most delicate transparent tissue, traversed by certain arteries, and lined by a crimson network of blood-vessels, the interlacing of which, with their rapid currents and dancing globules, forms one of the most beautiful and dazzling exhibitions of vascularity." Of the three arterial trunks which arise on each side from the *truncus arteriosus*, *b*, the first, or *cephalic*, *e*, is distributed entirely to the head, running first along the upper edge of the gill, and giving off a branch, *f*, to the thick-fringed lip which surrounds the mouth; after which it suddenly curves upwards and backwards, so as to reach the upper surface of the head, where it dips between the eye and the brain. The second main trunk, *h*, seems to be chiefly distributed to the gill, although it freely communicates by a network of vessels both with the first or cephalic and with the third or abdominal trunk. The latter also enters the gill and gives off branches; but it continues its course as a large trunk, bending downwards and curving towards the spine, where it meets its fellow to form the *abdominal aorta*, *i*, which, after giving-off branches to the abdominal viscera, is continued, as the *caudal artery*, *k*, to the extremity of the tail. The blood is returned from the tail by the *caudal vein*, *l*, which is gradually increased in size by its successive tributaries as it passes towards the abdominal cavity; here it approaches the kidney, *m*, and sends off a branch which incloses that organ on one side, while the main trunk continues its course on the other, receiving tributaries from the kidney as it passes. (This supply of the kidney by *venous* blood is a peculiarity of the lower Vertebrata.) The venous blood returned from the abdominal viscera, on the other hand, is collected into a trunk *p*, known as the *portal vein*, which distributes it through the substance of the liver, *o*, as in Man; and after traversing that organ it is discharged by numerous fine channels, which converge towards the great abdominal trunk, or *vena cava*, *n*, as it passes in close proximity to the liver, onwards to the *sinus venosus*, *q*, or rudimentary auricle of the heart. This also receives the *jugular vein*, *r*, from the head, which first, however, passes downwards in front of the gill close to its inner edge, and meets a vein, *t*, coming up from the abdomen, after which it turns abruptly in the direction of the heart. Two other abdominal veins, *u*, meet and pour their blood direct into the sinus venosus; and into this cavity is also poured the aërated blood returned from the gill by the *branchial vein*, *v*, of which only the one on the right side can be distinguished.—The lungs may be detected in a rudimentary state, even in the very young tadpole; being in that stage a pair of minute tubular sacs, united at the upper extremities, and lying behind the intestine and close to the spine. They may be best brought into view by immersing the tadpole for a few days in a weak solution of chromic acid, which renders the tissue friable, so that the parts that conceal them may be more readily peeled away. Their gradual enlargement may be traced during the period of the tadpole's transparence; but they can only be brought into view by dissection when the metamorphosis has been completed. The following are Mr. Whitney's directions for displaying the Circulation in these organs:—" Put the young Frog into a wineglass, and drop on him a single drop of chloroform. This suffices to extinguish sensibility. Then lay him on the back on a piece of cork, and fix him with small pins passed through the web of each foot. Remove the skin of the abdomen with a fine pair of sharp scissors and forceps. Turn aside the intestines from the *left* side, and thus expose the left lung, which may now be seen as a glistening transparent sac, containing air bubbles. With a fine camel-hair pencil the lung may now be turned-out, so as to enable the operator to see a large part of it by *transmitted* light. Unpin the frog, and place him on a slip of glass, and then transmit the light through the everted portion of lung. Remember that the lung is very elastic, and is emptied and collapsed by very slight pressure. Therefore, to succeed with this experiment, the lung should be touched as little as possible, and in the lightest manner, with the brush. If the heart is acting feebly, you will see simply a transparent sac, shaped according to the quantity of air-bubbles it may happen to contain, but void of red vascularity and circulation. But should the operator succeed in getting the lung well placed, full of air, and have the heart still beating vigorously, he will see before him a brilliant picture of crimson network, alive with the dance and dazzle of blood-globules, in rapid chase of one another through the delicate and living lace-work which lines the chamber of the lung." The position of the lungs in relation to the heart and the great vascular trunks, is shown in Plate XXIV., fig. 6.

687. *Injected Preparations.*—Next to the Circulation of the Blood in the living body, the varied distribution of the Capillaries in its several organs, as shown by means of 'injections' of coloring matter thrown into their principal vessels, is one of the most interesting subjects of Microscopic examination. The art of making successful preparations of this kind is one in which perfection can usually be attained only by long practice, and by attention to a great number of minute particulars; and better specimens may be obtained, therefore, from those who have made it a business to produce them, than are likely to be prepared by amateurs for themselves. For this reason, no more than a general account of the process will be here offered; the minute details which need to be attended-to, in order to attain successful results, being readily accessible elsewhere to such as desire to put it in practice.[1] Injections may be either *opaque*

---

[1] See especially the article 'Injection,' in the " Micrographic Dictionary;" M. Robin's work, " Du Microscope et des Injections;" Prof. H. Frey's Treatise " Das Mikroscop und die Mikroskopische Technik;" Dr. Beale's " How to Work with the

or *transparent*, each method having its special advantages. The former is most suitable where *solid form* and *inequalities of surface* are especially to be displayed, as in Figs. 479 and 485; the latter is preferable where the injected tissue is so thin as to be transparent (as in the case of the retina and other membranes of the eye), or where the distribution of its blood-vessels and their relation to other parts may be displayed by sections thin enough to be made transparent by mounting either in Canada balsam or Dammar (Plate xxv.).—The injection is usually thrown into the vessels by means of a brass syringe expressly constructed for the purpose, which has several jet-pipes of different sizes, adapted to the different dimensions of the vessels to be injected; and these should either be furnished with a stop-cock to prevent the return of the injection when the syringe is withdrawn, or a set of small corks of different sizes should be kept in readiness, with which they may be plugged. The pipe should be inserted into the cut end of the trunk which is to be injected, and should be tied therein by a silk thread. In injecting the vessels of Fish, Mollusks, etc., the softness of the vessels renders them liable to break in the attempt to tie them; and it is therefore better for the operator to satisfy himself with introducing a pipe as large as he can insert, and with passing it into the vessel as far as he can without violence. All the vessels from which the injection might escape should be tied, and sometimes it is better to put a ligature round a part of the organ or tissue itself; thus, for example, when a portion of the Intestinal tube is to be injected through its branch of the Mesenteric artery, not only should ligatures be put round any divided vessels of the mesentery, but the cut ends of the intestinal tube should be firmly tied.—For making those minute injections, however, which are needed for the purposes of anatomical investigation, rather than to furnish 'preparations' to be looked-at, the Author has found the glass-syringe (Fig. 106), so frequently alluded-to, the most efficient instrument; since the Microscopist can himself draw its point to the utmost fineness that will admit of the passage of the injection, and can push this point without ligature, under the Simple Microscope, into the narrowest orifice, or into the substance of the part into which the injection is to be thrown.—Save in the cases in which the operation has to be practised on living animals, it should either be performed when the body or organ is as fresh as possible, or after the expiry of sufficient time to allow the *rigor mortis* to pass-off; the presence of this being very inimical to the success of the injection. The part should be thoroughly warmed, by soaking in warm water for a time proportionate to its bulk; and the injection, the syringe, and the pipes should also have been subjected to a temperature sufficiently high to insure the free flow of the liquid. The force used in pressing-down the piston should be very moderate at first, but should be gradually increased as the vessels become filled; and it is better to keep up a steady pressure for some time, than to attempt to distend them by a more powerful pressure, which will be certain to cause extravasation. This pressure should be maintained[1] until the injection begins to flow from the large veins, and the tissue is thoroughly reddened, and if one syringeful of injection after another be required for this purpose, the return of the injection should be prevented

---

Microscope;" the "Handbook to the Physiological Laboratory;" and Rutherford's and Schäfer's treatises on "Practical Histology."

[1] Simple mechanical arrangements for this purpose, by which the fatigue of maintaining this pressure with his hand is saved to the operator, are described in the works referred-to in the preceding note.

by stopping the nozzle of the jet-pipe when the syringe is removed for refilling. When the injection has been completed, any openings by which it can escape should be secured, and the preparation should then be placed for some hours in cold water, for the sake of causing the size to 'set.'[1]

688. For *opaque* injections, the best coloring-matter, when only one set of vessels is to be injected, is Chinese vermilion. This, however, as commonly sold, contains numerous particles of far too large a size; and it is necessary first to reduce it to a greater fineness by continued trituration in a mortar (an agate or a steel mortar is the best) with a small quantity of water, and then to get rid of the larger particles by a process of 'levigation,' exactly corresponding to that by which the particles of coarse sand, etc., are separated from the *Diatomaceæ* (§ 300). The fine powder thus obtained, ought not, when examined under a magnifying power of 200 diameters, to exhibit particles of any appreciable dimensions. The size or gelatine should be of a fine and pure quality, and should be of sufficient strength to form a tolerable firm jelly when cold, whilst quite limpid when warm. It should be strained, whilst hot, through a piece of new flannel; and great care should be taken to preserve it free from dust, which may be best done by putting it into clean jars, and covering its surface with a thin layer of alcohol. The proportion of levigated vermilion to be mixed with it for injection, is about 2 oz. to a pint; and this is to be stirred in the melted size, until the two are thoroughly incorporated, after which the mixture should be strained through muslin.—Although no injections look so well by reflected light as those which are made with vermilion, yet other coloring substances may be advantageously employed for particular purposes. Thus a bright *yellow* is given by the yellow chromate of lead, which is precipitated when a solution of acetate of lead is mixed with a solution of chromate of potass; this is an extremely fine powder, which 'runs' with great facility in an injection, and has the advantage of being very cheaply prepared. The best method of obtaining it is to dissolve 200 grains of acetate of lead and 105 grains of chromate of potass in separate quantities of water, to mix these, and then, after the subsidence of the precipitate, to pour-off the supernatant fluid so as to get-rid of the acetate of potash which it contains, since this is apt to corrode the walls of the vessels if the preparation be kept moist. The solutions should be mixed cold, and the precipitate should not be allowed to dry before being incorporated with the size, four ounces of which will be the proportion appropriate to the quantity of the coloring-substance produced by the above process. The same materials may be used in such a manner that the decomposition takes-place within the vessels themselves, one of the solutions being thrown-in first, and then the other; and this process involves so little trouble or expense, that it may be considered the best for those who are novices in the operation, and who are desirous of perfecting themselves in the practice of the easier methods, before attempting the more costly. By M. Doyère, who first devised this method, it was simply recommended to throw-in saturated solutions of the two salts, one

---

[1] The Kidney of a Sheep or Pig is a very advantageous organ for the learner to practise-on; and he should first master the filling of the vessels from the arterial trunk alone, and then, when he has succeeded in this, he should fill the tubuli uriniferi with white injection, before sending colored injection into the renal artery. The entire systemic circulation of small animals, as Mice, Rats, Frogs, etc., may be injected from the aorta; and the pulmony vessels from the pulmonary artery.

after the other; but Dr. Goadby, who had much experience in the use of it, advised that gelatine should be employed in the proportion of 2 oz. dissolved in 8 oz. of water, to 8 oz. of the saturated solutions of each salt. This method answers very well for the preparations that are to be mounted dry; but for such as are to be preserved in fluid, it is subject to the disadvantage of retaining in the vessels the solution of acetate of potash, which exerts a gradual corrosive action upon them. Dr. Goadby has met this objection, however, by suggesting the substitution of nitrate for acetate of lead; the resulting nitrate of potash having rather a preservative than a corrosive action on the vessels.—When it is desired to inject two or more sets of vessels (as the arteries, veins, and gland-ducts) of the same preparation, different coloring substances should be employed. For a *white* injection, the carbonate of lead (prepared by mixing solutions of acetate of lead and carbonate of soda, and pouring-off the supernatant liquid when the precipitate has fallen) is the best material. No *blue* injections can be much recommended, as they do not reflect light well, so that the vessels filled with them seem almost black; the best is freshly precipitated prussian blue (formed by mixing solutions of persulphate of iron and ferrocyanide of potassium), which, to avoid the alteration of its color by the free alkali of the blood, should be triturated with its own weight of oxalic acid and a litte water, and the mixture should then be combined with size, in the proportion of 146 grains of the former to 4 oz. of the latter.

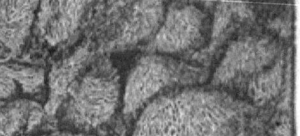

Fig. 479.

Villi of Small Intestine of Monkey.

689. Opaque injections may be preserved either dry or in fluid. The former method is well suited to sections of many solid organs, in which the disposition of the vessels does not sustain much alteration by drying; for the colors of the vessels are displayed with greater brilliancy than by any other method, when such slices, after being well dried, are moistened with turpentine and mounted in Canada balsam. But for such an injection as that shown in Fig. 479, in which the form and disposition of the intestinal *villi* would be completely altered by drying, it is indispensable that the preparation should be mounted in fluid, in a cell deep enough to prevent any pressure on its surface. Either Goadby's solution or weak Spirit answers the purpose very well; or by careful management even such may be mounted in Canada balsam or Dammar.

690. Within the last few years, the art of making *transparent* Injection has been much cultivated, especially in Germany; and beautiful preparations of this description have been sent over from that country in large numbers. The coloring-matter is chiefly employed is Carmine, which is dissolved in liquid ammonia; the solution (after careful filtration) being added in the requisite amount to liquid gelatine.

The following is given by Dr. Carter as a formula for a carmine injection which will run freely through the most minute capillaries, and which will not tint the tissues beyond the vessels themselves, a point of much importance:—Dissolve 60 grains of pure carmine in 120 grains of strong liquor ammoniæ (Pharm. Brit.), and filter if necessary; with this mix thoroughly 1½ oz. of a hot solution

PLATE XXV.

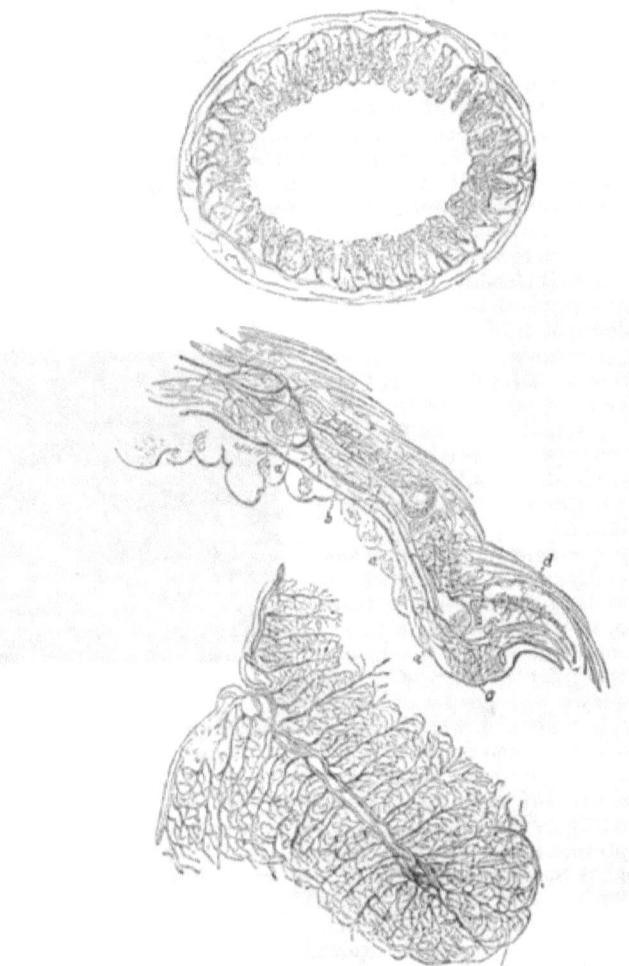

DISTRIBUTION OF CAPILLARY BLOOD-VESSELS AS SHOWN IN TRANSPARENT INJECTIONS (Original).

Fig. 1. Transverse section of Small Intestine of Rat, showing the villi *in situ*.
2. Section of the Toe of a Mouse:—*a, a, a,* tarsal bones; *b,* digital artery; *c,* vascular loops in the papillæ forming the thick epidermic cushion on the under surface, *d,* distribution of vessels in the matrix of the claw.
3. Distribution of Blood-vessels in the cortical of layer of the Brain, showing the manner in which the arteries, carried-in by the pia mater, dip-down into the furrows of the convolutions.

of gelatine (1 to 6 of water); mix another ½ oz. of the gelatine solution with 86 minims of glacial acetic acid; and drop this, little by little, into the solution of carmine, stirring briskly the whole time. After the part has been injected, and has been hardened either by partial drying or by immersion in the Chromic acid solution or in Alcohol, thin sections are cut with a sharp razor; and these are usually dried and mounted in Canada balsam.

Many of these transparent injections (Plate XXV.) are peculiarly well seen under the Binocular Microscope, which shows the capillary network not only in two dimensions (length and breadth), but also in its third dimension, that of its thickness; this is especially interesting in such injections as that (fig. 1) of the villi of the Intestine (seen *in situ* in a transverse section of its tube), a thin section of the Mouse's toe (fig. 2), or the convoluted layer of the Brain (fig. 3). The Stereoscopic effect is best seen, if the light reflected through the object be moderated by a ground-glass, or even by a piece of tissue-paper, placed behind it. —This method, however, does not serve to display anything *well*, save the distribution of the Capillary vessels; the structures they traverse being imperfectly shown. For the purpose of scientific research, therefore, the method followed by Dr. Beale (for full details of which the reader is referred to his Treatise) is much to be preferred.

The material recommended by him for the finest injections is prepared as follows:—Mix 10 drops of the tincture of perchloride of iron (Pharm. Brit.) with 1 oz. of glycerine: and mix 3 grains of ferrocyanide of potassium, previously dissolved in a little water, with another 1 oz. of glycerine. Add the first solution very gradually to the second, shaking them well together; and lastly, add 1 oz. of water, and 3 drops of strong hydrochloric acid. This 'prussian blue fluid,' though not a solution, deposits very little sediment by keeping; and it appears like a solution even when examined under high magnifying powers, in consequence of the minuteness of the particles of the coloring matter. Where a second color is required, a carmine injection may be used, which is to be prepared as follows:—Mix 5 grains of carmine with a few drops of water, and, when they are well incorporated, add about 5 drops of strong liquor ammoniæ. To this dark-red solution add about ½ oz. of glycerine, shaking the bottle so as to mix the two fluids thoroughly; and then very gradually pour in another ½ oz. of glycerine acidulated with 8 or 10 drops of acetic or hydrochloric acid, frequently shaking the bottle. Test the mixture with blue litmus paper; and mix with it another ½ oz. of glycerine, to which a few drops more acid should be added, if the acid reaction of the liquid should not have previously been decided. Finally, add gradually 2 drachms of alcohol previously well mixed with 6 drachms of water, and incorporate the whole by thorough shaking after the addition of each successive portion.

The *staining* process (§ 202) may be combined with the injecting; but Dr. Beale has now come to prefer the following method, when such a combination is desired. An alkaline carmine fluid rather stronger than that ordinarily employed (carmine 15 grs., strong liq. ammoniæ ½ drachm, glycerine 2 oz., alcohol 6 drachms) is first to be injected carefully with very slight pressure; the ammonia having a tendency to soften the walls of the vessels. When they are fully distended, the preparation is to be left for from 12 to 24 hours, in order that time may be allowed for the carmine liquid which has permeated the capillaries, to soak through the different tissues and stain the germinal matter fully. Next a little pure glycerine is to be injected, to get rid of the carmine liquid; and the prussian blue fluid is then to be injected with the utmost care. When the vessels have been fully distended, the injected preparation is to be divided into very small pieces; and these are to be soaked for an hour or two in a mixture of 2 parts of glycerine and 1 of water, and then for three or four days in strong glycerine acidulated with acetic acid (5 drops to 1 oz.). Preparations thus made are best mounted in Glycerine jelly; and may then be examined with the highest powers of the Microscope.

A well-injected preparation should have its vessels completely filled through every part; the particles of the coloring matter should be so closely compacted together, that they should not be distinguishable unless carefully looked for; and there should be no patches of pale uninjected tissue. Still, although the beauty of a specimen, as a Microscopic object, is much impaired by any deficiency in the filling of its vessels, yet to the Anatomist the disposition of the vessels will be as apparent when they are only filled in part, as it is when they are fully distended; and in thin sections mounted as transparent objects, imperfectly injected capillaries may often be better seen than such as have been completely filled.

691. A relation may generally be traced between the disposition of the Capillary vessels, and the functions they subserve; but that relation is obviously (so to speak) of a mechanical kind; the arrangement of the vessels not in any way determining the function, but merely administer-

FIG. 480.

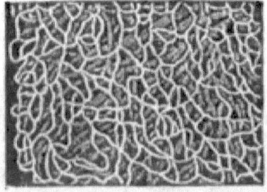

Capillary network around *Fat-cells*.

FIG. 481.

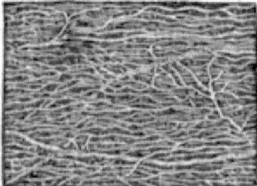

Capillary network of *Muscle*.

ing to it, like the arrangement of water or gas-pipes in a manufactory. Thus in Fig. 480 we see that the capillaries of adipose substance are disposed in a network with rounded meshes, so as to distribute the blood among the Fat-cells (§ 674); whilst in Fig. 481 we see the meshes

FIG. 482.

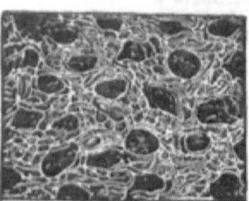

Distribution of Capillaries in *Mucous Membrane*.

FIG. 483.

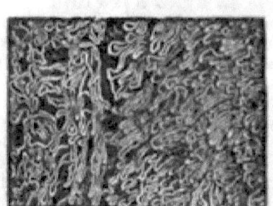

Distribution of Capillaries in *Skin of Finger*.

enormously elongated, so as to permit the Muscular fibres (§ 677) to lie in them. Again, in Fig. 482 we observe the disposition of the Capillaries around the orifices of the follicles of a Mucous membrane; whilst in Fig. 483 we see the looped arrangement which exists in the papillary surface of the Skin, and which is subservient to the nutrition of the epidermis and to the activity of the sensory nerves (§ 682).

692. In no part of the Circulating apparatus, however, does the disposition of the capillaries present more points of interest, than it does in

the Respiratory organs. In Fishes the respiratory surface is formed by an outward extension into fringes of *gills*, each of which consists of an arch with straight laminæ hanging down from it; and every one of these laminæ (Fig. 484) is furnished with a double row of leaflets, which is most minutely supplied with blood-vessels, their network (as seen at A) being so close that its meshes (indicated by the dots in the figure) cover less space than the vessels themselves. The gills of Fish are not ciliated on their surface, like those of Mollusks and of the larva of the Water-Newt; the necessity for such a mode of renewing the fluid in contact with them being superseded by the muscular apparatus with which their gill-chamber is furnished.—But in Reptiles the respiratory surface is formed by the walls of an internal cavity, that of the *lungs:* these organs, however, are constructed on a plan very different from that which they present in higher

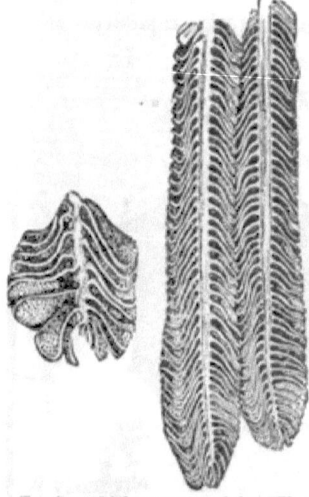

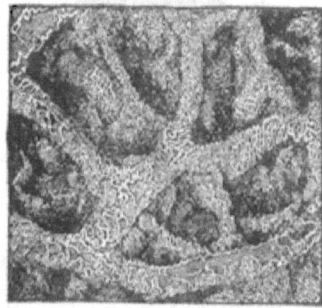

Two branchial processes of the *Gill of the Eel*, showing the branchial lamellæ: —A, portion of one of these processes enlarged, showing the capillary network of the lamellæ.

Interior of upper part of *Lung of Frog*.

Vertebrata, the great extension of surface which is effected in the latter by the minute subdivision of the cavity not being here necessary. In the Frog (for example) the cavity of each lung is undivided; its walls, which are thin and membranous at the lower part, there present a simple smooth expanse; and it is only at the upper part where the extensions of the tracheal cartilage form a network over the interior, that its surface is depressed into sacculi, whose lining is crowded with blood-vessels (Fig. 485). In this manner a set of air-cells is formed in the thickness of the upper wall of the lung, which communicate with the general cavity, and very much increase the surface over which the blood comes into relation with the air; but each air-cell has a capillary network of its own, which lies on one side against its wall, so as only to be exposed to the air on its free surface. In the elongated lung of the Snake the

same general arrangement prevails; but the cartilaginous reticulation of its upper part projects much further into the cavity, and incloses in its meshes (which are usually square, or nearly so) several layers of air-cells, which communicate, one through another, with the general cavity.
—The structure of the lungs of Birds presents us with an arrangement of a very different kind, the purpose of which is to expose a very large amount of capillary surface to the influence of the air. The entire mass of each lung may be considered as subdivided into an immense number of 'lobules' or 'lunglets' (Fig. 486, B), each of which has its own bronchial tube (or subdivision of the windpipe), and its own system of blood-vessels, which have very little communication with those of other lobules. Each lobule has a central cavity, which closely resembles that of a Frog's lung in miniature, having its walls strengthened by a network of cartilage derived from the bronchial tube, A, in the interspaces of which are openings leading to sacculi in their substance. But each of these cavities is surrounded by a solid plexus of blood-vessels, which does not seem to be covered by any limiting membrane, but which admits air from the central cavity freely between its meshes; and thus its capillaries are in

Fig. 486.

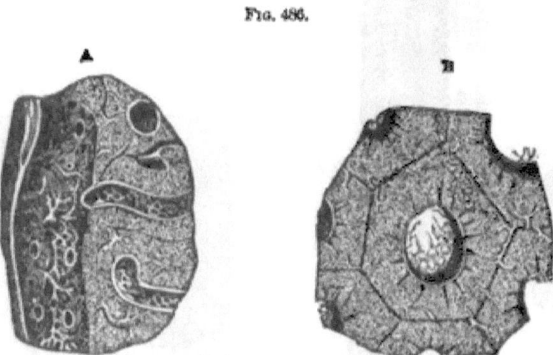

Interior structure of *Lung of Fowl*, as displayed by a section, A, passing in the direction of a bronchial tube, and by another section, B, cutting it across.

immediate relation with air on all sides, a provision that is obviously very favorable to the complete and rapid aëration of the blood they contain.
—In the lung of Man and Mammals, again, the plan of structure differs from the foregoing, though the general effect of it is the same. For its whole interior is divided-up into minute air-cells, which freely communicate with each other, and with the ultimate ramifications of the air-tubes into which the trachea subdivides; and the network of blood-vessels (Fig. 487) is so disposed in the partitions between these cavities, that the blood is exposed to the air on both sides. It has been calculated that the number of these air-cells grouped around the termination of each air-tube in Man is not less than 18,000; and that the total number in the entire lungs is *six hundred millions*.

693. The following list of the parts of the bodies of Vertebrata, of which injected preparations are most interesting as Microscopic objects, may be of service to those who may be inclined to apply themselves to their production.—*Alimentary Canal;* stomach, showing the orifices of the gastric follicles, and the rudimentary villi near the pylorus; small intestine, showing the villi and the orifices of the follicles of Lieberkühn,

and at its lower part the Peyerian glands ; large intestine, showing the various glandular follicles :—*Respiratory Organs ;* lungs of **Mammals, Birds,** and **Reptiles** ; gills and swimming-bladder of **fish** ; *Glandular Organs;* liver, gall-bladder, **kidney,** parotid :—*Generative Organs;* ovary of Toad ; oviduct of Bird and Frog ; Mammalian placenta ; uterine and fœtal cotyledons of Ruminants :—*Organs of Sense ;* retina, iris, choroid,

FIG. 487.

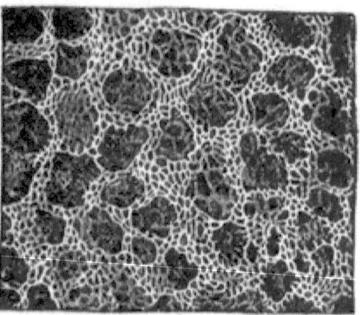

Arrangement of the Capillaries on the walls of the Air-cells of the *Human Lung.*

and ciliary processes of eye, pupillary membrane of fœtus ; papillæ of tongue ; mucous membrane of nose, papillæ of skin or finger ; *Tegumentary Organs ;* skin of different parts, hairy and smooth, with vertical sections showing the vessels of the hair-follicles, sebaceous glands, and papillæ; matrix of nails, hoofs, etc. :—*Tissues;* fibrous, muscular, adipose, sheath of tendon :—*Nervous Centres ;* sections of brain and spinal cord.

CHAPTER XXI.

APPLICATION OF THE MICROSCOPE TO GEOLOGICAL INVESTIGATION.

694. THE utility of the Microscope is by no means limited to the determination of the structure and actions of the Organized beings at present living on the surface of the Earth; for a vast amount of information is afforded by its means to the Geological inquirer, not only with regard to the minute characters of the many Vegetable and Animal remains that are entombed in the successive strata of which its crust is composed, but also with regard to the essential nature and composition of many of those strata themselves.—We cannot have a better example of its value in both these respects, than that which is afforded by the results of Microscopic examination of *lignite* or fossilized wood, and of ordinary *coal*, which we now assuredly know to be a product of the decay of wood.

695. Specimens of *fossilized wood*, in a state of more or less complete preservation, are found in numerous strata of very different ages,—more frequently, of course, in those whose materials were directly furnished by the dry land, and were deposited in its immediate proximity, than in those which were formed by the deposition of sediments at the bottom of a deep ocean. Generally speaking, it is only when the wood is found to have been penetrated by *silex*, that its organic structure is well preserved; but instances occur every now and then, in which penetration by *carbonate of lime* has proved equally favorable. In either case, transparent sections are needed for the full display of the organization; but such sections, though made with great facility when lime is the fossilizing material, require much labor and skill when silex has to be dealt with. Occasionally, however, it has happened that the infiltration has filled the cavities of the cells and vessels, without consolidating their walls; and as the latter have undergone decay without being replaced by any cementing material, the lignite, thus composed of the internal 'casts' of the woody tissues, is very friable, its fibres separating from each other like those of asbestos; and laminæ split asunder with a knife, or isolated fibres separated by rubbing-down between the fingers, exhibit the characters of the woody structure extremely well, when mounted in Canada balsam.—Generally speaking, the lignites of the Tertiary strata present a tolerably close resemblance to the woods of the existing period: thus the ordinary structure of *dicotyledonous* and *monocotyledonous* stems may be discovered in such lignites in the utmost perfection; and the peculiar modification presented by *coniferous* wood is also most distinctly exhibited (Fig. 259). As we go back, however, through the strata of the Secondary period, we more and more rarely meet with the ordinary dicotyledonous structure

and the lignites of the earliest deposits of these series are, almost universally, either Gymnosperms[1] or Palms.

696. Descending into the Palæozoic series, we are presented in the vast *coal* formations of our own and other countries with an extraordinary proof of the prevalence of a most luxuriant vegetation in a comparatively-early period of the world's history; and the Microscope lends the Geologist essential assistance, not only in determining the nature of much of that vegetation, but also in demonstrating (what had been suspected on other grounds) that Coal itself is nothing else than a mass of decomposed vegetable matter, derived from the decay of an ancient vegetation. The determination of the characters of the *Ferns, Sigillariæ, Lepidodendra, Calamites*, and other kinds of vegetation whose forms are preserved in the shales or sandstones that are interposed between the strata of Coal, has been hitherto chiefly based on their external characters; since it is seldom that these specimens present any such traces of minute internal structure as can be subjected to Microscopic elucidation. But persevering search has recently brought to light numerous examples of Coal-plants, whose internal structure is sufficiently well preserved to allow of its being studied microscopically: and the careful researches of Prof. W. C. Williamson have shown that they formed a series of connecting links between Cryptogamia and Flowering plants; being obviously allied to *Equisetaceæ, Lycopodiaceæ*, etc., in the character of their fructifications whilst their stem-structure foreshadowed both the 'endogenous' and 'exogenous' types of the latter.[2] Notwithstanding the general absence of any definite *form* in the masses of decomposed wood of which Coal itself consists (these having apparently been reduced to a pulpy state by decay, before the process of consolidation by pressure, aided perhaps by heat, commenced), the traces of *structure* revealed by the Microscope are often sufficient—especially in the ordinary 'bituminous' coal—not only to determine its vegetable origin, but in some cases to justify the Botanist in assigning the character of the vegetation from which it must have been derived; and even where the stems and leaves are represented by nothing else than a structureless mass of black carbonaceous matter, there are found diffused through this a multitude of minute resinoid yellowish-brown granules, which are sometimes aggregated in clusters and inclosed in sacculi; and these may now be pretty certainly affirmed to represent the *spores*, while the sacculi represent the *sporangia*, of gigantic *Lycopodiaceæ* (§ 347) of the Carboniferous Flora. The larger the proportion of these granules, the brighter and stronger is the flame with which the coal burns; thus in some blazing *cannel*-coals they abound to such a degree as to make up the greater proportion of their substance; whilst in *anthracite* or 'stone-coal,' the want of them is shown by its dull and slow combustion. It is curious that the dispersion of these resinoid granules through the black carbonaceous matter is sometimes so regular, as to give to transparent sections very much the aspect of a section of vegetable cellular tissue, for which they have been mistaken even by experienced microscopists; but this resemblance disappears under a more extended scrutiny, which shows it to be altogether accidental.

697. In examining the structure of coal, various methods may be fol-

---

[1] Under this head are included the *Cycadeæ*, along with the ordinary *Coniferæ* or pine and fir tribe.
[2] See his succession of Memoirs on the Coal-plants, in the recent volumes of the "Philosophical Transactions."

lowed. Of those kinds which have sufficient tenacity, thin sections may be made; but the opacity of the substance requires that such sections should be ground extremely thin before they become transparent; and its friability renders this process one of great difficulty. It may, however, be facilitated by using Marine Glue, instead of Canada balsam, as the cement for attaching the smoothed surface of the coal to the slip of glass on which it is rubbed-down. Another method is recommended by the authors of the "Micrographic Dictionary" (2d edit., p. 178):—"The coal is macerated for about a week in a solution of carbonate of potass; at the end of that time, it is possible to cut tolerably thin slices with a razor. These slices are then placed in a watch glass with strong nitric acid, covered, and gently heated; they soon turn brownish, then yellow, when the process must be arrested by dropping the whole into a saucer of cold water, or else the coal would be dissolved. The slices thus treated appear of a darkish amber-color, very transparent, and exhibit the structure, when existing, most clearly. We have obtained longitudinal and transverse sections of Coniferous wood from various coals in this way. The specimens are best preserved in glycerine, in cells; we find that spirit renders them opaque, and even Canada balsam has the same defect."—When the coal is so friable that no sections can be made of it by either of these methods, it may be ground to fine powder, and the particles may then, after being mounted in Canada balsam, be subjected to Microscopic examination: the results which this method affords are by no means satisfactory in themselves, but they will often enable the organic structure to be sufficiently determined, by the comparison of the appearances presented by such fragments with those which are more distinctly exhibited elsewhere. Valuable information may often be obtained, too, by treating the ash of an ordinary coal-fire in the same manner, or (still better) by burning to a white ash a specimen of coal that has been previously boiled in nitric acid, and then carefully mounting the ash in Canada balsam; for mineral 'casts' of vegetable cells and fibres may often be distinctly recognized in such ash; and such casts are not unfrequently best afforded by samples of coal in which the method of section is least successful in bringing to light the traces of organic structure, as is the case, for example, with the anthracite of Wales.

698. Passing on now to the Animal kingdon, we shall first cite some parallel cases in which the essential nature of deposits that form a very important part of the Earth's crust, has been determinined by the assistance of the Microscope; and shall then select a few examples of the most important contributions which it has afforded to our acquaintance with types of Animal life long since extinct.—It is an admitted rule in Geological science, that the past history of the Earth is to be interpreted, so far as may be found possible, by the study of the changes which are still going on. Thus, when we meet with an extensive stratum of fossilized *Diatomaceæ* (§ 299) in what is now dry land, we can entertain no doubt that this siliceous deposit originally accumulated either at the bottom of a fresh-water lake or beneath the waters of the ocean; just as such deposits are formed at the present time by the production and death of successive generations of these bodies, whose indestructible casings accumulate in the lapse of ages, so as to form layers whose thickness is only limited by the time during which this process has been in action (§ 298). In like manner, when we meet with a Limestone-rock entirely composed of the calcareous shells of *Foraminifera*, some of them entire, others broken-up into minute particles (as in the case of the *Fusulina*-limestone

of the Carboniferous period, § 485, and the *Nummulitic* limestone of the Eocene, § 489), we interpret the phenomenon by the fact that the dredgings obtained from certain parts of the ocean-bottom consist almost entirely of remains of existing Foraminifera, in which entire shells, the animals of which may be yet alive, are mingled with the *débris* of others that have been reduced by the action of the waves to a fragmentary state. Such a deposit consisting chiefly of *Orbitolites*, § 466, is at present in the act of formation on certain parts of the shores of Australia, as the Author was informed by Mr. J. Beete Jukes; thus affording the exact parallel to the stratum of Orbitolites (belonging, as his own investigations have led him to believe, to the very same species), that forms part of the 'calcaire grossier' of the Paris basin. So in the fine white mud which is brought up from almost every part of the sea-bottom of the Levant, where it forms the stratum that is continually undergoing a slow but steady increase in thickness, the Microscopic researches of Prof. Williamson[1] have shown, not only that it contains multitudes of minute remains of living organisms, both Animal and Vegetable, but that it is **entirely or** almost wholly composed of such remains. Amongst these were about 26 species of Diatomaceæ (siliceous), 8 species of Foraminifera (calcareous), and a miscellaneous group of objects (Fig. 488), consisting of calcareous and siliceous spicules of Sponges and Gorgoniæ, and fragments of the calcareous skeletons of Echinoderms and Mollusks. A

Fig. 488.

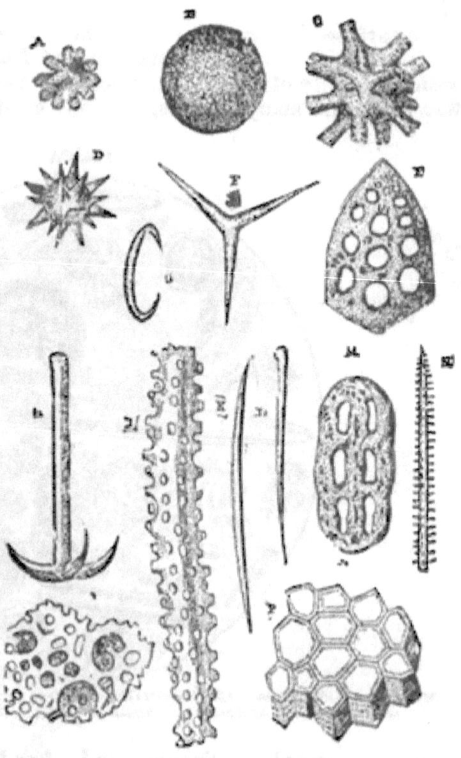

Microscopic Organisms in *Levant Mud:*—A, D, siliceous spicules of *Tethya;* B, H, spicules of *Geodia;* c, sponge-spicule (unknown); E, calcareous spicule of *Grantia;* F, G, M, O, portions of calcareous skeleton of *Echinodermata;* H, I, calcareous spicule of *Gorgonia;* K, L, N, siliceous spicules of *Halichondria;* P, portion of prismatic layer of shell of *Pinna.*

collection of forms strongly resembling that of the Levant mud, with the exception of the siliceous Diatomaceæ, is found in many parts of the 'calcaire grossier' of the Paris basin, as well as in other extensive deposits of the same early Tertiary period.

699. It is, however, in regard to the great *Chalk* Formation, that the information afforded by the Microscope has been most valuable

---

[1] "Memoirs of the Manchester Literary and Philosophical Society," Vol. vii.
20

Mention has already been made (§ 480) of the fact that a large proportion of the North Atlantic sea-bed has been found to be covered with an 'ooze' chiefly formed of the shells of *Globigerinæ;* and this fact, first determined by the examination of the small quantities brought up by the sounding apparatus, has been fully confirmed by the results of the recent exploration of the Deep-sea with the dredge; which, bringing up half a ton of this deposit at once, has shown that it is not a mere surface-film, but an enormous mass whose thickness cannot be even guessed at. "Under the Microscope," says Prof. Wyville Thomson[1] of a sample of 1½ cwt. obtained by the dredge from a depth of nearly three miles, "the surface-layer was found to consist chiefly of entire shells of *Globigerina bulloides,* large and small, and of fragments of such shells mixed with a quantity of amorphous calcareous matter in fine particles, a little fine sand, and many spicules, portions of spicules, and shells of Radio-

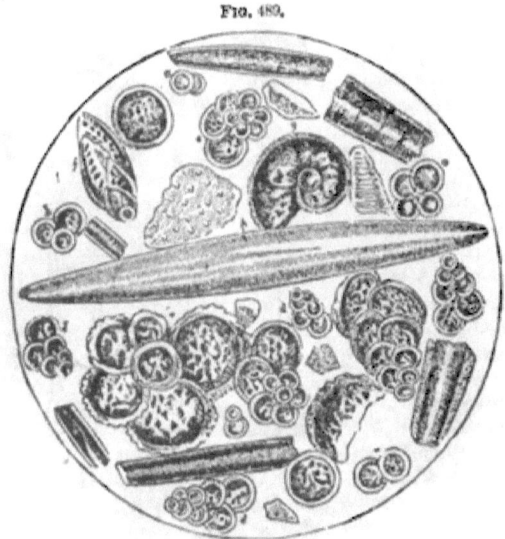

FIG. 489.

Microscopic Organisms in *Chalk* from Gravesend: *a, b, c, d,* Textularia globulosa; *e, e, e,* Rotalia aspera; *f,* Textularia aculeata; *g,* Planularia hexas; *h,* Navicula.

laria, a few spicules of Sponges, and a few frustules of Diatoms. Below the surface-layer the sediment becomes gradually more compact, and a slight gray color, due, probably, to the decomposing organic matter, becomes more pronounced, while perfect shells of Globigerina almost disappear, fragments become smaller, and calcareous mud, structureless, and in a fine state of division, is in greatly preponderating proportion. One can have no doubt, on examining this sediment, that it is formed in the main by the accumulation and disintegration of the shells of Globigerina; the shells fresh, whole, and living, in the surface-layer of the deposit; and in the lower layers dead, and gradually crumbling down by the decomposition of their organic cement, and by the pressure of the layers above." This white calcareous mud also contains in large amount

---

[1] "The Depths of the Sea," p. 410.

the 'coccoliths' and 'coccospheres' formerly described (§ 409).—Now the resemblance which this Globigerina-mud, when dried, bears to Chalk, is so close as at once to suggest the similar origin of the latter, and this is fully confirmed by Microscopic examination. For many samples of it consist in great part of the minuter kinds of Foraminifera, especially *Globigerinæ* (Figs. 489, 490), whose shells are imbedded in a mass of apparently amorphous particles, many of which, nevertheless, present indications of being the worn fragments of similar shells, or of larger calcareous organisms. In the Chalk of some localities, the disintegrated prisms of *Pinna* (§ 563), or of other large shells of the like structure (as *Inoceramus*), form the great bulk of the recognizable components; whilst, in other cases, again, the chief part is made up of the shells of *Cytherina*, a marine form of Entomostracous Crustacean (§ 604). Different specimens of Chalk vary greatly in the proportion which the distinctly organic remains bear to the amorphous residuum, and which

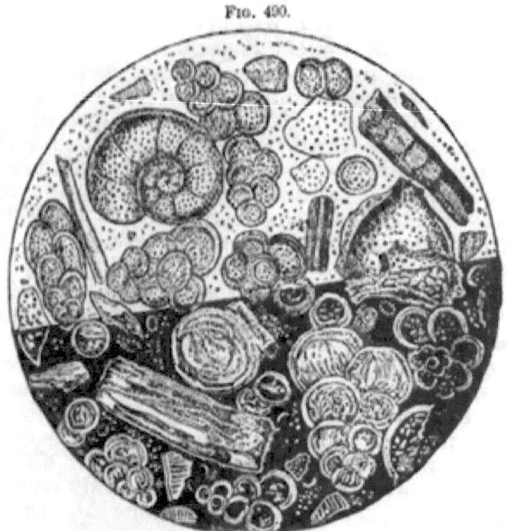

Fig. 490.

Microscopic Organisms in *Chalk* from Meudon; seen partly as opaque, and partly as transparent objects.

the different kinds of the former bear to each other; and this is quite what might be anticipated, when we bear in mind the predominance of one or another tribe of Animals in the several parts of a large area; but it may be fairly concluded from what has been already stated of the amorphous component of the Globigerina-mud, that the amorphous constituent of Chalk likewise is the disintegrated residuum of Foraminiferal shells. —But further, the Globigerina-mud now in process of formation is in some places literally crowded with Sponges having a complete *siliceous* skeleton (§ 511); and some of them bear such an extraordinarily close resemblance, alike in structure and in external form, to the *Ventriculites* which are well known as Chalk-fossils, as to leave no reasonable doubt that these also lived as siliceous sponges on the bottom of the Cretaceous sea. Other sponges, also, are found in the Globigerina-mud, the structure of whose horny skeleton corresponds so closely with the sponge-tissues which can be recognized in sections of nodular Flints, Agates,[1] etc., as to make it clear—when taken in connection with correspondence of external form—that such flints are really fossilized sponges, the silicifying material having been furnished by the solution of the skeletons of the siliceous sponges, or of deposits of Diatoms or Radiolaria. Further, in many sections of Flints there are found minute bodies termed *Xanthidia*, which bear a strong resemblance to the sporangia of certain *Desmidiaceæ* (Fig. 158, D); and the Author has found similar bodies in the midst of what appears to be sponge-tissue imbedded in the Globigerina-mud. And (as was first pointed out by Mr. Sorby) the coccoliths and coccospheres at present found on the sea-bottom (§ 409), are often to be discovered by the Microscopic examination of Chalk.[2] All these correspondences show that the formation of Chalk took place under conditions essentially similar to those under which the deposit of Globigerina-mud is being formed over the Atlantic sea-bed at the present time.

700. **In examining Chalk or other similar mixed aggregation, whose component particles are easily separable from each other, it is desirable to separate, with as little trouble as possible, the larger and more definitely organized bodies from the minute amorphous particles; and the mode of doing this will depend upon whether we are operating upon the large or upon the small scale. If the former, a quantity of soft Chalk should be rubbed to powder with water, by means of a soft brush; and this water should then be proceeded with according to the method of levigation already directed for separating the Diatomaceæ (§ 300). It will usually be found that the first deposits contain the larger Foraminifera, fragments of Shell, etc., and that the smaller Foraminifera and Sponge-spicules fall next; the fine amorphous particles remaining diffused through the water after it has been standing for some time, so that they may be poured-away. The organisms thus separated should be dried and mounted in Canada balsam.—If the smaller scale of preparation be preferred, as much Chalk scraped fine as will lie on the point of a knife is to be laid on a drop of water on the glass slide, and allowed to remain there for a few seconds; the water, with any particles still floating on it, should then be removed; and the sediment left on the glass should be dried and mounted in Balsam.**—For examining the structure of **Flints, such chips as may be obtained** with a hammer will commonly **serve very well**: a clear translucent flint being first selected, and the **chips that are obtained** being soaked for a short time in turpentine (which increases their transparence), those which show **organic structure**, whether Sponge-tissue or Xanthidia, are to **be selected and** mounted **in Canada** balsam. The most perfect specimens of Sponge-structure, **however, are** only to be obtained by slicing and polishing,—a **process** which **is best** performed by the lapidary.

701. There are various other deposits, of less extent and importance than the great Chalk-formation, which are, like it, composed in great part of Microscopic organisms, chiefly minute Foraminifera; and the presence of animals of this group may be largely recognized, by the assistance of this instrument, in sections of Calcareous rocks of various dates, whose other materials were fragments of Corals, Encrinite-stems,

---

[1] See Dr. Bowerbank's Memoirs in the "Trans. of the Geolog. Society," 1840, and in the "Ann. of Nat. Hist.," 1st Ser., Vols. vii., x.
[2] On the Organic origin of the so-called "Crystalloids" of Chalk; in "Ann. of Nat. Hist.," Ser. 3, Vol. viii. (1861), pp. 193-200.

or the shells of Mollusks. In the formation of the Coralline Crag' (Tertiary) of the eastern coast of England, *Polyzoaries* (§ 548) had the greatest share; but the Tertiary limestone of which Paris is chiefly built consists almost exclusively of the shells of *Miliolida* (§ 462), and is thus known as Miliolite (millet-seed) limestone. In the vast stratum of Numulitic limestone (Fig. 333), which was formed at the commencement of the Tertiary period, the Microscope enables us to see that the *matrix* in which the large entire Nummulites are imbedded, is itself composed of comminuted fragments and young of the same, together with minuter Foraminifera. In the Oolitic (Secondary) formation, again, there are many beds which are shown by the Microscope to have been chiefly composed of Foraminiferal shells; and in those portions which exhibit the 'roe-stone' arrangement from which the rock derives its name (such as is beautifully displayed in many specimens of Bath-stone and Portland-stone), it is found by Microscopic examination of transparent sections, that each rounded concretion is composed of a series of concentric spheres formed by successive calcareous deposits upon a central nucleus, which nucleus is often a Foraminiferal shell. In these and similar calcareous formations, the entire materials of which were obviously furnished by the accumulation of animal remains, it not unfrequently happens that all traces of their origin are obliterated by local 'metamorphic' action usually dependent upon neighboring Volcanic heat; and thus a crystalline marble, whose particles present not the least evidence of organic arrangement, may have been formed by the metamorphosis of Chalky, Oolitic, or Nummulitic limestone. Now there is very strong evidence that the vast mass of sub-crystalline 'Carboniferous' limestone, which forms our coal-basins, has had a similar origin in Foraminiferal and Zoophytic life; the traces of which have been for the most part removed by the metamorphic action involved in its upheaval. For where it has sustained but little disturbance, the evidences of its organic (chiefly Foraminiferal) origin are unmistakable. Thus in the great plains of Russia, there are certain bands of limestone of this epoch, varying in thickness from fifteen inches to five feet, and frequently repeated through a vertical depth of two hundred feet over very wide areas, which are almost entirely composed of the extinct genus *Fusulina* (Fig. 331). Again, those parts of the Carboniferous limestone of Ireland which have undergone least disturbance, can be plainly shown, by the examination of Microscopic sections, to consist of the remains of Foraminifera, Polyzoa, fragments of Coral, etc. And where, as not unfrequently happens, beds of this limestone are separated by clay seams, these are found to be loaded with 'Microzoa' of various kinds, particularly *Foraminifera* (of which the *Saccamina*, Fig. 319, a, has come down to the present time), and the beautiful *Polyzoaries* known as 'lace-corals.'

702. Mention has been already made (§ 487 *note*) of Prof. Ehrenberg's very remarkable discovery, that a large proportion (to say the least) of the *green sands* which present themselves in various stratified deposits, from the Silurian epoch to the Tertiary period, and which in certain localities constitute what is known as *the* Greensand formation (beneath the Chalk), is composed of the *casts* of the interior of minute shells of Foraminifera and Mollusca, the shells themselves having entirely disappeared. The mineral material of these casts has not merely filled the chambers and their communicating passages (Fig. 328, A, B), but has also penetrated, even to its minutest ramifications, the canal-system of the intermediate skeleton (Figs. 332, 337). The precise parallel to these

deposits presents itself in certain spots of the existing sea-bottom, such as the Agulhas bank near the Cape of Good Hope; where the dredge comes up laden with a green sand, which, on microscopic examination, proves to consist almost entirely of 'internal casts' of existing Foraminifera, that must have been formed by the chemical replacement of their protoplasmic bodies by ferruginous silicates precipitated from the Seawater. And this fact gives the clue to the interpretation of the conditions under which the 'Eozoic Limestone' of Canada (§ 497), formed on the sea-bottom of the Laurentian epoch by the extension of continuous Foraminiferal growth resembling a Coral reef, became interpenetrated with a like deposit of green silicate of magnesia (serpentine), of whose presence in large amount in the sea-water of that period there is ample evidence.—The determination of the organic nature of this Serpentinelimestone, which is one of the lowest members of a series of strata so far below those in which organic remains had previously been detected, that, to use the words of Sir William Logan, the appearance of the so-called 'Primordial Fauna' is a comparatively modern event,—may be regarded as the most remarkable achievement of Microscopic inquiry as applied to Geology.

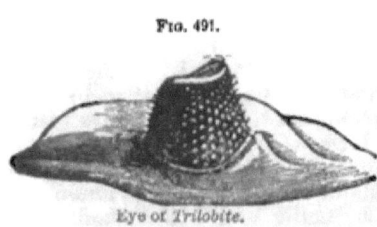

Fig. 491.

Eye of Trilobite.

703. It is obvious that, under ordinary circumstances, only the *hard* parts of the bodies of Animals that have been entombed in the depths of the earth are likely to be preserved; but from these a vast amount of information may be drawn; and the inspection of a microscopic fragment will often reveal, with the utmost certainty, the entire nature of the organism of which it formed part. Minute fragments of the tests or spines of all Echinodermata, and of all such Molluscous shells as present distinct appearances of structure (this being especially the case with the Brachiopods, and with certain families of Lamellibranchiate bivalves), may be unerringly identified by its means, when the external forms of these fragments would give no assistance whatever.—In the study of the important ancient group of *Trilobites*, not only does a Microscopic examination of the 'casts' which have been preserved of the surface of their Eyes (Fig. 491) serve to show the entire conformity in the structure of these organs to the 'composite' type which is so remarkable a characteristic of the higher Articulata (§ 626), but it also brings to light certain peculiarities which help to determine the division of the great Crustacean series with which this group has most alliance.'

704. It is, however, in the case of the Teeth, the Bones, and the Dermal skeleton of Vertebrated animals, that the value of Microscopic inquiry becomes most apparent; since their structure presents so many characteristics which are subject to well-marked variations in their several Classes, Orders, and Families, that a knowledge of these characters frequently enables the Microscopist to determine the nature of even the most fragmentary specimens, with a positiveness which must appear altogether misplaced to such as have not studied the evidence. It was in regard to *teeth*, that the possibility such determinations was first made

---

' See Prof. Burmeister "On the Organization of the Trilobites," published by the Ray Society, p. 19.

clear by the laborious researches of Prof. Owen;[1] and the following may be given as examples of their value:—A rock-formation extends over many parts of Russia, whose mineral characters might justify its being likened either to the *Old* or to the *New* Red sandstone of this country, and whose position relatively to other strata is such that there is great difficulty in obtaining evidence from the usual sources as to its place in the series. Hence the only hope of settling this question (which was one of great practical importance,—since, if the formation were *new* Red, Coal might be expected to underlie it, whilst if *old* Red, no reasonable hope of Coal could be entertained) lay in the determination of the Organic remains which this stratum might yield; but unfortunately these were few and fragmentary, consisting chiefly of teeth which are seldom perfectly preserved. From the gigantic size of these teeth, together with their form, it was at first inferred that they belonged to Saurian Reptiles, in which case the Sandstone would have been considered as New Red; but Microscopic examination of their intimate structure unmistakably proved them to belong to a genus of Fishes (*Dendrodus*) which is exclusively Palæozoic, and thus decided that the formation must be Old Red. — So again, the Microscopic examination of certain fragments of teeth found in a sandstance of Warwickshire, disclosed a most remarkable type of tooth-structure (shown in Fig. 492), which was also ascertained to exist in certain teeth that had been discovered in the 'Keupersandstein' of Wurtemberg; and the identity or close resemblance of the animals to which these teeth belonged having been thus established, it became

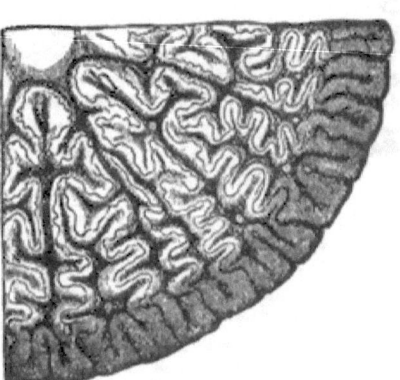

Fig. 492.

Section of Tooth of *Labyrinthodon*.

almost certain that the Warwickshire and Wurtemberg sandstones were equivalent formations, a point of much Geological importance. The next question arising out of this discovery, was the nature of the animal (provisionally termed *Labyrinthodon*, a name expressive of the most peculiar feature in its dental structure) to which these teeth belonged. They had been referred, from external characters merely, to the order of Saurian Reptiles; but it is now clear that they were gigantic Salamandroid *Amphibia*, having many points of relationship to *Ceratodus* (the Australian 'mud-fish'), which shows a similar though simpler dental organization.

705. The researches of Prof. Quekett on the minute structure of *bone*[2] have shown that from the average size and form of the lacunæ, their disposition in regard to each other and to the Haversian canals, and the

---

[1] See his magnificent "Odontography."
[2] See his Memoir on the 'Comparative Structure of Bone,' in the "Transact. of the Microsc. Soc.," Ser. 1, Vol. ii.; and the "Catalogue of the Histological Museum of the Roy. Coll. of Surgeons," Vol. ii.

number and course of the canaliculi (§ 653), the nature of even a minute fragment of Bone may often be determined with a considerable approach to certainty; as in the following examples, among many which might be cited:—Dr. Falconer, the distinguished investigator of the fossil remains of the Himalayan region, and the discoverer of the gigantic fossil Tortoise of the Sivalik hills, having met with certain small bones about which he was doubtful, placed them for minute examination in the hands of Prof. Quekett, who informed him, on Microscopic evidence, that they might certainly be pronounced Reptilian, and probably belonged to an animal of the Tortoise tribe : and this determination was fully borne-out by other evidence, which led Dr. Falconer to conclude that they were toe-bones of his great Tortoise.—Some fragments of Bone were found, many years since, in a Chalk-pit, which were considered by Prof. Owen to have formed part of the wing-bones of a long-winged sea-bird allied to the Albatross. This determination, founded solely on considerations derived from the very imperfectly-preserved external forms of these fragments, was called in question by some other Palæontologists; who thought it more probable that these bones belonged to a large species of the extinct genus *Pterodactylus*, a flying lizard whose wing was extended upon a single immensely-prolonged digit. No species of Pterodactyle, however, at all comparable to this in dimensions, was at that time known; and the characters furnished by the configuration of the bones not being in any degree decisive, the question would have long remained unsettled, had not an appeal been made to the Microscopic test. This appeal was so decisive, by showing that the minute structure of the bone in question corresponded exactly with that of Pterodactyle bone, and differed essentially from that of every known Bird, that no one who placed much reliance upon that evidence could entertain the slightest doubt on the matter. By Prof. Owen, however, the validity of that determination was questioned, and the bone was still maintained to be that of a Bird; until the question was finally set at rest, and the value of the Microscopic test triumphantly confirmed, by the discovery of undoubted Pterodactyle bones of corresponding and even of greater dimensions, in the same and other Chalk quarries.

706. The application of the Microscope to Geology is not, however, limited to the discovery or determination of Organic structure; for, as has been now satisfactorily demonstrated, very important information may be acquired by its means respecting the mineral composition of Rocks, and the mode of their formation. The Microscopic examination of the sediments now in course of deposition on various parts of the great Oceanic area, and especially of the large number of samples brought up in the 'Challenger' soundings, has led to this very remarkable conclusion,—that the *débris* resulting from the degradation of Continental land-masses are not carried far from their shores, being *entirely absent* from the bottom of the deep Ocean-basins. The sediments *there* found, where not of Organic origin, mainly consist of volcanic *sands* and *ashes*, which are found in Volcanic areas, and of *clay* that seems to have been produced by the disintegration of masses of pumice (vesicular lava), which, after long floating, and dispersion by surface-drift or ocean currents, have become water-logged and have sunk to the bottom. As no ordinary siliceous sand is found anywhere save in the neighborhood of Continents and Continental islands, and as all Oceanic islands are the products of local Volcanic outbursts, this absence of all trace of submerged Continental land over the great Oceanic area, affords strong confirmation

to the belief which Geological evidence has been gradually tending to establish, that the sedimentary rocks which form the existing land, were deposited in the immediate neighborhood of pre-existing land, whose degradation furnished their materials; and consequently that the *original* disposition of the great Continental and Oceanic areas was not very different from what it now is.[1] Further, the microscopic examination of these Oceanic sediments reveals the presence of extremely minute particles, which seem to correspond in composition to *meteorites*, and which there is strong reason for regarding as 'cosmic dust' pervading the interplanetary spaces.—Thus the application of the Microscope to the study of these deposits, brings us in contact with the greatest questions not only of Terrestrial but also of Cosmical Physics; and furnishes evidence of the highest value for their solution.

707. The application of the Microscope to the determination of the materials of the sediments now in process of deposition on the Ocean-bottom, leads us to another great department of Microscopic inquiry now being extensively prosecuted,—namely, *Microscopic Petrology*, or the study of the Mineral materials and Physical structure of Rocks. For although the Geologist has no difficulty in determining by his unaided eye, with the use of simple chemical tests, the mineral composition of rocks of coarse texture, and in distinguishing the fragments of previously existing rocks of which they have been built-up, the case is different with those of extremely fine grain, still more with such as present an apparently homogeneous, compact, and glassy character. For it is only by the microscopic study of these, that any trustworthy conclusions can be arrived at in regard to the mode in which they have originated, and the changes they have subsequently undergone; and such study often reveals facts of the most unexpected kind and the most striking significance.— Thus, many compact *sedimentary* rocks, whose homogeneous appearance to the eye or the hand-magnifier gives no clue to their origin, are found, when thin sections of them are examined microscopically, to be aggregations of minute rounded and water-worn grains (often less than 1-1000th of an inch in diameter) of Quartz, Felspar, Mica, soft and hard Clays, Clayslate, Oxide of Iron, Iron-pyrites, Carbonate of Lime, fragments of fossil Organisms, etc., arranged without any trace of decided structure or crystallization. In rocks exhibiting *slaty cleavage*, again, the direction in which the pressure has been applied is indicated in a microscopic section by the elongation or flattening out of some of the particles, with a sliding movement of others. In regard to *eruptive* or *igneous* rocks, on the other hand, the results of microscopic examination enable it to be stated that whether possessing the hardest and most compact substance, and presenting the most homogeneous and even glassy aspect, or existing under the form of the softest and finest powder (like the dust-ash of volcanoes), the rocks of this class are characterized—as a rule—by the minutely-crystalline character of their mineral conponents; and this even when their vitrification seems to the eye so complete, as to forbid the expectation of any such recognition. And in this manner a clue is obtained to the *sources* of these rocks; which (there is now strong reason to believe) have been formrd for the most part, if not universally, by the melting-down of the rocks pre-existing in the neighborhood, and not ejected (as according to the older theory) from the general molten inte-

---

[1] See Prof. Geikie's Lecture on 'Geographical Evolution,' in the "Proceedings of the Royal Geograpical Society," July, 1879.

rior of the earth.—Again, we are often enabled by the same means to trace-out the 'metamorphic' action by which one kind of rock has been converted into another subsequently to its first deposition as a sediment. Of this the change of a calcareous deposit made-up of the remains of Foraminifera with fragments of shells, corals, etc., into a crystalline Limestone, is one of the most common; occurring wherever the rock has been subjected to pressure and contortion, and especially in the near neighborhood of igneous outbursts. And there can now be little hesitation in attributing much of this conversion to the solvent action of water raised to a very high temperature under enormous pressure. A very curious piece of evidence, moreover, has now been furnished by Microscopic study, in support of the doctrine which other considerations render probable, that *some* forms of Granite (to say the least) have been generated from sedimentary rocks by metamorphic agency of a like nature. For it has been shown by Mr. Sorby that the quartz-crystals of Granite often inclose water or other liquids (sometimes liquid carbonic acid) in cavities in their interior; which cavities, however, are not filled with the liquid, the remaining spaces being occupied by vapor. This fact cannot be otherwise accounted for, than by supposing that the crystallization must have taken place in the presence of water; and that this water, though liquid, must have been so hot as at that time to *fill* the cavities which it now occupies only partially, the size of the present vacuity marking the amount of its subsequent shrinkage during the cooling of the mass.

708. As this study, however, can only be successfully prosecuted by such as have previously obtained a considerable knowledge of Mineralogy, further details would obviously be unsuitable to our present purpose; which is only to excite an interest in these researches, and to give such general directions as will be of service to beginners who may be disposed to follow them out.—The mode in which Rock-sections are to be cut, is essentially the same as that for which directions have already been given (§§ 192–196); but it will be found desirable to use broader and thicker glasses than the ordinary 3×1 inch size, so that the sections may be about an inch square. The emery-plate should only be used for the hardest rocks, as the softer will be disintegrated when rubbed upon it. For these last, a fine corundum-file, or a piece of pumice-stone, is to be preferred in the first instance, and a fine Water-of-Air stone for finishing. When the rock is very friable, it may be saturated with hardened Canada balsam before rubbing down. As sections of the thinness usually required may not bear being transferred from the glasses to which they are cemented, it will be desirable that the attachment of a flattened and polished surface to the glass on which any section is to remain, should be finally made before the reduction of its thickness has been such as to involve the risk of its fracture in the process.[1]

---

[1] An "Elementary Text-book of Petrology" has lately been published by Mr. F Rutley, of H. M. Geological Survey. The more advanced Student should have recourse to the successive Memoirs published by Mr. Sorby in the Journal of the Geological Society, the Proceedings of the Yorkshire Geological Society and elsewhere, especially the following:—'On some Peculiarities in the Microscopic Structure of Crystals,' in "Journ. of Geolog. Society," Vol. xiv., p. 242; 'On the Microscopic Structure of Crystals, indicating the Origin of Minerals and Rocks,' *Op. cit.*, p. 453; 'On the Original Nature and subsequent Alteration of Mica-Schist,' *Op. cit.*, Vol. xix., p. 401; 'Sur l'Application du Microscope à l'Etude de la Géologie Physique,' in "Bull. Soc. Géol. de Paris," 1859–60, p. 568; and his Presidential Addresses to the Geological Society, 1879 and 1880.—Also the Memoir by Mr. David Forbes, 'The Microscope in Geology,' in the "Popular Science Review,"

709. In the application of the Microscope to Petrological and Mineralogical research, the employment of Polarized light is constantly required; and various means and appliances are needful for its most advantageous application, which are not required by the ordinary Microscopist.[1] An instrument having been recently brought out by M. Nachet, which combines all that the large experience of MM. Fouqué and Michel Lévy has led them to think desirable for Mineralogical and Petrological investigation, an account of it is here subjoined.—In all Microscopes previously constructed for this purpose, the rotation of the object on the Stage between the Polarizing and the Analyzing prisms was liable to put it out of position in regard to the cross-threads in the eye-piece; as the centering of the Objective is scarcely ever so perfect as not to produce *some* displacement, and, if the centering be adjusted so as to be perfect for one Objective, it is likely to be faulty for another. Now, the peculiarity of M. Nachet's construction is, that the Eye-piece, with its cross-threads and analyzing prism, remains fixed above (being carried upon a separate arm), whilst the Body and Stage (with the object it carries) can be made to rotate altogether around the optic axis, above the Polarizing prism which remains fixed beneath; the angular amount of this rotation being measured by a graduated ring, and a vernier attached to the stage. By this arrangement, the object is made to rotate between the two prisms of the Polarizing apparatus, without changing its position beneath the Objective, and therefore without displacing its image from its contact with the cross-threads of the Eye-piece. The mode in which this plan is worked-

FIG. 493.

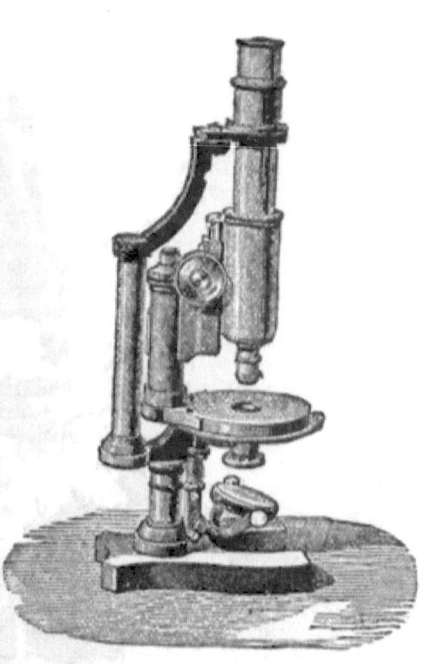

Nachet's Small Mineralogical Microscope.

Oct., 1867; the Treatise of Vogelsang, "Philosophie der Geologie und Mikroskopische Gesteinsstudien," Bonn, 1867; various subsequent Memoirs by the same; the Treatises of Zirkel, "Mikroskopische Beschaffenheit der Mineralien u. Gesteine," 1873, and "Microscopic Petrography" (U. S. Geological Exploration of Fortieth Parallel), 1876; the Treatises of Rosenbusch, "Mikroskopische Physiographie der petrographisch-wichtigen Mineralien," 1873 and "Mikroskopische Physiographie der massigen Gesteine," 1877; that of Jenzsch, "Mikroskopische Flora u. Fauna Krystallinischer Massengesteine," 1868; that of Von Lasaulx, "Elemente der Petrographie," 1875; and the great work of MM. Fouqué and Lévy, "Minéralogie Micrographique, Roches Eruptives Françaises," Paris, 1879.

[1] The description of a Microscope specially devised for this purpose by Mr. Rutley, and made by Mr Watson (of Pall Mall), will be found at p. 307 of his Textbook.

out in the ordinary small Continental model, is shown in Fig. 493; whilst, on the other hand, Fig. 494 represents the largest and most complete form of the instrument. In this last, the upper part of the body, carrying the eye-piece and analyzing prism, can be raised or lowered by the pinion attached to the fixed arm that carries it. At M, immediately beneath the eye-piece, is a small mirror, so placed as to illuminate the cross-wires when the field is dark. The analyzing prism is

FIG. 494.

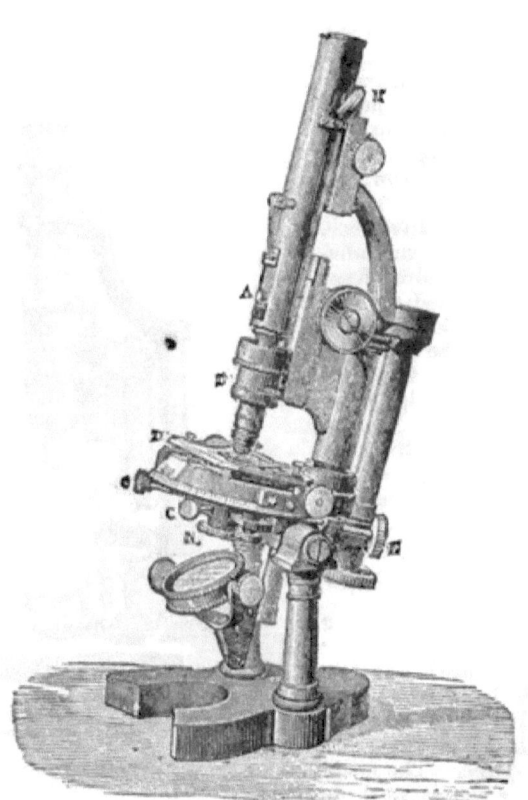

Nachet's Large Mineralogical Microscope.

inserted at A, in such a manner as to allow of being readily withdrawn when its action is not required. The Stage, with its traversing object-platform D, is made to rotate in the optic axis by the pinion E; which can be thrown out of gear so as to enable the rotation to be made by hand; and the object-platform which is graduated in both directions, is fitted with a square against which the slide abuts, so that any particular point in a section, whose place has been once noted by the scales, can be readily found again. The Polarizing prism N, is mounted quite independently of the stage, and can be precisely centered by the two milled-heads, c and c'. In the lower (rotating) part of the body, there is a horizontal slit at B for the introduction of laminæ of gypsum, quartz, etc., and into the lower end of the ocular tube can be fitted a cone that carries the converging lenses necessary to transform the instrument into an Amici microscope, its distance from the objective being regulated by the rack near the top of the eye-tube.

CHAPTER XX.

CRYSTALLIZATION.—POLARIZATION.—MOLECULAR COALESCENCE.

710. ALTHOUGH by far the most numerous and most important applications of the Microscope are those by which the structure and actions of Organized beings are made known to us, yet there are many Mineral substances which constitute both interesting and beautiful objects; being remarkable either for the elegance of their forms or for the beauty of their colors, or for both combined. The natural forms of Inorganic substances, when in any way symmetrical, are so in virtue of that peculiar arrangement of their particles which is termed *crystallization;* and each substance which crystallizes at all, does so after a certain type or plan,—the identity or difference of these types furnishing characters of primary value to the Mineralogist. It does not follow, however, that the form of the crystal shall be constantly the same for each substance; on the contrary, the same plan of crystallization may **exhibit** itself under a great variety of forms; and the study of these in such minute crystals as **are** appropriate subjects for observation **by the** microscope, is not only a **very** interesting application of **its powers, but** is capable of affording **some** valuable hints to the **designer.** This is particularly the case with crystals of *Snow,* which **belong to the** 'hexagonal system,' the basis of every figure **being a** hexagon **of six** rays; for these rays "become incrusted with an **endless** variety of secondary formations of the same kind, **some** consisting **of thin** laminæ alone, others **of** solid but translucent **prisms** heaped one upon another, and others gorgeously combining **laminæ** and prisms in the richest profusion;"[1] the angles by which these figures **are** bounded being invariably 60° or 120°. Beautiful arborescent forms are not unfrequently produced by the **peculiar** mode of aggregation of individual crystals: of this we have often an example on a large scale on a frosted window; but microscopic crystallizations sometimes present the same curious phenomenon (Fig. 495).—In the following list are enumerated some of the most interesting natural specimens which the Mineral kingdom affords as Microscopic objects; these should be viewed by reflected light, under a very low power:—

Antimony, sulphuret	Iron, ilvaite or Elba-ore
Asbestos	—— pyrites (sulphuret)
Aventurine	Lapis lazuli
Ditto, artificial	Lead, oxide (minium)
Copper, native	—— sulphuret (galena)
—— arseniate	Silver, crystallized
—— malachite-ore	Tin, crystallized
—— peacock-ore	—— oxide
—— pyrites (sulphuret)	—— sulphuret
—— ruby-ore	Zinc, crystallized.

---

[1] See Mr. **Glaisher's** Memoir on 'Snow-Crystals in 1855,' with numerous beautiful figures, in "Quart. Journ. of Microsc. Sci.," Vol. iii. (1855), p. 179.

Thin sections of Granite and other rocks of the more or less regularly-crystalline structure adverted to in the preceding paragraph, also of Agate, Arragonite, Tremolite, Zeolite, and other Minerals, are very beautiful objects for the Polariscope.

711. The actual process of the *Formation of Crystals* may be watched under the Microscope with the greatest facility; all that is necessary being to lay on a slip of glass, previously warmed, a saturated solution of the Salt, and to incline the stage in a slight degree, so that the drop shall be thicker at its lower than at its upper edge. The crystallization will speedily begin at the upper edge, where the proportion of liquid to solid is most quickly reduced by evaporation, and will gradually extend downwards. If it should go on too slowly, or should cease altogether, whilst yet a large proportion of the liquid remains, the slide may be again warmed, and the part already solidified may be re-dissolved, after which the process will recommence with increased rapidity. — This interesting spectacle may be watched under any Microscope; and the works of Adams and others among the older observers testify to the great interest which it had for them. It becomes far more striking, however, when the crystals, as they come into being, are made to stand out bright upon a dark ground, by the use of the Spot lens, the Paraboloid, or any other form of Black-ground illumination; still more beautiful is the spectacle when the Polarizing apparatus is employed, so as to invest the crystals with the most gorgeous variety of hues. Very interesting results may often be obtained from a mixture of two or more Salts; and some of the Double Salts give forms of peculiar beauty.[1] A further variety may be produced by *fusing* the film of the substance which has crystallized from its

FIG. 495.

Crystallized Silver.

---

[1] The following directions have been given by Mr. Davies ("Quart. Journ. of Microsc. Sci.," Vol. ii., 1862, p. 128, and Vol. v., p. 205) for obtaining these. "He makes a nearly saturated solution, say of the double Sulphate of Copper and Magnesia; he dries rapidly a portion on a glass slide, allowing it to become hot, so as to fuse the salt in its water of crystallization; there then remains an amorphous film on the hot glass. On allowing the slide to cool slowly, the particles of the salt will absorb moisture from the atmosphere, and begin to arrange themselves on the glass, commencing from points. If then placed under the Microscope, the points will be seen starting up here and there; and from those centres the crystals may be watched as they burst into blossom and spread their petals on the plate. Starting-points may be made at pleasure, by touching the film with a fine needle, to enable the moisture to get under it; but this treatment renders the centres imperfect. If allowed to go on, the crystal would slowly cover the plate, or if breathed-on they form immediately; whereas if it is desired to preserve the flower-like forms on a plain ground, as soon as they are large enough development is suspended by again applying gentle heat; the crystals are then covered with *pure* Canada balsam and thin glass, to be finished off as usual. The balsam must cover the edges of the film, or moisture will probably get under it, and crystallization go creeping on."

solution; since on the temperature of the glass slide during the solidification will depend the size and arrangement of the crystals. Thus *Santo-*

Fig. 496.

Radiating Crystallization of Santonine.

nine, when crystallizing rapidly on a very hot plate, forms large crystals radiating from centres without any undulations; when the heat is less

Fig. 497.

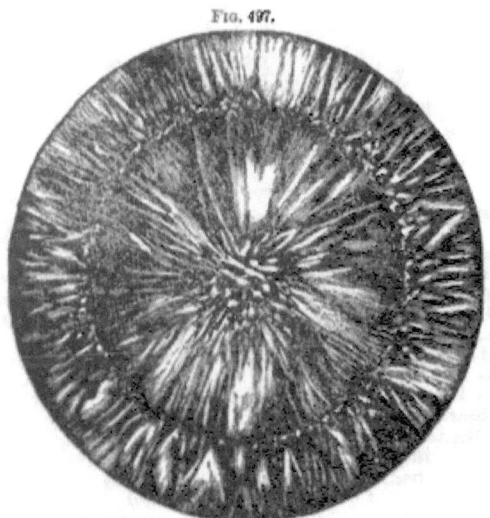

Radiating Crystallization of Sulphate of Copper and Magnesia.

considerable, the crystals are smaller, and show concentric waves of very decided form (Fig. 496); but when the slip of glass is cool, the crytals are exceedingly minute. It would seem as if these last results were due to interruptions in the formative process at certain points, consequent upon the hardening influence of cold, and the starting of a fresh formation at those points.[1] A curious example of the like kind in the crystallization of Sulphate of Copper to which a *small* quantity of Sulphate of Magnesia has been added, is shown in Fig. 497. The same principle has been carried out to a still greater extent in the case of Sulphate of Copper alone, by Mr. R. Thomas,[2] who has succeeded, by keeping the slide at a temperature of from 80° to 90°, in obtaining most singular and beautiful forms of *spiral* crystallization, such as that represented in Fig. 498. Mr. Slack has shown that a great variety of spiral and curved forms can be obtained by dissolving metallic salts, or Salicine, Santonine, etc., in water containing 3 or 4 per cent of colloid Silica. The nature of the action that takes place may be understood by allowing a drop of the Silica-solution to dry upon a slide; the result of which will be the pro-

Spiral Crystallization of Sulphate of Copper.

duction of a complicated series of cracks, many of them curvilinear. When a group of crystals in formation tend to radiate from a centre, the contractions of the Silica will often give them a trangential pull. Another action of the Silica is to introduce a very slight curling with just enough elevation above the slide to exhibit fragments of Newton's rings, when it is illuminated with Powell and Lealand's modification of Prof. Smith's dark-ground illuminator for high powers, and viewed with a 1-8th Objective. With crystalline bodies, these actions add to the variety of colors to be obtained with the Polariscope, the best slides

---

[1] See Davies on 'Crystallization and the Microscope,' in "Quart. Journ. of Microsc. Sci.," Vol. iv., p. 251.
[2] See his paper 'On the Crystallization at various Temperatures of the Double Salt, Sulphate of Magnesia and Sulphate of Zinc,' in "Quart. Journ. of Microsc. Sci.," N.S., Vol. vi., pp. 137, 177. See also H. N. Draper on 'Crystals for the Micro-Polariscope,' in "Intellectual Observer," Vol. vi. (1865), p. 437.

exhibiting a series of tertiary tints.[1]—The following List specifies the Salts and other substances whose crystalline forms are most interesting. When these are viewed with Polarized light, some of them exhibit a beautiful variety of colors of their own, whilst others require the interposition of the Selenite plate for the development of color. The substances marked *d* are distinguished by the curious property termed *dichroism*, which was first noticed by Dr. Wollaston, but has been specially investigated by Sir D. Brewster.[2] This property consists in the exhibition of different colors by these crystals, according to the direction in which the light is transmitted through them; a crystal of Chloride of Platinum, for example, appearing of a deep red when the light passes along its axis, and of a vivid green when the light is transmitted in the opposite direction, with various intermediate shades. It is only possessed by doubly-refracting substances; and it depends on the absorption of some of the colored rays of the light which is polarized during its passage through the crystal, so that the two pencils formed by double refraction become differently colored,—the degree of difference being regulated by the inclination of the incident ray to the axis of double refraction.

Acetate of Copper, *d*
——— of Manganese
——— of Soda
——— of Zinc
Alum
Arseniate of **Potass**
Asparagine
Aspartic Acid
Bicarbonate of Potass
Bichromate of Potass
Bichloride of Mercury
Binoxalate of Chromium and Potass
Bitartrate of Ammonia
——— of Lime
——— of Potass
Boracic Acid
Borate of **Ammonia**
——— of Soda (borax)
Carbonate of Lime (from urine of horse)
Carbonate of Potass
——— of Soda
Chlorate of **Potass**
Chloride of **Barium**
——— of Cobalt
——— of Copper and Ammonia
——— of Palladium, *d*
——— of Sodium
Cholesterine
Chromate of Potass
Cinchonoidine
Citric Acid
Cyanide of Mercury
Hippuric Acid
Hypermanganate of Potass
Iodide of Potassium
——— of Quinine
Mannite
Margarine
Murexide
Muriate of Ammonia
Nitrate of Ammonia
——— of Barytes
——— of Bismuth
——— of Copper
——— of Potass
——— of Soda
——— of Strontian
——— of Uranium
Oxalic Acid
Oxalate of Ammonia
——— of Chromium
——— of Chromium and Ammonia, *d*
——— of Chromium and Potass, *d*
——— of Lime
——— of Potass
——— of Soda
Oxalurate of Ammonia
Phosphate of Ammonia
——— Ammoniaco-Magnesian (triple of urine)
——— of Lead, *d*
——— of Soda
Platino-chloride of Thallium
Platino-cyanide of Ammonia, *d*
Prussiate of Potass (red)
Ditto   ditto   (yellow)
Quinidine
Salicine
Saliginine
Santonine
Stearine
Sugar
Sulphate of Ammonia
——— of Cadmium
——— of Copper
——— of Copper and Ammonia

---

[1] 'On the Employment of Colloid Silica in the preparation of Crystals for the Polariscope,' in " Monthly Microscopical Journal," Vol. v., p. 50.
[2] " Philosophical Transactions," 1819.

CRYSTALLIZATION.—POLARIZATION. 323

Sulphate of Copper and Magnesia
——— ——— of Copper and Potass
——— ——— of Iron
——— ——— of Iron and Cobalt
——— ——— of Magnesia
——— ——— of Nickel
——— ——— of Potassa

Sulphate of Soda
——— ——— of Zinc
Tartaric Acid
Tartrate of Soda
Uric Acid
Urate of Ammonia
——— ——— of Soda

It not unfrequently happens that a remarkably-beautiful specimen of Crystallization develops itself, which the observer desires to keep for display. In order to do this successsully, it is necessary to exclude the air; and Mr. Warrington recommends Castor-oil as the best preservative. A small quantity of this should be poured on the crystallized surface, a gentle warmth applied, and a thin glass cover then laid upon the drop and gradually pressed down; and after the superfluous oil has been removed from the margin, a coat of Gold-size or other varnish is to be applied.—Although most of the objects furnished by Vegetable and Animal structures, which are advantageously shown by Polarized light, have been already noticed in their appropriate places, it will be useful here to recapitulate the principal, with some additions.

Vegetable.

Cuticles, Hairs, and Scales, from Leaves (§§ 377–380)
Fibres of Cotton and Flax
Raphides (§ 359)
Spiral cells and vessels (§§ 357–362)
Starch-grains (§ 358)
Wood, longitudinal sections of, mounted in balsam (§ 368)

Animal.

Fibres and Spicules of Sponges (§ 510)
Polypidoms of Hydrozoa (§ 521)

Spicules of Gorgoniæ (§ 529)
Polyzoaries (§ 248)
Tongues (Palates) **of Gasteropods mounted in balsam (§§ 576–579)**
Cuttle-fish bone (§ 575)
Scales of Fishes (§§ **657, 658**)
Sections of Egg-shells (§ 712)
——— ——— of Hairs (§§ **661, 662**)
——— ——— of Quills (§ 660)
——— ——— of Horns (§ 664)
——— ——— of Shells (§§ 563–574)
——— ——— of Skin (§ 670)
——— ——— of Teeth (§§ 655, 656)
——— ——— **of Tendon, longitudinal (§ 668)**

712. *Molecular Coalescence.*—Remarkable modifications are shown in the ordinary forms of crystallizable substances, when the aggregation of the inorganic particles takes place in the presence of certain kinds of organic matter; and a class of facts of great interest in their bearing upon the mode of formation of various calcified structures in the bodies of Animals, was brought to light by the ingenious researches of Mr. Rainey,[1] whose method of experimenting essentially consisted in bringing-about a slow decomposition of the salts of Lime contained in Gum-arabic, by the agency of Subcarbonate of Potash. The result is the formation of spheroidal concretions of Carbonate of Lime, which progressively increase in diameter at the expense of an amorphous deposit which at first intervenes between them; two such spherules sometimes coalescing to produce 'dumb-bells,' whilst the coalescence of a larger number gives rise to the mulberry-like body shown in Fig. 499, *b*. The particles of such composite spherules appear subsequently to undergo re-arrangement according to a definite plan, of which the stages are shown at *c* and *d;* and it is upon this plan that the further increase takes place, by which such

---

[1] See his Treatise "On the Mode of Formation of the Shells of Animals, of Bone, and of several other structures, by a process of Molecular Coalescence, demonstrable in certain artificially-formed products" (1858); and his 'Further Experiments and Observations,' in "Quart. Journ. of Microsc. Sci.," N.S., Vol. i. (1861), p. 23.

larger concretions as are shown at *a, a,* are gradually produced. The structure of these, especially when examined by Polarized light, is found to correspond very closely with that of the small calculous concretions which are common in the urine of the Horse, and which were at one time supposed to have a matrix of cellular structure. The small calcareous concretions termed '*otoliths,*' or ear-stones, found in the auditory sacs of Fishes, present an arrangement of their particles essentially the same. Similar concretionary spheroids have already been mentioned (§ 613) as occurring in the skin of the Shrimp and other imperfectly-calcified shells of Crustacea; they occur also in certain imperfect layers of the shells of Mollusca; and we have a very good example of them in the outer layer of the envelope of what is commonly known as a 'soft egg,' or an 'egg without shell,' the calcareous deposit in the fibrous matting already described (§ 668) being here insufficient to solidify it. In the external layer of an ordinary egg-shell, on the other hand, the concretions have enlarged themselves by the progressive accretion of calcareous particles, so as to form a continuous layer, which consists of a series of polygonal plates resembling those of a tessellated pavement. In the solid 'shells'

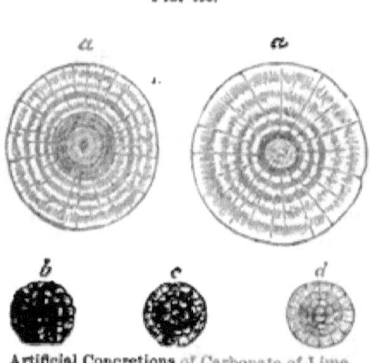

Fig. 499.

Artificial Concretions of Carbonate of Lime.

of the eggs of the Ostrich and Cassowary, this concretionary layer is of considerable thickness; and vertical as well as horizontal sections of it are very interesting objects, showing also beautiful effects of color under Polarized light. And from the researches of Prof. W. C. Williamson on the scales of Fishes (§ 657), there can be no doubt that much of the calcareous deposit which they contain is formed upon the same plan.

713. This line of inquiry has been contemporaneously pursued by Prof. Harting, of Utrecht, who, working on a plan fundamentally the same as that of Mr. Rainey (viz., the slow precipitation of insoluble salts of Lime in the presence of an Organic 'colloid'), has not only confirmed but greatly extended his results; showing that with *animal* colloids (such as egg-albumen, blood-serum, or a solution of gelatine) a much greater variety of forms may be thus produced, many of them having a strong resemblance to Calcareous structures hitherto known only as occurring in the bodies of Animals of various classes. The mode of experimenting usually followed by Prof. Harting, was to cover the hollow of an ordinary porcelain plate with a layer of the organic liquid, to the depth of from 0.4 to 0.6 of an inch; and then to immerse in the border of the

liquid, but at diametrically opposite points, the solid salts intended to act on one another by double decomposition, such Muriate, Nitrate, or Acetate of Lime, and Carbonate of Potass or Soda; so that, being very gradually dissolved, the two substances may come slowly to act upon each other, and may throw down their precipitate in the midst of the 'colloid.' The whole is then covered with a plate of glass, and left for some days in a state of perfect tranquillity; when there begin to appear at various spots on the surface, minute points reflecting light, which gradually increase and coalesce, so as to form a crust that comes to adhere to the border of the plate; whilst another portion of the precipitate subsides, and covers the bottom of the plate. Round the two spots where the salts are placed in the first instance, the calcareous deposits have a different character; so that in the same experiment several very distinct products are generally obtained, each in some particular spot. The length of time requisite is found to vary with the temperature, being generally from two to eight weeks. By the introduction of such a coloring matter as madder, log-wood, or carmine, the concretions take the hue of the one employed. When these concretions are treated with dilute acid, so that their calcareous particles are wholly dissolved-out, there is found to remain a basis-substance which preserves the form of each; this, which consists of the 'colloid' somewhat modified, is termed by Harting *calco-globuline.*—Besides the globular concretions with the peculiar concentric and radiating arrangement obtained by Mr. Rainey (Fig. 499), Prof. Harting obtained a great variety of forms bearing a more or less close resemblance to the following:—1. The 'discoliths' and 'cyatholiths' of Prof. Huxley (Fig. 293). 2. The tuberculated 'spicules' of *Alcyonaria* (Figs. 362, 363), and the very similar spicules in the mantle of some species of *Doris* (§ 573). Lamellæ of 'prismatic shell-substance' (§ 363), which are very closely imitated by crusts formed of flattened polyhedra, found on the surface of the 'colloid.' 4. The spheroidal concretions which form a sort of rudimentary shell within the body of *Limax* (§ 573). 5. The sinuous lamellæ which intervene between the parallel plates of the 'sepiostaire' of the *Cuttle-fish* (§ 575); the imitation of this being singularly exact. 6. The calcareous concretions that give solidity to the 'shell' of the bird's egg: the semblance of which Prof. Harting was able to produce *in situ*, by dissolving away the calcareous component of the egg-shell by dilute acid, then immersing the entire egg in a concentrated solution of chloride of calcium, and transferring it thence to a concentrated solution of carbonate of potass, with which, in some cases, a little phosphate of soda was mixed.[1] Other forms of remarkable regularity and definiteness, differing entirely from anything that ordinary crystallization would produce, but not known to have their parallels in living bodies, have been obtained by Prof. Harting. Looking to the relations between the calcareous deposits in the scales of Fishes (§§ 657–659) and those by which Bones and Teeth are solidified, it can scarcely be doubted that the principle of 'molecular coalescence' is applicable to the latter, as well as to the former; and that an extension and variation of this method of experimenting would throw much light on the process of ossification and tooth-formation.—The inquiry has been

---

[1] See Prof. Harting's "Recherches de Morphologie Synthétique sur la production artificielle de quelques Formations Calcaires Inorganiques, publiées par l'Académie Royale Néderlandaise des Sciences," Amsterdam, 1872; and "Quart. Journ. of Microsc. Sci.," Vol. xii., p. 118.

further prosecuted by Dr. W. M. Ord, with express reference to the formation of Urinary and other Calculi.[1]

714. *Micro-Chemistry of Poisons.*—By a judicious combination of Microscopical with Chemical research, the application of re-agents may be made effectual for the detection of Poisonous or other substances, in quantities far more minute than have been previously supposed to be recognizable. Thus it is stated by Dr. Wormley,[2] that Micro-Chemical analysis enables us by a very few minutes' labor to recognize with unerring certainty the reaction of the 100,000th part of a grain of either Hydrocyanic Acid, Mercury, or Arsenic; and that in many other instances we can easily detect by its means the presence of very minute quantities of substances, the true nature of which could only be otherwise determined in comparatively large quantity, and by considerable labor. This inquiry may be prosecuted, however, not only by the application of ordinary Chemical Tests under the Microscope, but also by the use of other means of recognition which the use of the Microscope affords. Thus it was originally shown by Dr. Guy[3] that by the careful sublimation of Arsenic and Arsenious Acid,—the sublimates being deposited upon small disks of thin-glass,—these are distinctly recognizable by the forms they present under the Microscope (especially the Binocular) in extremely minute quantities; and that the same method of procedure may be applied to the volatile metals, Mercury, Cadmium, Selenium, Tellurium, and some of their Salts, and to some other volatile bodies, as Sal-Ammoniac, Camphor, and Sulphur. The method of sublimation was afterwards extended by Dr. Helwig[4] to the Vegetable Alkaloids, such as Morphine, Strychnine, Veratrine, etc. And subsequently Dr. Guy, repeating and confirming Dr. Helwig's observations, has shown that the same method may be further extended to such Animal products as the constituents of the Blood and of Urine, and to volatile and decomposable Organic substances generally.[5] By the careful prosecution of Micro-Chemical inquiry, especially with the aid of the Spectroscope (where admissible), the detection of Poisons and other substances in very minute quantity can be accomplished with such facility and certainty as were formerly scarcely conceivable.

---

[1] See his Treatise "On the Influence of Colloids upon Crystalline Form and Cohesion," London, 1879.

[2] "Micro-Chemistry of Poisons," New York, 1867.

[3] 'On the Microscopic Characters of the Crystals of Arsenious Acid,' in "Trans. of Microsc. Society," Vol. ix. (1861), p. 50.

[4] "Das Mikroskop in der Toxikologie," 1865.

[5] 'On Microscopic Sublimates; and especially on the Sublimates of the Alkaloids,' in "Trans of Royal Microsc. Soc.," Vol. xvi. (1868), p. 1; also "Pharmaceutical Journal," June to September, 1867.

APPENDIX.

'NUMERICAL APERTURE' AND 'ANGULAR APERTURE.'

THE introduction of the 'immersion system' has rendered necessary a considerable modification in the mode of determining the real 'Apertures' of Achromatic Objectives; which were formerly estimated entirely by their respective '*angles* of aperture,'—such angles being (as formerly explained, § 10), those contained, in each case, between the most diverging of the rays issuing from the axial point of an object, that can enter the lens and take part in the formation of an image. A careful investigation of the whole subject of 'Aperture,' both theoretically and practically, has of late been carried out with the greatest ability by Prof. Abbe, of Jena; of whose important discovery of the dependence of 'resolving power' upon *dif* fraction—not *re*fraction—an account has been already given (§ 157). This investigation has enabled him to place the question on an exact basis; and not only to clear up a great deal that was formerly obscure, but to formulate a definite principle for the comparison of 'immersion' with 'dry' or 'air' objectives, which shows that the advantages obtainable from the use of the former are much greater than had been previously conceived.

Prof. Abbe has also made an important contribution to the practical part of this inquiry, by the invention of an 'Apertometer' for the precise measurement of angular apertures,[1] by which more exact and definite results can be obtained than by any of the methods previously in use; and he has further shown that a comparison of 'dry' and of 'immersion' lenses by their respective 'angles' alone is so completely fallacious, as to necessitate the introduction of a new scale of 'numerical apertures,' to which, as to a common standard, both could be referred.—It is the object of this Addendum, in the first place, to explain to the readers of this treatise the precise meaning of Prof. Abbe's term; and then to put before them the new views in regard to the capacities of 'immersion' Objectives, to which his investigations have led him. As (for obvious reasons) conclusions only can be here stated, those who desire to master the train of reasoning by which those conclusions have been worked-out, are recommended to study the two most recent expositions of the doctrine; one given by Prof. Abbe himself in his Paper 'On the Estimation of Aperture,' and the other by his disciple, Mr. Frank Crisp (one of the Secretaries of the Royal Microscopical Society) in his 'Notes on Aper-

---

[1] "Journ. of Roy. Microsc. Soc.," Vol. i. (1878), p. 19. Another method devised by Prof. Hamilton Smith (Op. cit., Vol. ii., 1879, p. 775), gives nearly the same results as that of Prof. Abbe. And yet another has been proposed by Mr. Tolles (Op. cit., Vol. iii., 1880, p. 887), who does not, however, give any reason to question the accuracy of Prof. Abbe's instrument.

ture, Microscopical Vision, and the Value of Wide-angled Immersion Objectives;' contained in the "Journal of the Royal Microscopical Society," for April and June, 1881.

It can be easily demonstrated mathematically, that the 'aperture' of a single lens used as a magnifying glass—that is, its capacity for receiving, and bringing to a remote conjugate focus, the rays emanating from the axial point of an object brought very near to it—is determined by the ratio between its absolute diameter (or clear 'opening') and its focal length; while that of an ordinary Achromatic Objective, composed of several lenses, is determined by the ratio of the diameter of its *back* lens (so far as this is really utilized) to its focal length. This ratio is most simply expressed, when the medium is the same, by the sine of its semi-angle of aperture (sin u); and we hence see how different are the proportionate 'apertures' of different lenses from their proportionate 'angles of aperture.' For as the sine of half 180°,—the largest possible *theoretical* angle, whose two boundaries lie in the same straight line,—is equal to radius, and as the sine of half 60° is equal to $\frac{1}{2}$ radius, it follows that a lens having an angle of 60° has an aperture equal to *half* (instead of being only *one-third*) of the theoretical maximum. And as the sines of angles beyond 60° increase very slowly, an objective whose angle is 120° will have (instead of only two-thirds) as much as about 87-100ths of the aperture given by the theoretical maximum.

When, however, the medium in which the Objective works is not air, but a liquid of higher refractive index—such as water or oil—an additional circumstance has to be taken into consideration; for we may now have three *angles* of aperture expressed by the *same* number of degrees, which yet denote quite *different* 'apertures.' For instance, an 'angle' of 90° in oil will give a greater 'aperture' than one of 90° in water; and the latter a greater aperture than 90° in air. For since, when light is transmitted from any medium into another of greater refractive index (§ 1), its rays are bent towards the perpendicular, the rays forming a pencil of given angular extension in air, will, when they pass into water or oil, be closed-together or *compressed;* so that in comparing (for instance) an object mounted in balsam with one mounted dry, the balsam angle, though much reduced, may nevertheless contain all the rays that were spread-out over the whole hemisphere when the object was in the less dense medium. It follows, therefore, that a given 'angle' in oil or water represents an *increase* in 'aperture' over the same angle in air. The amount of this increase having been determined by Prof. Abbe to be proportional, in each case, to the index of refraction of the interposed medium, the comparative 'apertures' of lenses working in different media are in the compound ratio of two factors,—the sines of their respective semi-angles of aperture, and the refractive indices of the interposed fluids.

It is the product of these (n sin u) that gives what is termed by Prof. Abbe the *Numerical Aperture;* which serves, therefore, as the standard of comparison not only between 'immersion' and 'dry' objectives, but also between objects of like kind. For, when the medium is the same, the factor (n) which represents the refractive index may, of course, be neglected; the 'numerical apertures' of such objectives then being simply the sines of their respective semi-angles.

Thus, taking as a standard of comparison a 'dry' objective of the maximum theoretical angle of 180,' whose 'numerical aperture' is the sine of 90°, =radius or 100, we find this standard to be equalled by a 'water' immersion objective of only 96°, and by an 'oil' or 'homogeneous' immersion lens of only 82°; the 'numerical apertures' of these, obtained by multiplying the sines of their respective semi-angles by the refractive index of water in one case and of oil in the other, being 1.00 in both. Each, therefore, will have as great a power of receiving and utilizing divergent rays, as any 'dry' lens can even theoretically possess, —an angle of nearly 70° being the limit of what is practically attainable. But as the actual angle of an 'immersion' Objective can be opened-out to the same extent as that of an 'air' objective, it follows that the 'aperture' of the former can be augmented *far beyond* even the theoretical maximum of the latter; the maxima of numerical aperture being 1.52 for Oil-immersion, and 1.33 for Water-immersion objectives, as against 1.00 for 'dry;' and these being nearly attainable in practice.[1]

So, if we have four Objectives, two of which are 'dry,' the third a water-immersion, and the fourth an oil-immersion, their apertures have hitherto been designated, on the angular aperture notation, by (for instance) 47° and 74° air-angle; 85° water-angle; and 117° oil-angle; so that it is difficult without calculation to judge of their relative apertures. By the numerical notation, however, the apertures of the four are seen to be as .40, .60, .90 and 1.30; so that a comparison is readily made, and it is seen whether the two latter have larger or smaller apertures than the maximum of a dry objective.

This important doctrine may be best made practically intelligible by a comparison (Fig. 500) of the relative diameters of the *back* lenses of 'dry' with those of 'water' and 'oil' immersion Objectives of *the same power*, from an 'air-angle' of 60° to an 'oil-angle' of 180°; these diameters expressing in each case, the opening between the extreme pencil-forming rays at their issue from the posterior surface of the combination, to meet in its conjugate focus for the formation of the image; *the extent of which opening in relation to focal length* (not that of the rays entering the Objective), *is the real measure of the Aperture of the combination*. The dotted circles in the interior of 1 and 2 are of the same diameter as 3; and therefore show the excess in the diameters of the back lenses of the 'oil' and 'water' immersion-objectives, over that of the 'dry' at their respective theoretical limits.

Now this difference is capable of being practically tested by a simple experiment originally suggested by Mr. Stephenson, and thus described by Prof. Abbe:—"Take any immersion-objective of balsam angle exceeding the critical angle, and focus it on a balsam-mounted object, which is illuminated by any kind of immersion-condenser, in such a way that the whole range of the aperture-angle is filled by the incident rays. Remove the eye-piece, and place the pupil of the eye at the place where the air-image is projected by the objective, and look down on the lens. You see a uniformly bright circle of well-defined diameter, which is the true cross section of the image-forming pencil emerging from the Microscope (for the eye receives now all rays which have been transmitted through a small central portion of the object—that portion which is conjugate to the pupil—and receives *no other* rays). After this, focus the *same* objec-

---

[1] At p. 325 of Vol. i., Ser. 2 (1881) of the "Journ. of the Roy. Microsc. Soc.," will be found a Table calculated by Mr. Stephenson of the Equivalent Angles of Aperture of Dry, Water-immersion, and Oil (or homogeneous) immersion Objectives, with their respective Illuminating powers, and Theoretical Resolving powers, for every 0.02 of Numerical Aperture, from 0.40 to 1.52.

tive on an ordinary dry-mounted preparation (or on one which is connected with the slide, the cover-glass being put on dry), and repeat the observation; you will now see a well-defined circle, a cross section of the emergent pencil, but of *less* diameter than in the former case, surrounded by a dark annulus, visible by faint diffused light only."[1] The explanation of this experiment is, that in focussing an immersion-objective on an object with air above it (*i. e.*, between itself and the cover-glass), the under-surface of the cover-glass acts as the plane front surface of the system, converting it into a true 'dry' lens of 180° angular aperture, which gathers-in almost the *whole hemisphere* of light from the radiant in air; and yet the emergent pencil of rays is *much narrower* than when the same objective is used as an immersion, and focussed on an object in balsam, the extreme divergence of whose rays is not more than 138°.

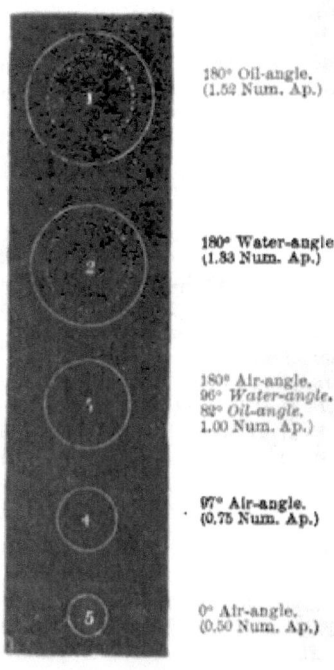

Fig. 500.

180° Oil-angle.
(1.52 Num. Ap.)

180° Water-angle.
(1.33 Num. Ap.)

180° Air-angle.
96° *Water-angle.*
82° *Oil-angle.*
1.00 Num. Ap.)

97° Air-angle.
(0.75 Num. Ap.)

0° Air-angle.
(0.50 Num. Ap.)

A wide-angled 'immersion' Objective can therefore utilize rays from an object mounted in a dense medium, such as balsam, which are entirely *lost* for the image (since they do not exist, physically) when the same object is in air, or is observed through a film of air. And this loss cannot be compensated-for by an increase of illumination; because the rays which are lost are *different* rays, physically, from those obtained by any illumination, however intense, in a medium like air.

It is by increasing the number of 'diffraction-spectra,' that the rays admitted from the object contribute to the 'resolving power' of the Objective for lined and dotted objects; the truth of the image formed by the recombination of these spectra, being, as formerly shown (§ 157), essentially dependent upon the augmentation of the number which the objective can be made to receive.

Upon the 'aperture' of an objective are dependent (1) its illuminating power, (2) its resolving power, and (3) its penetrating power;—the first varying as the square of the numerical aperture, the second being in *direct*, and the third in *inverse* proportion to the numerical aperture.

Whilst Prof. Abbe's investigation has made it clear that the 'aperture' of an immersion objective may exceed the maximum of that of a dry objective, it is hardly necessary to point out that the *act* of the excess is a distinct question from that of the *value* of the excess for particular cases. As the penetrating power of the objective is diminished in proportion as the aperture is increased, it is seen that large apertures can

---

[1] The diameter of the *emergent* pencil may be accurately measured by the use of an eye-piece Micrometer with the "auxiliary Microscope" of Prof. Abbe's Apertometric apparatus, already referred to.

only be obtained at the expense of a great reduction of penetration or focal depth, and consequently also of working distance,—qualities which are essential in some of the most important kinds of Biological investiga-

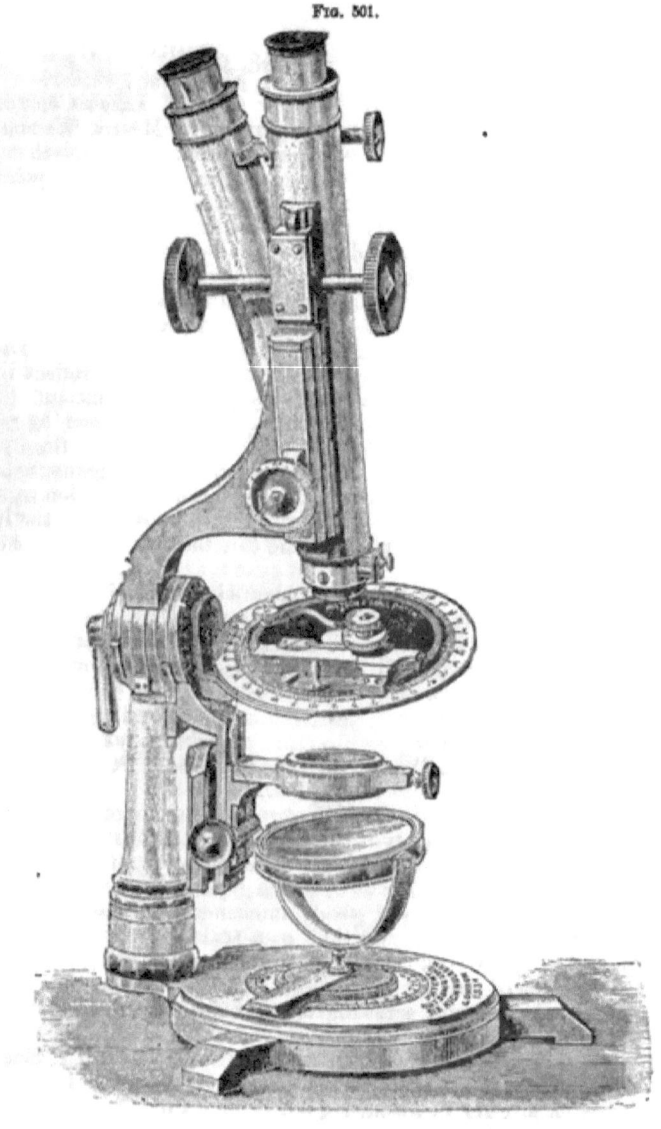

Fig. 501.

tion; and the Author, therefore, still holds to the opinion, that for objectives intended to be used for such purposes, 'moderate angles' are preferable; objectives of wide angle being kept for 'critical' investigations upon objects specially demanding their use. In this view he is entirely supported by Mr. Dallinger, whose unrivalled experience in Biological work of the highest kind, entitles his opinion on such a point to the highest respect. See Preface, pp. v., vi.

MICROSCOPES, ETC.

Messrs. Watson's New Models.—A new form of Large Compound Microscope (Fig. 501) has lately been brought out by Messrs. Watson (of Holborn), the peculiarity of which essentially consists in this,—that the horizontal axis on which it is suspended passes through the axial point of the plane in which the object lies; so that by inclination of the body and stage—the source of light remaining fixed,—illuminating rays may be made to fall on the object at any degree of obliquity. The mechanical stage is so constructed (by placing the entire movement *above* the object platform) as to give it a thinness not otherwise attainable with the power of making a complete revolution. The mirror with its frame may be slipped off the swinging arm that ordinarily carries it, and slid into a fitting on the foot, on which it can be readily centred so as to reflect light upon the centre of the stage, whatever may be the inclination of the latter. And a further variety of illumination may be obtained by rotating the whole instrument on its foot, the mirror retaining its fixed position in the centre.—The principle of these ingenious arrangements is to give to the stage, and all that is above it, every variety of position in relation to a fixed source of light, instead of varying the position of the light in relation to the object.—Experience alone can test its advantages over the old models.

The above-named Makers have also adapted a 'swinging sub-stage,' not merely to this large instrument, but to a smaller one on the scale of the 'Student's Microscope' of Messrs. Ross (Fig. 43); which is furnished, in addition, with a graduated disk for the precise measurement of the obliquity given to the illuminating apparatus. Having carefully examined this instrument, with the Objectives supplied by the makers, the Author is able to speak favorably of its workmanship; and would desire to add the name of Messrs. Watson to those of whose Students' microscopes he has spoken with approval at p. 67.

Messrs. Swift's New Students' Microscope.—These excellent Makers, having adopted the general plan of the 'Wale' model (Fig. 44), of which the Author has spoken in terms of high commendation, have applied to it a new fine adjustment of their own, which gives to the ring that carries the objective a very delicate and steady movement;[1] replacing the iris-diaphragm of the Wale model with their own 'calotte' diaphragm.

M. Nachet's Objective-carrier.—Every working Microscopist has desired a ready means of varying his 'powers,' without the trouble of unscrewing one Objective and screwing on another. This difficulty has been partly met by the use of the 'nose-piece;' but this cannot be conveniently made (at least, in the case of the heavily-mounted English objectives) to carry more than two powers. By Messrs. Parkes, of Birmingham, as already mentioned § 53, sliding tubes are substituted for screws; but the use of them requires the withdrawal of the nose of the microscope to a considerable distance above the stage.—The attention

---

[1] See "Journ. of Roy. Microsc. Soc.," Vol. i., N.S. (1881), p. 297.

of M. Nachet having been long directed to this point, he has recently brought out a form of 'porte-objectif' (an improvement on a suggestion originally made by Prof. Thury) which allows the change of objectives to be readily made without as much raising of the body from the stage as is required in screwing and unscrewing. It consists (Fig. 502) of a fixed inner cylinder, whose top screws into the bottom of the body; this being embraced by a movable outer cylinder (A), that is kept closely pressed up to its lower end by a strong spiral spring between the two. The bottom of this outer cylinder is formed by a shoulder that is cut away for about one-fourth of its circumference, so as to allow a collar (B) at the top of the objective to be slipped into the opening as shown at c. When this is

Fig. 502.

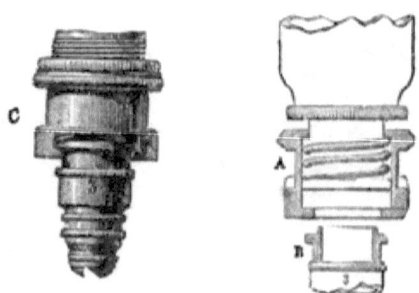

Nachet's Objective carrier.

done, the objective is held firmly in place by the pressure of the spring; and all that is needed to remove it is a slight pulling down of the outer cylinder, which enables the collar of the objective to be slipped out again. The inner cylinder is supplied by M. Nachet (when desired) with the Society's screw, and the 'collar' can be adapted to receive either M. Nachet's or any other Objectives.—Having been enabled, by the kindness of M. Nachet, to make a trial of this little apparatus, the Author is glad to be able to speak most favorably both of its simplicity and its effectiveness.

INDEX.

ABBE, Prof., on Homogeneous immersion, i. 17; on Diffraction spectra, i. 156-160; on Penetration of objectives, i. 163 *note;* on Numerical aperture of objectives, ii. 327; on Apertometer, ii. 327.
Aberration, Chromatic, i. 8, 9.
——— Spherical, i. 6, 7.
——— means of reducing and correcting, i. 7-14.
Absorption bands, i. 89-93.
Acalephs, see *Medusæ.*
Acanthometrina, ii. 113.
Acarida, ii. 248, 249.
Achlya prolifera, i. 251, 253.
Achnanthes, i. 296, 297.
Achromatic Condenser, i. 102, 103; use of, i. 144, 145.
Achromatic Correction, i. 10, 11.
Achromatic Objectives, see Objectives.
Acinetina, ii. 39, 40.
Acrocladia, spines of, ii. 143.
Actinia, ii. 135; thread-cells of, ii. 135, 136.
Actinocyclus, i. 291.
Actinomma, ii. 113.
Actinophrys, ii. 10-12.
Actinoptychus, i. 292.
Actinosphærium, ii. 13.
Actinotrocha, ii. 198.
ACTINOZOA, ii. 134-137.
Adjustment of Focus, i. 82, i. 137-139.
Adjustment of Object-glass, i. 11-15; i. 139-141.
Æcidium tussilaginis, i. 324.
Agamic eggs, of Rotifera, ii. 58-60; of Entomostraca, ii. 210, 211; of Insects, ii. 246, 247.
Agrion, circulation in larva of, ii. 238.
Air-bubbles, microscopic appearances of, i. 152; in microscopic preparations, i. 198, i. 216-217.
Albuminous substances, tests for, i. 208, 209.
Alburnum, i. 364, 370.
Alcohol, as hardening agent, i. 202; as test, i. 209.
Alcyonian Zoophytes, ii. 136, 137.
Alcyonidium, ii. 161.
ALGÆ, higher, microscopic structure of, i. 331-335 (see Protophyta).

Allman, Prof., on Sarcode organisms, ii. 22 *note;* on Noctiluca, ii. 34; on Peridinium, ii. 37; on Myriothela, ii. 122; on Tubularida, ii. 129 *note;* on fresh-water Polyzoa, ii. 162 *note;* on Appendicularia, ii. 169, 170.
Alternating Circulation of Ascidians, ii. 166, 168.
Alternation of Generations, ii. 134.
Alveolina, ii. 73.
Amaranthus, seeds of, i. 386.
Ambulacral disks of Echinida, ii. 141.
Amici, Prof., his early construction of Achromatic lenses, i. 11, 15; his invention of the immersion system, i. 16; his drawing Camera, i. 97; his Prism for oblique illumination, i. 106.
Amœba, ii. 15-17.
Amœboids of Volvox, i. 244; of protoplasm of Chara, i. 262 *note;* of protoplasm of roots of Mosses, i. 339; of Myxomycetes, i. 326; of Sponges, ii. 117-119; of Polypes, etc., ii. 122, 123; of colorless Blood-corpuscles, ii. 271.
Amoroucium, ii. 165.
Amphipleura pellucida, resolution of, i. 171.
Amphistegina, ii. 91.
Amphitetras, i. 295.
Amplifiers, i. 86.
Anacharis alsinastrum, formation of cells in, i. 356; cyclosis in, i. 356, 357.
Anagallis, petal of, i. 383.
Androspores of Œdogonium, i. 258.
Angle of Aperture, i. 8; limitation of, for Binocular, i. 35-38; its relation to Angular Aperture, i. 160 *note;* to Numerical Aperture, ii. 327.
Anguillulæ, ii. 193.
Angular Aperture of Object Glasses, i. 160 *note;* its relation to resolving power, i. 157, 163; its real meaning, ii. 327; limits to its value, Preface, v.
Anguliferæ, i. 295.
Aniline dyes, as staining agents, i. 206.
Animal Tissues, formation of, ii. 252-255.
Animalcule-cage, i. 123.
ANIMALCULES, ii. 24 (see Infusoria, Monerozoa, Rhizopoda, and Rotifera).
Animals, distinction of, from Plants,

i. 222, 223; links connecting with Plants, i. 325-328.
ANNELIDA, ii. 192-204; marine, circulation in, ii. 196, 197; metamorphoses of, ii. 197, 198; remarkable forms of, ii. 199-202; luminosity of, ii. 202; fresh-water, ii. 202, 203.
Annual layers of Wood, i. 369-371.
Annular Ducts, i. 365, 366.
ANNULOSA, ii. 192—see Entozoa, Turbellaria, and Annelida.
Anodon, shell of, ii. 174; parasitic embryo of, ii. 183; ciliary action on gills of, ii. 189.
Anomia, fungi in shell of, i. 321.
Ant, red, integument of, ii. 220.
Antedon, development of, ii. 152-154.
Antennæ of Insects, ii. 232, 233.
Antheridia, of Chara, i. 262; of Marchantia, i. 337; of Mosses, i. 340; of Ferns, i. 347;—see Antherozoids.
Antherozoids, i. 229; of Volvox, i. 242; of Vaucheria, i. 251; of Sphæroplea, i. 255, 256; of Œdogonium, i. 258; of Characeæ, i. 261, 266; of Fuci, i. 333; of Marchantia, i. 337; of Mosses, i. 339; of Ferns, i. 347.
Anthers, structure of, i. 383, 384.
Anthony, Dr., on scale of Gnat, i. 155; on battledoor scales, ii. 223; on tongue of Fly, ii. 235 note.
Antirrhinum, seeds of, i. 386.
Apertometers, ii. 327.
Aperture, Angular, see Angular Aperture; Numerical, ii. 327.
Aphides, agamic reproduction of, ii. 246.
Aphthæ, fungus of, i. 320.
Aplanatic Searcher, i. 8 note.
Apothecia of Lichens, i. 330.
Appendicularia, ii. 169, 170.
Apple, cuticle of, i. 376.
Apus, ii. 209.
Aquatic Box, i. 123.
ARACHNIDA, microscopic forms of, ii. 248, 249; eyes of, ii. 250; respiratory organs of, ii. 250; feet of, ii. 250; spinning apparatus of, ii. 250, 251.
Arachnoidiscus, i. 293.
Arachnosphæra, ii. 114.
Aralia, cellular parenchyma of, i. 354.
Arcella, ii. 18, 19.
Archegonia, of Marchantia, i. 336; of Mosses, i. 340; of Ferns, i. 347.
Archer, Mr., on zoöspores of Desmidiaceæ, i. 267 note; on Chlamidomyxis, i. 327-328; on Clathrulina, ii. 14.
Arenaceous Foraminifera, ii. 77-85.
Arenicola, ii. 196.
Areolar tissue, ii. 273.
Argulus, ii. 212.
Aristolochia, stem of, i. 375.
Artemia, ii. 207, 210.
Ascaris, ii. 193; fungous vegetation on, i. 319.

Asci, of Lichens, i. 330; of Fungi, i. 322, 323.
Ascidia parallelogramma, ii. 164.
Ascidians, solitary, ii. 164; compound ii. 165-167; social, 166-167; development of, ii. 168-170.
Ascogonia, of Fungi, i. 322; of Lichens, i. 330.
Ascomycetes, i. 322.
Asphalte-varnish, i. 178.
Aspidisca-form of Trichoda, ii. 48.
Aspidium, fructification of, i. 345.
Asplanchna, ii. 57, 58.
Astasia, ii. 33.
Asteriada, skeleton of, ii. 145; metamorphoses of, ii. 150, 151.
Asterolampra, i. 291.
Asteromphalus, i. 291.
Astromma, ii. 112.
Astrophyton, ii. 145.
Astrorhiza, ii. 78.
Auditory vesicles of Mollusks, ii. 190; development of, ii. 186.
Aulacodiscus, i. 293.
Auxospores of Diatomaceæ, i. 283.
Avicula, nacre of, ii. 174.
Avicularia of Polyzoa, ii. 162.
Axile bodies of sensory papillæ, ii. 285.
Axis-cylinder of Nerve-fibres, ii. 284-286.
Azure-blue butterfly, scales of, ii. 222, 223.

Bacillaria paradoxa, i. 287; movements of, i. 283.
Bacillus, i. 308-311.
Bacteria, i. 308-315.
Bacteriastrum, i. 296.
Badcock, Mr., on metamorphosis of Acinetina, ii. 40.
Bailey, Prof., his Diatomaceous tests, i. 171; on siliceous cuticle, i. 349; on internal casts of Foraminifera, ii. 93 note.
Baker, Mr., his Students' Microscope, i. 59; his Students' Binocular, i. 6?; his Travelling Microscope, i. 81, 8?; his Pond-stick, i. 219.
Balanus, metamorphosis of, ii. 213, 214.
Balbiani, M., on generation of Infusoria, ii. 49-52.
Balsam, Canada, see Canada Balsam.
Banksia, stomata of, i. 380.
Barbadoes, Polycystina of, ii. 109, 114.
Bark, structure of, i. 373.
Barnacle, metamorphosis of, ii. 213, 214.
Basidia of Fungi, i. 322.
Bat, hair of, ii. 264; cartilage of ear of, ii. 279.
Batrachospermeæ, i. 258, 259.
Battledoor scale of Polyommatus, ii. 222 223.
Bathybius, ii. 20.
Beading of Diatom-valves, i. 277-279; of Insect-scales, Dr. Royston Pigott on, ii. 224, 226.

Beale, Prof., his Pocket Microscope, i. 80; his Demonstrating Microscope, i. 81; his views of viscid media, i. 210-212; his Views of Tissue-formation, ii. 253-255.
Beck, Messrs., their Economic Microscopes, i. 67; their Popular Microscope, i. 68; their Large Compound Microscope, i. 77; their improved ditto, i. 80; their Achromatic Condensers, i. 102, 103; their arrangement of Polarizing apparatus, i. 113; their Compressors, i. 126, 127; their Binocular Magnifier, i. 187 *note;* their Microtome, i. 192.
—— Mr. Joseph, on **scales of Thysanuræ**, ii. 223-226.
—— Mr. Richd., his Dissecting Microscope, **i.** 48; his Disk-holder, i. 121; his Side-Reflector, i. 116; his Vertical Illuminator, i. 118, 119; on scales of of Thysanuræ, ii. 224; on Spider's threads, ii. 251.
Bee, eyes of, ii. 229, 230; hairs of, ii. 227; proboscis of, ii. 235; wings of, ii. 242; sting of, ii. 245; reproduction of, ii. 247.
Berg-mehl, i. 302.
Bermuda-earth, i. 292, 302.
Beroë, ii. 137, 138.
Biddulphia, i. 293; growth of, i. 275 *note;* surface-marking of, i. 276; self-division of, i. 280, 281.
Bignonia, seed of, i. 386.
Biliary Follicles, ii. 280.
Biloculina, ii. 71.
Binary subdivision of Vegetable Cells, i. 227, 228; of Animal Cells, ii. 253; see Cells, Animal and Vegetable.
Binocular Eye-piece, i. 33.
—————— Magnifier, Nachet's, i. 48, **49;** Beck's, i. 187 *note.*
—————— - Microscopes, Stereoscopic, principles of construction of, i. 25-28; advantages of, i. 35-38; Objectives suitable for, i. 35-37; different forms of, Nachet's, i. 28-29; Wenham's, i. 29-31; Stephenson's, i. 30-33.
—————— Non-Stereoscopic, Powell and Lealand's, i. 84; Wenham's, i. 85.
—————— - Stereo-Pseudoscopic, Nachet's, i. **33-35.**
—————— **Vision, i.** 25-28.
Bipinnaria-larva of Star-fish, ii. 151.
Bird, Dr. Golding, on preparation of Zoophytes, ii. 131.
BIRDS, bone of, ii. 257; feathers of, ii. 266; blood of, ii. 269; lungs of, ii. 300.
Bird's-head processes of Polyzoa, ii. 162.
Bisulphide of Carbon, mounting Diatoms in, i. 291.
Bivalve Mollusks. shells of, **ii.** 171-178.
Black ground illuminators, **i.** 106-110, i. 147.
Blackham, Dr., on Focal Depth, i. 163 *note.*

Blankley, Mr., his Selenite Stage, i. 113.
Blenny, viviparous, scales of, ii. 261.
Blights of Corn, i. 323.
Blood, Absorption-bands of, i. 92, 93.
Blood-disks of Vertebrata, ii. 267-271; mode of preserving, ii. 271; circulation of, see Circulation.
Blood-vessels, injection of, ii. 292-297; disposition of, in different parts, ii. 297-301.
Bocket Lamp, **i.** 131.
Bone, structure of, ii. 255-258; mode of making sections of, i. 196-199, ii. 258.
Bones, fossil, examination of, ii. 311.
Botryllians, ii. 160.
Botrytis, of Silkworms, i. 317-326.
Botterill, Mr., his Growing-slide, i. 122, 123; his Zoophyte-trough, i. 125.
Bowerbankia, ii. 161.
Brachionus, ii. 54, 62.
BRACHIOPODA, Shell-structure of, ii. 177, 178.
Brady, Mr. H. B., on Saccammina, ii. 78; on Loftusia, ii. 84; on Globigerina, ii. 87.
Braithwaite, Dr., on Sphagnaceæ, i. 344.
Branchiopoda, ii. 207-209.
Branchipus, ii. 209.
Braun, Prof., on development of Pediastreæ, i. 270-272.
Brittan, Dr., on Fungus-germs, i. 321.
Brownian Movement, i. 153, 154.
Browning, Mr., his Platyscopic Lens, i. 21; his smaller Stephenson Binocular, i. 72-73; his Rotating Microscope, i. 65; his Micro-Spectroscope, i. 89-91.
Bryozoa, see POLYZOA.
Buccinum, palate of, ii. 182; egg capsules of, ii. 184; development of, ii. 187.
Buckthorn, stem of, i. 369.
Bugs, ii. 219; wings of, ii. 242.
Bugula avicularia, ii. 162, 163.
Built-up Cells, i. 182.
Bulbels, of Chara, i. 261; of Marchantia, i. 336, 337.
Bulimina, ii. 89.
Bull's-eye Condenser, **i.** 114, 115; use of, 148-150.
Burdock, stem of, i. 375.
Busk, Mr. G., on Volvox, i. 238-243; on Polyzoa, ii. **162,** 163.
Butterflies, **see** Lepidoptera.

Cabinet for Microscopic Apparatus, i. 130; for Objects, i. 218.
Cacao-butter, for imbedding, i. 195.
Cactus, raphides of, i. 363.
Calcaire Grossier, ii. 305, 307.
Calcareous Deposits, organic origin of, ii. 308, 309.
Calcareous Sponges, ii. 120.
Calcarina, ii. 90.
Calycanthus, stem of, i. 374.
Calyptra of Mosses, i. 339.

338 INDEX.

Cambium-layer i. 373, 374.
Camera Lucida, i. 96-98; **use of, in** Micrometry, i. 99.
Campanularidæ, ii. 129.
Campylodiscus, i. 288.
Canada Balsam, use of, as Cement, i. 176, i. 197-199; mounting of objects in, i. 209, 214, 215.
Canaliculi of Bone, ii. 256, 257.
Canal-system of Foraminifera, ii. 70, 90-99.
Capillaries, circulation in, ii. 286-292; injection of, ii. 293-298; distribution of, ii. 298-300.
Capsule of Mosses, i. 340.
Carbolic Acid, as preservative, i. 210; use of for dehydration, i. 215.
Carmine, as staining agent, i. 205; injection with, ii. 295, 297.
Carp, scales of, ii. 262.
Carpenteria, ii. 88.
Carrot, seeds of, i. 387.
Cartilage, structure of, ii. 278, 279.
Caryophyllia, ii. 135.
Caryophyllum, seeds of, i. 386.
Caterpillars, feet of, ii. 245.
Cedar, stem of, i. 371.
Cells for mounting objects, i. 179-183; mounting objects in, i. 215-217.
Cells, Animal, formation, of, ii. 253; binary subdivision of, ii. 253, 279; in Protozoa, ii. 1, 9, 12, 16, 26-30, 37, 38, 45, 47.
—— Vegetable, i. 224-227; origin and multiplication of, i. 227, 228; binary subdivision of, in protophyta, i. 230, 233, 241, 246, 254, 264-266, 279; in Phanerogamia, i. 352-356; cyclosis in, i. 260, 263, 356-359; thickening deposits in, i. 359-361; spiral deposits in, i. 361; starch-grains in, i. 361, 362; raphides in, i. 363.
Cellular Tissue, Animal, ii. 273; Vegetable, ordinary forms of, i. 353-356; stellate, i. 354, 355; formation of, i. 356.
Cellulose, i. **225**; tests for, i. 208, 209.
Cements, Microscopic, i. 178, 179.
Cement-Cells, i. 180.
Cementum of Teeth, ii. **259**.
CEPHALOPODS, shell of, ii. **180; chromatophores** of, ii. 191.
Ceramiaceæ, i. 334.
Ceratium, ii. 38, 39.
Cercomonas, development of, ii. **29, 30**.
Cestoid Entozoa, ii. 192, 193.
Chætocereæ, i. 295.
Chætophoraceæ, i. 257, 258.
Chalk, formation of, ii. 305-308.
'Challenger' Expedition, use of tow-net in, i. 221 *note*; collection of Globigerinæ in, ii. 86, 87; observations in, on Bathybius, ii. 20; on deep-sea sediments, ii. 312, 313.
Characeæ, i. 259-262; cyclosis of fluid in, i. 260, 261; multiplication of, by gonidia, i. 262; sexual apparatus of, i. 260-262.
Cheilostomata, ii. 162.
Chemical Microscope, i. 82-83.
—— Re-agents, i. 208, 209.
Chemistry, microscopic, ii 326.
Cherry-stone, cells of, i. 360.
Chilodon, teeth of, ii. 43; self-division of, ii. 46.
Chirodota, calcareous skeleton of, ii. 149.
Chitine of Insects, ii. 220.
Chlamidomyxis, i. 327, 328.
Choroid, pigment of, ii. 275.
Chromatic Aberration, i. 8, 9; means of reducing and correcting, i. 10-16; residual, in high-angled Objectives, i. 173.
Chromatophores of Cephalopods, ii. 191.
Chromic acid, as solvent, i. 201; use of, for hardening, i. 203.
Chyle, corpuscles of, ii. 270.
Cienkowski, on Myxomycetes, i. 327; on Noctiluca, ii. 37.
Cidaris, spines of, ii. 135.
Ciliary action, nature of, ii. 41; in Protophytes, i. 250; in Infusoria, ii. 41; on gills of Mollusks, ii. 189; on epithelium of Vertebrata, ii. 277.
Ciliate Infusoria, ii. 41.
Cilio-flagellata, ii. 37-39.
Circulation of Blood, in Vertebrata, ii. 286-292; in Insects, ii. 237, 238; alternating, in Tunicata, ii. 164, 167
Circulation, Vegetable, see Cyclosis.
Cirrhipeds, metamorphosis of, ii. 213, 214.
Cladocera, ii. 209.
Clark, Prof. H. James, on flagellate Infusoria, ii. 32; on Sponges, ii. 118 *note*.
Clathrulina elegans, ii. 13, 15.
Clavellinidæ, ii. 166-168.
Cleanliness, importance of, to Microscope, i. 134; in mounting objects, i. 217.
Clematis, stem of, i. 368, 376.
Closterium, cyclosis in, i. 264; binary subdivision of, i. 264, 265; conjugation of, i. 267-269.
Clypeaster, spines of, ii. 144.
Coal, nature of, ii. 303, 304.
Coalescence, molecular, ii. 323-325.
Cobweb-Micrometer, i. 92, 93.
Coccoliths and Coccospheres, ii. 19, 20, 325.
Cocconeidæ, i. 296.
Cockchafer, cellular integument of, ii. 220; eyes of, ii. 229; antenna of, ii. 232, 233; spiracle of larva of, ii. 240.
Cockle of Wheat, ii. 193.
Coddington lens, i. 20.
Codosiga, life history of, **ii. 32, 33**.
Cœlenterata, ii. 2.
Cœnurus, **ii.** 192, 193.

Cohn, Dr., his researches on Protococcus, i. 232-236; on Volvox, i. 242,243; on Stephanosphæra, i. 242 note; on Sphæroplea, i. 255; on Schizomycetes, i. 307, 308; on reproduction of Rotifera, ii. 59; his cultivation-solution, i. 307 note.
Coleoptera, integument of, ii. 220; antennæ of, ii. 232, 233; mouth of, ii. 233, 234.
Collection of Objects, general directions for, i. 219-221.
Collins, Mr., his Harley Binocular, i. 70-71; his Eye-piece caps, i. 70; his Students' Microscope, i. 59; his Graduating Diaphragm, i. 102, 104.
Collomia, spiral fibres of, i. 361.
Collozoa, ii. 115.
Colonial nervous system of Polyzoa, ii. 160.
Colorless corpuscles of Blood, ii. 270, 271.
Columella of Mosses, i. 342.
Comatula, metamorphosis of, ii. 152-156; nervous system of, ii. 285.
Compound Microscope, optical principles of, i. 22-25; mechanical construction of, i. 40-42, 51, 52; Educational, i. 52-55; Students', i. 55-67; Second Class, i. 67-73; First class, i. 73-80; for special purposes, i. 80-85.
Compressor, i. 126, 127; use of, i. 142.
Concave lenses, refraction by, i. 5, 6; use of, in Achromatic combinations, i. 10-15.
Conceptacles of Marchantia, i. 337.
Concretions, calcareous, ii. 323-325.
Condensers, Achromatic, i. 102-104; Webster, i. 103; Swift's new combination, i. 113.
——— for Opaque objects, ordinary, i. 114; Bull's-eye, i. 115; mode of using, i. 147-150.
Confervaceæ, i. 253; self-division of, i. 254; zoospores of, i. 255; sexual reproduction of, i. 254, 255.
Conidia of Fungi, i. 322.
Coniferæ, peculiar woody fibre of, i. 364; absence of ducts in, i. 369; structure of stem in, i. 371, 372; pollen-grains of, i. 385 note; fossil, ii. 302.
Conjugateæ, i. 236, 237.
Conjugation, of Palmoglæa, i. 232; of Desmidiaceæ, i. 267, 268; of Diatomaceæ, i. 281, 282; of Conjugateæ, i. 236, 237; of Monadina, ii. 28-31; of Noctiluca, ii. 37; of Vorticellina, ii. 52;—see Zygosis.
Connective Tissue, ii. 273; corpuscles of, ii. 254, 273, 274.
Contractile vesicle, of Volvox, i. 238; of Actinophrys, ii. 11; of Amoeba, ii. 16; of Infusoria, ii. 25, 45.
Conversion of Relief, i. 27, 28, 33, 34.
Convex lenses, refraction by, i. 3-5; formation of images by, i. 6.

Copepoda, ii. 208.
Coquilla-nut, cells of, i. 360.
Coral, cutting sections of, with animal. i. 200.
Corallines, true, i. 335; Zoophytic, ii. 129.
Cork, i. 373.
Corn, blights of, i. 323; ii. 193.
Corn-grains, husk of, i. 388.
Corns, structure of, ii. 276.
Cornuspira, ii. 70.
Corpuscles of Blood, ii. 267-271.
Correction of Object-glasses, for Spherical Aberration, i. 8, 9, 173; for Chromatic Aberration, i. 10, 11, 173; for thickness of covering glass, i. 11, 14, 135-141.
Coscinodisceæ, i. 289-291.
Cosmarium, binary subdivision of, i. 265; conjugation of, i. 267; development of, i. 268.
Cover-correction of Objectives, i. 11, 14, 135-141.
Covering-glass, i. 176, 177.
Crab, shell-structure of, ii. 214, 215; metamorphosis of, ii. 216.
Crabro, integument of, ii. 220.
Crag-Formation, ii. 309.
Cricket, gastric teeth of, ii. 237; sounds produced by, ii. 242.
Crinoidea, skeleton of, ii. 146; metamorphosis of, ii. 153-156.
Cristatella, ii. 162.
Cristellaria, ii. 85.
Crouch, Mr., his Educational Microscope, i. 53; his Students' Binocular, i. 65, 66; his stage-centering adjustment, i. 80; his adapter for Beck's side-reflector, i. 116.
Crusta Petrosa of Teeth, ii. 259.
CRUSTACEA, ii. 205-217; lower forms of, ii. 205-206; Entomostracous, ii. 206-211; Suctorial, ii. 212; Cirrhiped, ii. 213, 214; Decapod, shell of, ii. 214, 215; metamorphosis of, ii. 215, 216.
CRYPTOGAMIA, general plan of structure of, i. 331, 351, 352;—see Protophyta, Algæ, Lichens, Fungi, Hepaticæ, Mosses, Ferns, etc.
Crystallization, Microscopic, ii. 318, 323.
Ctenoid scales of Fish, ii. 261, 262.
Ctenophora, ii. 137-139.
Culture of Protophytic Fungi, i. 123 note, i. 307; of Flagellate Infusoria, ii. 31.
Curculionidæ, scales of, ii. 220, 228; elytra of, ii. 128; foot of, ii. 244.
Cuticle of Animals, ii. 275-277.
——— of Equisetaceæ, i. 349; of leaves, i. 379.
Cutis Vera, ii. 274.
Cuttle-fish, shell of, ii. 180; chromatophore of, ii. 191.
Cyanthus, seeds of, i. 387.
Cyclammina, ii. 82.
Cycloclypeus, ii. 70.

Cycloid scales of Fish, i. 261, 262.
Cyclops, ii. 208, 209; fertility of, ii. 210.
Cyclosis, in Vegetable cells, i. 226; in Closterium, i. 263; in Diatomaceæ, i. 273; in Chara, i. 259, 260; in cells of Phanerogamia, i. 356–359; in Rhizopods, ii. 6.
Cydippe, ii. 137, 138.
Cymbelleæ, i. 297
Cynipidæ, ovipositor of, ii. 245.
Cypris, ii. 207.
Cypræa, structure of shell of, ii. 179.
Cystic Entozoa, ii. 192, 193.
Cysticercus, ii. 192, 193.
Cytherina, ii. 207.

Dactylocalyx, ii. 121.
Dallinger, Mr., on flagellum of Bacterium termo, i. 309; his Microscope Lamp, i. 133; on qualities of Objectives, i. 162, 165; Preface, v., vi.
Dallinger and Drysdale, their researches on Monadina, ii. 26–32.
Dallingeria Drysdali, ii. 26–28.
Dalyell, Sir J. G., on development of Medusæ, ii. 132–134.
Dammar-Varnish, i. 178, 213.
Daphnia, ii. 209, 210.
Darker's Selenites, i. 113.
Davies, Mr., on Microscopic Crystallization, ii. 319–321 *note*.
Dawson, Dr., on Eozoön, ii. 102.
Deane's Gelatine, i. 211.
De Bary, Dr., on Myxomycetes i. 327 *note*.
Decalcification, i. 201.
Decapod Crustacea, shell of, ii. 214, 215; metamorphosis of, ii. 215, 216.
Defining power of Objectives, i. 161–163, 173.
Dehydration, by Alcohol, i. 195; by Carbolic Acid, i. 215.
Delsaux, Rev. J., on Brownian movements, i. 154 *note*.
Demodex folliculorum, ii. 249.
Demonstrating Microscope, Beale's, i. 81.
Dendritina, ii. 72.
Dendrodus, teeth of, ii. 311.
Dentine of Teeth, ii. 258–260.
Depressions, distinction of, from elevations, i. 151.
Dermestes, hair of, ii. 228.
Desiccation, tolerance of, by Protophytes, i. 235; by Infusoria, ii. 31, 49; by Rotifera, ii. 59, 60; by Entomostraca, ii. 210.
Desmidiaceæ, general structure of, i. 262–264; cyclosis in, i. 263; binary subdivision of, i. 264–266; formation of gonidia in, i. 267; conjugation in, i. 267, 268; classification of, i. 268, 269; collection of, i. 269, 270.
Deutzia, stellate hairs of, i. 379.
Development, of Annelids, ii. 197–201; of Anodon, ii. 183, of Ascidians, ii. 168; of Cirrhipeds, ii. 213, 214; of Crab, ii. 215; of Desmidiaceæ, i. 267; of Diatomaceæ, i. 279–281; of Echinodermata, ii. 150–156; of Embryo(Animal), ii. 1, 122, of Embryo (Vegetable), i. 351, 352; of Entomostraca, ii. 211–213; of Ferns, i. 348; of Gasteropods, ii. 184–189; of Insects, ii. 247, 248; of Leaves, i. 356; of Medusæ, ii. 126, 127; of Mosses, i. 342; of Nudibranchiata, ii. 185; of Palmoglæa, i. 230; of Pollen-grains, i. 384; of Protococcus, i. 232; of Sponges, ii. 118; of Stem, i. 374; of Vegetable-cell, i. 227, 228; of Volvox, i. 238–241.
Diagonal Scales, i. 94, 99.
Diamond-beetle, scales of, ii. 220; elytra of, ii. 228; foot of, ii. 243.
Diaphragm Eye-piece, Slack's, i. 95.
Diaphragm-Plate, i. 101–103.
Diatoma, i. 286.
Diatomaceæ, general structure of, i. 273, 274; silicified valves of, i. 274–276; surface-markings of, i. 276–279; binary subdivision of, i. 279, 280; conjugation of, i. 281, 282; gonidia of, i. 282; auxospores of, i. 282; movements of, i. 283; classification of, i. 284; general habits of, i. 301, 302; fossilized deposits of, i. 302, 303, ii. 304; collection of, i. 302–304; mounting of, i. 305, 306.
Diatoms, as Tests, i. 169–172, i. 276–279.
Dichroism, ii. 322.
Dicotyledonous Stems, structure of, i. 369–375.
Dictyoloma, seeds of, i. 387.
Didemnians, ii. 166.
Didymoprium, i. 269; self-division of, i. 265; conjugation of, i. 269.
Difflugia, ii. 18.
Diffraction, errors arising from, i. 154–157; production of microscopic images by, i. 157–160.
Diphtheria, fungus of, i. 320.
Dipping-tubes, i. 127.
Diptera, mouth of, ii. 235; halteres of, ii. 243; ovipositors of, ii. 246.
Discorbina, ii. 89.
Disk-holder, Beck's, i. 121; Morris's, i. 122.
Disk-illuminator, Wenham's, i. 105.
Dispersion, chromatic, i. 8, 9.
Dissecting Instruments, i. 187; Trough, i. 187; Microscopes, i. 44–50.
Dissection, Microscopic, i. 186–188.
Distoma, ii. 194.
Dog, epidermis of foot of, ii. 276.
Doris, palate of, ii. 182; spicules of, ii. 179; development of, ii. 185.
Dorsal Vessel of Insects, ii. 237.
Double-staining, i. 207.
Doublet, Wollaston's, i. 20; Steinheil's, i. 21.
Dragon-fly, eyes of, ii. 230; larva of, ii. 238–240.

INDEX. 341

Drawing Apparatus, i. 96-99.
Draw-Tube, i. 87.
Dropping Bottle, i. 211.
Drosera, hairs of, i. 379.
Dry-mounting of Objects, i. 179, 183.
Drysdale, Dr., see Dallinger.
Ducts, of Plants, i. 365, 366.
Dujardin, M., on Sarcode of Foraminifera, etc., i. 222 *note*; on Rotifera, ii. 60-62.
Dunning's Turn-Table, i. 184.
Duncan, Dr., on Fungi in coral, i. 321.
Duramen, i. 370.
Dytiscus, foot of, ii. 244; trachea and spiracle of, ii. 339.

Eagle-Ray, teeth of, ii. 260.
Earwig wings of, ii. 242.
Eccremocarpus seeds of, i. 386.
Echinida, shell of, ii. 140, 141; ambulacral disks of, ii. 141, 142; spines of, ii. 142, 143; pedicellariæ of, ii. 144; teeth of, ii. 144, 145; metamorphosis of, ii. 151, 152.
ECHINODERMATA, skeleton of, ii. 140-145; metamorphoses of, ii. 150-153.
Echinus-spines, cutting sections of, i. 196-200, ii. 146-148.
Ectocarpaceæ, i. 332.
Ectoderm, ii. 1.
Ectosarc of Rhizopods, ii. 7, 15.
Ectoplasm of Vegetable cell, i. 225.
Edmunds, Dr., his immersion-paraboloid, i. 109; his parabolized gas-slide, i. 125.
Educational Microscopes, i. 53-55.
Eel, scale of, ii. 262; gills of, ii. 299.
Eels, of paste and vinegar, ii. 193.
Eggs of Insects, ii. 246;—see Winter-Eggs.
Egg-shell, fibrous structure of, ii. 272; calcareous deposit in, ii. 324.
Ehrenberg, Prof., his researches on Infusoria, ii. 24, 25; on Rotifera, ii. 24; on Polycystina, ii. 109, 116; on composition of Greensands, ii. 93 *note*, ii. 309
Elastic Ligaments, ii. 272.
Elaters of Marchantia, i. 338.
Elementary Parts of Animal body, ii. 253-255;—see Tissues.
Elevations, distinction of, from depressions, i. 151.
Elytra of Beetles, ii. 242.
Embryo, see Development.
Embryo-sac of Phanerogamia, i. 352.
Empusa musci, i. 318.
Enamel of Teeth, ii. 259.
Encrinites, see Crinoidea.
Encysting process, of Protophytes, i. 230-236; of Infusoria, ii. 47-49.
End-bulbs of sensory Nerves, ii. 285.
Endochrome, of Vegetable cell, i. 225; of Diatomaceæ, i. 273.
Endoderm, ii. 1.

Endogenous Stems, structure of, i. 367, 375.
Endoplasm of Vegetable cell, i. 225.
Endosarc of Rhizopods ii. 7, 15.
Endosperm of Phanerogams, i. 353.
Enterobryus, i. 319, 320.
Entomostraca (Crustacea), ii. 205-211; classification of, ii. 207-210; reproduction of, ii. 210-212.
Entophytic Fungi, i. 323, 324.
Entozoa, ii. 192-194; Cestoid, ii. 192; Cystic, ii. 192; Nematoid, ii. 193, 194; Trematode, ii. 194.
Eosin, as staining agent, i. 206.
Eozoic Limestone, ii. 310.
Eozoön Canadense, ii. 101-106.
Ephemera, larva of, ii. 219, 237, 240.
Ephippium of Daphnia, ii. 211.
Epidermis, Animal, ii. 275, 276.
——— Vegetable, i. 377-380.
Epithelium, ii. 276; ciliated, ii. 277.
Epithemia, i. 285; conjugation of, i. 281.
Equisetaceæ, cuticle of, i. 349; spores of, i. 349.
Erecting Binocular, see Stephenson.
Erecting Prism, Nachet's, i. 88.
Erector, Lister's, i. 87.
Errors of Interpretation, i. 150-156.
Euglena, ii. 33.
Eunotieæ, i. 285.
Euplectella, ii. 121.
Euryale, skeleton of, ii. 145.
Ewart, Prof., on Bacillus, i. 311-313.
Exogenous Stems, structure of, i. 369-375.
Eyes, care of, i. 134; Preface, vi., vii.
Eyes of Mollusks, ii. 190, 191; of Insects, ii. 229-231; of Trilobite, ii. 310.
Eye-piece, i. 22; Huyghenian, i. 23, 24; Kellner's, i. 24, 25; solid, i. 25; Ramsden's, i. 25; Binocular i. 33; Erecting, i. 88; Spectroscopic, i. 89; Micrometric, i. 92-95; Diaphragm i. 95.
——— Collins's shades for, i. 70.
Eye-piecing, deep, disadvantage of, i. 136, 137, Preface, vi.

Falconer, Dr., on bones of fossil Tortoise, ii. 312.
Fallacies of Microscopy, i. 150-156.
Farrant's Medium, i. 211, 213.
Farre, Dr. Arthur, his researches on Bowerbankia, ii. 161.
Fat-cells, ii. 278; capillaries of, ii. 298.
Feathers, structure of, ii. 263, 266.
Feet of Insects, ii. 243-245; of Spiders, ii. 250.
Fermentation, influence of vegetation on, ii. 315.
FERNS, i. 344-349; scalariform ducts of, i. 344; fructification of, i. 344-346; spores of, i. 346; prothallium of, i. 347; antheridia and archegonia of, i. 347; generation and development of, i. 347, 348.

Fertilization of ovule, in Flowering plants, i. 352, 385.
Fibre-cells of anthers, i. 384; of seeds, i. 360, 361.
Fibres, Muscular, ii. 281-283.
——— Nervous, ii. 284-286.
——— Spiral, of Plants, i. 360, 361, i. 364-366.
Fibrillæ of Muscle, structure of ii. 281.
Fibro-Cartilage, ii. 279.
Fibro-Vascular Tissue of Plants, i. 363.
Fibrous Tissues of Animals, ii. 271-273; formation of, ii. 254.
Field's Dissecting and Mounting Microscope, i. 49, 50; his Educational Microscope, i. 53.
Filiferous Capsules of Zoophytes, ii. 135.
Finders, i. 99; Maltwood's, i. 100.
Fine Adjustment, i. 40; uses of, i. 137-139.
FISHES, bone of, ii. 257, 258; teeth of, ii. 258, 259; scales of ii. 261-263; blood of, ii. 268, 269; circulation in, ii. 288; gills of, ii. 299.
Fishing tubes, i. 127.
Flagella, of Protococcus, i. 233; of Vorvox, i. 237; of Bacteria, i. 308.
Flagellata (Infusoria), ii. 26-37; their relation to Sponges, ii. 117.
Flatness of field of Objectives, i. 164.
Flints, organic structure in, ii. 308; examination of, ii. 308.
Flint Glass, dispersive power of, i. 10; use of, in Objectives, i. 15, 16.
Floridæ, i. 334.
Floscularians, ii. 60, 61.
Flowers, small, as Microscopic objects, i. 382; structure of parts of, i. 382-385.
Fluid, mounting objects in, i. 215-217.
Fluke, ii. 194.
Flustra, ii. 157-161.
Fly, fungous disease of, i. 318; number of objects furnished by, ii. 218; eye of ii. 230; circulation in, ii. 238; tongue of, ii. 234; spiracle of, ii. 239; wing of, ii. 241; foot of, ii. 243; development of, ii. 248.
Focal Adjustment, i. 137-139; errors arising from imperfection of, i. 151-153.
Focal Depth of Objectives, i. 163; increase of, with Binocular, i. 38.
Follicles of Glands, ii. 280.
Foot of Fly, ii. 243; of Dytiscus, ii. 244; of Spider, ii. 250.
FORAMINIFERA, ii. 64-106; their relation to Rhizopods, ii. 7, 66; their general structure, ii. 66-70; porcellanous, ii. 70-77; arenaceous, ii. 77-85; vitreous, ii. 85-106; collection and mounting of, ii. 107-109; fossil deposits of, see Fossil Foraminifera; mode of making sections of, i. 198 *note*.
Forceps, i. 123; Stage, i. 120; Slider, i. 185.

Forficulidæ, wings of, ii. 242.
Formed Material, Dr. Beale on, ii. 253-255.
Fossil Bone, ii. 311.
——— Diatoms, i. 302, 303, ii. 304, 305.
——— Foraminifera, ii. 72, 73, 78, 82-84, 90, 96-106, 304-310.
——— Radiolaria, ii. 109.
——— Sponges, ii. 307, 308.
——— Teeth, ii. 310, 311.
——— Wood, i. 371, 373, ii. 302-304.
Fowl, lung of, ii. 300.
Fragillariæ, i. 286.
Free Cell formation in Plants, i. 227, 228.
Freezing Microtome, i. 191, 192.
Frog, blood of, ii. 268-271; pigment cells of, ii. 275, 276; circulation in web of, ii. 286-288; in tongue of, ii. 288; in lung of, ii. 288; structure of lung of, ii. 299, 300.
Fructification, of Chara, i. 260-262; of Fuci, i. 332-334; of Florideæ, i. 334-335; of Lichens, i. 329; of Fungi, i. 321, 324; of Marchantia, i. 337, 338; of Mosses, i. 339-342; of Ferns, i. 344-348; of Equisetaceæ, i. 349; of Lycopodiaceæ, i. 350.
Fucaceæ, i. 331-334; sexual apparatus of, i. 332-334; development of, i. 334.
FUNGI, relation of, to Algæ, i. 229, 307; to Animals, i. 307, 325-329; to Lichens, i. 329; simplest forms of, i. 307-316; in bodies of living Animals, i. 316-320; in substance, or on surface, of Plants, i. 323, 324; amœboid states of, i. 325, 326; universal diffusion of sporules of, i. 313-321; culture of, i. 23 *note*, i. 307.
Furcularians, ii. 62.
Fusulina, ii. 90, 91.
Gad-flies, ovipositor of, ii. 246.
Gall-flies, ovipositor of, ii. 245.

Galls of Plants, ii. 245.
Ganglion-Cells, ii. 284.
Ganoid scales of Fish, ii. 263.
GASTEROPODA, structure of shells of, ii. 178; palates of, ii. 180-183; development of, ii. 183-189; organs of sense of, ii. 190, 191.
Gastrula, ii. 1.
Geikie, Prof., on Geographical evolution, ii. 313 *note*.
Gelatine, see Glycerine jelly.
Gelatinous Nerve fibres, ii. 284.
Generation, distinguished from Growth, i. 227-229; in Cryptogams, i. 350; in Phanerogams, i. 352.
Geology, applications of Microscope to, ii. 302-314.
Geranium-petal, peculiar cells of, i. 383.
Germ-cell of Cryptogams, i. 350; of Phanerogams, i. 352.
Germinal Matter, Dr. Beale on, ii. 253.

INDEX.

Gills, of Mollusks, ciliary motion on, ii. 189, 190; of Fishes, distribution of vessels in, ii. 299; of Water-newt, circulation in, ii. 288.
Gizzard of Insects, ii. 237.
Glands, structure of, ii. 279, 280.
Glandular woody fibre of Conifers, i. 364.
Glass Slides, i. 175, 176.
—— —— Stage-plate, i. 122.
—— —— Thin, i. 176, 177.
Glaucium, cyclosis in hairs of, i. 359.
Globigerina, ii. 86, 87.
Globigerina-mud, ii. 86; its relation to Chalk-formation, ii. 305-309.
Globigerinida, ii. 86-91.
Glochidium, ii. 183.
Glue, Liquid, i. 179.
——, Marine, uses of, i. 179, 181.
Glycerine, for mounting objects, i. 210-213.
Glycerine Jelly, i. 211; mounting in, i. 212.
Glycerine and Gum medium, i. 211, 213.
Gnat, scale of, i. 155; transparent larva of, ii. 237.
Gold-Size, use of, i. 178.
Gomphonemeæ, i. 298.
Goniometer, i. 95.
Gonidia, i. 229 *note*, i. 230; multiplication by, in Desmidiaceæ, i. 267; in Pediastreæ, i. 270; in Diatomaceæ, i. 280; in Hydrodictyon, i. 253; in Chara, i. 261; in Lichens, i. 329; in Fungi, i. 322, 324; in Volvox, i. 241.
Gordius, ii. 193.
Gorgonia, spicules of, ii. 136, 137.
Gosse, Mr., on mastax of Rotifers, ii. 56, 57; on sexes of Rotifers, ii. 58; on Melicerta, ii. 61; on thread-cells of Zoophytes, ii. 136.
Grammatophora, i. 288; its use as test, i. 171.
Grantia, structure of, ii. 120, 121.
Grasses, silicified cuticle of, i. 379.
Gray, Dr., on palates of Gasteropods, ii. 182; on development of Buccinum, ii. 187.
Green Sands, Foraminiferal origin of, ii. 309, 310; Prof. Ehrenberg on composition of, ii. 93 *note*, ii. 309.
Gregarinida, ii. 21, 22.
Gromia, ii. 8-10.
Growing-Slide, i. 122.
Growth, distinguished from Generation, i. 227-229.
Guano, Diatomaceæ of, i. 303.
Gulliver, Mr., on Raphides, i. 363; on sizes of Blood-disks, ii. 269-271.
Gum Arabic, i. 179.
Guy, Dr., on sublimation of Alkaloids, ii. 326.
Gymnosperms, i. 352, 353.

Haeckel, Prof., on Gastræa theory, ii.
22 *note;* on Monerozoa, ii. 2; on Bathybius, ii. 19; on Radiolaria, ii. 110; on Infusoria, ii. 26 *note;* on Calcareous Sponges, ii. 120 *note.*
Hæmatococcus, i. 245; its relations to Protococcus, i. 233.
Hæmatoxylin, as staining agent, i. 206.
Hairs, of Insects, ii. 227; of Mammals, ii. 263-266.
—— of Vegetable cuticles, i. 378; rotation of fluid in, i. 358, 359.
Halichondria, spicules of, ii. 119.
Halifax, Dr., on making Sections of Insects, ii. 219.
Haliomma, ii. 113.
Haliotis, palate of, ii. 182.
Haliphysema, ii. 80.
Halodactylus, ii. 161.
Halophragmium, ii. 81.
Halteres of Diptera, ii. 243.
Hand-Magnifiers, i. 19-22, 43, 44.
Hard Substances, cutting Sections of, i. 196-199.
Hardening of Animal Substances, i. 203.
Harley Binocular, i. 70.
Harting, Prof., on Calcareous Concretions, ii. 324.
Hartnack, M., his diagonal Micrometer, i. 94; on Surirella, i. 171.
Harvest-bug, ii. 249.
Haversian Canals of Bone, ii. 256.
Haustellate Mouth, ii. 236.
Haycraft, Mr., on Muscular fibre, ii. 286.
Hazel, stem of, i. 370.
Hearing, organs of (?), in Insects, ii. 233.
Heart-wood, i. 370.
Heat, tolerance of, by Bacteria, etc., i. 313; by Infusoria, ii. 31, 32.
Heliopelta, i. 292.
Heliozoa, ii. 8, 11-14.
Helix, palate of, ii. 181.
Hemiptera, wings of, ii. 242.
Hepaticæ, i. 335-338; see Marchantia.
Hepworth, Mr., on feet of Insects, ii. 244.
Hertwig, Dr., on Rhizopods, ii. 8 *note*, ii. 9; on Foraminifera, ii. 64 *note.*
Heteromita, ii. 30.
Heterostegina, ii. 99.
Hexiradiate Sponges, ii. 121.
Hicks, Dr., on Volvox, i. 244; on Amœboid production in root-fibres of Mosses, i. 339; on eyes of Insects, ii. 230; on peculiar organs of sense in Insects, ii. 233 *note*, ii. 243.
Hincks, Rev. T., on Hydroid Zoophytes, ii. 126; on Polyzoa, ii. 162 *note.*
Hippocrepian Polyzoa, ii. 161, 162.
Hogg, Mr., on development of Lymnæus, ii. 186.
Hollyhock, pollen-grains of, i. 37, 167, i. 385.
Holothurida, skeletons of, ii. 148-150.
Holtenia, ii. 121.
Homogeneous Immersion, i. 17 18.

Hoofs, structure of, ii. 267.
Hooker, Sir J. D., on Antartic Diatoms, i. 301.
Hoop, of Diatoms, i. 275, 279, 280.
Hormosina, ii. 81.
Hornet, wings of, ii. 242.
Horns, structure of, ii. 267.
Houghton, Rev. W., on Glochidium, ii. 183.
Hudson, Dr., on in Pedalio, ii. 58, 63.
Huxley, Prof., on Protoplasm, i. 222; on cell formation in Sphagnaceæ, i. 342; **on** Bathybius, ii. 19; on Coccoliths, **ii.** 19, 20; on Rotifera, ii. 59, 63, on Thalassicolla, ii. 116 *note;* on Noctiluca, ii. 34 *note;* on Shell of Mollusca, ii. 173; **on** Appendicularia, ii. 169; on Blood **of** Annelida, ii. 197; on Shell **of** Crustacea, ii. 214 *note;* on Reproduction of Aphides, ii. 246.
Huyghenian eye piece, i. 23, 24.
Hyalodiscus, i. 171, **i.** 289.
Hydatina, ii. 62; reproduction **of, ii.** 58.
Hydra, life-history **of, ii.** 122–126.
Hydra tuba, developments of Acalephs from, ii. 132–134.
Hydrodictyon, i. 252, 253.
Hydrozoa, simple, ii. 123; composite, ii. 126–131; their relation to Medusæ, ii. 126, 131–134.
Hyla, preparation of nerves of, ii. **286.**

Ice-Plant, cuticle of, i. 378.
Ichneumonidæ, ovipositor of, ii. 245.
Illumination of Opaque objects, i. 147–150; of Transparent objects, i. 144–147; diverse effects of, on lined objects, i. 146–147.
Illuminators, Black-ground, i. 106–110, i. 147.
——————— Oblique, i. 104, 105, i. 145–147.
——————— Parabolic, i. 107, 108.
——————— Reflex, i. 109, 110.
——————— Side, i. 114–117.
——————— Vertical, i. 117–119.
——————— Wenham's Disk, i. 105; his Reflex, i. 109, 110.
——————— White Cloud, i. 111.
Imbedding processes, i. 194–196.
Immersion-Lenses, i. 16–18.
Images, formation of, by convex lenses, i. 6.
Index of Refraction, i. 1–3.
Indigo-carmine, as staining agent, i. 207.
Indian Corn, cuticle of, i. 377, 380.
Indicator, Quekett's, i. 96.
Indusium of Ferns, i. 346.
INFUSORIA, ii. 25–53; Flagellate, ii. 26; Suctorial, ii. 39; Ciliate, ii. 41; movements of, ii. 43, 44; internal structure of, ii. 44, 45; binary subdivision of, ii. 46; encysting process of, ii. 47–49; sexual generation (?) of, ii. 49–52.
Infusorial Earths, i. 302.

Injections of Blood-vessels, mode of making, ii. 292–298.
INSECTS, great numbers of objects furnished by, ii. 218; microscopic forms of, ii. 219; antennæ of, ii. 232, 233; circulation of blood in, ii. 237, 238; eggs of, ii. 246, 247; eyes of, ii. 229–231; feet of, ii. 243–245; gastric **teeth** of, **ii.** 237; hairs of, ii. 227; integument of, ii. 219, 229; mouth of, ii. 233–236; organs of hearing in, ii. 233; of smell in, ii. 243; of taste in, ii. 236; ovipositors of, ii. 245, 246; scales of, ii. 220–228; spiracles of, ii. 239, 240; stings of, ii. 245; tracheæ of, ii. 238–240; wings of, ii. 241–243.
Interference-spectra, i. 156–160.
Intermediate Skeleton of Foraminifera, ii. 70, 86, 91, 94, 103.
Internal Casts of Foraminifera, ii. 89, 92, 93, 99, 100, 105, 309.
Interpretation, errors of, i. 150–155.
Inverted Microscope, Dr. L. Smith's, i. 82.
Iodine, as test, i. 208.
Iris, structure of leaf of, **i.** 380, 381.
Iris-diaphragm, i. 102.
Isthmia, i. 293; markings on, i. 276; self-division of, i. 280.
Itch-Acarus, ii. 249.
Iulus, fungous vegetation in, i. 319.

Jackson, Mr. G., his model for Compound Microscope, i. 52; his Eye-piece Micrometer, i. 93, 94.
Jevons, Prof., on Brownian movement, **i.** 154.
Jukes, Prof., **on** Foraminiferal reef, ii. 305.

Kellner's Eye-piece, i. 24, **25.**
Kerona silurus, ii. 42.
Kent, Mr. S., on Flagellate Infusoria, ii. 32, 33; on Sponges, ii. 118 *note*.
Kidney, structure of, ii. 280.
Klein, Dr., on Cells and Nuclei, ii. 252 *note*.
Kleinenberg, Prof., on Hydra, ii. 126 *note*, ii. 253 *note;* his preparing fluid, i. 203; his staining fluid, i. 206.
Koch, on Sections of hard and soft substances, i. 200.
Kölliker, Prof., on Fungi in Shells, etc., **i.** 320 *note*.
Kovalevsky, on development of Ascidians, ii. 169 *note*.
Kühne, on contraction of Vorticellastalk, ii. 44.

Labelling of Objects, i. 218, 219.
Laboratory Dissecting Microscope, i. 47.
Labyrinthodon, tooth of, ii. 311.
Lachmann, see Claparède and Lachmann.
Lacinularia, Prof. Huxley on, ii. 58 *note*.

INDEX. 345

Lacunæ of Bone, ii. **256**, 257.
Lagena, ii. 66, 85.
Laguncula, ii. 157-160.
Lamellicornes, antennæ of, ii. 232.
Lamps for Microscope, i. 131, 132.
Lankester, Prof. E. Ray, on amœboids in fresh-water Medusa, ii. 122; on development of Limnæus, ii. 186.
Larvæ of Echinoderms, ii. 150-156.
Laurentian Formation of Canada, ii. 101, ii. 310; of Europe, ii. 101 *note*.
Leaves, structure of, i. 380-382.
Leech, teeth of, ii. 203.
Leeson, Dr., his double-refracting Goniometer, i. 95; his Selenite-plate, i. 112.
Legg, Mr., on collection of Foraminifera, ii. 107, 108.
Leidy, Dr., on Enterobryus, i. 319; on Rhizopods, ii. 19 *note*.
Lenses, refraction by, i. 3, 4.
Lepidocyrtus, scales of, see Podura.
Lepidoptera, scales of, ii. 220-229; proboscis of, 236, 237; wings of, ii. 220, 242; eggs of, ii. 247.
Lepidosteus, bony scales of, ii. **257**, 263.
Lepidostrobi, i. 350.
Lepisma, scales of, ii. 223, **224**; diffraction-spectrum of, i. 159.
Lepralia, ii. 158, 162.
Lernœa, ii. 213.
Levant-Mud, microscopic organisms of, ii. 305.
Lever of Contact, i. 177.
Lewis, Mr. B., his freezing Microtome, i. 192.
Libellula, eyes of, ii. 230; respiration of larva of, ii. 240.
Liber, i. 374.
Lichens, composite nature of, ii. 329, 330.
Lichmophoreæ i. 286.
Lieberkühn (speculum), i. **117**, 118; mode of using, i. 150.
Lieberkühnia, ii. 6, 7.
Ligaments, structure of, ii. 272.
Light, for Microscope, i. 131-134; arrangement of, for Transparent objects, i. 143-147; for Opaque objects, i. 147-150.
Light-modifiers, i. 110.
Ligneous Tissue, i. 363, 364.
Limax, shell of, **ii.** 179; palate of, ii. 181.
Limestones, organic origin of, ii. 308, 309; Fusuline, ii. 90, 91, 309; Nummulitic, ii. 94, 96, 309; Milioline, ii. 309; Orbitoidal, ii. 100; Eozoic, ii. 101, 102, 310.
Limiting Angle, i. 3.
Limpet, palate of, ii. 182.
Liquid Glue, i. 179.
Lined Objects, diverse effects of Illumination on, i. 146, 147.
—— Tests, resolution of, i. 169-172.

Lister, Mr. J. J., his improvements in Achromatic lenses, i. 14; his Erector, i. 87; his observations on Zoophytes, ii. 127; on Social Ascidians, ii. 167, 168.
Lister, Prof., on Bacteria, etc., i. 314.
Lituolida, ii. 81-85.
Live-box, i. 123.
Liver, structure of, ii. 280, 281.
Liverwort, i. 335-338.
Lobb, Mr., on binary **subdivision in** Micrasterias, i. 266.
Lobosa, **ii.** 8, 14-19.
Loftusia, ii. **84**.
Logan, Sir W., on Laurentian **Formation**, ii. 101 *note*, ii. 310.
Lophophore of Polyzoa, ii. 158.
Lophyropoda, ii. 207, 208.
Lowne, Mr., on feet of Insects, ii. 244 *note;* on eyes of Insects, ii. 231 *note;* on development of Insects, ii. 248.
Lubbock, Sir J., on Daphnia, ii. 211; on Thysanura, ii. 223.
Lüders, Mad., on fermentation, i. 316.
Luminosity of Noctiluca, ii. 33, 36; of Anelida, ii. 202.
Lungs of Reptiles, ii. 299; of Birds, ii. 300; of Mammals, ii. 300.
Lycænidæ, scales of, ii. 221, 223.
Lycopodiaceæ, i. 350.
Lymnæus, development of, ii. **184**, 186.
Lymph, corpuscles of, ii. 270.

Machilis, scale of, ii. 224.
Macro-gonidia, i. 229 *note;* of Volvox, i. 241; of Pediastreæ, i. 271; of Hydrodictyon, i. 252.
Maddox, Dr., his Growing-Slide, i. 123; on cultivation of Microscopic Fungi, i. 123 *note*.
Magnifying power, augmentation of, i. 136, 137; determination of, i. 173, 174.
Magenta, as staining agent, i. 206.
Mahogany, section of, i. 372.
Malpighian bodies of Kidney, ii. 280.
—— layer of Skin, ii. 275.
Maltwood's Finder, i. 100.
Malvaceæ, pollen-grains of, **i. 385; their** use as tests, i. 37, 167.
Mammals, bone of, ii. 255-258; teeth of, ii. 259-261; hairs, hoofs, etc., of, ii. 263-267; blood of, ii. 267-271; **lungs** of, ii. 300.
Man, teeth **of,** 259-261; **hair of,** ii. 265, 266; blood **of,** ii. 267-271.
Mandibulate mouth **of** Insects, ii. **233**.
Marchantia, general structure of, **i. 335**; stomata of, i. 336; conceptacles **of, i.** 337; sexual apparatus of, i. 338.
Margaritaceæ, shells of, 173, 174.
Marine Glue, uses of, i. 179, 181.
Marsh, Dr. S., his section-lifter, i. 205; on Section-cutting, etc., i. 194, 213.
Mastax of Rotifera, ii. 56.
Mastogloia, i. 299, 300.

Matthews, Dr., his Micro-megascope, i. 88; his saw for Section-cutting, i. 197.
Media, Preservative, i. 209-211.
Medullary Rays, i. 354, 371-373.
——— Sheath, i. 364, 369.
Medusæ, their relation to Polypes, ii. 126, 131-134; fresh-water, amœboids in, 122.
Megalopa-larva of Crab, ii. 216.
Megatherium, teeth of, 261.
Melanospermeæ, i. 332.
Melicertians, ii. 60, 61.
Melolontha, see Cockchafer.
Melosira, i. 289; auxospores of, i. 282.
Menelaus, scale of, ii. 221, 222.
Meniscus Lenses, refraction by, i. 6.
Meridion circulare, i. 285.
Mesembryanthemum, cuticle of, i. 378.
Mesocarpus, i. 236.
Metamorphosis, of Annelids, ii. 198-201; of Ascidians, ii. 168-170; of Cirrhipeds, ii. 213, 214; of higher Crustacea, ii. 215, 216; of Entomostraca, ii. 211; of Echinoderms, ii. 150-153; of Infusoria, ii. 47-49; of Insects, ii. 248; of Mollusks, ii. 183-189.
Metazoa, ii. 2.
Mica-Selenite Stage, i. 113.
Micrasterias, binary sub-division of, i. 266; stato-spores of, i. 267.
Micro-Chemistry, ii. 326.
Micrococcus, i. 308.
Micro-gonidia, i. 229 *note;* of Protoccus, i. 234; of Desmidiaceæ, i. 267; of Hydrodictyon, i 253.
Micro-megascope, i. 88.
Micrometers, Ramsden's, i. 92, 93; Jackson's, i. 93, 94; Hartnack's, i. 95.
Micrometry, by Micrometer, i. 93-95; by Camera Lucida, i. 99.
Micropyle of Vegetable Ovule, i. 353, 386.
MICROSCOPE, support required for, i. 130, 131; care of, i. 134, 135; focal adjustment of, i. 137-141; arrangement of, for Transparent objects, i. 141-147; for Opaque objects, i. 147-150.
——— *Binocular,* see Binocular Microscope.
——— *Compound,* see Compound Microscope.
——— *Simple,* see **Simple Microscope.**
——— ——— Chemical, i. 82, 83.
——— ——— Demonstrating, i. 81.
——— ——— Dissecting, i. 44-50.
——— ——— Educational, i. 53-55.
——— ——— Inverted, i. 82-84.
——— ——— Mineralogical, ii. 315-317.
——— ——— Pocket, i. 8.
——— ——— Popular, i. 68.
——— ——— Portable, Binocular, i. 83.
——— ——— Student's, i. 55-67, ii. 332.
——— ——— Travelling, i. 81, 82.
Microscopic Dissection, i. 186-188.

Micro-Spectroscope, i. 89-92.
Microtome, Simple, i. 189; Hailes's, i. 190, 191; Strassburg, i. 191; freezing, i. 192; Rivet-Leiser, i. 192, 193.
Microzymes, i. 314.
Mildew of Corn, i. 323.
Miliolida, ii. 70.
Millon's test for Albuminous substances, i. 208.
Milne-Edwards, M., on Compound Ascidians, ii. 166, *note.*
Mineral Objects, ii. 313-326.
Minnow, circulation in, ii. 288.
Misinterpretation of microscopic appearances, causes of, i. 150-156.
Mites, ii. 249.
Mivart, Prof., on Radiolaria, ii. 110.
Moderator, Rainey's, i. 110.
Molecular Coalescence, ii. 323-325.
——— Movement, i. 153, 154.
MOLLUSCA, shells of, ii. 171-180; palates of, ii. 180-183; development of, ii. 183-189; ciliary motion on gills of, ii. 189, 190; organs of sense of, ii. 190, 191.
Molybdate of Ammonia, i. 207.
Monadina, ii. 26-32.
Monerozoa, ii. 2-7.
Monocotyledonous Stems, structure **of,** i. 367, 368.
Monothalamous Foraminifera, ii. 66.
Morehouse, Mr., on Lepisma-scale, **ii.** 224, ii. 227 *note.*
Morris, Mr., his Object-holder, i. 121, 122; on mounting Zoophytes, ii. 131.
MOSSES, structure of, i. 338, 339; sexual apparatus of, i. 340-342; development of spores of, i. 341.
Mother-of-Pearl, structure of, ii. **174.**
Moths, see Lepidoptera.
Moulds, fungous, i. 321, 322.
Mounting of objects, i. 211; in Canada Balsam, i. 214, 215; in cement cells, i. 215; in deep cells, i. 216.
Mounting-Instrument, Smith's, i. 186.
——— ——— Microscope, Field's, i. 49, 50.
Mounting-Plate, i. 185.
Mouse, hair of, ii. 264; cartilage of ear of, ii. 278; vessels of toe of, ii. 297.
Mouth of Insects, ii. 233-236.
Mucor, i. 323.
Mucous Membranes, structure of, ii. 275; capillaries of, ii. 298.
Müller, Dr. Fritz, on Polyzoa, ii. 160.
Müller, Prof. J., on Radiolaria, i. 110; on Echinoderm-larvæ, ii. 150-153.
Müller's fluid, for hardening, i. 203.
Muscardine, of Silk-worms, i. 317, 318.
Muscular Fibre, structure of, ii. 281-284; mode of examining and preparing, ii. 282; capillaries of, ii. 298.
Musk-deer, hair of, ii. 264; minute blood-corpuscles of, ii. 269.
Mussel, ciliary action on gills of, ii. 189; development of, ii. 183.

Mya, structure of hinge-tooth of, ii. 175.
Mycelium of Fungi, i. 320-325.
Myliobates, teeth of, ii. 258, 259.
Myriapods, hairs of, ii. 228.
Myriothela, amœboids in, ii. 122.
Myxomycetes, i. 325-327.

Nachet, M., his Stereoscopic Binocular, i. 28, 29; Stereo-pseudoscopic Binocular, i. 33-35; Binocular Magnifier, i. 48, 49; Student's Microscope, i. 63, 64; Chemical Microscope, i. 82, 83; Mineralogical Microscope, ii. 315-317; Erecting Prism, i. 88; Camera, i. 98. Porte-Objectif, ii. 333.
Nacre, structure of, ii. 173, 174.
Nais, ii. 202, 203.
Nassula, teeth of, ii. 43.
Navicellæ of Gregarinida, ii. 22.
Naviculæ, i. 298; movements of, i. 283.
Needles for Dissection, i. 188.
Nematoid Entozoa, ii. 193, 194.
Nemertes, larva of, ii. 198, 199.
Nepa, tracheal system of, ii. 239.
Nepenthes, spiral vessels of, i. 364.
Nervous Tissue, structure of, ii. 284, 285; mode of examining, ii. 286.
Net, Collector's, i. 219-221.
Nettle, sting of, i. 379.
Neuroptera, circulation in, ii. 237, 240; wings of, ii. 241.
Neutral-tint Reflector, i. 98.
Newt, circulation in larva of, ii. 288.
Nicol-Prism, i. 111.
Nitella, i. 259.
Nitzschieæ, i. 287.
Nobert's Test, i. 169, 170.
Noctiluca, ii. 33-37.
Nodosaria, ii. 85.
Nonionina, ii. 94.
Nose piece, i. 99.
Nostochaceæ, i. 249.
Nucleus, of Vegetable cells, i. 225-228; of Animal cells, ii. 254.
Nudibranchs, development of, ii. 185, 186.
Numerical Aperture of Objectives, ii. 327.
Nummulinida, ii. 69, 91-101.
Nummulite, structure of, ii. 94-98.
Nummulitic Limestone, ii. 94, ii. 309.
Nuphar lutea, parenchyma of, i. 354, 355.

Oak, galls of, ii. 245.
Object-Glasses, Achromatic, principle of, i. 7-10; Angular aperture of, i. 8, i. 161-163; ii. 327; Numerical aperture of, ii. 327; construction of, i. 11-16; immersion, i. 16-18; adjustment of, for covering glass, i. 14, i. 139-141; adaptation of, to Binocular, i. 35-38; working distance of, i. 161; defining power of, 161-163; focal depth of, i. 163; increase of, with Binocular, i. 38; resolving power of, i. 164; flatness of field of, i. 164; comparative value of i. 161-165; Preface, v., vi.; different powers of, tests for, i. 165-173; determination of magnifying power of, i. 173, 174.
Object-Holder, i. 121, 122.
Objects, mode of mounting, dry, i. 179, 183; in Canada balsam, i. 213, 214; in preservative media, i. 209-213; in cells, i. 215-217; see Opaque and Transparent Objects.
——— labelling and preserving of, i. 218, 219.
——— collection of, i. 219-221.
Oblique Illuminators, i. 104, 105.
Ocelli of Insects, ii. 229-231.
Octospores of Fuci, i. 333.
Œdogonieæ, i. 257, 258.
Oidium, i. 323.
Oil-globules, microscopic appearances of, i. 152, 153.
Oil-immersion Objectives, i. 17, 18; ii. 329.
Oleander, cuticle of, i. 378; stomata of, i. 380.
Oncidium, spiral cells of, i. 361.
Onion, raphides of, i. 363.
Oögonia of Fucaceæ, i. 333.
Oolite, structure of, ii. 282.
Oöspores, i. 228, 229; of Volvox, i. 243; of Achlya, i. 251; of Sphæroplea, i. 255; of Œdogonium, i. 258; of Batrachospermeæ, i. 259; of Chara, i. 262; of Fucaceæ, i. 333.
Opaque Objects, arrangement of Microscope for, i. 147-150; modes of mounting, i. 179, 183.
Operculina, ii. 94-96.
Ophiocoma, teeth and spines of, ii. 145.
Ophioglosseæ, prothallium of, i. 348.
Ophiurida, skeleton of, ii. 145; development of, ii. 151.
Ophrydinæ, ii. 46.
Orbiculina, ii. 72, 73.
Orbitoides, structure of, ii. 100, 101.
Orbitolina, ii. 90.
Orbitolites, structure and development of, ii. 70, 73-77; fossil, ii. 305.
Orbulina, ii. 86.
Orchideous Plants, i. 360.
Ord, Dr. W. M., on Calculi, ii. 326.
Ornithorynchus, hair of, ii. 265.
Orthoptera, wings of, ii. 242.
Osmic acid, uses of, i. 203.
Osmunda, prothallium of, i. 348 note.
Oscillatoriaceæ, i. 247, 249.
Ostraceæ, shells of, ii. 175, 176.
Ostracoda, ii. 207.
Otoliths of Gasteropods, ii. 190; of Fishes, ii. 324.
Ovipositors of Insects, ii. 245, 246.
Ovules of Phanerogamia, i. 352; fertilization of, i. 385; mode of studying, i. 385, 386.
Owen, Prof., on fossil Teeth, ii. 311; on fossil Bone, ii. 312.

Oxytricha form of Trichoda, ii. 47–49.
Oyster, shell of, ii. 176, 177

Pachymatisma, spicules of, ii. 120.
Pacinian corpuscles, ii. 286.
Palates of Gasteropods, ii. 180–183.
Palm, stem of, i. 367, 368.
Palmellaceæ, i. 245, 246.
Palmodictyon, i. 246.
Palmoglœa macrococca, life-history of, i. 230, 231.
Papillæ of Skin, structure of, ii. 274. 285; capillaries of, ii. 298; of Tongue, ii. 285.
Parabolic Speculum, i. 116.
Parabolized Gas-Slide. i. 125.
Paraboloid, i. 107, 108; immersion, i. 108, 109.
Paraffin, imbedding in, i. 194–196.
Paramecium, ii. 42; contractile vesicles of, ii. 45; binary subdivision of, ii. 46; sexual generation (?) of, ii. 49.
Parasitic Fungi in Animal bodies, i. 317–321; in Plants, i. 323, 324.
Parker, Mr. Jeffery, on Hydra, ii. 122.
Parkeria, ii. 83–85.
Passulus, fungous vegetation in, i. 321.
Paste, Eels of, ii. 93.
Pasteur, M., his researches on ferments, i. 313; on pébrine, i. 315.
Patella, palatal tube of, ii. 182.
Pearls, structure of, ii. 174.
Pébrine, i. 315.
Pecari, hair of, ii. 265.
Pecten, eyes of, ii. 190; tentacles of, ii. 191.
Pedalion, ii. 58. 63.
Pedesis, Prof. Jevons on, i. 154.
Pediastreæ, structure of, i. 270–273; multiplication and development of, i. 270, 271; varieties of, i. 272.
Pedicellariæ of Echinoderms, ii. 144.
Pedicellina, ii. 162.
Pelargonium, cells of petal of, i. 383.
Pelomyxa palustris, ii. 17.
Peneroplis, ii. 66, 71.
Penetrating power of Object-glasses, i. 163, increase of, with Binocular, i. 38.
Penicillium, i. 323.
Pentacrinoid larva of Comatula, ii. 152–156.
Pentacrinus, skeleton of, ii. 146.
Perennibranchiata, bone of, ii. 257; blood-corpuscles of, ii. 269, 270.
Peridinium, ii. 37, 38.
Peristome of Mosses, i. 389–342.
Peronospora, i. 323.
Perophora, ii. 167, 168.
Petals of Flowers, structure of, i. 383.
Petrology, Microscopic, ii. 312–317.
Pettenkofer's test, i. 208.
PHANEROGAMIA, distinctive peculiarities of. i. 352, 353; elementary tissues of, i. 353–367 (see Tissues of Plants); Stems and Roots of, i. 367–377; Cuticles and Leaves of, i. 377–382; Flowers of, i. 382–386; Seeds of, i. 386–388.
Phyllopoda. ii. 209.
Picric acid, for hardening, i. 203.
Picro-aniline, as staining agent, i. 207.
Picro-carmine, as staining agent, i. 206.
Pieridæ, scales of, ii. 221, 222.
Pigott, Dr. Royston, his Aplanatic Searcher, i. 8 *note;* his Micrometers, i. 92 *note;* on angle of aperture, i. 162; on scales of Insects, ii. 224, 226.
Pigment-cells, ii. 275, 276; of Cuttlefish, ii. 191; of Crustacea, ii. 215.
Pigmentum nigrum, ii. 275.
Pilidium-larva of Nemertes, ii. 199.
Pillischer, Mr., his International Microscope, i. 59, 60.
Pilulina, ii. 79.
Pinna, structure of shell of, ii. 171–173; fossil, in Chalk, ii. 307.
Pinnularia, i. 298.
Pistillidia, see Archegonia.
Pith, structure of, i. 354, 368.
Placoid scales of Fish, ii. 263.
Planaria, ii. 194, 195.
Planorbulina, ii. 89.
Plantago, cyclosis in hairs of, i. 359.
Plants, distinction of, from Animals, i. 222–224.
Plasmodium, of Myxomocetes, i. 326; of Protomyxa, ii. 3.
Plate-glass Cells, i. 182.
Pleurosigma, i. 298; nature of markings on, i. 274–279; value of, as Test, i. 170–172; diverse aspects of, i. 146–151; diffraction-spectrum of, i. 160.
Pluteus-larva of Echinus, ii. 152.
Plumules of Butterflies, i. 221.
Pocket Microscope, Beale's, i. 80.
Podophrya quadripartita, ii. 39–41.
Podura, scale of, ii. 223–227; use of, as Test object, i. 172, 173
Poisons, detection of, ii. 326.
Polarization, Objects suitable for, i. 318–323.
Polarizing Apparatus, i. 111–114.
Polistes, fungous vegetation in, i. 318.
Pollen-grains, development of, i. 383; structure and markings of, i. 383–385.
Pollen-tubes, **fertilizing** action of, i. 386.
Polycelis, ii. 195.
Polyclinians, ii. 165.
Polycystina, ii. 109, 113–116.
Polygastrica, see Infusoria.
Polymorphina, ii. 85.
Polyommatus argus, scale of, ii. 222, 223.
Polypes, see *Hydra* and *Zoophytes*.
Polypide of Polyzoa, ii. 157.
Polypodium, fructification of, i. 345.
Polystomella, ii. 92–94.
Polythalamous Foraminifera, ii. 66–68.
Polytoma uvella, ii. 29.
Polytrema, ii. 90.

POLYZOA, general structure of, ii. 157-163; classification of, ii. 162.
Polyzoary, ii. 157.
Pond-Stick, Baker's, i. 219.
Poppy, seeds of, i. 386, 387.
Popular Microscope, Beck's, i. 68.
Porcellanous Foraminifera, ii. 68, 70-77.
Porcellanous shells of Gasteropods, ii. 178.
Porcupine, quill of, ii. 265.
PORIFERA, see Sponges.
Portable Binocular, i. 83.
Potato-disease, i. 323.
Powell and Lealand's Microscopes, i. 67, 68, 77, 79; their non-stereoscopic Binocular, i. 85; their Achromatic Condenser, i. 103; their Light-modifier, i. 110; their Oil-immersion objectives, i. 18; their Vertical Illuminator, i. 116.
Prawn, shell of, ii. 215.
Preservative Media, i. 209-211
Primordial Utricle, i. 225, 356.
Pringsheim, Dr., his observations on Vaucheria, i. 251; on Hydrodictyon, i. 252; on Œdogonium, i. 258.
Prismatic Shell-substance, ii. 171, 172.
Prism, Amici's, i. 106; Nachet's Erecting, i. 88; Wenham's Binocular, i. 30, 85; Stephenson's Binocular, i. 31; Camera Lucida, i. 96-98; Spectroscope, i. 90; Polarizing, 111, 112.
Proboscis, of Bee, ii. 234, 235; of Butterfly, ii. 236; of Fly, ii. 234.
Proteus, blood-corpuscles of, ii. 269, 270.
Prothallium of Ferns, i. 346-348.
Protococcus, life-history of, i. 231-236.
Protomyxa, ii. 2, 3.
Protoplasm, i. 222; of Vegetable cell, i. 224-228; of Animals, ii. 253-255.
PROTOPHYTA, general characters of, i. 222-228.
Protophytic Algæ, i. 229-306.
Protophytic Fungi, i. 229, 307; relation of, to Protozoa, i. 307; cultivation of, i. 123, 307.
PROTOZOA, ii. 1, 2; their relations to Protophyta, i. 224.
Pseud-embryo of Echinoderms, ii. 150.
Pseudo-navicellæ of Gregarinida, ii. 21.
Pseudopodia of Rhizopods, ii. 2-19, different forms of, ii. 7.
Pseudoscope, i. 27, 28.
Pseudoscopic Microscope of MM. Nachet, i. 33-36.
Pteris, fructification of, i. 345; prothallium of, i. 346.
Pterodactyle, bone of, ii. 312.
Puccinia, i. 323.
Purpura, egg-capsules of, ii. 184; development of, ii. 187-189.
Pycnogonidæ, ii. 205-207.

Quadrula symmetrica, ii. 19.

Quatrefages, M. de, on **luminosity of** Annelids, ii. 202.
Quekett, Prof. J., his Dissecting Microscope, i. 45; his Indicator, i. 96; on Raphides, i. 363; on structure of Bone, ii. 258, 311.
Quinqueloculina, ii. 71.

Radiating Crystallization, ii. 320, 321.
Radiolaria, ii. 109, 110; their relation to Heliozoa, ii. 109; their general structure, ii. 110, 111; their classification, ii. 112, 113; collection and mounting of, ii. 115, 116.
Rainey, Mr., his Light modifier, i. 110; on Molecular coalescence, ii. 313-325.
Ralfs, Mr., on Desmidiaceæ, i. 263 *note*; on Diatomaceæ, i. 284 *note*.
Ralph, Dr., his mode of mounting, i. 215.
Ramsden's Micrometer, i. 92, 93.
Raphides, i. 363.
Re-agents, Chemical, use of, in Microscopic research, i. 208, 209, ii. 326.
Red Corpuscles of blood, ii. 267-270.
Red Snow, i. 245.
Reflection by Prisms, i. 2, 3.
Reflex Illuminator, Wenham's, i. 109.
Refraction, laws of, i. 1-3; by convex lenses, i. 3-5; by concave and meniscus lenses, i. 5-6.
Reindeer, hair of, ii. 264.
Reophax, ii. 82.
REPTILES, bone of, ii. 257, 258, 311; teeth of, ii. 259, 311; scales of, ii. 263; blood of, ii. 268-271; lungs of, ii. 299, 300.
Resolving **power of** Object-glasses, i. 158, 164.
Reticularia, ii. 7-11.
Reticulated Ducts, i. 366.
Rhabdammina, ii. 80.
Rhinoceros, horn of, ii. 267.
Rhizocarpeæ, i. 350.
RHIZOPODA, ii. 7-19; their subdivions, ii. 7, 8; their relation to higher Animals, ii. 252, 253.
Rhizosolenia, i. 296.
Rhizostoma, ii. 134.
Rhodospermeæ, i. 334.
Rhubarb, raphides of, i. 363.
Rhynchonellidæ, structure of Shell of, ii. 178.
Rice, starch-grains of, i. 362.
Rice-paper, i. 354, 355.
Ricinæ, ii. 249.
Ring-Cells, i. 181.
Ring-Net, i. 219-221.
Rivet-Leiser Microtome, i. 192, 193.
Roasted Corn, detection of, in Chicory, i. 388.
Robin, M., on Noctiluca, ii. 34 *note*, ii. 37 *note*.
Rochea, epidermis of, i. 378.
Rocks, structure of, ii. 304-310 313-317.

Roots, structure of, i. 375, 376; **mode of** making sections of, i. 376.
Ross. Mr., on correction of Object-glass, i. 14, 15; his First-class Microscopes, i. 73-76; his Achromatic Condenser, i. 103; his Students' Microscope, i. 60, 61; his Simple Microscope, i. 43-45; his Lever of contact, i. 177; his Compressor, i. 126.
Ross-Model for Compound Microscope, i. 51, 52.
Rotalia, ii. 67, 68, 89, 90.
Rotaline Foraminifera, ii. 67, 89, 90.
Rotating Microscope, Browning's, i. 64, 65.
Rotifer, anatomy of, ii. 55-58; reproduction of, ii. 58, 59; tenacity of life of, ii. 59; occurrence of, in leaves of Sphagnum, i. 343, ii. 53.
ROTIFERA, general structure of. ii. 53-63; reproduction of, ii. 58, 59; desiccation of, ii. 59; classification of, ii. 60-63.
Royston-Pigott, Dr., see Pigott.
Rush, stellate parenchyma of, **i. 354**, 355.
Rust, of Corn, i. 323.
Rutherford, Prof., his **freezing Microtome**, i. 192, 196.

Sable, hair of ii 264.
Saccammina, ii. 78, 79.
Saccharomyces, i 315.
Saccolobium, spiral cells of, i. 361.
Safety-stage, Stephenson's, i. 120.
Salpingœca, ii. 32.
Salter, Mr. Jas., on teeth of Echinida, ii. 144.
Salts, crystallization of, ii. 318-323.
Salvia, spiral fibres of seed of, i. 361.
Salicylic Acid, as preservative, i. 210.
Sand-wasp, integument of. ii. 220.
Sandy tests of Foraminifera, ii. 77-85.
Sarcina ventriculi, i. 316.
Sarcode, of Protozoa, i. 222 *note*, i. 222.
Sarcoptes scabiei, ii. 249.
Sarsia, ii. 127.
Saw-flies, ovipositor of, ii. 245. 246.
Scalariform ducts of Ferns, i. 344, 366.
Scales, of cuticle of Plants, i. 378, 379.
——— of Fish, ii. 261-263.
——— of Insects, ii. 220-228; their **use** as Test-objects, i. 167-173.
——— of Reptiles and Mammals, **ii.** 263.
Schiek's Compressor, i. 126.
Schizomycetes, i 307-313: their Zymotic action, i. 313-315.
Schizonemeæ, i. 299.
Schultz's test, i. 208.
Schultze, Prof. Max, on Protoplasm, i. 222 *note*; on movement of fluid in Diatoms, i. 273; on surface-markings of Diatoms, i. 277 *note*.
Schulze, Mr. A., on use of Illuminators, i. 110.

Schwann, doctrines of, ii. 252.
Schwendener, on Lichens, i. 329.
Scissors, for microscopic dissection, i. 188; for cutting thin sections, ii. 188.
Sclerogen, deposit of, on walls of Cells, i. 359, 360.
Scolopendrum, sori of, **i. 345**.
Sea Anemone, ii. 135, 136.
Section-cutting Instruments, **i.** 189-193.
Section-lifter, Marsh's, i. 205.
Sections, thin, mode of making, of Soft substances, i. 188-196; modes of mounting, i. 212-214; of Hard Substances, i. 196-200. of Foraminifera, i. 198 *note*; of Leaves, i. 382; of Wood, i. 376; of Echinus-spines, ii. 146, 147; of Insects. ii. 219; of Bones and Teeth, ii. 258; of Hairs, ii. 266.
Seeds, testæ of, i. 386-388; spiral cells in, i. 361.
Segmentation of Yolk-mass, **ii. 185, 187**.
Selaginelleæ, i. 350.
Selenite Stages, i. 112-114.
Sepiola, eggs of, ii. 191.
Sepiostaire of Cuttle-fish, ii. 180, 325.
Serialaria, colonial nervous system of, ii. 160.
Serous Membranes, structure of, ii. 274.
Serpentine-Limestone, **ii. 101-107**, 310.
Sertularidæ, ii. 129-131.
Sexual Generation, lowest forms of, in Protophytes, i. 229, 230, 236, 237; in Infusoria, ii. 26-30.
Shadbolt, Mr., on Arachnoidiscus, i. 293: his Annular condenser, i. 107 *note;* his Turn-table, i. 184.
Shark, teeth of, ii. 258, 259; scales, etc., of, ii. 263.
Shell, of Crustacea, ii. 214, 215; of Echinida, ii. 140, 141; of Foraminifera, ii. 68-70; of Mollusca, ii. 172-180; Fungi in, i. 321.
Shrimp, shell of, ii. 215.
Side Illuminators, i. 114-116.
Side-Reflector, Beck's, i. 116, 117.
Siebert and Kraft's Dissecting Microscope, i. 46.
Siebold, Prof., on reproduction of Bee, ii. 247.
Silica crack-slide, **i.** 152, 162, ii. 321.
Siliceous Epiderms, i. 349, 379.
——— Sponges, ii. 120, 121.
Silk-worm diseases, i. 315-317.
Silver, crystallized, ii. 319.
Simple Microscope, optical principles of, i. 18-22; various forms of, i. 43-51.
Siphonaceæ, i. 250, 251.
Siricidæ, ovipositors of, ii. 245, 246.
Skin, structure of, ii. 274, 275; papillæ of, ii. 284, 285, 298.
Slack, Mr., on Pinnularia, i. 298; on artificial Diatoms, i. 277 *note*; his Diaphragm-Eyepiece, i. 95; his Light-modifier, i. 111; his Stage-vice, i. 121; his Compressors, i. 126, 127; his Silica

INDEX. 351

crack-slide, i. 152, 162; his crystallizations from silicated solutions, ii. 321.
Sladen, Mr. P., on preserving Echinoderm larvæ, ii. 153.
Slider-Forceps, i. 185.
Slides, Glass, i. 175, 176.
——— Wooden, i. 183.
Slug, rudimentary shell of, ii. 179; palate of, ii. 181, 182; eyes of, ii. 190.
Smith, Mr. Jas., his Mounting Instrument, i. 186; his use of Bull's-eye Condenser, i. 118.
Smith, Dr. Lawrence (U. S.), **his** Inverted Microscope, i. 82.
Smith, Prof. H. L. (U. S.), on Binocular Eyepiece, i. 33; his vertical Illuminator, i. 118; his cells for dry-mounting, i. 180; on mounting Diatoms, i. 306.
Smith, Prof. J. Edwards (U. S.), on development of Œdogonium, i. 257; **on** wide-angled Objectives, Preface, vi., vii.
Smith, Prof. W., on Diatomaceæ, i. 170, 273, 391 *note*.
Smith and Beck, see Beck, Messrs.
Smut, of Wheat, i. 323.
Snail, palate of, ii. 181, 182; eyes of, ii. 190.
Snake, lung of, ii. 299.
Snow crystals, ii. 318.
Social Ascidians, ii. 166-168.
Soemmering's speculum, i. 97.
Sole, skin **and scales** of, ii. 261, 262.
Sollitt, Mr., on Diatom-tests, i. 170.
Sorby, Mr., on skeleton of Echinoderms, ii. **146** *note;* his Spectroscope Eyepiece, i. 90; his Microscopic examination of Rocks, ii. 314, 315 *note*.
Soredia of Lichens, i. 329.
Sori of Ferns, i. 344-346.
Spatangidium, i. 291.
Spatangus, spines of, ii. 143.
Spectacles, for Dissection, i. 187.
Spectro-Micrometer, Browning's, i. 91.
Spectroscope Eye-piece, i. 90.
Spectroscopic Analysis, principles of, i. 89-92.
Speculum, Parabolic, i. 116, 117.
Spermogonia of Fungi, i. 322; of Lichens, i. 330.
Sphacelaria, i. 332.
Sphæria, development of, within Animals, i. 318.
Sphæroplea, sexual reproduction of, i. 255.
Sphærosira volvox, i. 243.
Sphærozosma, i. 266.
Sphærozoum, ii. 115.
Sphagnaceæ, peculiarities **of,** i. 342-344; occurrence of parasites **in** leaf-cells of, i. 327, ii. 53.
Spherical Aberration, i. 6, **7;** means of reducing and correcting, i. 7, 8.
Spicules of Sponges, ii. 120-122; Alcyonian Zoophytes, ii. 136; of Doris, ii. 179.

Spiders, eyes of, ii. 250; respiratory organs of, ii. 250; feet of, ii. 250; spinning apparatus of, ii. 251.
Spines of Echinida. ii. 142, **143; mode** of making sections of, ii. 146, 147; of Spatangus, ii. 143.
Spinning apparatus of Spiders, ii. 251.
Spiracles of Insects, ii. 239-241.
Spiral Cells of Sphagnum, i. 343; of Orchideæ, i. 360; of anthers, i. 384.
——— Crystallization, ii. 321.
——— **Ducts**, i. 366.
——— **Fibres**, i. 361.
——— **Vessels**, i. 364; **in petals,** i. 383.
Spiriferidæ, **Shell-structure of,** ii. 178.
Spirillina, ii. 85.
Spirillum, i. 312.
Spirolina, ii. 72.
Spiroloculina, ii. 71.
SPONGES, general structure and relations of, ii. 117, 118; reproduction of, ii. 118; skeleton of, ii. 119-122; fossil, ii. 307, 308.
Spongilla, ii. 118, 121.
Spongiole of Root, i. 375.
Spores, different kinds of, i. 228-230; of Fungi, general diffusion of, i. 321-323; of Hepaticæ, i. 338; of Mosses, i. 342; of Ferns, i. 345-348; of Equisetaceæ, i. 349;—see Oöspores and Zygospores.
Spot-Lens, i. 107.
Spring-Clip, i. 186.
——— Press, i. 186.
——— Scissors, i. 188.
Squirrel, hair of, ii. 264.
Stage-centering adjustment, **i. 80**.
Stage, Glass, i. 63.
Stage, Safety, **i. 120**.
Stage-Forceps, **i. 120**.
Stage-Plate, glass, **i. 122**.
Stage-Vice, i. 121.
Staining Processes, i. 204-208.
Stanhope Lens, i. 21.
Stanhoscope, i. 21.
Star-Anise, seed-coat of, i. 360.
Starch-granules, in Cells, i. 361, 362; appearance of, by Polarized light, i. 362.
Star-fish, Bipinnarian larva of, ii. 150, 151.
Stato-spores, i. 230 *note;* of Volvox, **i.** 244; of Hydrodictyon, i. 253.
Staurastrum, prominences of, i. 263; self-division of, i. 265; varieties of, i. 272.
Stauroneis, i. 299.
Steenstrup, Prof., on **Alternation of** generations, ii. 134.
Stein, Dr., his doctrine of Acineta forms, ii. 41 *note*, ii. 52 *note;* his researches on Infusoria, ii. 63 *note*.
Steinheil Doublet, i. 21.
Stellaria, petal of, i. 343.
Stellate cells, of Rush, i. 354, **355; of Water lily,** i. 354, 355.

Stemmata of Insects, ii. 231.
Stem, i. 367; Monocotyledonous, structure of, i. 367, 368; Exogenous, structure of, i. 368-374; development of, i. 374, 375; mode of making sections of, i. 376.
Stentor, ii. 44; its conjugation, ii. 52.
Stephanoceros Eichornii, ii. 60, 61.
Stephanosphæra, i. 243 note, i. 244 note.
Stereoscope, i. 25.
Stereoscopic Spectacles, i. 187.
———— Vision, principles of, i. 25-28; application of, to Compound Microscope, i. 27-39; to Simple Microscope, i. 48, 49.
Stephenson, Mr., his suggestion of homogeneous immersion Objectives, i. 17; on diffraction-doctrine, i. 157-160; his Binocular Microscope, i. 31-33; his safety-stage, i. 120; on mounting in bisulphide of carbon, i. 279; on Coscinodiscus, i. 290.
Stewart, Mr., on internal skeleton of Echinodermata, ii. 148.
Stick-net, i. 220.
Stigmata of Insects, ii. 239, 240.
Stings of Plants, structure of, i. 379; of Insects, ii. 245, 246.
Stokes, Prof., on Absorption bands of blood, i. 91, 92.
Stomata, of Marchantia, i. 336; of Flowering Plants, i. 379, 380.
Stones, for polishing Sections, i. 198.
Stones, of Fruit, structure of, i. 360.
Strassburger, Dr., on cell-division, ii. 254 note.
Striatelleæ, i. 288.
Student's Microscopes, principles of construction of, i. 55-57; Objectives suitable for, i. 57-58; various forms, of, i. 59-67.
Suctorial Crustacea, ii. 212, 213.
Suctorial Infusoria, ii. 39.
Sulphate of Copper and Magnesia, radiating crystallization of, ii. 320.
Sulphate of Copper, spiral crystallization of, ii. 321.
Sulphuric Acid, as test, i. 208.
Sundew, hairs of, i. 379.
Sunk Cells, i. 182.
Surirella, i. 287; conjugation of, i. 281; use of, as test, i. 171.
Swift, Mr., his Challenge Microscope, i. 71; his Portable Binocular, i. 82, 83; his swinging Sub-stage, i. 71 note, his combination Sub-stage, i. 113, 114; his Aquatic box, i. 124; his Microscope lamp, i. 122; his Wale Students' Microscope, ii. 332.
Synapta, calcareous skeleton of, ii. 149.
Syncoryne, ii. 127.
Syncrypta, i. 244.
Synedreæ, i. 287.
Syringe, small glass, i. 128; uses of, i. 142, 204, 208, 212, 217, ii. 188 note, ii. 293.

Syringe, injecting, ii. 293.

Tabanus, ovipositor of, ii. 246.
Table for Microscope, i. 130.
Tadpole, pigment cells of, ii. 276; circulation in, ii. 288-292.
Tænia, ii. 192, 193.
Tardigrada, ii. 62.
Teeth, of Echinida, ii. 144, 145; of Ophiocoma, ii. 145; of Mollusks, ii. 181-183; of Leech, ii. 203; of Vertebrata, structure of, ii. 258-261; fossil, ii. 311, 312; mode of making sections of, ii. 258.
Tendon, structure of, ii. 273.
Tenthredinidæ, ovipositor of, ii. 245.
Terebella, circulation and respiration in, ii. 196, 197.
Terebratula, shell-structure of, ii. 177, 178; muscular fibre of, ii. 283.
Terpsinoë, i. 288.
Tests, of Rhizopods, ii. 18, 19; of Foraminifera, ii. 77-85.
Test-Liquids, i. 208, 209.
Test-Objects, i. 165; for low powers, i. 166, 167; for medium powers, i. 167-169; for high powers, i. 169-173.
Tetramitus rostratus, ii. 30, 31.
Tetraspores of Ceramiaceæ, i. 334.
Textularia, ii. 68, 88.
Thalassicolla, ii. 115.
Thallus of lower Cryptogamia, i. 246, 329, 331.
Thaumantias, ii. 131.
Thecæ, of Ferns, i. 345; of Equisetaceæ, i. 349.
Thin Glass, i. 176, 177.
Thomas, Mrs. H., on Cosmarium, i. 266.
Thomas, Mr. R., on microscopic crystallization, ii. 321.
Thompson, Mr. J. V., on development of Comatula, ii. 153; on metamorphosis of Cirrhipeds, ii. 213; on metamorphosis of Crustacea, ii. 216.
Thomson, Sir Wyville, on Globigerina, ii. 87; on Siliceous Sponges, ii. 121; on development of Pentacrinoid larva, ii. 156; on Chalk-formation, ii. 306.
Thread-cells of Zoophytes, ii. 137, 138.
Thrush, fungous vegetation of, i. 320.
Thurammina, ii. 81.
Thwaites, Mr., on conjugation of Diatoms, i. 281, 282; on filamentous extensions of Palmelleæ, i. 246 note.
Ticks, ii. 248.
Tinea favosa, fungus of, i. 320.
Tinoporus, ii. 89, 90.
Tipula, larva of, ii. 240.
Tissues, Elementary, of Animals, microscopic study of, ii. 252; formation of, ii. 253-255; see Blood, Bone, Capillaries, Cartilage, Epidermis, Epithelium, Fat, Feathers, Fibrous Tissues, Glands, Hair, Horn, Mucous Membranes, Muscle, Nervous Tissue,

INDEX. 353

Pigment-cells, Scales, Serous Membranes, Teeth.
Tissues, Elementary, of Plants, i. 353; Cellular, i. 353-363; Woody, i. 363, 364; Vascular, i. 365, 366; dissection of, i. 366; preparation of, i. 201.
Tolles, Mr., his Binocular Eye-piece, i. 33; his Amplifier, i. 86; his vertical Illuminator, i. 119.
Tomopteris, ii. 199-201.
Tongues of Gasteropods, ii. 181-183; of Insects, ii. 234-236.
Torula cerevisiæ, i. 315.
Tous-les-mois, Starch-grains of, i. 362.
Tow-net, i. 220.
Tracheæ of Insects, ii. 238-240; mode of preparing, ii. 240, 241.
Tradescantia, cyclosis in hairs of, i. 358.
Transparent Objects, arrangement of Microscope for, i. 141-145; various modes of Illuminating, i. 141-146.
Travelling Microscopes, i. 81, 82.
Trematode, Entozoa, ii. 194.
Triceratium, i. 295; markings on, i. 277.
Trichoda, bristles of, ii. 43; metamorphosis of, ii. 47-49.
Trichogyne, of Lichens, i. 330; of Florideæ, i. 335.
Trilobite, eye of, ii. 310.
Triloculina, ii 71.
Triple Staining, i. 207.
Trochus, palate of, ii. 181.
Trout, circulation in young of, ii. 289.
Tube-cells, i. 180.
Tubular Nerve-substances, ii. 284, 285.
Tubularia, ii. 127.
TUNICATA, general organization of, ii. 163, 164; see *Ascidians.*
Turbellaria, ii. 194-196.
Turn-tables, i. 184.
Tyndall, Prof., on Bacteria, etc., i. 313, 314, 321.

Ulvaceæ, i. 246, 247.
Unicellular nature of Infusoria, ii. 25.
Unicellular Plants, i. 229.
Unionidæ, shells of, ii. 174-176.
Uredo, i. 323.
Urns of Mosses, i. 340.
Urella, i. 235.

Vacuoles, i. 225; microscopic appearances of, i. 153.
Vallisneria, cyclosis in, i. 356, 357.
Vampyrella, ii. 3-5.
Van Beneden, Prof. Ed., on gigantic Gregarina, ii. 21.
Vanessa, haustellium of, ii. 236.
Variation, tendency to, in Desmideaceæ, i. 271; in Diatomaceæ, i. 282; in Foraminifera, ii. 72, 77, 96; in Polycystina, ii. 112 *note.*
Varnishes and Cements, i. 178, 179.
Vaucheria, zoospores of, i. 250; sexual reproduction of, i. 251.

Vegetable Ivory, i. 360.
Vegetable Kingdom, differentiated from Animal, i. 222-233.
Vegetable substances, preparation of, i. 201.
Ventriculites, ii. 307.
Vermilion injections, ii. 294, 295.
VERTEBRATA, elementary structure of, ii. 252 (see Tissues); blood of, ii. 267-271 circulation in, ii. 286-292.
Vertical Illuminators, i. 118, 119;
Vesicular Nerve Substance, ii. 284.
Vessels of Plants, i. 365, 366.
Vibracula of Polyzoa, ii. 163.
Vibrio, i. 311, 312.
Villi of intestine, injections of, ii. 295.
Vine-disease, i. 323.
Vinegar, Eels of, ii. 193.
Vitreous Foraminifera, ii. 69, 85-107.
Volvox, structure of, i. 237-240; development and multiplication of, i. 240, 241; generation of, i. 241-243; amœboid state of, i. 241-243.
Vorticella, ii. 43, 46; encysting process in, ii. 47; conjugation of, ii. 52.

Wale's New Working Microscope, i. 61-63.
Wallich, Dr., on making sections of Foraminifera, i. 198 *note;* on Diatoms, i. 275 *note,* i. 277 *note;* on Coccospheres, ii. 19; on nucleus in Gromia, ii. 9; on Globigerinæ, ii. 86; on Polycystina, ii. 113 *note.*
Warts, structure of, ii. 276.
Water-Bath, i. 185.
Water-immersion Objectives, i. 16, 17; ii. 327, 330.
Water-Lily, stellate cells of, i. 354, 355; leaf of, i. 382.
Water-newt, circulation in larva of, ii. 288.
Water-Vascular system, of Rotifera, ii. 57; of Planaria, ii. 195.
Watson, Messrs., their new form of Microscope, ii. 331, 332.
Weber's Annular Cell, i. 124.
Webster-Condenser, 1. 103, 104.
Wenham, Mr., his new Achromatic combination, i. 15; his suggestion of homogeneous immersion, i. 17; his Binocular Microscope, i. 19-31; his Non-Stereoscopic Binocular, i. 84; his Disk-illuminator, i. 105; his Parabolic Illuminator, i. 107, 108; his Reflex Illuminator, i. 109; on adjustment of Object-glasses, i. 140; his observations on Pleurosigma, i. 277 *note;* on Cyclosis, i. 358, 359; on Podura scale, ii. 226.
Whalebone, structure of, ii. 267.
Wheat, blights of, i. 323, ii. 193.
Wheatstone, Sir C., his invention of the Stereoscope, i. 25-27; of the Pseudoscope, i. 27, 28.
Wheel-animalcules, see Rotifera.

White-cloud Illuminator, i. 111.
White Corpuscles of blood, ii. 270, 271.
White Fibrous tissue, ii. 272.
Whitney, Mr., on circulation in Tadpole, ii. 289-292.
Williamson, Prof. W. C., on Volvox, i. 243 *note;* on shells of Crustacea, ii. 215 *note;* on scales of Fishes, ii. 261, 262; on Coal-plants, ii. 303; on Levantmud, ii. 305.
Wings of Insects, ii. 241-243.
Winter-eggs, of Rotifera, ii. 59; of Hydra, ii. 125; of Entomostraca, ii. 210.
Wollaston, Dr., his Camera Lucida, i. 96.
Wood, of Exogenous stems, i. 369, 370.
Woodward, Col. Dr., his Prism, i. 105; his resolution of Amphipleura pellucida, i. 171; of Surirella gemma, i. 172; on scale of Gnat, i. 155, on Podura-scale, ii. 227.
Woody Fibre, i. 363; glandular, of Conifers, i. 364.
Working-distance of Objectives, i. 161.
Wormley, Dr., on Micro-Chemistry, ii. 326.
Wyth's Amplifier, i. 86.

Xanthidia of Flints, i. 267 *note*, ii. 308.

Yeast-plant, i. 315.
Yellow Fibrous tissue, ii. 273.
Yucca, epidermis of, i. 377; stomata of, i. 380.

Zeiss's oil-immersion Objectives, i. 17, 18; his adjusting Low-power, i. 136, 166; his Sub-stage Condenser, i. 104 *note*.
Zentmayer, Mr., on defining power, i. 162 *note;* his swinging tail-piece, i. 61, 75; his glass stage, i. 63.
Zoea-larva of Crab, ii. 216.
Zoantharia, ii. 135.
Zooglœa, i. 308.
Zoophyte-Trough, i. 125.
ZOOPHYTES, ii. 122-137; see *Actinozoa, Alcyonaria,* and *Hydrozoa*.
Zoöspores, i. 230; *note;* of Protococcus, i. 233, 234; of Ulvaceæ, i. 247; of Vaucheria, i. 250; of Achlya, i. 251; of Confervaceæ, i. 254; of Chætophora, i. 258; of Pediastreæ, i. 271; of Fucaceæ, i. 334.
Zygnemaceæ, i. 236, 237.
Zygospores, i. 229; of Conjugateæ, i. 232, 236; of Desmidiaceæ, i. 267, 268; of Diatomaceæ, 281, 282.
Zygosis of Actinophrys, ii. 12; of Amœba; ii. 17; of Gregarina, ii. 22.
Zymotic action of Bacillus-organisms, 313-315.

ERRATUM.

The first sentence in the Note to p. 163. vol. i., should run thus:—

The Author is informed by Prof. Abbe, that the 'penetration' of Objectives decreases in a corresponding ratio with the increase of their respective Numerical Apertures; or, when Objectives of the same class are compared, with the increase in the sines of their respective semi-angles of aperture.

www.ingramcontent.com/pod-product-compliance
Lightning Source LLC
Chambersburg PA
CBHW020325240426
43673CB00039B/918